European Workshop on Advanced Control and Diagnosis 2014

Journal of Physics: Conference Series Volume 570

Berlin, Germany
13 – 14 November 2014

ISBN: 978-1-63439-755-1
ISSN: 1742-6588

Printed from e-media with permission by:

Curran Associates, Inc.
57 Morehouse Lane
Red Hook, NY 12571

Some format issues inherent in the e-media version may also appear in this print version.

Printed by Curran Associates, Inc. (2015)

For permission requests, please contact the Institute of Physics at the address below.

Institute of Physics
Dirac House, Temple Back
Bristol BS1 6BE UK

Phone: 44 1 17 929 7481
Fax: 44 1 17 920 0979

techtracking@iop.org

Additional copies of this publication are available from:

Curran Associates, Inc.
57 Morehouse Lane
Red Hook, NY 12571 USA
Phone: 845-758-0400
Fax: 845-758-2634
Email: curran@proceedings.com
Web: www.proceedings.com

TABLE OF CONTENTS

SYSTEM IDENTIFICATION

NONLINEAR AND DISCRETE CONTROL

MODEL PREDICTIVE CONTROL

APPLICATIONS OF FUZZY LOGIC

CONTROL APPLICATIONS

MODELLING AND SIMULATION

FAULT DIAGNOSIS

FAULT-TOLERANT CONTROL

11th European Workshop on Advanced Control and Diagnosis 2014

Preface

The European Workshop on Advanced Control and Diagnosis is an annual event that has been organised since 2003 by Control Engineering departments of several European universities in Germany, France, the UK, Poland, Italy, Hungary and Denmark. The overall planning of the workshops is conducted by the Intelligent Control and Diagnosis (ICD) steering committee. This year's ACD workshop took place at HTW Berlin (University of Applied Sciences) and was organised by the Control Engineering group of School of Engineering I of HTW Berlin.

38 papers were presented at ACD 2014, with contributions spanning a variety of fields in modern control science: Discrete control, nonlinear control, model predictive control, system identification, fault diagnosis and fault-tolerant control, control applications, applications of fuzzy logic, as well as modelling and simulation, the latter two forming a basis for all tasks in modern control.

Three interesting and high-quality plenary lectures were delivered. The first plenary speaker was Wolfgang Weber from Pepperl+Fuchs, a German manufacturer of state-of-the-art industrial sensors and process interfaces. The second and third plenary speakers were two internationally high-ranked researchers in their respective fields, Prof. Didier Theilliol from Université de Lorraine and Prof. Carsten Scherer from Universität Stuttgart. Taken together, the three plenary lectures sought to contribute to closing the gap between theory and applications.

On behalf of the whole ACD 2014 organising committee, we would like to thank all those who submitted papers and participated in the workshop. We hope it was a fruitful and memorable event for all.
Together we are looking forward to the next ACD workshop in 2015 in Pilsen, Czech Republic.

Horst Schulte (General Chair), Sören Georg (Programme Chair)

Organisation

General Chair

Horst Schulte, HTW Berlin

Programme Chair

Sören Georg, HTW Berlin

Organising Committee

A special thanks to the following students, who greatly helped with the organisation on the two workshop days:
Moritz Andrejewski, Stefan Heyde, Jonas Jungjohann, Philipp Kammerer, Daniel Kleimann, Ramy Abdelmonem Rashad

International Programme Committee

Steering Committee

Hosting Institution

Control Group of School of Engineering I, HTW Berlin (University of Applied Sciences)

htw. Hochschule für Technik und Wirtschaft Berlin
University of Applied Sciences

Sponsor

PEPPERL+FUCHS
SENSING YOUR NEEDS

Journal of Physics: Conference Series **570** (2014) 001001 doi:10.1088/1742-6596/570/0/001001

URL

www.acd2014.htw-berlin.de

Group Photo

Participants and organisers of ACD 2014 in front of the workshop venue at HTW Berlin

Peer Review Statement

All papers (except paper 082004) published in this volume of *Journal of Physics: Conference Series* have been peer reviewed through processes administered by the proceedings Editors. Reviews were conducted by expert referees to the professional and scientific standards expected of a proceedings journal published by IOP Publishing. No external reviews were received for paper 082004. It was included at the discretion of the editors.

A Simple Method for Estimation of Parameters in First order Systems

Henrik Niemann and Robert Miklos

Dept. of Electrical Engineering, Automation and Control, Technical University of Denmark, DK-2800 Kgs. Lyngby, Denmark

E-mail: hhn@elektro.dtu.dk, s112994@student.dtu.dk

Abstract. A simple method for estimation of parameters in first order systems with time delays is presented in this paper. The parameter estimation approach is based on a step response for the open loop system. It is shown that the estimation method does not require a complete step response, only a part of the response and the steady state value of the system before the step is applied. Further, for calculation of the time delay, it is also required that the time for the step is known.

Keywords:
First order systems, parameter estimation, time delay, step response.

1. Introduction

First order models with a time delay are used in a large number of different applications, where the main dynamic is reasonable damped. First order models are also the basic for a number of controller tuning methods as e.g. [4, 5, 7, 8, 10, 11]. Some of these tuning methods are also applied in a number of different auto-tuning methods for simple feedback controllers, see e.g. in [2, 12].

There exists a number of methods for estimation of the parameters in a first order system as described in e.g. [1, 3, 9, 11]. Some of these methods are standard methods in basic control text books, see e.g. [4, 6].

One of the most used methods for estimation of parameters in stable first order systems is to apply a step response. A step is shown in Fig. 1 for a system with a gain $k = 1$, a time constant $\tau = 1$, a time delay $\theta = 1$ and stepsize 1.

Based on the step response, it is easy to determine the gain, the time constant and the time delay, [4, 6]. However, the step response will not always be as nice and smooth as shown in Fig. 1. The steady value of the system before the step input might not be known exactly because disturbance might be included as well as other issues that might affect the steady state value. Further, the system will not in general exactly be a first order system, but a higher order system with a first order dynamic as the main dynamic. The result of this is that it might be difficult to calculate a tangent in the beginning of the step response. The tangent is one way to determine the time constant in the system. Further, when the exact starting point of the step response is unclear, it is also difficult to calculate the time delay in the system.

Another problem is that the steady state value after the step input might not necessarily occur or it might be unclear. The steady state output will not occur if the step input is not active for a time that is at least

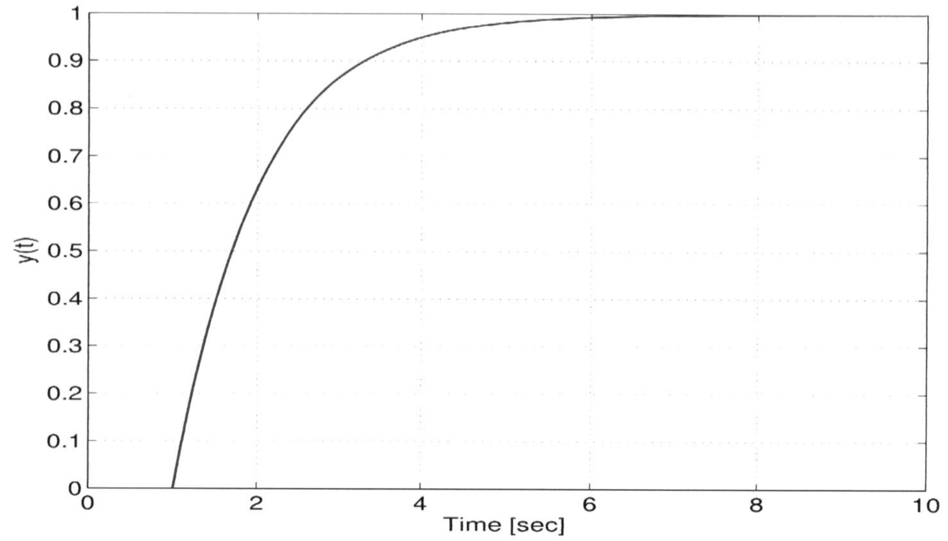

Figure 1. The step response for a first order system.

$5-6$ times the time constant τ. Another reason is that the input might not be a pure step input and therefore the steady state output might not be correct. If the steady state value occurs, it might again be disturbed by other known or unknown inputs to the system. A more realistic step response is shown in Fig. 2. The step response is from a dynamic system in an industrial environment. Considering the response, it is reasonable to describe the dynamic by using a first order model with a time delay.

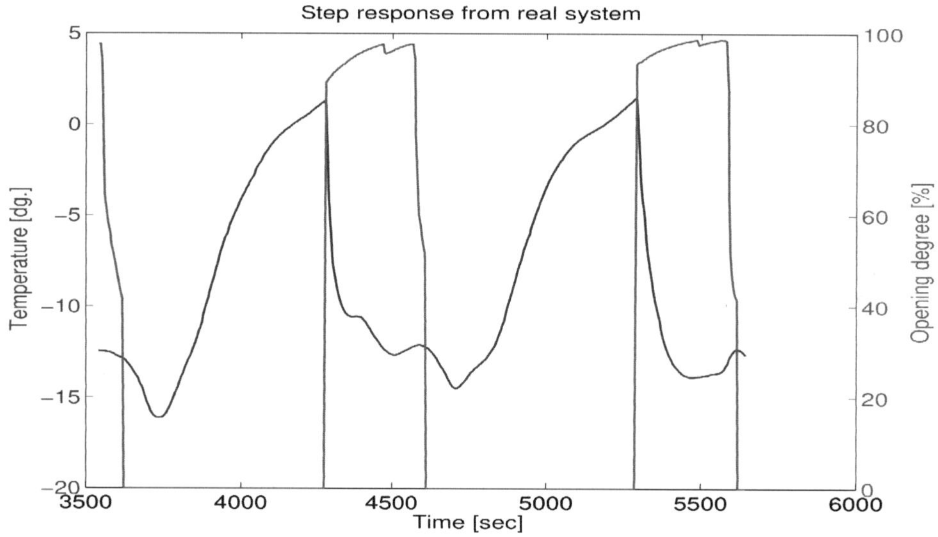

Figure 2. A step response from a refrigeration system. The opening degree is the input to the valve and temperature is the output of the system.

As shown in Fig. 2, we will typical have limited information about the step response. Therefore, we need

Journal of Physics: Conference Series **570** (2014) 012001 doi:10.1088/1742-6596/570/1/012001

to base the estimation of the system parameters directly on the available information. In the following, it will be shown how it is possible to calculate the system parameters based on a part of a step response. The parameter estimation is based on the following information about the step response:

(i) A part of the step response is known.
(ii) The gain of the step input is known.
(iii) An estimate of the steady state value of the output before the step is needed, a mean value.
(iv) The time for the step input is applied, if the time delay is going to be estimated.

It will be shown in the following that based on the above, it is possible to estimate the parameters in a first order system including the time delay. It should be pointed out that it is possible to calculate the steady-state output value of the system as well as the time constant only based on a part of the step response.

The rest of this paper is organized as follows. In Section 2, the system is described together with some preliminary results. The main results are given in Section 3 followed by an example in Section 4. The paper is closed with a conclusion in Section 5.

2. System description

Let a first order system $G(s)$ be given by:

$$G(s) = k\frac{1}{\tau s + 1}e^{-\theta s} \tag{1}$$

where k is the gain, τ is the time constant and θ is the time delay in the system.

The problem is to estimate the three parameters in (1) based on a step response of the system.

Assume that a unit step is applied on the system at time $t = 0$.

It will be assumed that the time delay θ is zero in the following. A non-zero time delay is handled later in this paper.

The step response for the first order system with a unit step input is given by:

$$y(t) = k(1 - e^{-t/\tau}) \tag{2}$$

The steady state value is given by

$$y_{ss} = k \tag{3}$$

Further, we have some standard results with respect to the step response for the first order system, [4, 6].

$$y(\tau) = k(1 - e^{-1}) = k \times 0,632 \tag{4}$$

The derivative of the step response is given by:

$$y'(t) = k\frac{1}{\tau}e^{-t/\tau} \tag{5}$$

and at $t = 0$, we have that

$$y'(0) = k\frac{1}{\tau} \tag{6}$$

Multiplying the tangent in (6) by τ gives

$$y'(0)\tau = k = y_{ss} \tag{7}$$

i.e. it is equal to the steady state value.

The relation between the derivative given by (5) and the derivative for $t = 0$ given by (6) is:

$$\frac{y'(t)}{y'(0)} = e^{-t/\tau}$$

i.e. it depends only of the rate t/τ.

Based on these equations, there are some interesting observations that can be used in the following. First of all, we have that the value of the output is $63,2\%$ of the steady state value after $t = \tau$, see (4). Further, the tangent slope times τ will be equal to the steady state value of the output, see (7). These two equations are normally used for the calculation of the parameters in the first order system. This is shown in Fig. 3, where the tangent given by (6) has been included. Further, both (4) and (7) can be seen from Fig. 3.

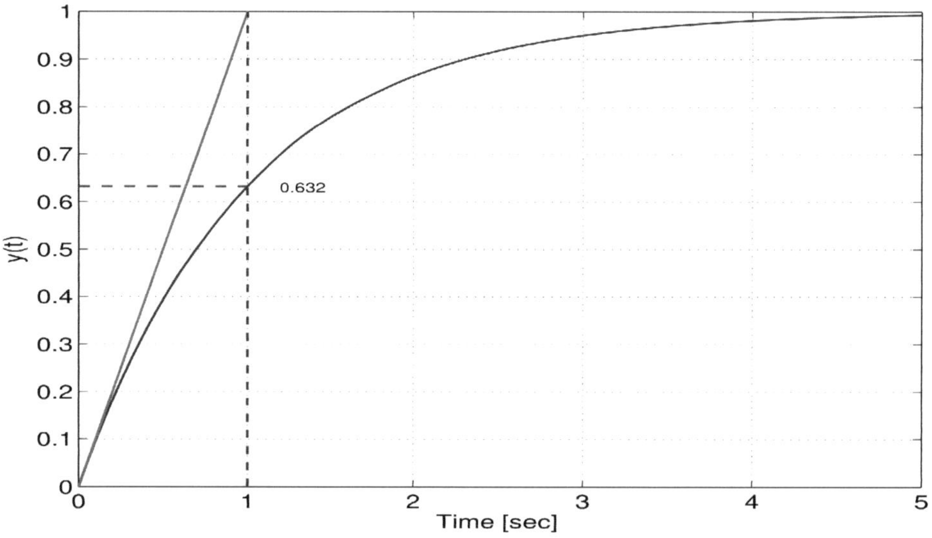

Figure 3. The step response for a first order system with $k = \tau = 1$.

3. Main Results

To be able to estimate the parameters in a first order system, we will have the following assumptions:

(i) The step response is measure in a time interval that does not need to include the start of the response as well as the steady state value of the response.

(ii) The step input is known, i.e. the gain of the step.

(iii) The value of the output $y(t)$ is known before the step is applied. This can be a mean value of the output in a given period before the step.

Note that it is not required that the time for the step input is known. If the system include a time delay that need to be estimated, then we need to know the time for the step input.

Based on these assumptions, we are able to estimate the system parameters.

3.1. Estimation of τ

It is assumed that the step is applied at time t_0. Let's consider the step response in the time interval $[t_1,\ t_2] = [t_0 + \Delta t_1,\ t_0 + \Delta t_2]$. The time interval will not necessarily include the beginning of the step response ($\Delta t_1 = 0$) as well as the steady state values of the output ($\Delta t_2 \to \infty$). The step response that is available is given by:

$$y(t) = y_0 + k(1 - e^{-(t-t_0)/\tau}),\ t \in [t_1,\ t_2] \tag{8}$$

where y_0 is the off-set or steady state value before the step input. The derivative of $y(t)$ is given by:

$$y'(t) = k\frac{1}{\tau}e^{-(t-t_0)/\tau},\ t \in [t_1,\ t_2] \tag{9}$$

Consider the tangent slope given by (9) at time $t = t_1 = t_0 + \Delta t_1$, i.e. $y'(t_1)$ multiplied by the time interval Δt:

$$y'(t_1)\Delta t = k\frac{1}{\tau}e^{-(t_1-t_0)/\tau}\Delta t \tag{10}$$

As in the previous section, consider the tangent slope with a time interval $\Delta t = \tau$. This gives:

$$y'(t_1)\tau = ke^{-(t_1-t_0)/\tau} \tag{11}$$

This tangent start in $y(t_1) = y_0 + k(1 - e^{-(t_1-t_0)/\tau})$. If this is added to the tangent, we get:

$$\begin{aligned} y &= y(t_1) + y'(t_1)\tau \\ &= y_0 + k(1 - e^{-(t_1-t_0)/\tau}) + ke^{-(t_1-t_0)/\tau} \\ &= y_{ss} \end{aligned} \tag{12}$$

(12) together with (10) shows that any tangent at the time length τ will end at the steady state value y_{ss} of the output. This means that it is always possible to find the steady state value of the output if it is possible to calculate a tangent and the time constant τ is known.

At this point, τ is unknown, but it will be shown how τ can be calculated based on the part of the step response in the time interval $[t_1,\ t_2]$. Consider the difference in output for the time $t = t_1$ and $t = t_1 + \Delta t$. It is here assumed that $\Delta t \le t_2 - t_1$. The difference is given by:

$$\begin{aligned} y(t_1 + \Delta t) - y(t_1) &= y_0 + k(1 - e^{-(t_1+\Delta t-t_0)/\tau}) - y_0 - k(1 - e^{-(t_1-t_0)/\tau}) \\ &= k(e^{-(\Delta t_1)/\tau} - e^{-(\Delta t_1+\Delta t)/\tau}) \\ &= k(1 - e^{-\Delta t/\tau})e^{-(\Delta t_1)/\tau} \end{aligned} \tag{13}$$

Further, consider the ratio between the tangent line and the real value of the output in the interval $[t_1,\ t_1 + \Delta t]$. Let the ratio be given by:

$$R_1(\Delta t) = \frac{y(t_1 + \Delta t) - y(t_1)}{y'(t_1)\Delta t} \tag{14}$$

Using (10) and (13) in (14) gives:

$$\begin{aligned} R_1(\Delta t) &= \frac{k(1-e^{-\Delta t/\tau})e^{-(\Delta t_1)/\tau}}{k\frac{1}{\tau}e^{-(\Delta t_1)/\tau}\Delta t} \\ &= \frac{\tau(1-e^{-\Delta t/\tau})}{\Delta t} \end{aligned} \tag{15}$$

Journal of Physics: Conference Series **570** (2014) 012001 doi:10.1088/1742-6596/570/1/012001

Note that $R_1(\Delta t)$ is independent of t_0 and t_1. This mean that we do not need to know the time for the step injected in the system as well as we do not need to know the exact time for the t_1 and t_2. We only need to know the time difference Δt.

Using $\Delta t = \tau$ in (15) gives directly:

$$R_1(\tau) = (1 - e^{-1}) = 0.632 \quad (16)$$

This is the same result (with $k = 1$) as given by (4) when the calculation is based on the whole step response. (15) and (16) shows that we do not need to have the complete step response for calculation of the time constant τ.

A part of the step response in Fig. 3 is shown in Fig. 4. Here, the tangent has been calculated at $t = 1.5sec$. It is shown that the tangent is equal to the steady state value of the system for $t = 1.5sec + \tau = 2.5sec$. Further, we can also calculate $R_1(\tau)$ from Fig. 4. Based on the data in Fig. 4, we get:

$$\begin{aligned} R_1(\tau) &= \frac{y(1.5+\tau)-y(1.5)}{y'(1.5)\tau} \\ &= \frac{0.918-0.777}{0.223} \\ &= 0.632 \end{aligned}$$

where $y'(1.5) = 0.223$.

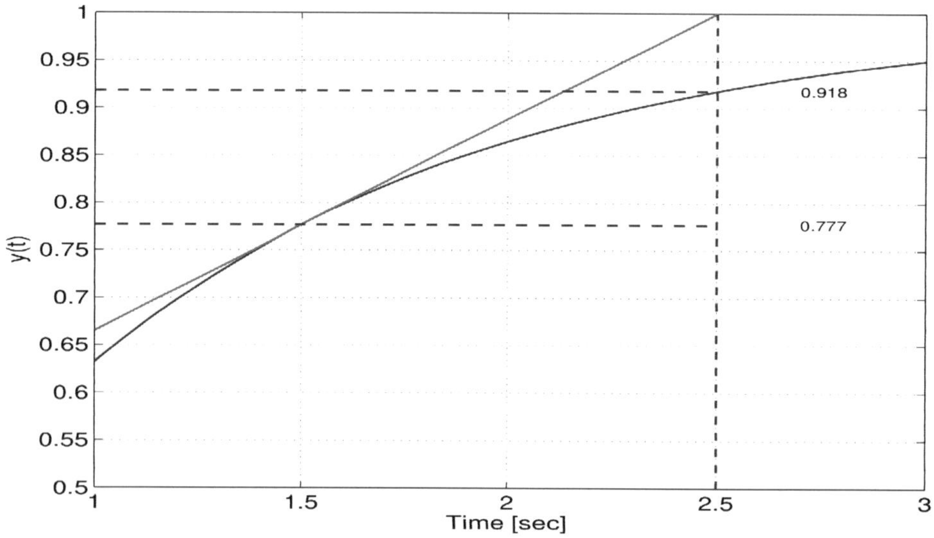

Figure 4. Part of step response for a first order system with $k = \tau = 1$.

In the above results, it is shown that it is easy to find τ where $R_1(\Delta t)$ is equal to $0,632$. However, τ might be larger than $t_2 - t_1$, so it is not possible to get the ratio directly given by (16). Instead, it is possible to calculate the ratio for different values of Δt. Now, let (15) be independent of the time by using

$$\Delta t = \alpha\tau$$

where α is a scaling factor with respect to the time constant τ. Including this in (15) gives directly:

$$R_1(\alpha\tau) = \frac{1 - e^{-\alpha}}{\alpha} \quad (17)$$

Note that the ratio R_1 is independent of τ but depend only on α, i.e. $R_1(\alpha\tau) = R_1(\alpha)$. The ratio $R_1(\alpha)$ as function of α is shown in Fig. 5.

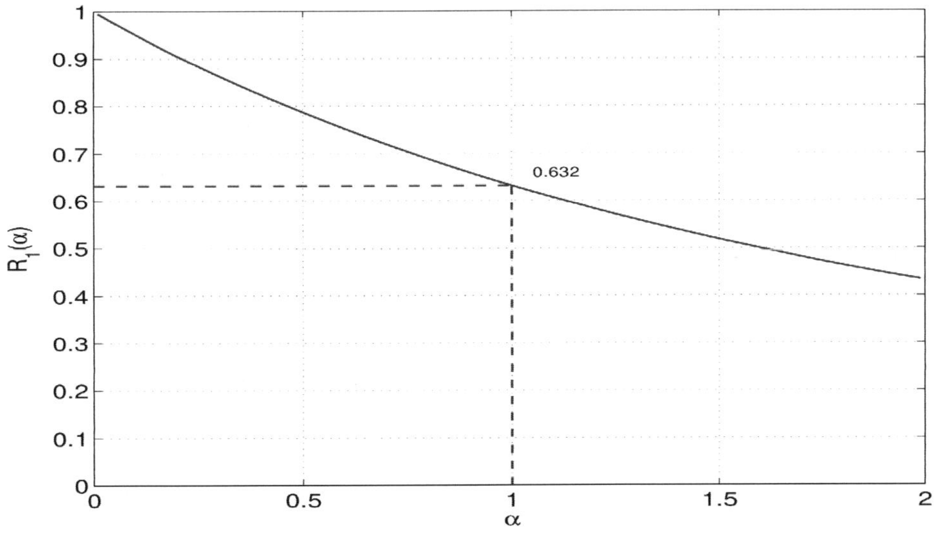

Figure 5. The ratio $R_1(\alpha)$ is shown as function of of α.

Based on Fig. 5, the time constant τ is then given by:

$$\tau = \frac{\Delta t}{\alpha} \tag{18}$$

Again a part of the step response from Fig. 3 is shown in Fig. 6. The same tangent as shown in Fig. 4 has been shown, but the calculation of τ is now based on $0.5sec$. Based on the data from the figure, $R_1(\Delta t)$ in (14) is then given by:

$$\begin{aligned} R_1(0.5) &= \tfrac{y(2.0)-y(1.5)}{y'(1.5)0.5} \\ &= \tfrac{0.865-0.777}{0.223\times 0.5} \\ &= 0.789 \end{aligned}$$

From Fig. 5, we have that $R_1 = 0.789$ gives an α around 0.5. From (18) gives directly that $\tau = 1$.

Until now, we have only used a part of the step response to calculate the steady state value y_{ss} and the time constant τ. We still need to calculate the gain k and the time t_0 when the step is injected in the system. In general, we will know t_0, but we need to be able to calculate the start of the step response in the case when the system includes a time delay. It will then be possible to calculate the time delay as the difference between the time for the start of the step response and t_0. So based on this, assume that we know y_0 and calculate k and t_0 based on this information.

3.2. Estimation of gain k

It has been assumed that the amplitude of the input is 1. This gives directly that the gain k is given by:

$$k = y_{ss} - y_0 \tag{19}$$

where the steady state output y_{ss} is calculated from (12) when the time constant τ is known.

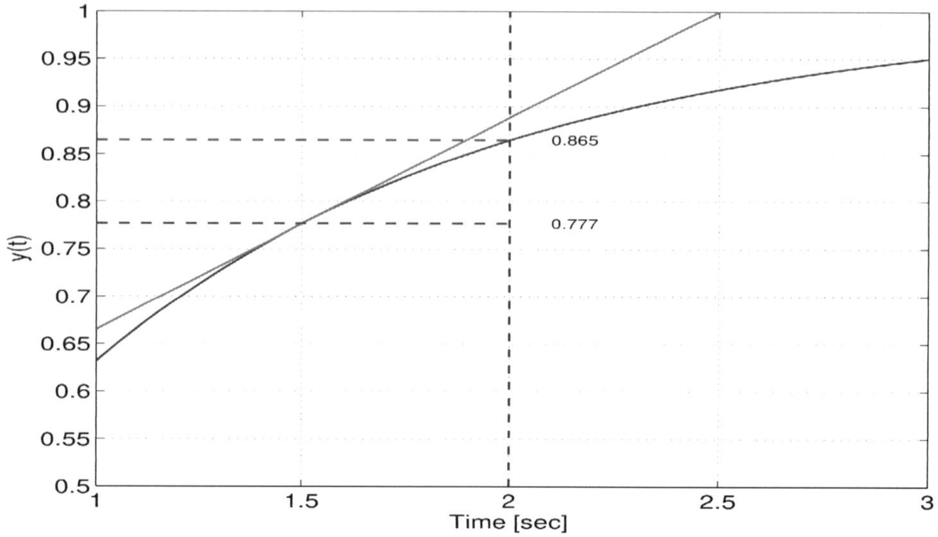

Figure 6. Part of step response for a first order system with $k = \tau = 1$.

If a step is applied with an amplitude different from 1, the term $y_{ss} - y_0$ need to be divided by this amplitude.

3.3. Calculation of the step time t_0

For calculation of t_0, consider the step response at time $t = t_1$. From (8) we have directly:

$$\begin{aligned} y(t_1) &= y_0 + k(1 - e^{-(t_1 - t_0)/\tau}) \\ &= y_0 + k(1 - e^{-(\Delta t_1)/\tau}) \end{aligned} \quad (20)$$

Rewriting (20) gives the following index $R_2(\Delta t_1)$:

$$R_2(\Delta t_1) = \frac{y(t_1) - y_0}{k} = (1 - e^{-(\Delta t_1)/\tau}) \quad (21)$$

The index $R_2(\Delta t_1)$ can be calculated based on the output $y(t_1)$, the off-set y_0 and the gain k. The index $R_2(\Delta t_1)$ will always be in the interval from 0 to 1. $R_2(\Delta t_1)$ can now be used for the calculation of Δt_1. Again, let's introduce a scaling factor β:

$$\Delta t_1 = \beta \tau \quad (22)$$

This gives the following index $R_2(\beta\tau)$:

$$R_2(\beta\tau) = (1 - e^{-\beta}) \quad (23)$$

As in connection with R_1, R_2 is also independent of τ.

The index $R_2(\beta)$ as function of β is shown in Fig. 7.

Based on Fig. 7, β can be found and then Δt_1 can be calculated by using (22). Let t_0 be the time when the response starts. With t_1 known, the time that the t_0 is given by:

$$t_0 = t_1 - \Delta t_1 \quad (24)$$

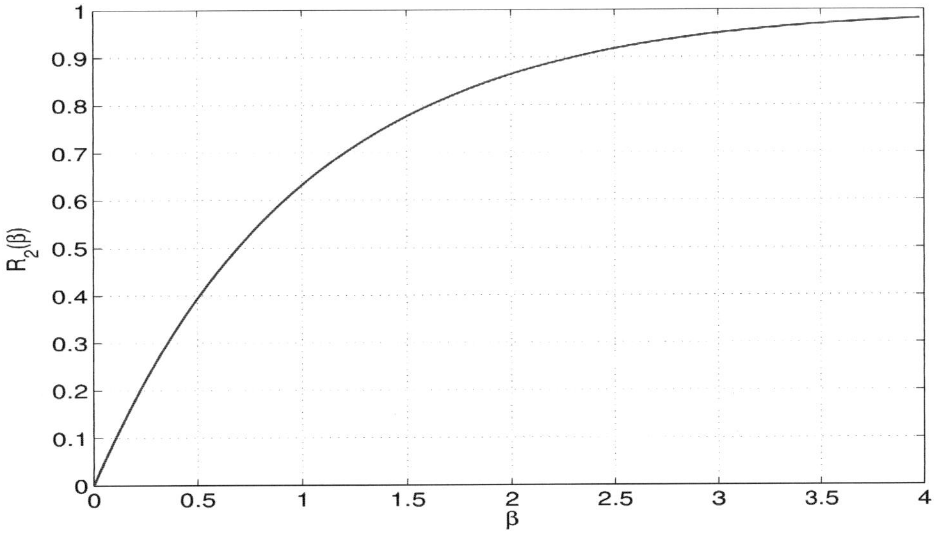

Figure 7. The ratio $R_2(\beta)$ is shown as function of of β.

Consider again the step response shown in Fig. 6. From the data in the figure, we can calculate R_2:

$$R_2 = \frac{y(2) - y_0}{k} = \frac{0.865 - 0}{1} = 0.865$$

From Fig. 7, β can then be found to 2. This gives directly t_0:

$$t_0 = t_1 - \Delta t_1 = t_1 - \beta\tau = 2 - 2 \times 1 = 0$$

3.4. System with time delay

In the case when the system includes a time delay as given by (1), the time delay θ also needs to be calculated. Again, it is assumed that the step is applied in the system at t_0. Let t_θ be the time when the system react. (24) take then the following form:

$$t_\theta = t_1 - \Delta t_1 \tag{25}$$

The system time delay is the difference between t_0 and t_θ, i.e.

$$\theta = t_\theta - t_0 = t_1 - t_0 - \Delta t_1 \tag{26}$$

4. Example

Let's consider again a step response from a refrigation system shown in Fig. 8.

As it can be seen from Fig 8, the system has different dynamics depending of the opening degree of the valve as expected. Further, it is also clear that it is not possible to describe the dynamic only by using first order systems. However, it is in general assumed that the main dynamic in refrigeration systems can be described by using first order dynamics.

In this example, let's consider the step response for an opening degree of 100%. In Fig. 9, a part of the step response from Fig. 8 for t = 1430 sec. to t = 1670 sec. is shown.

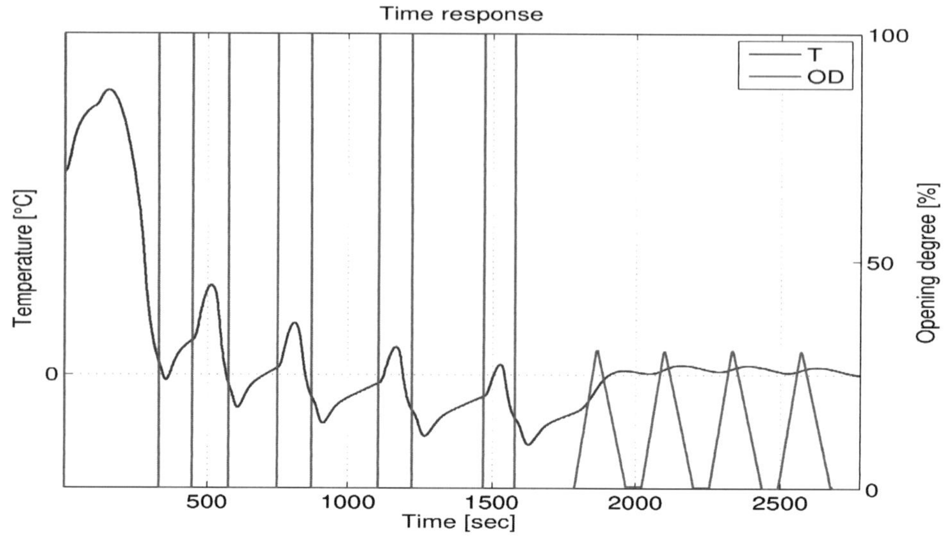

Figure 8. A step response from a refrigeration system. The opening degree is the input to the valve and temperature is the output of the system.

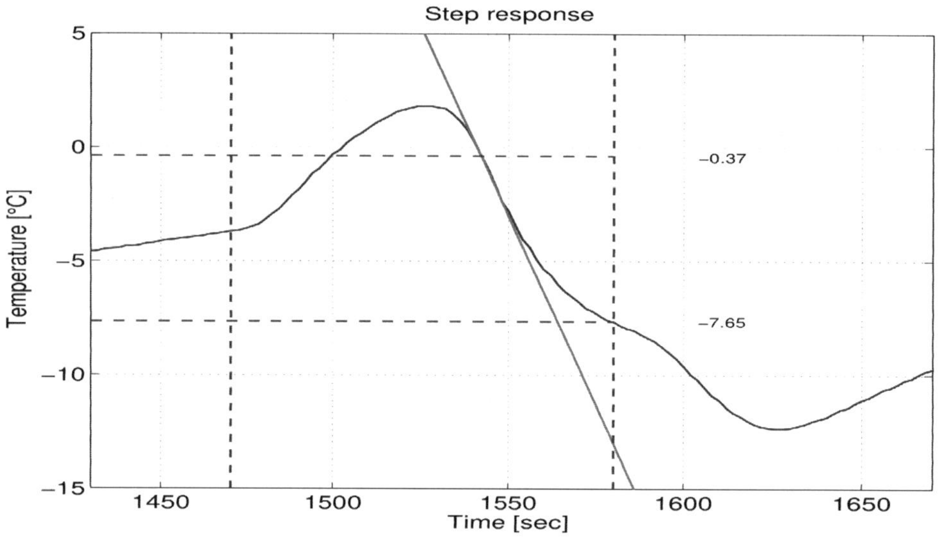

Figure 9. Part of the step response from Fig 8. The two vertical lines indicate the start and stop of the step input. Further, the calculation of the tangent around $t = 1542$ sec.

The step input starts at $t = 1470$ sec. and stops at $t = 1580$ sec. The opening degree is 100% in this time interval. The temperature is maximal at $t = 1524$ sec. with $T = 1.81^oC$. Further, it can also be seen from Fig. 9 that the response fits reasonable with a first order response when the temperature gets below zero degree around $t = 1542$ sec. to around $t = 1590$ sec. with $T = -8.36^oC$.

Calculating a tangent to the step response curve around $t = 1542$ sec. gives the equation for the tangent as:

$$T_{tangent}(t) = -0.32 * t + 493.1$$

The tangent is also shown in Fig. 9.

Based on the tangent in Fig. 9, the ratio $R_1(\Delta t)$ given by (14) can be calculated. From the data given in Fig. 9 together with $y'(t_1) = -0.32$ and $\Delta t = 38$ sec. gives

$$R_1(\Delta t) = \frac{y(t_1 + \Delta t) - y(t_1)}{y'(t_1)\Delta t} = \frac{-7.65 - (-0.37)}{-0.32 \times 38} = 0.599$$

From Fig. 5, α can be calculated. This gives (can also be calculated by using (17)):

$$\alpha = 1.13$$

The time constant is then given by:

$$\tau = \frac{\Delta t}{\alpha} = \frac{38}{1.13} = 33.6 \approx 34\ sec.$$

The steady state value of the temperature can then be calculated based on the tangent. The tangent crosses the steady state value at $t_1 + \tau = 1542 + 34$ sec. $= 1576$ sec. Using the equation for the tangent gives that the steady state value for the temperature is:

$$T_{ss} = -11.25$$

From Fig. 9, it is clear that the system is not in steady state when the step starts, so we need to decide the steady state value before the step. We select the maximal temperature in the interval as the steady state value, i.e. $T_0 = 1.81^oC$. The reason is that at this point it starts decreasing due to the step input. From this, the gain of the system, k, is given by:

$$k = T_{ss} - T_0 = -11.25 - 1.81 = -13.06 \approx -13$$

The time when the system react t_θ can now be calculated. This is done by using the ratio $R_2(\Delta t)$ given by (21). Using $T(1542) = -0.37^oC$ gives:

$$R_2(\Delta t_1) = \frac{-0.368 - 1.81}{-13} = 0.17$$

Using Fig. 7, β is found to:

$$\beta \approx 0.2$$

which gives that

$$t_\theta = t_1 - \beta\tau = 1542 - 0.2 \times 34 \approx 1535 sec.$$

The time delay is then given by:

$$\theta = t_\theta - t_0 = 1535 - 1470 = 65 sec.$$

The complete system is then given by

$$G(s) = -13\frac{1}{34s + 1}e^{-65s}$$

The step response from the system and the model is shown in Fig. 10. As expected, the step response from the model does not follow the system exactly, but it will in general be good enough.

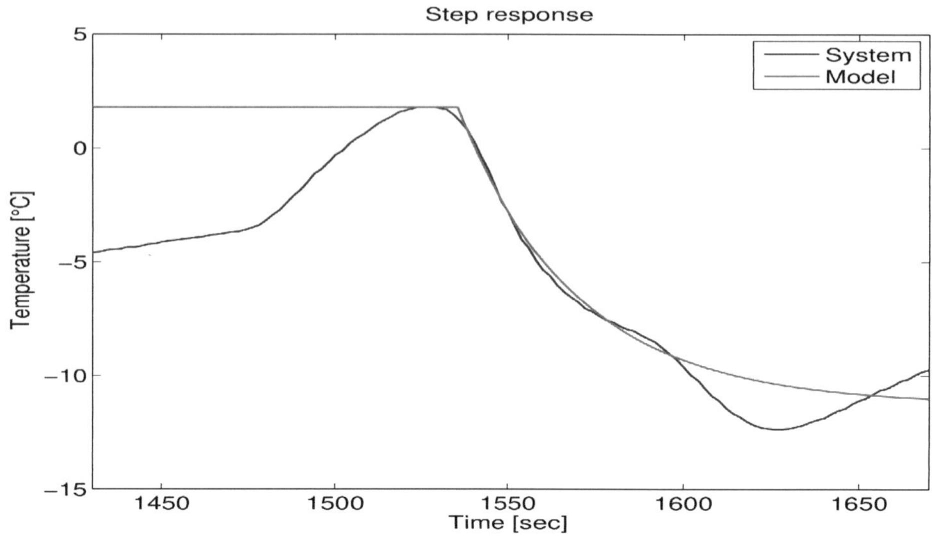

Figure 10. Step response from the real system compared with a step response from the model.

5. Conclusion

A simple method for estimation of parameters in a first order system with time delay has been considered. It is shown that it is possible to estimate the parameters based only on a part of a step response. This is very relevant in connection with data from real systems, where it will not always be possible to get a complete step response, but only part of a step response will be available.

References

[1] S. Ahmed, B. Huang, and S.L. Shah. Novel identification method from step response. *Control Engineering practice*, 15:545 – 556, 2007.

[2] K.J. Åström and T. Hagglund. *PID controllers: Theory, Design and Tuning*. Instrument Society of America, 1995.

[3] A. Eisinberg, G. Fedele, and D. Frascino. An analytic optimization procedure to estimate a first-order plus time delay model from step response. In *16th Mediterranean Conference on Control and Automation*, pages 729 – 734, Congress Centre, Ajaccio, France, June 2008.

[4] G.F. Franklin, J.D. Powell, and A. Emami-Naeini. *Feedback control of dynamic systems*. Pearson, 2010.

[5] D.T. Korsane, V. Yadav, and K.H. Raut. Pid tuning rules for first order plus time delay system. *Int. Journal of Innovative Research in Electrical, Instrumentation and Control Engineering*, 2(1):582 – 586, January 2014.

[6] K. Ogata. *Modern systems engineering*. Pearson, 2010.

[7] D.D. Ruscio. On system pi controllers for integrating plus time delay systems. *Modelling, Identification and Control*, 31(4):145 – 164, 2010.

[8] F.A. Salem. New efficient model-based pid design method. *European Scientific Journal*, 9(15):181 – 199, May 2013.

[9] J.B.M. Santos and P.R. Barros. Simple identification techniques for first-order and second-order plus time-delay systems. In *XVIII Congresso Brasileiro de Automatica*, pages 1432 – 1439, Bonito-MS, September 2010.

[10] S. Skogestasd. Probably the best simple pid tuning rules in the world.

[11] S.W. Sung, J. Lee, and I.B. Lee. *Process identification and PID control*. John Wiley & Sons (Asia), 2009.

[12] Cheng-Ching Yu. *Autotuning of PID controllers - A relay feedback approach*. Springer Verlag, London, 2006.

Journal of Physics: Conference Series **570** (2014) 012002 doi:10.1088/1742-6596/570/1/012002

Gaussian process based recursive system identification

Jakub Prüher and Miroslav Šimandl

University of West Bohemia, Czech Republic

E-mail: `jacobnzw@ntis.zcu.cz`, `simandl@ntis.zcu.cz`

Abstract. This paper is concerned with the problem of recursive system identification using non-parametric Gaussian process model. Non-linear stochastic system in consideration is affine in control and given in the input-output form. The use of recursive Gaussian process algorithm for non-linear system identification is proposed to alleviate the computational burden of full Gaussian process. The problem of an online hyper-parameter estimation is handled using proposed ad-hoc procedure. The approach to system identification using recursive Gaussian process is compared with full Gaussian process in terms of model error and uncertainty as well as computational demands. Using Monte Carlo simulations it is shown, that the use of recursive Gaussian process with an ad-hoc learning procedure offers converging estimates of hyper-parameters and constant computational demands.

1. Introduction

Throughout the sciences the task of building models from observed data is of fundamental importance. In the area of control this task is known as system identification. The way of obtaining the model is in direct contrast to mathematical modelling, where the knowledge from wide variety of disciplines, such as physics, biology, economy and engineering, is exploited. In control engineering, the need for system identification arises whenever there is uncertainty in the knowledge of the system central to the task.

Identification methods can be divided according to the way they process data into: *batch* methods and *recursive* methods. By the batch identification we will understand a method that processes all the measured data at once to produce the system model. On the contrary, recursive identification algorithms process the measurements as they become available during the course of the system operation and gradually refine the system model. These algorithms are at the core of adaptive control mechanisms.

Models of the system can be *parametric* or *non-parametric*. Parametric models typically assume fixed structure with finite number of free parameters, which need to be estimated. Well known model structures include for example ARX and ARMAX models for linear systems. For non-linear systems various types of neural networks can be named. On the other hand, non-parametric models assume no fixed structure and, from the viewpoint of parametric models, their number of 'parameters' can be potentially infinite. Spectral and correlation approaches fall into this category along with increasingly popular Gaussian process regression models.

Contrary to the parametric models, such as the neural networks, Gaussian process (GP) models do not possess a fixed structure and thus are much more flexible. Neal showed [1], that a neural network, where the number of hidden units goes to infinity, is equivalent to GP model with certain covariance function. Specifying a GP can be effectively thought of as putting a prior distribution over functions themselves, where functions can be informally regarded as vectors of function values of infinite dimension. Various communities pay increasing attention to the GP models due to their effectiveness in modelling of highly non-linear phenomena. They are attractive, also because they combine tractable Bayesian inference with non-parametric nature of the model. GP models have been utilised in many applications. These include non-linear filtering [2–5], model predictive control [6], non-linear system identification [7], time series forecasting [8], reinforcement learning [9, 10], numerical quadrature [11] and many others.

Great advantage of parametric models is that they lend themselves nicely to the possibility of devising a recursive estimation algorithm. As GPs are non-parametric models, the situation is much more complicated. Typically, it is necessary to consider some kind of GP approximation to counteract the problem of increasing data size. Särkkä [12] proposed a solution to this problem based on Kalman filtering and smoothing applied to the state-space representation of the GP with specific covariance functions. This solution, however, works for one-dimensional inputs only and is thus inappropriate for our purposes. Very frequently used alternative is the Sparse Online Gaussian Process (SOGP) [13–15]. SOGP however does not solve the problem of online learning of hyper-parameters.

Our goal is the investigation of the recently proposed *Recursive Gaussian Process* (RGP) algorithm [16] for the purpose of non-linear system identification. Ultimately, we envision the future incorporation of RGP system identification algorithm into the functional dual adaptive control framework [17, 18].

The rest of the paper will present a description of the problem formulation in section 2, followed by an RGP based non-linear system identification in section 3. In section 4, a simple numerical illustration comparing the full GP and the recursive GP in non-linear system identification is presented. Finally, section 5 concludes the paper.

2. Problem formulation

In this work, we are dealing with the non-linear stochastic discrete time-invariant system given in an input-output representation

$$\mathcal{S}: \qquad y_{k+1} = f\left(\boldsymbol{x}_k^a\right) + g\left(\boldsymbol{x}_k^a\right) u_k + e_{k+1}, \tag{1}$$

where $f, g : \mathbb{R}^{n_y+n_u+1} \to \mathbb{R}$ are unknown non-linear functions, $\boldsymbol{x}_k^a = \left[y_k, \dots, y_{k-n_y}, u_{k-1}, \dots, u_{k-n_u}\right]^\mathrm{T} \in \mathbb{R}^{n_y+n_u+1}$ is the state vector, u_k and y_k are input and output signals respectively at discrete time instants $k \in 0, 1, \dots,$ N-1 and e_k is the noise term. In addition, the following assumptions are considered:

A1: The non-linear functions are smooth, i.e. $f\left(\boldsymbol{x}_k^a\right), g\left(\boldsymbol{x}_k^a\right) \in C^\infty$.

A2: The structural parameters n_y and n_u of the system are known.

A3: $e_k \in \mathbb{R}$ is a white zero-mean Gaussian noise with known variance σ_e^2.

The unknown non-linear system is modelled by a non-parametric GP model as an alternative to the parametric neural network model [19]. The structural information about affinity of the system input u_k is tackled by employing a specific form of covariance function. We postpone discussion of the specifics to the section 3, where modelling of the non-linear system with the help of GPs is described in detail in system identification setting. For now, the focus will be given to the GP regression in it's original formulation to illustrate the shortcomings, when used for sequential data processing. This approach is termed *full GP* (FGP) in section 4.

To motivate the need for recursive GP regression algorithm, the Gaussian process regression problem will be described. *Gaussian process regression* refers to the estimation of an underlying function $h : \mathbb{R}^D \to \mathbb{R}$ of a noisy static mapping

$$y_i = h(\boldsymbol{x}_i) + \varepsilon_i \tag{2}$$

from a given set of input vectors $\boldsymbol{X} = \left\{\boldsymbol{x}_i \in \mathbb{R}^D\right\}_{i=1}^n$ and corresponding set of observations $\boldsymbol{y} = \left\{y_i \in \mathbb{R}\right\}_{i=1}^n$, where $\{\varepsilon_i \in \mathbb{R}\}_{i=1}^n$ is the noise sequence. The key concept here is that the estimated unknown function h is modelled using a GP prior

$$h \sim \mathcal{GP}(m, k) \tag{3}$$

which, akin to the Gaussian distribution, is fully determined by its first two moments defined as

$$m(\boldsymbol{x}_p \mid \boldsymbol{\theta}) \triangleq \mathrm{E}_h\Big[h(\boldsymbol{x}_p) \mid \boldsymbol{\theta}\Big], \tag{4}$$

$$k(\boldsymbol{x}_p, \boldsymbol{x}_q \mid \boldsymbol{\theta}) \triangleq \mathrm{E}_h\Big[\big(h(\boldsymbol{x}_p) - m(\boldsymbol{x}_p)\big)\big(h(\boldsymbol{x}_q) - m(\boldsymbol{x}_q)\big) \mid \boldsymbol{\theta}\Big], \tag{5}$$

where E_h is expectation over function values h, $m(\boldsymbol{x}_p \mid \boldsymbol{\theta}) : \mathbb{R}^D \to \mathbb{R}$ is a *mean function* and $k(\boldsymbol{x}_p,\ \boldsymbol{x}_q \mid \boldsymbol{\theta}) : \mathbb{R}^D \times \mathbb{R}^D \to \mathbb{R}$ is a *covariance function* for all $\boldsymbol{x}_p, \boldsymbol{x}_q \in \mathbb{R}^D$. The covariance function is parametrised by a vector of free *hyper-parameters* $\boldsymbol{\theta}$. To keep the notation uncluttered, dependence of $m(\cdot)$ and $k(\cdot)$ on $\boldsymbol{\theta}$ will be omitted from now on. Designers have an incredible variety of covariance functions at their disposal each expressing different assumptions about the modelled phenomena. For comprehensive account of the wide range of covariance functions and their relationship to other models reader should consult [20]. Compared to parametric regression models, where the assumptions on the modelled function are typically expressed in the form of fixed model structure, the assumptions of non-parametric GP models, expressed by the choice of covariance function, are much less restrictive.

With a GP prior in hand, Bayesian techniques of inference can further be employed to refine the GP prior $p(h \mid \boldsymbol{\theta})$ by incorporating the evidence from the set of input vectors $\boldsymbol{X}$ and observations $\boldsymbol{y}$ to produce the GP posterior

$$p(h \mid \boldsymbol{y}, \boldsymbol{X}, \boldsymbol{\theta}) = \frac{p(\boldsymbol{y} \mid h, \boldsymbol{X}, \boldsymbol{\theta})\, p(h \mid \boldsymbol{\theta})}{p(\boldsymbol{y} \mid \boldsymbol{X}, \boldsymbol{\theta})}, \tag{6}$$

where the GP posterior $h \mid \boldsymbol{y}, \boldsymbol{X}, \boldsymbol{\theta} \sim \mathcal{GP}(m^+, k^+)$ is fully described by a posteriori mean function m^+ and covariance function k^+ given by

$$m^+(\boldsymbol{x}) = m(\boldsymbol{x}) + k(\boldsymbol{x}, \boldsymbol{X}) k(\boldsymbol{X}, \boldsymbol{X})^{-1} \boldsymbol{y}, \tag{7}$$

$$k^+(\boldsymbol{x}, \boldsymbol{x}) = k(\boldsymbol{x}, \boldsymbol{x}) - k(\boldsymbol{x}, \boldsymbol{X}) k(\boldsymbol{X}, \boldsymbol{X})^{-1} k(\boldsymbol{X}, \boldsymbol{x}), \tag{8}$$

where $k(\boldsymbol{x}, \boldsymbol{X}) = \big[k(\boldsymbol{x}, \boldsymbol{x}_1), \ldots, k(\boldsymbol{x}, \boldsymbol{x}_n)\big]$, $k(\boldsymbol{X}, \boldsymbol{x}) = k(\boldsymbol{x}, \boldsymbol{X})^\top$ and $k(\boldsymbol{X}, \boldsymbol{X})$ is an $n \times n$ prior covariance matrix of elements $k(\boldsymbol{x}_i, \boldsymbol{x}_j)$. In other words, for any fixed $\boldsymbol{x} \in \mathbb{R}^D$ the equations (7) and (8) define the Gaussian posterior predictive distribution over the values of function h at this point.

Looking at equations (7), (8) one can notice the fact that the prediction is calculated using all available data, which is the very feature of non-parametric GP modelling. In offline settings, where all the data are available beforehand, this results in the necessity to invert large matrices. Moreover, in situations, where the real-time data processing is required, the size of the matrix $k(\boldsymbol{X}, \boldsymbol{X})$ grows, which leads to ever increasing computational demands for matrix inversion in each time step as more and more data are being processed. To alleviate this problem, we propose the use of RGP regression algorithm for system identification task, which is described in the next section.

3. Recursive Gaussian process regression

This section covers the identification of a non-linear stochastic system by means of the RGP regression algorithm. RGP algorithm is briefly outlined along with the ad-hoc procedure for hyper-parameter learning. Last section covers the application of RGP to the system identification task.

3.1. Recursive Gaussian Process algorithm

Recursive Gaussian process regression algorithm (RGP) was proposed by Huber [16] as a solution to the problem of GP regression, where the data arrive sequentially. The main idea employed in RGP is to use a predefined set of *basis vectors* as a sparsification element [21], from which the predictions are computed. The number of basis vectors $s \ll N$ (where N is the number of data points) is fixed throughout the operation of the algorithm, which enables to keep the computational demands in check. On the other hand, this acts as an approximation to the full GP approach characterized by the equations (7) and (8) as the prediction is no longer a function of all the previously seen data.

Keeping with the notation of [16], let $\boldsymbol{X} = [\boldsymbol{x}_1, \boldsymbol{x}_2, \ldots, \boldsymbol{x}_s]$ denote the matrix of locations of the basis vectors and $\boldsymbol{g} = g(\boldsymbol{X})$ the corresponding vector of values of the unknown latent function g. Because g is considered to be a GP, the distribution $p_0(\boldsymbol{g}) = \mathcal{N}(\boldsymbol{g}; \boldsymbol{\mu}_0^g, \boldsymbol{C}_0^g)$ at initial time step $k = 0$ is Gaussian. For any time step $k > 0$, new set of nk observations $\boldsymbol{y}_k = [y_{k,1}, y_{k,2}, \ldots, y_{k,nk}]$ at input locations $\boldsymbol{X}_k = [\boldsymbol{x}_{k,1}, \boldsymbol{x}_{k,2}, \ldots, \boldsymbol{x}_{k,nk}]$ is processed. The goal is then to calculate posterior distribution

$$p_k(\boldsymbol{g} | \boldsymbol{y}_{1:k}) = \mathcal{N}(\boldsymbol{g}; \boldsymbol{\mu}_k^g, \boldsymbol{C}_k^g) \tag{9}$$

at time k, where $\boldsymbol{y}_{1:k} = [\boldsymbol{y}_1, \boldsymbol{y}_2, \ldots \boldsymbol{y}_k]$, by combining the new observations $\boldsymbol{y}_k$ with the distribution

$$p_{k-1}(\boldsymbol{g} | \boldsymbol{y}_{1:k-1}) = \mathcal{N}(\boldsymbol{g}; \boldsymbol{\mu}_{k-1}^g, \boldsymbol{C}_{k-1}^g) \tag{10}$$

from the previous time step, which functions as a prior in Bayesian setting. Calculation of the posterior $p_k(\boldsymbol{g} | \boldsymbol{y}_{1:k})$ is performed in two steps:

Inference: calculating the joint prior $p_{k-1}(\boldsymbol{g}, \boldsymbol{g}_k | \boldsymbol{y}_{1:k-1})$ given the prior (10),
Update: updating the joint prior with new observations and integrating out $\boldsymbol{g}_k$.

In summary, RGP algorithm operates by means of the following set of equations

$$\boldsymbol{J}_k = k(\boldsymbol{X}_k, \boldsymbol{X}) k(\boldsymbol{X}, \boldsymbol{X})^{-1}, \tag{11}$$
$$\boldsymbol{\mu}_k^p = m(\boldsymbol{X}_k) + \boldsymbol{J}_k \left(\boldsymbol{\mu}_{k-1}^g - m(\boldsymbol{X})\right), \tag{12}$$
$$\boldsymbol{C}_k^p = k(\boldsymbol{X}_k, \boldsymbol{X}_k) + \boldsymbol{J}_k \left(\boldsymbol{C}_{k-1}^g - k(\boldsymbol{X}, \boldsymbol{X})\right) \boldsymbol{J}_k^\top, \tag{13}$$

$$\boldsymbol{G}_k = \boldsymbol{C}_{k-1}^g \boldsymbol{J}_k \left(\boldsymbol{C}_k^p + \sigma^2 \boldsymbol{I}\right)^{-1}, \tag{14}$$
$$\boldsymbol{\mu}_k^g = \boldsymbol{\mu}_{k-1}^g + \boldsymbol{G}_k \left(\boldsymbol{y}_k - \boldsymbol{\mu}_k^p\right), \tag{15}$$
$$\boldsymbol{C}_k^g = \boldsymbol{C}_{k-1}^g - \boldsymbol{G}_k \boldsymbol{J}_k \boldsymbol{C}_{k-1}^g. \tag{16}$$

The equations (11) – (13) perform the inference step and serve for function value predictions $\boldsymbol{g}_k$ at locations $\boldsymbol{X}_k$ using previous estimate $\boldsymbol{\mu}_{k-1}^g$. Equation (13) provides covariance matrix of the predictions. The equations (14) – (16) perform the update step, where the first two moments of the posterior (9) are updated using the last observations $\boldsymbol{y}_k$. For detailed explanations and derivation of the equations (11) – (16) reader is referred to the original article [16].

3.2. Ad-hoc online learning procedure

RGP algorithm is able to perform GP regression recursively. However, the problem of online hyper-parameter learning still stands. One possibility would be utilization of the RGP* algorithm [22], which is able to learn the hyper-parameters in recursive manner in addition to recursively performing the GP regression. Nonetheless, RGP* algorithm proved to be challenging for implementation. For this reason we resort to tackling the online hyper-parameter learning problem by the following ad-hoc procedure.

Typically, to find an optimal setting of the hyper-parameters $\boldsymbol{\theta}$, the logarithm of marginal likelihood

$$\log p(\boldsymbol{Y}_k | \boldsymbol{X}_k, \boldsymbol{\theta}) = -\frac{1}{2}\left[\boldsymbol{Y}_k^\top \left(\boldsymbol{K}_\theta + \sigma_e^2 \boldsymbol{I}\right)^{-1} \boldsymbol{Y}_k + \log\left|\boldsymbol{K}_\theta + \sigma_e^2 \boldsymbol{I}\right| + n \log(2\pi)\right] \tag{17}$$

is maximised, where $\boldsymbol{K}_\theta = k(\boldsymbol{X}_k, \boldsymbol{X}_k)$. The advantage of the marginal likelihood maximisation is that it produces optimal values that do not result in the model over-fitting or under-fitting. The disadvantage is that the optimisation problem is non-linear which makes finding of the global maxima practically impossible. It is helpful to point out, that the objective function (17) is a function of $\boldsymbol{\theta}$, where the data (noisy observations $\boldsymbol{Y}_k$ at locations $\boldsymbol{X}_k$) are given. In sequential framework these quantities represent all the currently available data up to a certain time step.

The idea of the proposed ad-hoc approach lies in the treatment of the current estimate of the latent function values $\boldsymbol{\mu}_k^g$ as perfect observations. In practical terms this entails modification of the marginal likelihood objective function (17), whereby the noisy observations $\boldsymbol{Y}_k$ are replaced with the perfect observations $\boldsymbol{\mu}_k^g$ at basis vector locations $\boldsymbol{X}$. Thus we obtain the following objective function

$$\log p\left(\boldsymbol{\mu}_k^g \mid \boldsymbol{X}, \boldsymbol{\theta}\right) = -\frac{1}{2}\left[\left(\boldsymbol{\mu}_k^g\right)^\top \boldsymbol{K}_\theta^{-1} \boldsymbol{\mu}_k^g + \log\left|\boldsymbol{K}_\theta\right| + s\log(2\pi)\right], \tag{18}$$

which is maximized in every time step using iterative optimization solver, with the initial starting point set to the hyper-parameter estimate obtained in the previous time step. With maximum number of solver iterations fixed, the computational burden associated with hyper-parameter estimation is kept bounded. The notation $\boldsymbol{K}_\theta$ emphasizes the fact that the prior covariance matrix is a function of hyper-parameters $\boldsymbol{\theta}$.

3.3. Non-linear system identification with RGP

The unknown non-linear stochastic system to be identified (1) is modelled by the GP prior. This can be written in the following way

$$\mathcal{M}: \qquad \hat{y}_{k+1} \sim \mathcal{GP}(m, k), \tag{19}$$

where $\hat{y}_{k+1}$ is the prediction of system output at the next time step. Let $\boldsymbol{X}^a = \left[\boldsymbol{x}_1^a, \ldots, \boldsymbol{x}_s^a\right]$ be a $D \times s$ matrix, where each column is a vector of regressors of dimension $D = ny + nu + 1$ and let $\boldsymbol{X} = \left[(\boldsymbol{X}^a)^\top, \boldsymbol{U}^\top\right]^\top$, where $\boldsymbol{U} = [u_1, \ldots, u_s]$ is the last row of $(D+1) \times s$ basis vector matrix $\boldsymbol{X}$. The first two moments of the GP prior were chosen as

$$m\left(\boldsymbol{x}_p\right) \equiv 0, \tag{20}$$

$$k\left(\boldsymbol{x}_p, \boldsymbol{x}_q\right) = k_f\left(\boldsymbol{x}_p^a, \boldsymbol{x}_q^a\right) + k_g\left(\boldsymbol{x}_p^a, \boldsymbol{x}_q^a\right)u_p u_q + k_n\left(\boldsymbol{x}_p^a, \boldsymbol{x}_q^a\right), \tag{21}$$

where

$$k_f\left(\boldsymbol{x}_p^a, \boldsymbol{x}_q^a\right) = s_f^2 \exp\left(-\frac{1}{2}\left(\boldsymbol{x}_p^a - \boldsymbol{x}_q^a\right)^\top \boldsymbol{\Lambda}_f^{-1}\left(\boldsymbol{x}_p^a - \boldsymbol{x}_q^a\right)\right), \tag{22}$$

$$k_g\left(\boldsymbol{x}_p^a, \boldsymbol{x}_q^a\right) = s_g^2 \exp\left(-\frac{1}{2}\left(\boldsymbol{x}_p^a - \boldsymbol{x}_q^a\right)^\top \boldsymbol{\Lambda}_g^{-1}\left(\boldsymbol{x}_p^a - \boldsymbol{x}_q^a\right)\right), \tag{23}$$

$$k_n\left(\boldsymbol{x}_p^a, \boldsymbol{x}_q^a\right) = \sigma_e^2 \delta_{pq}. \tag{24}$$

The quantities s_f, s_g are vertical lengthscales, $\boldsymbol{\Lambda}_f = \operatorname{diag}\left(\lambda_{f1}^2, \ldots, \lambda_{fD}^2\right)$ are horizontal lengthscales and σ_e^2 is the system output noise variance. The symbol δ_{pq} stands for Kronecker delta. The above mentioned quantities are GP hyper-parameters forming a vector $\boldsymbol{\theta} = \left[s_f^2, \lambda_{f1}, \ldots, \lambda_{fD}, s_g^2, \lambda_{g1}, \ldots, \lambda_{gD}\right]^\top$. Choice of the form of the covariance function $k\left(\boldsymbol{x}_p, \boldsymbol{x}_q\right)$ follows from the system description (1) and is inspired by [23], where this particular structure of covariance was exploited to ease the derivation of control law. The term k_n models the system output disturbance as a white Gaussian noise with variance σ_e^2. The form of the covariance functions k_f and k_g is often called *squared exponential* with *automatic relevance determination*, which is a common choice in a number of applications [4, 7]. This choice expresses the assumption that the functions f and g involved in the system equation (1) are smooth, i.e. they meet the assumption A1 from section 2.

With GP prior specified by (20) – (24), we can now derive expressions for system output prediction and prediction variance. Using the equations (11) – (16) we obtain

$$\mathrm{E}\left[\hat{y}_{k+1}\right] = \boldsymbol{\mu}^{p}_{k+1} = \boldsymbol{J}_k \boldsymbol{\mu}^{g}_{k}, \tag{25}$$

$$\mathrm{var}\left[\hat{y}_{k+1}\right] = \boldsymbol{C}^{p}_{k+1} = k_{ss} + \boldsymbol{J}_k\left(\boldsymbol{C}^{g}_{k} - k(\boldsymbol{X},\boldsymbol{X})\right)\boldsymbol{J}_k, \tag{26}$$

where the quantities $\boldsymbol{J}_k$, $\boldsymbol{k}_s$ and k_{ss} are computed as

$$\boldsymbol{J}_k = \boldsymbol{k}_s \boldsymbol{K}^{-1}, \tag{27}$$

$$\boldsymbol{K} = k_f\left(\boldsymbol{X}^a, \boldsymbol{X}^a\right) + k_g\left(\boldsymbol{X}^a, \boldsymbol{X}^a\right) \circ \boldsymbol{U}, \tag{28}$$

$$\boldsymbol{k}_s = \boldsymbol{k}_{sf} + \boldsymbol{k}_{sg}\, u_k + \boldsymbol{k}_{sn}, \tag{29}$$

$$k_{ss} = k_{ssf} + k_{ssg}\, u_k^2 + k_{ssn} \tag{30}$$

with

$$\boldsymbol{k}_{sf} = k_f\left(\boldsymbol{X}^a, \boldsymbol{x}^a_k\right), \qquad k_{ssf} = k_f\left(\boldsymbol{x}^a_k, \boldsymbol{x}^a_k\right), \tag{31}$$

$$\boldsymbol{k}_{sg} = k_g\left(\boldsymbol{X}^a, \boldsymbol{x}^a_k\right) \circ \boldsymbol{U}, \qquad k_{ssg} = k_g\left(\boldsymbol{x}^a_k, \boldsymbol{x}^a_k\right), \tag{32}$$

$$\boldsymbol{k}_{sn} = k_n\left(\boldsymbol{X}^a, \boldsymbol{x}^a_k\right), \qquad k_{ssn} = k_n\left(\boldsymbol{x}^a_k, \boldsymbol{x}^a_k\right), \tag{33}$$

where the symbol $\circ$ denotes the element-wise product of two vectors. Finally, update of $\boldsymbol{\mu}^g$ and $\boldsymbol{C}^g$ is performed by the already familiar equations (14) – (16). Note, that we also get an estimate of the uncertainty associated with the system output prediction, which can be of use in the control design. Also note, that the RGP algorithm was formulated in section 3.1 as being able to process nk observations at a time. For the system identification we only consider processing of one observation at a time; thus $nk = 1$.

To fully elucidate the workings of RGP regression algorithm with ad-hoc hyper-parameter learning in the system identification process, the following summary is provided.

RGP identification algorithm:	
Step 0:	Set $k \leftarrow 0$, initialise hyper-parameters $\boldsymbol{\theta}_0$, choose basis vector locations $\boldsymbol{X}$, set $\boldsymbol{\mu}^g_0 = \boldsymbol{0}$, $\boldsymbol{C}^g_0 = k(\boldsymbol{X}, \boldsymbol{X}; \boldsymbol{\theta}_0)$,
Step 1:	Generate input u_k and measure the system output y_{k+1},
Step 2:	Calculate system output prediction $\boldsymbol{\mu}^p_{k+1}$ by (25) and (26), utilizing (27) – (33).
Step 3:	Calculate updated $\boldsymbol{\mu}^g_{k+1}$, $\boldsymbol{C}^g_{k+1}$ from $\boldsymbol{\mu}^p_{k+1}$, $\boldsymbol{C}^p_{k+1}$ using (14) – (16),
Step 4:	Optimise hyper-parameters: $\boldsymbol{\theta}_{k+1} = \arg\max_{\boldsymbol{\theta}} \log\left(p\left(\boldsymbol{\mu}^g_{k+1} \mid \boldsymbol{X}, \boldsymbol{\theta}\right)\right)$.
Step 5:	$k \leftarrow k+1$, go to Step 1.

4. Numerical simulations

In numerical simulations the RGP was compared with the performance of the FGP on the following benchmarking example of non-linear stochastic discrete-time system

$$y_{k+1} = \frac{1.5\, y_k}{1 + y_k^2} + \left(2 + \cos\left(y_k\right)\right) u_k + e_{k+1}, \tag{34}$$

where the sequence e_k is a zero-mean white Gaussian noise with variance $\sigma_e^2 = 0.025$. In this case the vector of regressors $\boldsymbol{x}_k = \left[y_{k-1}, u_{k-1}\right]$ is two dimensional. Hyper-parameters were initialised with values $\boldsymbol{\theta}_0 = [0, 0, 0, 0]^{\top}$ for both algorithms. As the excitation signal u_k we considered uniformly random noise within the interval $[-3, 3]$.

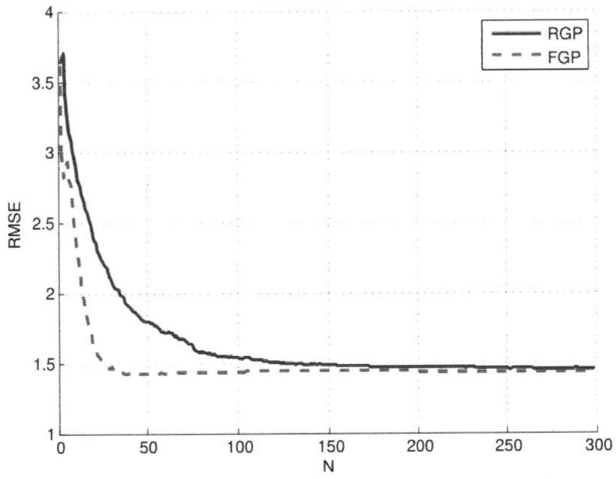

Figure 1: Root mean square error.

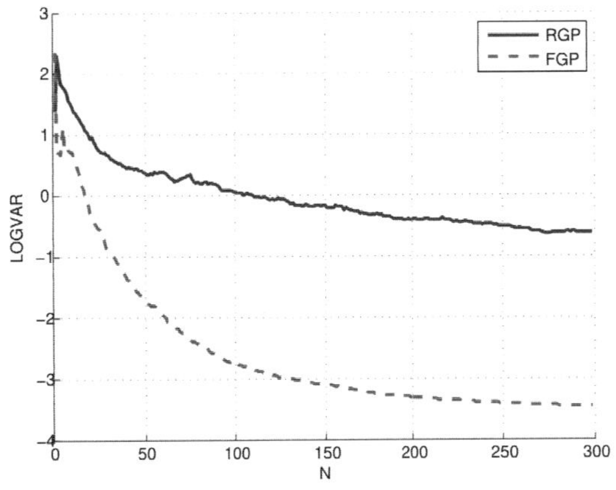

Figure 2: Logarithm of average predictive variance.

In case of RGP algorithm, 75 basis vectors were used to create a 15×5 grid, that covered the area of state space, where the system operates. Initialisation $\boldsymbol{\mu}_0^g = \mathbf{0}$ and $\boldsymbol{C}_0^g = k(\boldsymbol{X}, \boldsymbol{X})$ was performed with initial hyper-parameter values in mind. Ad-hoc hyper-parameter optimisation routine was started with the estimate obtained in previous iteration. The FGP was trained using all available data up to the current time step, where the hyper-parameter optimisation routine was always started with initial values $\boldsymbol{\theta}_0$. Hyper-parameter optimisation was realised using MATLAB's `fminunc()` solver with the maximum number of iterations set to 100.

The experiments performed were focused on the comparison of the two approaches in terms of obtained GP model error and computational demands for prediction in every time step. Root mean square error (RMSE)

$$\text{RMSE} = \sqrt{\frac{1}{N_{ts}} \sum_{i=1}^{N_{ts}} \left(y_i - \hat{y}_i\right)^2} \tag{35}$$

was used to measure the error in model validation process, where y_i is the measurement of system output and $\hat{y}_i$ is the GP model prediction. Furthermore, the logarithm of average predictive variance

$$\text{LOGVAR} = \log\left(\frac{1}{N_{ts}} \sum_{i=1}^{N_{ts}} \text{var}\left(\hat{y}_i\right)\right) \tag{36}$$

was used to quantify the overall model prediction uncertainty. Both metrics were calculated at each of the $N = 300$ time steps using $N_{ts} = 1600$ test points, which were obtained from the system operation. The computational demand is assumed to be the sum of the time required for obtaining the current hyper-parameter estimate and the time required for computing the one-step ahead prediction (model response). This was implemented by employing MATLAB's `tic/toc` command functionality and results recorded for every time step. 100 MC simulations were run with the results shown in figures.

As seen in figure 1, RMSE of the FGP is slightly lower than that of the RGP algorithm. This can be caused by the fact, that the RGP uses for prediction a *fixed amount* of predefined basis vectors. The FGP, however, makes predictions based on *all* available past data points. The main drawback of the RGP is that, when prediction is to be made further away from the area covered by the basis vectors, it's quality will be significantly worse than that of the FGP. Development of the LOGVAR in figure 2 suggests, that predictions of the RGP algorithm are less confident than those of the FGP. Figure 3 compares the system and model response during training. Note how the model response approaches the true system response

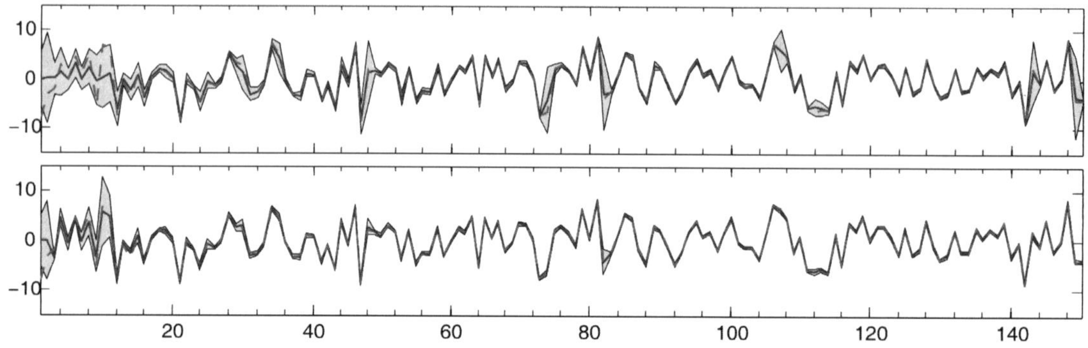

Figure 3: System response (dashed) compared to the RGP (top) and FGP (bottom) model response with the increasing amount of data used for system identification (GP model training). The grey band represents the model uncertainty.

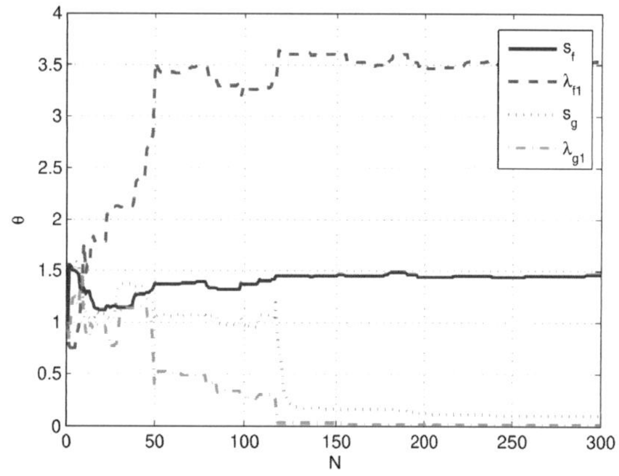

Figure 4: Hyper-parameter development in time.

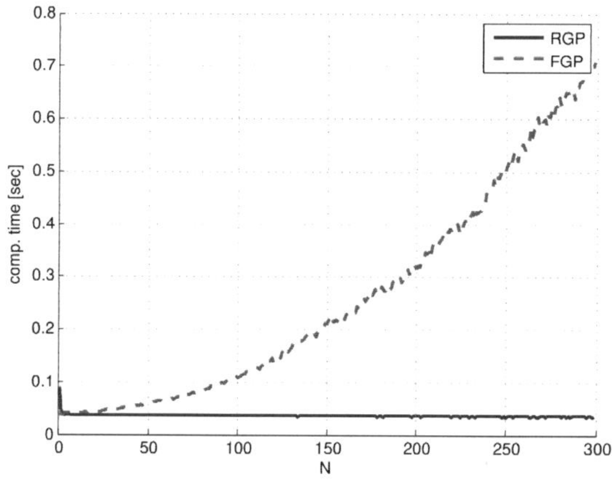

Figure 5: Computation time per iteration.

and the model uncertainty is reduced as the amount of data used for identification increases. Figure 4 shows converging estimates of each of the hyper-parameters as the number of available data for system identification (GP model training) increases. Estimates may not always converge to the same value since the marginal likelihood has multiple local extrema. The figure 5 demonstrates the unsurprising fact, that the RGP algorithm, owing to it's recursive nature and limited number of solver iterations, has constant computational demands per time step. On the contrary, time requirements of the FGP algorithm for prediction increase with every time step regardless of the limited number of solver iterations.

5. Conclusions

In this paper, we proposed use of the RGP algorithm with an ad-hoc hyper-parameter learning procedure for the non-linear system identification. The RGP algorithm was compared with the FGP approach in terms of computational demands per iteration as well as the RMSE and the logarithm of average predictive variance. The ad-hoc learning procedure provides converging hyper-parameter estimates. In performed simulations it was demonstrated that use of the RGP algorithm is clearly superior to the FGP approach in terms computational demands. Compared to the FGP, the RGP algorithm provides more conservative predictions and achieves comparable RMSE, though exhibits slower convergence. Main disadvantage of the RGP algorithm is that the number of basis vectors increases exponentially with the dimension of the state space (vector of regressors). Also, quality of predictions at the locations further away from the area of state space covered by the basis vectors is worse than those of the FGP.

Future research activity will be aimed at incorporating the RGP identification algorithm into the functional dual adaptive control loop.

Acknowledgments

The work was supported by the European Regional Development Fund (ERDF), project "NTIS - New Technologies for the Information Society", European Centre of Excellence, CZ.1.05/1.1.00/02.0090 and by the Czech Science Foundation, project GACR P103/13/07058J.

References

[1] Neal R M 1995 *Bayesian learning for neural networks* Ph.D. thesis University of Toronto
[2] Ko J and Fox D 2008 *Intelligent Robots and Systems, 2008. IROS 2008. IEEE/RSJ International Conference on* vol 27 (Springer) pp 3471–3476 ISBN 9781424420582
[3] Ko J, Klein D J, Fox D and Haehnel D 2007 *Intelligent Robots and Systems, 2007. IROS 2007. IEEE/RSJ International Conference on* vol 54 (IEEE) pp 1901–1907 ISBN 1424409128
[4] Deisenroth M P, Turner R D, Huber M F, Hanebeck U D and Rasmussen C E 2012 *IEEE Trans. on Automatic Control* **57** 1865–1871
[5] Särkkä S, Hartikainen J, Svensson L and Sandblom F 2014 *Information Fusion (FUSION), 2014 17th International Conference on* pp 1–8
[6] Kocijan J, Murray-Smith R, Rasmussen C and Girard A 2004 *Proc. of the American Control Conf.* vol 3 pp 2214–2219
[7] Kocijan J, Girard A, Banko B and Murray-Smith R 2005 *Mathematical and Computer Modelling of Dynamic Systems* **11** 411–424
[8] Girard A, Rasmussen C E, Quiñonero Candela J and Murray-Smith R 2003 *Advances in Neural Information Processing Systems 15* ed Becker S, Thrun S and Obermayer K (MIT Press) pp 545–552
[9] Deisenroth M and Rasmussen C E 2011 *Proc. of the 28th Int. Conf. on Machine Learning (ICML-11)* pp 465–472
[10] Park S, Mustafa S K and Shimada K 2013 *Robotic Intelligence In Informationally Structured Space (RiiSS), 2013 IEEE Workshop on* (IEEE) pp 120–127 ISBN 978-1-4673-5877-4
[11] Rasmussen C E and Ghahramani Z 2003 *Advances in Neural Information Processing Systems 15* 1 ed Becker S, Thrun S and Obermayer K (MIT Press) pp 505—-512
[12] Särkkä S, Solin A and Hartikainen J 2013 *IEEE Signal Processing Magazine* **30** 51–61 ISSN 1053-5888
[13] Csató L and Opper M 2002 *Neural Computation* **14** 641–668
[14] Chowdhary G, Kingravi H A, How J P and Vela P A 2014 *Neural Networks and Learning Systems, IEEE Transactions on* **PP** 1–1 ISSN 2162-237X
[15] Ranganathan A, Yang M H and Ho J 2011 *IEEE Transactions on Image Processing* **20** 391–404
[16] Huber M 2013 *Acoustics, Speech and Signal Processing (ICASSP), 2013 IEEE Int. Conf. on* pp 3362–3366 ISSN 1520-6149
[17] Fabri S and Kadirkamanathan V 2001 *Functional Adaptive Control: An Intelligent Systems Approach* (Springer London) ISBN 978-1-4471-1090-3
[18] Šimandl M, Král L and Hering P 2005 *IFAC Proc. Volumes (IFAC-PapersOnline)* vol 16 pp 58–63
[19] Sbarbaro D, Murray-Smith R and Valdes A 2004 *Proc. of Int. Symp. on Neural Networks* pp 52–58
[20] Rasmussen C E and Williams C K 2006 *Gaussian Processes for Machine Learning* (The MIT Press) ISBN 978-262-18253-9
[21] Quiñonero-Candela J and Rasmussen C E 2005 *The Journal of Machine Learning Research* **6** 1939–1959
[22] Huber M F 2014 *Pattern Recognition Letters* **45** 85–91 ISSN 01678655
[23] Sbarbaro D and Murray-Smith R 2005 *Switching and Learning in Feedback Systems* vol 3355 (Springer) pp 140–157

Modelling and data-based identification of heating element in continuous-time domain

Ivan Zajic[1], Muriel Iten[2] and Keith J Burnham[1]

[1] Control Theory and Applications Centre, Coventry University, Coventry, UK
[2] Civil Engineering, Architecture and Building, Coventry University, Coventry, UK

E-mail: zajici@uni.coventry.ac.uk

Abstract.
A unique data-based and physically meaningful nonlinear continuous-time model of heating element is presented. The model is considered to be of low complexity yet achieving high simulation performance. The physical meaningfulness of the model provides enhanced insight into the performance and functionality of the system. In return, this information can be used during the system simulation and improved model based control designs for tight temperature regulation. The second contribution presented in this work is the parameter estimation of the derived nonlinear model in continuous-time domain itself. For this purpose the application of refined instrumental variable methods has been found to be particularly suitable.

1. Introduction

The paper reports on modelling and data-based identification of a heating element, which preheats the inlet air used by a testbed for testing and modelling of phase change materials (PCM) used in passive air conditioning for sustainable housing applications, for more details see [1]. The main motivation for the work is the actual modelling of the phase change materials itself where a clear parallel between the air being conditioned by a heating element or the PCM based heat exchanger can be found. The later heat exchange process additionally includes nonlinear effects caused by the phase change of the PCM. Therefore the emphasis is given on physical meaningfulness of the derived heating element model with the view of extending this model for later modelling of PCM based heat exchangers.

The presented modelling procedure has been motivated by, but not necessarily strictly follow, the 'data-based mechanistic' approach to identification and modelling of systems [2, 3, 4, 5]. In this approach a parsimonious data-based model is inferred first using statistical system identification tools. Once the model parameters, order and structure are inferred from measured data the obtained data-based model is interpreted in a physical sense. Furthermore, this system identification procedure can be re-iterated such that the physical laws can assist in priming the black-box model structure and *vice versa.* The search for the system model stops when the identified model explains the measured data well and also the model must provide meaningful interpretation of the system in physical terms. In this regard, the final obtained heating element model is unique in its own right, where up to the best knowledge of authors no alike model has been reported in the literature up to date.

The system identification in continuous-time domain is essential so that a direct link between the data-based identified model and physical laws can be established in a straight forward manner. For this purpose a simplified refined instrumental variable method for linear continuous-time transfer function model identification (SRIVC) has been adopted, see [6, 7] and references therein. The identified heating element model is, however, not linear in structure and resemble a bilinear system models. Therefore, modified SRIVC algorithm capable of parameter estimation of such nonlinear models is used instead [8]. Finally, the nonlinear dynamic behaviour of the identified heating element model throughout the operational envelope is discussed, together with the implications for parameter estimation and simulation of such model, see Sub-section 3.5. Motivated by [9], it will be shown that the dynamic behaviour of the considered nonlinear model at given operating point can be replicated at zero operating point, which has further applications in system identification and simulation of such system models (which are, however, not shown here).

2. Heating element details and system identification set-up

The heating element is part of a larger testbed system for testing and numerical modelling of phase change materials, see [1] for further references. The speed and temperature of air passing around the tested PCM can be freely regulated to meet the demanded test environmental conditions. The air is conditioned by means of the considered heating element which is in a down stream series connection with a cooling unit. Schematic diagram of the heating element installed in the supply duct is provided in Figure 1.

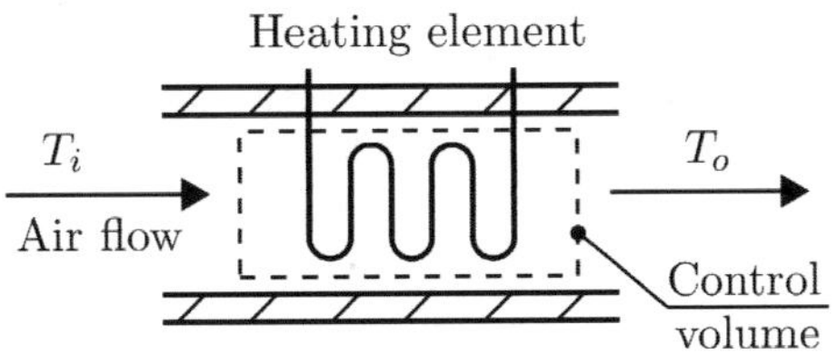

Figure 1. Schematic description of the heating element.

For the system identification purposes the following system inputs are considered: measured inlet air temperature, denoted $T_i(t_k)$ $[K]$, and the air velocity, denoted $v(t_k)$ $[m/s]$. Both system inputs are measured in the centre of the cross sectional area of the supply duct. The outlet air temperature, denoted $T_o(t_k)$ $[K]$, is the assumed system output. The data acquisition experimental set-up has been designed such that the supplied power, denoted $q(t_k)$ $[W]$, to the heating element is constant with the average reading of $q(t_k) = 830$ $[W]$. The inlet air temperate has been altered at random intervals by means of up stream cooling unit; simultaneously the air speed has been stepped up or down by means of altering the supply fan power. The test bed is installed in a large indoor space with approximately constant ambient air temperature, denoted $T_a(t_k)$ $[K]$. The time index (t_k) is chosen to emphasise that the continuous time signals are measured at discrete time index k. The chosen sampling interval is $\tau = 1$ $[s]$, where selection of faster sampling has not been feasible due to the limitations of the data acquisition device.

It is expected to observe that for increasing inlet air temperature the outlet air temperature will also increase and *vice versa*. On the other hand, for increasing air speeds the outlet air temperature should decrease (and the other way around). This can be explained such that for increasing air speeds the air volumetric flow also increases, hence the need for heating, while the heating power is kept constant. For example, the regulation of supply air volume is a common method used in air conditioning of offices and alike spaces [10].

3. Data-based model identification

Having obtained the measured input and output data, being the inlet and outlet air temperatures together with the air velocity, a continuous-time model is identified in a data-based mechanistic manner [3, 4, 5]. The first principles modelling approach, provided in subsequent Section 4, is used to prime the data-based model structure and *vice versa.* The data-based model is used to prime the first principles model, where the suggested model terms are supported by the evidence based on measured data. In overall this iterative procedure confirms the 'appropriateness' of the selected model structure for the data-based identified model and allows a meaningful interpretation in physical terms.

The main outcome of applying the physical laws to prime the data-based model structure has been the selection of two model inputs being the product between the inlet air temperature and air velocity and also the product between the outlet air temperature and air velocity, i.e. $T_i(t_k)v(t_k)$ and $T_o(t_k)v(t_k)$, respectively. This makes the suggested model nonlinear in structure, while still being linear in parameters so that least-squares based model parameter estimation methods can be adopted. However, it should be noted that despite the insight offered by physical laws a purely black-box modelling approach has been also adopted, where various linear and nonlinear regression terms and model orders have been tested with no clear advantage gained in terms of model fit.

3.1. Parameter estimation stage

In order to estimate the model parameters a simplified refined instrumental variable method for linear continuous-time transfer function model identification, SRIVC, has been adopted. The SRIVC algorithm is a direct parameter estimation method, which uses sampled (discrete), noisy, input-output signals and is based on the state variable filter (SVF) approach. The purpose of the optimally selected SVF is twofold, first, to obtain the filtered unknown time derivatives, second, to noise prefilter the measured input-output signals. In the case of white, zero mean, additive measurement noise with Gaussian amplitude distribution the SRIVC algorithm yields asymptotically unbiased statistically efficient (minimum variance) parameter estimates. Another feature of the SRIVC algorithm is the use of instrumental variables (IV), which together with the optimal prefilters form the core of the algorithm. The source of the instrumental variables is the simulated system output. Additionally, the SRIVC algorithm is known for its rapid convergence properties and ease of implementation.

It is important to note that the SRIVC algorithm is designed for linear transfer function model estimation, while the adopted model is nonlinear in structure due to the selection of input signals, cf. (1). Special attention must be paid to the use of instrumental variables, where one of the model inputs contains the measured output. The issue of using SRIVC algorithm in a presence of such a signal product is in detail addressed in [8]. Additionally, the nonlinear model needs to be estimated at its original point of operation, which then implies that a constant offset term must be included in the regression vector of the SRIVC algorithm - this method has been used in this work. Alternatively, in the same manner as in the case of estimation of linear models, it is advantageous to subtract the initial values of measured input-output signal so that the model is estimated at zero operating point. This is not an issue in the case of linear models, however the values of nonlinear model parameters depend on the point of operation. The procedure of tackling this issue is discussed in [8, 9] for the case of identification of bilinear and Hammerstein-bilinear models. Due to the close connection of the considered nonlinear model, cf. (1), with the bilinear model class the same procedure could be adopted.

It should be emphasized, that both system inputs are measurement noise contaminated, i.e. $T_i(t_k)$ and $v(t_k)$. This in effect creates an errors-in-variables identification conceptual scenario, [11], causing a bias in the parameter estimates. However, the SRIVC method uses optimal prefilters which help to attenuate the influence of this input noise on the parameter estimates,

Table 1. Estimated model parameters for model with orders ($n = 2, m = 2$) are provided on the first two lines, while the parameters for the model ($n = 2, m = 1$) are provided on the last two lines. The corresponding estimated standard errors are provided in the brackets. The offset terms are $\hat{o} = -279.25$ and $\hat{o} = -122.95$ for models ($n = 2, m = 2$) and ($n = 2, m = 1$), respectively.

$\hat{\alpha}_1 \times 10^{-3}$	$\hat{\alpha}_2 \times 10^{-6}$	$\hat{\beta}_0 \times 10^{0}$	$\hat{\beta}_1 \times 10^{-2}$	$\hat{\beta}_2 \times 10^{-4}$	$\hat{\eta}_0 \times 10^{0}$	$\hat{\eta}_1 \times 10^{-2}$	$\hat{\eta}_2 \times 10^{-4}$
8.5257	−9.2335	1.1581	3.8449	1.2031	−0.9620	−3.6187	−1.4548
(0.6627)	(2.3038)	(0.0231)	(0.1194)	(0.1095)	(0.0186)	(0.1004)	(0.1060)
55.026	−92.835	−	9.9837	7.7661	−	−8.5689	−8.3067
(2.0110)	(3.3983)	−	(0.2512)	(0.2835)	−	(0.2070)	(0.2984)

see [7], so that the estimated model may suffice for practical purposes.

The final selected model has two integral terms and a feed-through term, i.e. a second order-like model $n = 2$ with two zeros $m = 2$. The estimated pure time delay is null. It is assumed that any pure time delay has been effectively captured by the dominant dynamics of the model. The final estimated nonlinear model in transfer function-like form is given by

$$T_o(t_k) = \frac{\hat{\beta}_0 s^2 + \hat{\beta}_1 s + \hat{\beta}_2}{s^2 + \hat{\alpha}_1 s + \hat{\alpha}_2}\Big\{T_i(t_k)v(t_k)\Big\} + \frac{\hat{\eta}_0 s^2 + \hat{\eta}_1 s + \hat{\eta}_2}{s^2 + \hat{\alpha}_1 s + \hat{\alpha}_2}\Big\{T_o(t_k)v(t_k)\Big\} + \hat{o} \tag{1}$$

where s denotes a differential operator defined as $s^p x(t) = \frac{\mathrm{d}^p x(t)}{\mathrm{d}t^p}$. The hat symbol over the parameters indicates that these are estimated. The estimated parameters are given in Table 1, where the estimated corresponding standard errors are provided in the parentheses and the estimated offset term is in the table's caption.

3.2. Simulation results and observations

The simulation results are presented in Figure 2. The upper plot shows measured (grey solid line) and simulated (black solid line) outlet air temperature, denoted by $\hat{x}(t_k)$. The middle plot shows the simulation error $e(t_k)$, defined as $e(t_k) = T_o(t_k) - \hat{x}(t_k)$. The lower plot shows inputs $v(t_k)$ (grey solid line) and $T_i(t_k)$ (black solid line). Two model performance criteria are evaluated, namely, the integral of absolute error, denoted IAE [°C], and the coefficient of determination, denoted R_T^2 [%]. The integral of absolute error is defined as

$$IAE = \frac{1}{N}\sum_{k=1}^{N} |e(t_k)| \tag{2}$$

where N is the number of data samples. The IAE criterion then provides an average simulation error in the units of the evaluated output signal (in this case [°C]). The coefficient of determination is defined as

$$R_T^2 = 100\left(1 - \frac{\sigma_e^2}{\sigma_{T_o}^2}\right) \tag{3}$$

where σ_e^2 and $\sigma_{T_o}^2$ are variances of simulation error and outlet temperature, respectively. The R_T^2 criterion then provides a measure of how much of the measured output variance is explained (or captured) by the simulated output variance.

Subsequently, the simulation results with $R_T^2 = 99.68$ [%] and $IAE = 0.1582$ [°C] have been obtained. The evaluated performance criteria indicate an acceptable predictive performance of

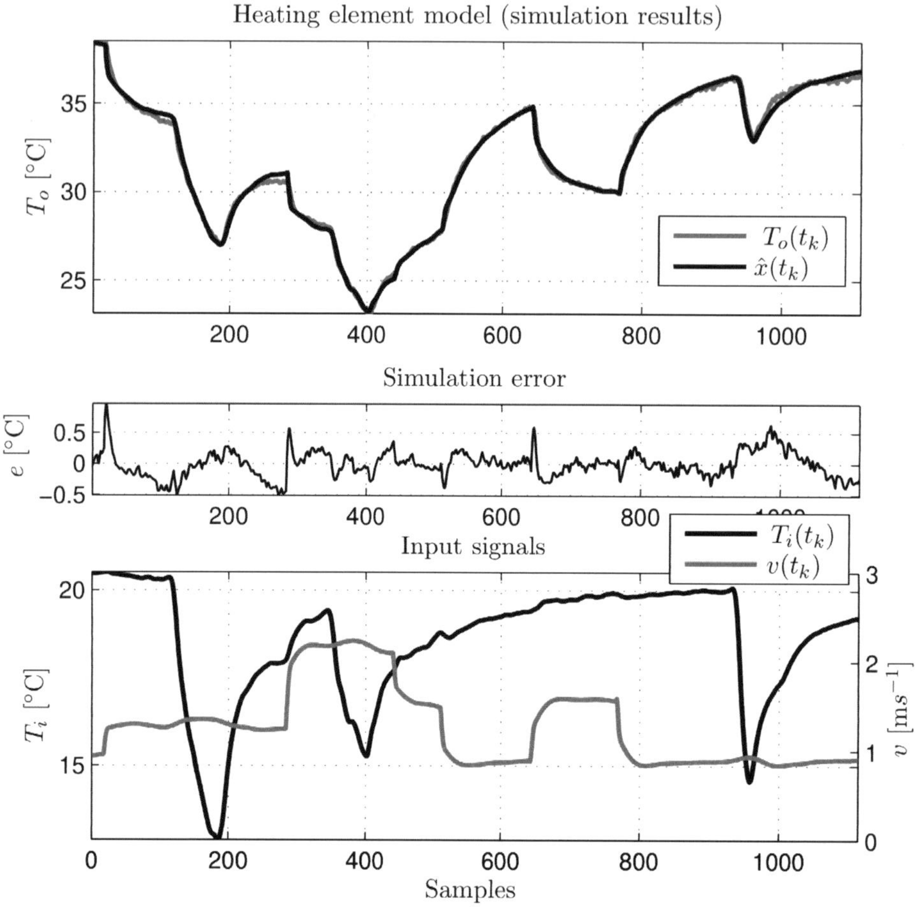

Figure 2. The upper plot shows measured (grey solid line) and simulated (black solid line) outlet air temperatures. The middle plot shows the simulation error defined as the difference between measured and simulated outlet air temperatures. The lower plot shows system inputs, i.e. air velocity (grey solid line) and inlet air temperature (black solid line). Sampling interval is one second.

the estimated heating element model. The simulation error (being also the model residuals), plotted in the middle part of Figure 2, is not a zero mean white noise signal, however that is to be expected in the case of nonlinear model identification. In overall this indicates that the model could not capture all of the deterministic processes caused by the system inputs. It is anticipated that this may have also affected the accuracy of the parameter estimates and that the reported standard errors are probably too optimistic.

Finally, it is instructive to evaluate the nonlinear model (1) dynamics at distinct point of operation. It has been found that the speed of the response depends only on air velocity while the achieved steady-state gain depends on both, the air velocity and the inlet air temperature. Subsequently, linearising the estimated nonlinear model (1) at two stead-state values of air velocity, low and high, respectively, yields two distinct time constants. Considering a low air velocity of $v^* = 0.84$ [ms^{-1}] the two time constants are $T_1 = 55.3$ [s] and $T_2 = 289$ [s]. For faster air velocity the system response becomes also faster with the two time constants being $T_1 = 41.2$ [s] and $T_2 = 242$ [s].

3.3. Relation to bilinear system models

It is interesting to note that if the inlet air temperature is constant then the identified nonlinear model (1) reduces to a so called bilinear model (a special class of nonlinear models). The characteristic feature of bilinear models is the input dependent dynamic behaviour, see the last paragraph in Sub-section 3.2.

Bilinear models are defined to be *linear* in both state and control when considered independently, with the *bilinearity* (or nonlinearity) arising from coupled terms involving products of the system state and control input. It is this close connection to linear systems, which makes the bilinear models particularly appealing so that many techniques developed for linear systems can be extended and applied to the bilinear case [12]. This has been also one of the reasons why it has been possible to use the SRIVC algorithm which has been originally developed for linear transfer function model estimation, see [8] for more details.

3.4. Reduced order model

The identified second best model in terms of IAE criterion is of the same form as model (1), however without the feed-through terms. The estimated model parameters are provided in Table 1. In order words the two feed-through parameters are $\beta_0 = 0$ and $\eta_0 = 0$ and the model order is $(n = 2, m = 1)$. The obtained performance criteria are $IAE = 0.2269$ [°C] and $RT^2 = 99.47$ [%], which indicates only a slight performance deterioration.

There are two main reasons why the reduced order model might be the preferred option for more practical considerations. The first reason is the reduced computational burden such that the feed-through terms of the original nonlinear model (1) have been inevitably causing an algebraic loop during the model simulation. Secondly, for model based control design purposes it is commonly required that at least one lag (or sample delay) exists between the control input and process output.

Physically meaningful interpretation of the originally estimated nonlinear model (1) is provided in next Section 4. Such interpretation is not provided for the considered reduced order model due to the space limitations. Nevertheless, the first principles model (7), which relates to the originally estimated data-based model (1), can be easily re-arranged to conform to the reduced order model instead. This could be done by combining the energy balance equations (7a) and (7b), hence

$$C\frac{\mathrm{d}T_o(t)}{\mathrm{d}t} = q(t) - v(t)\rho_a A_a c_a\left[T_o(t) - T_i(t)\right] - (UA)_{int}\left[T_o(t) - T_w(t)\right] \tag{4}$$

and leaving (7c) intact. The model parameters in (4) are introduced in the next Section 4 as well as the meaning of (4).

3.5. Non-linear model operation: Simulation at zero operating point

In general, the dynamic and steady-state behaviour of nonlinear models differ throughout the operation envelope. Therefore, if the nonlinear model, such as the identified model (1), is simulated at different operating points a distinct dynamical response will be obtained. This imposes interesting issues in the area of nonlinear system identification. For example, it is not possible to pre-process the measured input-output signals by means of subtracting mean values or initial values prior to parameter estimation and then attempt to estimate the nonlinear model parameters. Of course, this is a common practice when estimating linear transfer function model parameters, however identification of nonlinear models is, in general, not possible in this way (with some exceptions, see [9]).

It is going to be shown, however, that it is possible to simulate the identified nonlinear model (1) at zero operating point while the obtained dynamical response is identical to the one which would be obtained at the original point of operation. This has naturally implications for system identification but also for simulation of nonlinear systems, where the issue of computer simulation initialisation is elevated (initialisitaion of integral terms in particular).

To start with, the input-output signals adopted in the identified model (1) are re-defined or rather substituted with $T_o(t) := T_o(t) + \bar{T}_o$, $T_i(t) := T_i(t) + \bar{T}_i$ and $v(t) := v(t) + \bar{v}$. The input-output signals then comprise the time varying component (deviations) and the steady-state, base-line, component which is denoted by a bar. In the present context the adopted input signals start initially as constant signals and the modelled system is allowed to settle to reach the steady-state. Therefore, the initial values (first sample) of input-output signals are also the base-line components (denoted by the bar). To simplify the following discussion the reduced order model is assumed instead, i.e. $\beta_0 = 0$ and $\eta_0 = 0$ c.f. Section 3.4. Subsequently, if the base-line components are removed (subtracted) from the the input-output signals of model (1) then also the offset term o is effectively removed, see [9], and the following nonlinear, base-line *adjusted*, model can be derived

$$T_o(t) = \frac{B(s)}{A(s|\bar{v})}\Big\{T_i(t)v(t)\Big\} + \frac{E(s)}{A(s|\bar{v})}\Big\{T_o(t)v(t)\Big\} + \frac{B(s|\bar{v})}{A(s|\bar{v})}T_i(t) + \frac{B(s|\bar{T}_i,\bar{T}_o)}{A(s|\bar{v})}v(t) \quad (5)$$

where the constant coefficient polynomials are defined as

$$A(s|\bar{v}) = s^2 + (\alpha_1 - \eta_1\bar{v})\,s + (\alpha_2 - \eta_2\bar{v}) \quad (6a)$$
$$B(s) = \beta_1 s + \beta_2 \quad (6b)$$
$$E(s) = \eta_1 s + \eta_2 \quad (6c)$$
$$B(s|\bar{v}) = (\beta_1\bar{v})\,s + (\beta_2\bar{v}) \quad (6d)$$
$$B(s|\bar{T}_i,\bar{T}_o) = \left(\beta_1\bar{T}_o + \eta_1\bar{T}_i\right)s + \left(\beta_2\bar{T}_o + \eta_2\bar{T}_i\right) \quad (6e)$$

The obtained model (5) is then able to replicate the dynamic behavior of the original nonlinear model (1) simulated at the operating point $(\bar{v}, \bar{T}_i, \bar{T}_o)$. Comparing the base-line adjusted model (5) with the original model (1) it is noticed that two new transfer function terms have appeared on the right hand side of (5). These terms compensate for the discrepancy between the response of the original model simulated at operating point $(\bar{v}, \bar{T}_i, \bar{T}_o)$ and the response which would be obtained at zero operating point, i.e. $(\bar{v} = 0, \bar{T}_i = 0, \bar{T}_o = 0)$.

3.5.1. Simulation at zero operating point - An example: The originally identified nonlinear model (1) is simulated using the artificially designed input signals displayed in the lower left plot of Figure 3. The adopted parameters of the reduced order model are given in the lower part of Table 1 with the offset term provided in the caption. The obtained corresponding output is shown in the upper left plot of Figure 3. Consequently, the following base-line values are inferred from the obtained plots, i.e. $(\bar{v} = 1, \bar{T}_i = 20, \bar{T}_o = 36.52)$.

To be able to simulate the nonlinear model at zero operating point the base-line components of input signals are removed; the base-line compensated input signals are provided in the lower right plot of Figure 3. The compensated input signals are then used by the originally identified nonlinear model (1), where the obtained simulated output is shown in the upper right plot of Figure 3 as a grey solid line. The same compensated inputs are also used by the adjusted model (5), where the obtained output is shown as solid black line. The parameters of the adjusted model are calculated using the relations provided in (6) with the previously inferred values of $(\bar{v}, \bar{T}_i, \bar{T}_o)$.

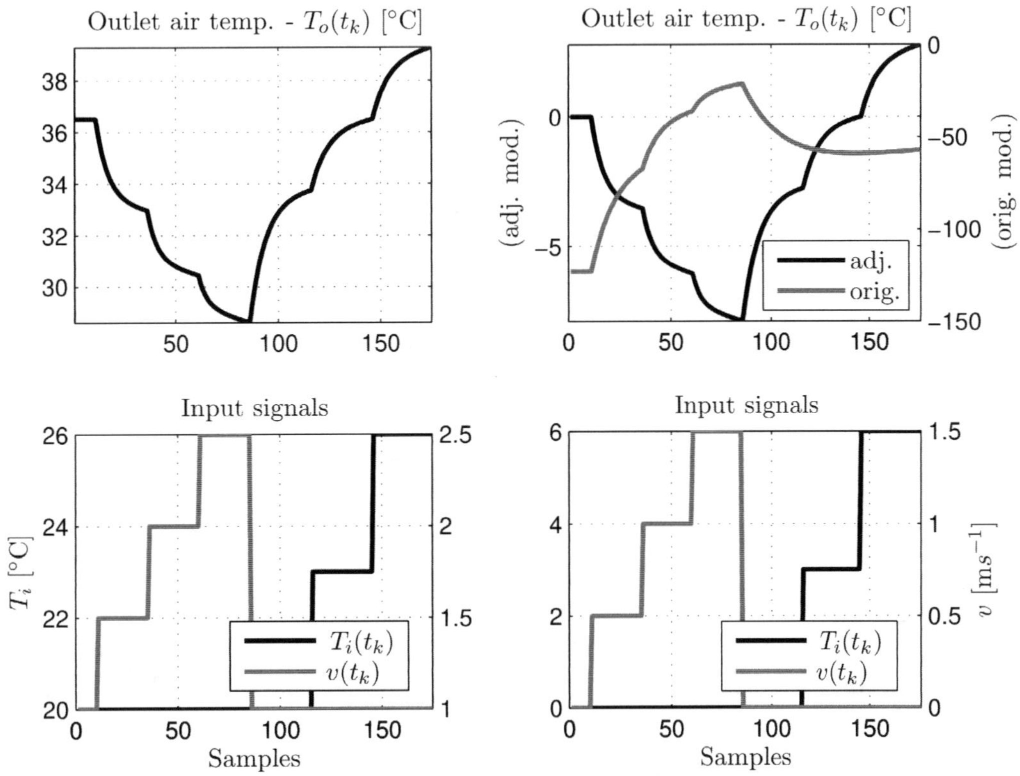

Figure 3. The upper left plot shows simulated outlet air temperature; originally estimated model is used. The lower left plot shows corresponding input signals, i.e. air velocity (grey solid line) and inlet air temperature (black solid line). The upper right plot shows simulated outlet air temperature, where original model (black solid line) and adjusted model (grey solid line) are used. Both simulated outputs are obtained using the base-line compensated input signals shown in the lower right plot, i.e. starting at zero value. The sampling interval is one second.

As expected, it is noticed, that the dynamical response of the original nonlinear model considerably differs at different operating points. It is also observed that the response of the adjusted model (5) is identical to the response of the original nonlinear model obtained at the original point of operation, i.e. ($\bar{v} = 1, \bar{T}_i = 20, \bar{T}_o = 36.52$), while the adjusted model is simulated at zero operating point, i.e. ($\bar{v} = 0, \bar{T}_i = 0, \bar{T}_o = 0$). Similarly to [9], it is possible to operate the base-line adjusted model at any operating point (not just zero base-line), while preserving the dynamical response of the nonlinear model from any other point of operation, but this is beyond the scope of the current article.

4. First principles considerations: Mechanistic model interpretation

From the system details provided in Section 2 it is known that the modelled system compose of three main elements, that is the heating element itself, the air control volume and the duct walls. Additionally, the supplied constant heat gain $q(t)$ causes a temperature rise over the heating element. The data-based model (1) indicates, however, that only two dynamical modes are present (fast and slow mode) and also a feed-through term is clearly present. Therefore, one of the main system components must have negligible thermal capacity.

The air surrounding the heating element is assumed to be perfectly mixed, where the schematic diagram of the system is shown in Figure 1, so that a lumped parameter modelling approach is considered. Under this assumption the outlet air temperature, $T_o(t)$, is assumed to be equal to the mean temperature of the whole control volume. Consequently, the energy balance equations for the heating element, the control volume of air surrounding the heating element and adjacent duct walls, are, respectively, given by

$$C_h \frac{\mathrm{d}T_h(t)}{\mathrm{d}t} = q(t) - (UA)_h \left[T_h(t) - T_o(t)\right] \tag{7a}$$

$$0 = (UA)_h \left[T_h(t) - T_o(t)\right] - v(t)\rho_a A_a c_a \left[T_o(t) - T_i(t)\right] - (UA)_{int} \left[T_o(t) - T_w(t)\right] \tag{7b}$$

$$C_w \frac{\mathrm{d}T_w(t)}{\mathrm{d}t} = (UA)_{int} \left[T_o(t) - T_w(t)\right] - (UA)_{ext} \left[T_w(t) - T_a(t)\right] \tag{7c}$$

where C_h $[J/K]$ is the thermal capacity of the heating element, C_w $[J/K]$ thermal wall capacity (insulated plywood), c_a $[J/kgK]$ is the air specific heat capacity, A_a $[m^2]$ denotes the cross sectional area of the duct, ρ_a $[kg/m^3]$ is the air density. The heat transfer coefficient is denoted by U $[J/m^2K]$ while the product of the heat transfer coefficient and the efficient surface area, A $[m^2]$, through which the heat is transmitted is denoted by $(UA)_h$ $[J/K]$ (heating element), $(UA)_{int}$ $[J/K]$ (inner duct wall) and $(UA)_{ext}$ $[J/K]$ (outer duct wall). The mean temperature of the heating element and wall temperature are denoted, respectively, by $T_h(t)$ $[K]$ and $T_w(t)$ $[K]$. The time index (t) is chosen, emphasising that the signals are not measured but generated by a physical laws based model.

4.1. Remarks on model derivation

The heating element system is represented by three governing nonlinear differential equations (7). The first equation (7a) relates to the energy balance of the heating element itself. It assumed that the inner temperature and the surface temperature of the element are the same and are equal to $T_h(t)$. This is surely simplifying assumption, however considering the small volume of the heating spiral and high temperatures involved, not so unrealistic. It is further assumed that the hating element can be represented as a lumped heat source dissipating all power $q(t)$ into the passing air.

It is assumed that the passing air has negligible thermal capacity so that the heat exchange between the heating element and the air is instantaneous, hence the left hand side of (7b) equals to zero, see e.g. [13]. It is this assumption which supports (implies) the presence of the feed-through terms in data-based model (1). Furthermore, the presence of feed-through terms in (1) can be also interpreted as if not all inlet air is in direct contact with the heating element and that the air just passes around, which surely does occur in practice. Subsequently, the heat transmitted into the passing air must be allowed to dissipate otherwise a first type integrating model would be derived - which is not supported by measured data. It is assumed that the heat loss solely occurs through the walls of the duct, which is expressed by the last energy balance equation (7c).

It should be emphasised that the derived (relatively simple) first principles model (7) is not the only physically feasible interpretation of the data-based model (1), however the model, and the mechanistic interpretation it provides, is considered to be sufficient for the given application. For example, the identified data-based model (1) has two dynamical modes, fast and slow mode, respectively. Based on the first principles model (7) the following mechanistic interpretation can be provided: It can be assumed that the slow mode (time constant is $T \approx 289$ [s]) corresponds to the heat loss through the insulated walls of the duct. Similarly, it can be assumed that the fast dynamical mode (time constant is $T \approx 55.3$ [s]) relates to the thermal mass of the heating element itself.

4.2. Nonlinear transfer function-like form

For the system identification purposes it is convenient to visualise the set of nonlinear differential equations (7) in a transfer function-like form, from which the interaction between the system inputs and outputs will be clearly visible. In return, this supports the model structure selection of the data-based model.

Using the differential operator s the energy balance equation for the heating element (7a) can be conveniently expressed as follows

$$\left[s + \frac{(UA)_h}{C_h}\right] T_h(t) = q(t) + \frac{(UA)_h}{C_h} T_o(t) \tag{8}$$

Similarly the energy balance equation for the wall (7c) can be arranged as

$$\left[s + \frac{(UA)_{int}}{C_w} + \frac{(UA)_{ext}}{C_w}\right] T_w(t) = \frac{(UA)_{int}}{C_w} T_o(t) + \frac{(UA)_{ext}}{C_w} T_a(t) \tag{9}$$

Subsequently, the set of differential equations (7) is combined by substituting (8) for the unknown heating element temperature $T_h(t)$ in the energy balance equation for air (7b), also the expression (9) is substituted for the unknown wall temperature $T_w(t)$ in (7b). Rewriting and rearranging the resulting expression in a transfer function-like form gives

$$\begin{aligned} T_o(t) = & \frac{\beta_0 s^2 + \beta_1 s + \beta_2}{s^2 + \alpha_1 s + \alpha_2}\Big\{T_i(t)v(t)\Big\} + \frac{\eta_0 s^2 + \eta_1 s + \eta_2}{s^2 + \alpha_1 s + \alpha_2}\Big\{T_o(t)v(t)\Big\} \\ & + \frac{\gamma_1 s + \gamma_2}{s^2 + \alpha_1 s + \alpha_2} q(t) + \frac{\delta_1 s + \delta_2}{s^2 + \alpha_1 s + \alpha_2} T_a(t) \end{aligned} \tag{10}$$

where the individual model parameters are defined as

$$\begin{aligned} \alpha_1 &= \frac{C_h[(UA)_h(UA)_{int} + (UA)_h(UA)_{ext} + (UA)_{int}(UA)_{ext}] + C_w(UA)_h(UA)_{int}}{[(UA)_h + (UA)_{int}]C_h C_w} \\ \alpha_2 &= \frac{(UA)_h(UA)_{int}(UA)_{ext}}{[(UA)_h + (UA)_{int}]C_h C_w} \\ \beta_0 &= -\eta_0 = \frac{C_h C_w \rho_a A_a c_a}{[(UA)_h + (UA)_{int}]C_h C_w} \\ \beta_1 &= -\eta_1 = \frac{[C_h(UA)_{int} + C_h(UA)_{ext} + C_w(UA)_h]\rho_a A_a c_a}{[(UA)_h + (UA)_{int}]C_h C_w} \\ \beta_2 &= -\eta_2 = \frac{(UA)_h[(UA)_{int} + (UA)_{ext}]\rho_a A_a c_a}{[(UA)_h + (UA)_{int}]C_h C_w} \end{aligned} \tag{11}$$

and

$$\begin{aligned} \delta_1 &= \frac{C_h(UA)_{int}(UA)_{ext}}{[(UA)_h + (UA)_{int}]C_h C_w} & \delta_2 &= \frac{(UA)_h(UA)_{int}(UA)_{ext}}{[(UA)_h + (UA)_{int}]C_h C_w} \\ \gamma_1 &= \frac{C_w(UA)_h}{[(UA)_h + (UA)_{int}]C_h C_w} & \gamma_2 &= \frac{(UA)_h[(UA)_{int} + (UA)_{ext}]}{[(UA)_h + (UA)_{int}]C_h C_w} \end{aligned} \tag{12}$$

It is interesting to note that the obtained nonlinear transfer function (10) clearly indicates that the signal products $T_i(t)v(t)$ and $T_o(t)v(t)$ should be considered as system inputs during the data-based identification stage. Indeed, it has been found in Section 3 that this is the case. The first system input, $T_i(t)v(t)$, directly relates to the amount of heat carried by the inlet air, while the second input, $T_o(t)v(t)$, can be viewed as an internal feedback which assures that as the inlet air speed goes to plus infinity the outlet air temperature converges to the inlet air temperature. Furthermore, considering the model parameters β_i and η_i, $i = 0, 1$ and 2, provided in (11), it can be noticed that assumed physical laws indicate that η_i parameters are equal to β_i parameters however being negative. The estimated $\hat{\beta}_i$ and $\hat{\eta}_i$ parameters provided in Table 1 do clearly confirm this (taking into account the standard error bounds provided by the estimated standard errors).

4.2.1. Mechanistic interpretation of the data-based model - An example: The first principles model (10) does not just provide the information on feasible model structure but also gives us an insight into the estimated data-based model (1). The first principles model suggests that the steady-state gains of the four transfer functions in (10) are related to each other as follows

$$\begin{aligned} \mathrm{SSG}_{vT_i} &= \frac{[(UA)_{int} + (UA)_{ext}]\rho_a A_a c_a}{(UA)_{int}(UA)_{ext}} \\ \mathrm{SSG}_{vT_o} = -\mathrm{SSG}_{vT_i}, \quad \mathrm{SSG}_q &= \frac{\mathrm{SSG}_{vT_i}}{\rho_a A_a c_a}, \quad \mathrm{SSG}_{T_a} = 1 \end{aligned} \tag{13}$$

where SSG denotes the steady-state gain, which has been obtained by setting the differential operator to zero, i.e. $s \to 0$. Considering the identified data-based model (1) it is then possible to infer the value of the steady-state gains (13) as follows

$$\widehat{\mathrm{SSG}}_{vT_i} = \frac{\hat{\beta}_2}{\hat{\alpha}_2}, \quad \widehat{\mathrm{SSG}}_{vT_o} = -\widehat{\mathrm{SSG}}_{vT_i}, \quad \widehat{\mathrm{SSG}}_q = \frac{\widehat{\mathrm{SSG}}_{vT_i}}{\rho_a A_a c_a}, \quad \widehat{\mathrm{SSG}}_{T_a} = 1 \tag{14}$$

Subsequently, recalling that the heating element power has been assumed to be constant as well as the ambient air temperature, i.e. $T_a(t) \approx T_a$ and $q(t) \approx q$, the following must hold

$$\hat{o} = \widehat{\mathrm{SSG}}_q q + \widehat{\mathrm{SSG}}_{T_a} T_a \tag{15}$$

For example, it can be now hypothesised that it is desirable to estimate the heating power q, which is not measured under standard circumstances. Rearranging (15) with respect to 'unknown' q and using the definitions of $\widehat{\mathrm{SSG}}_q$ and $\widehat{\mathrm{SSG}}_{T_a}$ from (14), then

$$\begin{aligned} q &= \frac{\hat{o} - \widehat{\mathrm{SSG}}_{T_a} T_a}{\widehat{\mathrm{SSG}}_q} \\ &= \frac{\hat{\alpha}_2 (\hat{o} - T_a)(\rho_a A_a c_a)}{\hat{\beta}_2} \end{aligned} \tag{16}$$

The average measured ambient air temperature was $T_a = 20.48$ [°C]; from the technical sheets the cross section area is $A_a = 0.17$ $[m^2]$, air density $\rho_a = 1.3$ $[kg/m^3]$ and air specific heat capacity $c_a = 1005$ $[J/kgK]$. Subsequently, the estimated heater power is $\hat{q} = 868.5$ $[W]$, which is well in alignment with the measured value of $q = 830$ $[W]$, see system details in Section 2.

5. Conclusions and further work

A unique physically meaningful nonlinear model of heating element has been identified in continuous-time time domain. A relatively high predictive performance of the estimated model has been observed despite its simplicity. Additionally a reduced order model has been introduced which finds its application in model based control designs. Interestingly, the identified nonlinear model reduces to a so called bilinear model class for constant inlet air temperatures, where such models have been found particularly useful in modelling of heating, ventilation and air conditioning (HVAC) systems by the authors.

The further work will follow two separate but related research streams. Firstly, it is intended to extend the obtained model to include the heating element power as another input. It is then anticipated that such a model will become more universal and will find its application in the area of simulation and control design of HVAC systems. The second work stream will follow modelling and data-based identification of phase change materials (PCM) for air conditioning applications. In these applications the air passes around a heat exchanger filled with PCM. The main difference between such heat exchanger and the considered heating element is that no constant heat power is provided into the heat exchanger. The PCM acts as a heat storage and depending on the air state the heat is either released into the passing air or the air is cooled down. It is assumed that the heat exchange process between the PCM and the passing air can be modeled by the same model as the one obtained in this work, while the inclusion of varying heating power (heat storage) will be required.

References

[1] M. Iten and S. Liu. A work procedure of utilising PCMs as thermal storage systems based on air-TES systems. *Energy conversion and management*, 77:608–627, 2014.

[2] P.C. Young. Data-based mechanistic modelling of environmental, ecological, economic and engineering systems. *Environmental Modelling and Software*, 13:105–122, 1998.

[3] P. C. Young. Data-based mechanistic modelling, generalised sensitivity and dominant mode analysis. *Computer Physics Communications*, 117:113–129, 1999.

[4] L. Price, P. Young, D. Berckmans, K. Janssens, and J. Taylor. Data-based mechanistic modelling (DBM) and control of mass and energy transfer in agricultural buildings. *Annual reviews in Control*, 23:71–82, 1999.

[5] A. Youssef, H. H. Yen, S. E. Özcan, and D. Berckmans. Data-based mechanistic modelling of indoor temperature distributions based on energy input. *Energy and buildings*, 43:2965–2972, 2011.

[6] X. Liu, J. Wang, and W. X. Zheng. Convergence analysis of refined instrumental variable method for continuous-time system identification. *IET Control Theory and Applications*, 5(7):868–877, 2011.

[7] P. C. Young. *Recursive Estimation and Time-Series Analysis: An Introduction for the Student and Practitioner.* Springer-Verlag, 2nd edition, 2011.

[8] I. Zajic. *A Hammerstein-bilinear approach with application to heating ventilation and air conditioning systems.* PhD thesis, Coventry University, UK, 2013.

[9] I. Zajic. and K. J. Burnham. Operating point adjustment procedure for a class of continuous-time bilinear models. In *UKACC Int. Control Conf.*, pages 183–188, Loughborough, UK, 2014.

[10] C. P. Underwood. *HVAC control systems: Modelling, analysis and design.* E.&F.N. Spon, London, 1999.

[11] T. Larkowski. *Extended algorithms for the errors-in-variables identification.* PhD thesis, Coventry University, UK, 2009.

[12] K. J. Burnham, I. Zajic, and T. Larkowski. Review of developments in bilinear systems modelling and control for efficient energy utilization on industrial plant. In *Proc. of 24th Int. Conf. on Efficiency, Cost, Optimization, Simulation and Environmental Impact of Energy Systems*, pages 1856–1870, Novi Sad, Serbia, 2011.

[13] C. P. Underwood. Robust control of HVAC plant I: modelling. *Building services engineering research and technology*, 21:53–61, 2000.

Performance diagnosis of local filters in state estimation of nonlinear systems

Ondřej Straka, Jindřich Duník, and Miroslav Šimandl
European Centre of Excellence - New Technologies for the Information Society, Department of Cybernetics, Faculty of Applied Sciences, University of West Bohemia, Univerzitní 8, 306 14 Pilsen, Czech Republic
E-mail: straka30@kky.zcu.cz, dunikj@kky.zcu.cz, simandl@kky.zcu.cz

Abstract. The paper deals with state estimation of nonlinear stochastic discrete-time systems with a special focus on filters providing a point estimate of the state. The aim is concentrated on performance diagnosis of the filters, which should inform about a possible decrease of estimate quality provided by the filter. The paper proposes a diagnostics of the filter performance using the non-Gaussianity measures. More specifically, the paper focuses on a third-order moment based non-Gaussianity measure and provides relations for its recursive computation. It is shown that this measure can effectively be used for the performance diagnosis of the filters, which is illustrated using a numerical example. The example considers tracking motion of a ship using the extended Kalman filter.

Keywords: state estimation, nonlinear filters, Kalman filtering, non-Gaussianity measures, performance diagnosis

1. Introduction

The problem of state estimation of discrete-time nonlinear stochastic dynamic systems from noisy measured data plays a key role in fields such as target tracking and localization [1], navigation, signal processing [2], fault detection [3], and adaptive and optimal control problems. Hence, it has been a subject of intensive research interest for the last several decades [4, 5, 6].

A general solution to the problem is represented by the Bayesian recursive relations (BRRs) [4]. The relations compute the probability density functions (PDFs) of the state conditioned by the available measurements. These PDFs represent a complete description of the state estimate.

In many cases, the complete description of the state estimate is superfluous and only a point estimate is searched often by optimizing the mean square error (MSE) criterion [7]. The resulting minimum MSE (MMSE) estimate is given A general solution to the problem is represented by the Bayesian recursive relations (BRRs) [8]. The relations compute the probability density functions (PDFs) of the state conditioned by the available measurements. These PDFs represent a complete description of the state estimate by the conditional mean of the state conditioned by the available measurements. The estimators (for coinciding time of the state estimate and time of the available measurement also called filters) providing a point estimate of the state will henceforth be called local, owing to the locally limited estimate validity in the state space$*$.

No matter which approach is chosen (either the Bayesian or the optimization), a closed-form solution to the relations is available only for a few special cases, such as the linear Gaussian system [9]. In other

$*$ The algorithms of local filters may also be derived from the BRRs when an assumption on noises distributions is imposed.

cases, approximate approaches must be utilized. In the case of the optimization approach, the structure of the filter is usually fixed to be a linear function of the measurement, which results in a linear MMSE (LMMSE) estimate.

For majority of nonlinear systems, computing an LMMSE estimate requires further approximations. The performed approximations then lead to the fact, that the estimate cannot be guaranteed to converge to the true state and also the estimate error may be unbounded. In [10] it has been proven that the estimate error can be bounded with probability one for systems with mild nonlinearities, exact initial condition, and small enough disturbing noises. Nonlinearity of a system function may result in initial Gaussian PDF of the state becoming a heavy-tailed or a multi-modal distribution. Subsequently, such a distribution might have unfavourable effect on the estimate quality.

Degree of nonlinearity of the functions appearing in the system description cannot be assessed in isolation, but must be tested in connection with the location of the state. As the state has stochastic nature, uncertainty of its location may be characterized by its mean and covariance matrix. Hence, these two moments of the state should be taken into account when assessing nonlinearity of the functions.

For the assessment, several nonlinearity measures (NLMs) [11, 12, 13] have been proposed. In [11], the NLM serves for a model assessment for selection of a suitable filter, in [13] for reliable assessments of the effectiveness of nonlinear mitigation techniques and in [12] for providing information for a decision whether a filter parameter requires an adaptation. The NLMs have, however, difficulty with capturing a cumulative effect of mild nonlinearities. The effect may appear due to several subsequent transformations of a Gaussian PDF of the state that are individually only mildly nonlinear, however, their joint effect is strongly nonlinear [14].

The cumulative effect of several nonlinear transformations can be effectively captured by non-Gaussianity measures (NGMs) [14]. The NGMs measure a distance of the PDF of a transformed random variable to a Gaussian PDF. The measurement is performed based on either the complete PDFs or on higher-order moments.

Besides measuring nonlinearity of the system functions, the NGMs may also be used for assessing Gaussianity of state conditional PDFs. This is useful especially for a number of recently developed local filters (LFs) based on the assumption of Gaussian conditional PDF of the state. These include, the filters based on various deterministic (cubature or quadrature) or stochastic integration rules or on the Fourier-Hermite expansion [15, 5, 16, 17, 18, 19].

The goal of the paper is to demonstrate utilization of the NGMs for diagnosing performance of the LFs. It will be shown that besides for the LFs assuming Gaussian distribution of the conditional state PDF, other LFs such as the extended Kalman filter (EKF) may utilize the NGMs for the diagnosis. The paper will focus on the moment-based NGMs which will monitor performance of a LF indicating a potential degradation of the estimate quality. The diagnosis will be illustrated using a multi-dimensional tracking of a maneuvering ship.

The rest of the paper is organized as follows. Section 2 provides system specification and a generic LF algorithm with a focus on the EKF. Section 3 then discusses NGMs and their utilization in the LF framework for performance diagnosis. Numerical illustration and concluding remarks are given in Sections 4 and 5, respectively.

2. System Description and Local Filters

2.1. System description

Let a discrete-time nonlinear stochastic system be considered given by the following state-space model

$$\mathbf{x}_{k+1} = \mathbf{f}_k(\mathbf{x}_k) + \mathbf{w}_k, k = 0, 1, 2, \ldots, \tag{1}$$

$$\mathbf{z}_k = \mathbf{h}_k(\mathbf{x}_k) + \mathbf{v}_k, k = 0, 1, 2, \ldots, \tag{2}$$

where the vectors $\mathbf{x}_k \in \mathbb{R}^{n_x}$ and $\mathbf{z}_k \in \mathbb{R}^{n_z}$ represent the immeasurable state of the system and measurement at time instant k, respectively, $\mathbf{f}_k : \mathbb{R}^{n_x} \to \mathbb{R}^{n_x}$ and $\mathbf{h}_k : \mathbb{R}^{n_x} \to \mathbb{R}^{n_z}$ are known vector

Journal of Physics: Conference Series **570** (2014) 012004 doi:10.1088/1742-6596/570/1/012004

functions, and $\mathbf{w}_k \in \mathbb{R}^{n_x}$ and $\mathbf{v}_k \in \mathbb{R}^{n_z}$ are the state and measurement white noises. The PDFs of the noises are supposed to be Gaussian with zero means and known covariance matrices $\boldsymbol{\Sigma}_k^{\mathbf{w}}$ and $\boldsymbol{\Sigma}_k^{\mathbf{v}}$, i.e., $p_{\mathbf{w}_k}(\mathbf{w}_k) = \mathcal{N}\{\mathbf{w}_k; \mathbf{0}, \boldsymbol{\Sigma}_k^{\mathbf{w}}\}$ and $p_{\mathbf{v}_k}(\mathbf{v}_k) = \mathcal{N}\{\mathbf{v}_k; \mathbf{0}, \boldsymbol{\Sigma}_k^{\mathbf{v}}\}$, respectively. The PDF of the initial state is Gaussian, i.e., $p_{\mathbf{x}_0}(\mathbf{x}_0) = \mathcal{N}\{\mathbf{x}_0; \bar{\mathbf{x}}_0, \mathbf{P}_0^{\mathbf{xx}}\}$ with known mean and covariance matrix; the initial state $\mathbf{x}_0$ is independent of the noises.

2.2. *State estimation by local filters*

The aim of a LF is to compute the first two moments of the state conditioned by the measurements, more specifically, the conditional mean $\hat{\mathbf{x}}_{k|k} = \mathsf{E}[\mathbf{x}_k|\mathbf{z}^k]$ and covariance matrix of the estimate error $\mathbf{P}_{k|k}^{\mathbf{xx}} = \mathsf{cov}[\mathbf{x}_k|\mathbf{z}^k]$ in which $\mathbf{z}^k = [\mathbf{z}_0, \mathbf{z}_1, \ldots, \mathbf{z}_k]$. The moments might be understood as a Gaussian approximation of the conditional PDF, i.e., $p(\mathbf{x}_k|\mathbf{z}^k) \approx \mathcal{N}\{\mathbf{x}_k : \hat{\mathbf{x}}_{k|k}, \mathbf{P}_{k|k}^{\mathbf{xx}}\}$ [5, 18], depending on the type of used approximation.

The group of LFs imposing the assumption of Gaussianity of the conditional PDF of the state consists of the quadrature, cubature, and stochastic integration based filters utilizing deterministic and stochastic integration rules [20, 5, 18], and the Fourier-Hermite Kalman filter [19]. The LFs providing the point estimate of the state without the assumption are e.g., the extended Kalman filter (EKF) or the second order filter [21], the divided difference filters (DDFs) utilizing the Stirling polynomial interpolation [22] or the unscented Kalman filter (UKF) using the unscented transform [23],

All the above mentioned LF algorithms share the following steps [6, 5]:

Algorithm 1: Generic Local Filter

Step 1: **(initialization)** Set the time instant $k = 0$ and define the initial condition by the predictive mean $\hat{\mathbf{x}}_{0|-1} = \mathsf{E}[\mathbf{x}_0] = \bar{\mathbf{x}}_0$ and the predictive covariance matrix $\mathbf{P}_{0|-1}^{\mathbf{xx}} = \mathrm{cov}[\mathbf{x}_0] = \mathbf{P}_0^{\mathbf{xx}}$.

Step 2: **(filtering)** The state filtering estimate is given by updating the state predictive estimate with respect to the last measurement $\mathbf{z}_k$ according to

$$\hat{\mathbf{x}}_{k|k} = \hat{\mathbf{x}}_{k|k-1} + \mathbf{K}_k(\mathbf{z}_k - \hat{\mathbf{z}}_{k|k-1}), \tag{3}$$

$$\mathbf{P}_{k|k}^{\mathbf{xx}} = \mathbf{P}_{k|k-1}^{\mathbf{xx}} - \mathbf{K}_k\mathbf{P}_{k|k-1}^{\mathbf{zz}}\mathbf{K}_k^T, \tag{4}$$

where $\mathbf{K}_k = \mathbf{P}_{k|k-1}^{\mathbf{xz}}(\mathbf{P}_{k|k-1}^{\mathbf{zz}})^{-1}$ is the filter gain and

$$\hat{\mathbf{z}}_{k|k-1} = \mathsf{E}[\mathbf{z}_k|\mathbf{z}^{k-1}] = E[\mathbf{h}_k(\mathbf{x}_k)|\mathbf{z}^{k-1}], \tag{5}$$

$$\begin{aligned}\mathbf{P}_{k|k-1}^{\mathbf{zz}} &= \mathsf{E}[(\mathbf{z}_k - \hat{\mathbf{z}}_{k|k-1})(\mathbf{z}_k - \hat{\mathbf{z}}_{k|k-1})^T|\mathbf{z}^{k-1}] = \\ &= \mathsf{E}[(\mathbf{h}_k(\mathbf{x}_k) - \hat{\mathbf{z}}_{k|k-1}) \times (\mathbf{h}_k(\mathbf{x}_k) - \hat{\mathbf{z}}_{k|k-1})^T|\mathbf{z}^{k-1}] + \boldsymbol{\Sigma}_k^{\mathbf{v}},\end{aligned} \tag{6}$$

$$\mathbf{P}_{k|k-1}^{\mathbf{xz}} = \mathsf{E}[(\mathbf{x}_k - \hat{\mathbf{x}}_{k|k-1})(\mathbf{z}_k - \hat{\mathbf{z}}_{k|k-1})^T|\mathbf{z}^{k-1}]. \tag{7}$$

Step 3: **(prediction)** The state predictive estimate is given by the relations

$$\hat{\mathbf{x}}_{k+1|k} = \mathsf{E}[\mathbf{x}_{k+1}|\mathbf{z}^k] = E[\mathbf{f}_k(\mathbf{x}_k)|\mathbf{z}^k], \tag{8}$$

$$\mathbf{P}_{k+1|k}^{\mathbf{xx}} = \mathsf{E}[(\mathbf{x}_{k+1} - \hat{\mathbf{x}}_{k+1|k})(\mathbf{x}_{k+1} - \hat{\mathbf{x}}_{k+1|k})^T|\mathbf{z}^k] = \mathsf{E}[(\mathbf{f}_k(\mathbf{x}_k) - \hat{\mathbf{x}}_{k|k})(\mathbf{f}_k(\mathbf{x}_k) - \hat{\mathbf{x}}_{k|k})^T|\mathbf{z}^k] + \boldsymbol{\Sigma}_k^{\mathbf{w}}. \tag{9}$$

Let $k = k + 1$. The algorithm then continues by **Step 2: (filtering)**.

The particular LFs differ in the approximation applied when calculating the measurement and state predictive statistics (5)–(7) and (8)–(9). Usage of the Taylor expansion by the EKF will be described in the following subsection.

2.2.1. The first order Taylor expansion The basic approach for an approximate computation of the predictive statistics is based on an approximation of the nonlinear function by the first order Taylor expansion (TE1) [9]. The TE1 of the nonlinear function in measurement equation under assumption of the known state predictive statistics $\hat{\mathbf{x}}_{k|k-1}$ and $\mathbf{P}^{xx}_{k|k-1}$ (defining the approximation point for the filtering step) is of the form

$$\mathbf{z}_k = \mathbf{h}_k(\mathbf{x}_k) + \mathbf{v}_k \approx \mathbf{h}_k(\hat{\mathbf{x}}_{k|k-1}) + \mathbf{H}_k(\hat{\mathbf{x}}_{k|k-1})\left(\mathbf{x}_k - \hat{\mathbf{x}}_{k|k-1}\right) + \mathbf{v}_k, \tag{10}$$

where $\mathbf{H}_k(\hat{\mathbf{x}}_{k|k-1}) = \frac{\partial \mathbf{h}(\mathbf{x}_k)}{\partial \mathbf{x}_k}|_{\mathbf{x}_k=\hat{\mathbf{x}}_{k|k-1}}$ is the Jacobian of $\mathbf{h}_k(\cdot)$. The desired statistics† are then computed as

$$\hat{\mathbf{z}}^{\mathrm{TE1}}_{k|k-1} = \mathbf{h}_k(\hat{\mathbf{x}}_{k|k-1}), \tag{11}$$
$$\mathbf{P}^{\mathbf{zz},\mathrm{TE1}}_{k|k-1} = \mathbf{H}_k\mathbf{P}^{xx}_{k|k-1}\mathbf{H}_k^T + \mathbf{\Sigma}^{\mathbf{v}}_k, \tag{12}$$
$$\mathbf{P}^{\mathbf{xz},\mathrm{TE1}}_{k|k-1} = \mathbf{P}^{xx}_{k|k-1}\mathbf{H}_k^T, \tag{13}$$

where $\mathbf{H}_k = \mathbf{H}_k(\hat{\mathbf{x}}_{k|k-1})$.

The TE1 of the nonlinear function in the state equation under assumption of the known state filtering statistics $\hat{\mathbf{x}}_{k|k}$ and $\mathbf{P}^{xx}_{k|k}$ (defining the approximation point for the prediction step) is of the form

$$\mathbf{x}_{k+1} = \mathbf{f}_k(\mathbf{x}_k) + \mathbf{w}_k \approx \mathbf{f}_k(\hat{\mathbf{x}}_{k|k}) + \mathbf{F}_k(\hat{\mathbf{x}}_{k|k})\left(\mathbf{x}_k - \hat{\mathbf{x}}_{k|k}\right) + \mathbf{w}_k, \tag{14}$$

where $\mathbf{F}_k(\hat{\mathbf{x}}_{k|k}) = \frac{\partial \mathbf{f}(\mathbf{x}_k)}{\partial \mathbf{x}_k}|_{\mathbf{x}_k=\hat{\mathbf{x}}_{k|k}}$ is the Jacobian of $\mathbf{f}_k(\cdot)$. The desired statistics are then computed as

$$\hat{\mathbf{x}}^{\mathrm{TE1}}_{k+1|k} = \mathbf{f}_k(\hat{\mathbf{x}}_{k|k}), \tag{15}$$
$$\mathbf{P}^{\mathbf{xx},\mathrm{TE1}}_{k+1|k} = \mathbf{F}_k\mathbf{P}^{xx}_{k|k}\mathbf{F}_k^T + \mathbf{\Sigma}^{\mathbf{w}}_k, \tag{16}$$

where $\mathbf{F}_k = \mathbf{F}_k(\hat{\mathbf{x}}_{k|k})$.

3. Non-Gaussianity Measures

The LFs provide reliable results if the approximation point is close to the true state. This usually happens if the nonlinearities in system description (1) and (2) are mild and if the state conditional PDFs have Gaussian-like shape (unimodal and not heavy-tailed).

Assessment of the system nonlinearity or the resulting state PDF properties is, therefore, a vital part of any filter design procedure. Hence, several NLMs have been proposed so far. A NLM can be utilized in two ways: i) as a global NLM assessing the overall nonlinearity of the given system without any reference to the usage of a specific filter [11], or ii) as a local NLM designed for an on-line monitoring of (local) filters.

While the global NLM provides an answer whether a LF is expected to perform well or a more complicated filter producing PDF estimates needs to be used, the local NLM is suitable for self-monitoring of the LFs. However, the local NLM has a very limited capability in monitoring the transformation cumulative effect (the NLMs in two subsequent time instants are independent; even the NLM in the filtering step is independent of the NLM in the prediction step at the same time). As was illustrated in [14], a sequence of mildly nonlinear transformations (transformations with very low values of NLMs) with, for example, an initially Gaussian PDF, might result in a heavy-tailed or multi-modal distribution. As a consequence, assumptions of some LFs are violated and the estimate quality deteriorates.

This cumulative effect of the nonlinear transformations can be monitored by NGMs, that evaluate influence of the nonlinear transformation on the PDF of a transformed state. Essentially, the NGMs are

† The statistics are generally approximate. Exact values are obtained for a few special functions $\mathbf{h}_k(\cdot)$ only, e.g., for a linear function $\mathbf{h}_k(\cdot)$.

based on a comparison of two PDFs, the estimated one and its Gaussian approximation. The comparison may consider the shape of the PDFs (e.g., based on the negentropy [14]) or be moment-based. The moment-based NGMs can, for example, compare the third and higher-order moments, skewness or kurtosis [14]. This paper will focus on the third moment, which is for a Gaussian random variable zero.

In this paper, the following parametrization of the filtering third moment of a vector state $\mathbf{x}_k$, denoted as $\mathbf{M}^{\mathbf{xxx}}_{k|k}$, is adopted

$$\mathbf{M}^{\mathbf{xxx}}_{k|k} \triangleq \mathsf{E}[(\mathbf{x}_k - \hat{\mathbf{x}}_{k|k})(\mathbf{x}_k - \hat{\mathbf{x}}_{k|k})^T \otimes (\mathbf{x}_k - \hat{\mathbf{x}}_{k|k})^T |\mathbf{z}^k], \tag{17}$$

where the symbol $\otimes$ stands for the Kronecker product [24]‡. The predictive third moment $\mathbf{M}^{\mathbf{xxx}}_{k+1|k}$ is defined analogously. It can be seen that the 3rd moment is described by a matrix of the dimension $n_x \times (n_x)^2$; thus, the number of elements of the moment, which will be compared, is cubic. To reduce the costs of moments computation and comparison for higher dimensions n_x, the mixed elements may be discarded and only n_x elements are compared.

As the third moment of a Gaussian distribution is zero, the NGM related to the third moment of n-th element of the state is given by

$$J_{3,k}(\mathbf{x}_{(n),k}) = |\mathbf{M}^{\mathbf{xxx}}_{k|k}(n, (n-1)n_x + n)|, n = 1, 2, \ldots, n_x, \tag{18}$$

where $\mathbf{M}^{\mathbf{xxx}}_{k|k}(i, j)$ is the (i, j)-th element of the matrix $\mathbf{M}^{\mathbf{xxx}}_{k|k}$.

When comparing the NGM (18) with a threshold $J_{3,thr}$ the inequality $J_{3,k}(\cdot) \geq J_{3,thr}$ indicates that the corresponding estimated PDF is far from being Gaussian. Thus, in this case the estimate quality is possibly low. The comparison is the main component of the performance diagnosis of the LF.

Note that the threshold value needs to be specified on the basis of an off-line analysis or a simulation study and might be different for each state component. Now, having the NGM specified based on the third moment, calculation of the moment will be pursued in the following subsection.

3.1. Recursive computation of the third moments

The *filtering third moment* of the state is given by (17). Substituting (3) into (17) and defining the error variables $\tilde{\mathbf{x}}_{k|k-1} = \mathbf{x}_k - \hat{\mathbf{x}}_{k|k-1}$ and $\tilde{\mathbf{z}}_{k|k-1} = \mathbf{z}_k - \hat{\mathbf{z}}_{k|k-1}$, the following relation can be obtained:

$$\begin{aligned}
\mathbf{M}^{\mathbf{xxx}}_{k|k} &= \mathsf{E}[(\tilde{\mathbf{x}}_{k|k-1} - \mathbf{K}_k\tilde{\mathbf{z}}_{k|k-1})(\tilde{\mathbf{x}}_{k|k-1} - \mathbf{K}_k\tilde{\mathbf{z}}_{k|k-1})^T \otimes (\tilde{\mathbf{x}}_{k|k-1} - \mathbf{K}_k\tilde{\mathbf{z}}_{k|k-1})^T] \\
&= \mathsf{E}[\underbrace{\tilde{\mathbf{x}}_{k|k-1}\tilde{\mathbf{x}}^T_{k|k-1} \otimes \tilde{\mathbf{x}}^T_{k|k-1}}_{a} - \underbrace{\mathbf{K}_k\tilde{\mathbf{z}}_{k|k-1}\tilde{\mathbf{x}}^T_{k|k-1} \otimes \tilde{\mathbf{x}}^T_{k|k-1}}_{b} - \underbrace{\tilde{\mathbf{x}}_{k|k-1}\tilde{\mathbf{z}}^T_{k|k-1}\mathbf{K}^T_k \otimes \tilde{\mathbf{x}}^T_{k|k-1}}_{c} \\
&+ \underbrace{\mathbf{K}_k\tilde{\mathbf{z}}_{k|k-1}\tilde{\mathbf{z}}^T_{k|k-1}\mathbf{K}^T_k \otimes \tilde{\mathbf{x}}^T_{k|k-1}}_{d} - \underbrace{\tilde{\mathbf{x}}_{k|k-1}\tilde{\mathbf{x}}^T_{k|k-1} \otimes \tilde{\mathbf{z}}^T_{k|k-1}\mathbf{K}^T_k}_{e} + \underbrace{\mathbf{K}_k\tilde{\mathbf{z}}_{k|k-1}\tilde{\mathbf{x}}^T_{k|k-1} \otimes \tilde{\mathbf{z}}^T_{k|k-1}\mathbf{K}^T_k}_{f} \\
&+ \underbrace{\tilde{\mathbf{x}}_{k|k-1}\tilde{\mathbf{z}}^T_{k|k-1}\mathbf{K}^T_k \otimes \tilde{\mathbf{z}}^T_{k|k-1}\mathbf{K}^T_k}_{g} - \underbrace{\mathbf{K}_k\tilde{\mathbf{z}}_{k|k-1}\tilde{\mathbf{z}}^T_{k|k-1}\mathbf{K}^T_k \otimes \tilde{\mathbf{z}}^T_{k|k-1}\mathbf{K}^T_k}_{h}].
\end{aligned} \tag{19}$$

‡ Alternative parametrizations can be found e.g., in [25].

The individual terms of (19) can be rewritten (for the ease of further derivations) according to the equality $\mathbf{AD} \otimes \mathbf{BG} = (\mathbf{A} \otimes \mathbf{B})(\mathbf{D} \otimes \mathbf{G})$ [24] as

$$\mathsf{E}[a] = \mathbf{M}^{\mathbf{xxx}}_{k|k-1}, \tag{20}$$

$$\mathsf{E}[b] = \mathbf{K}_k \mathbf{M}^{\mathbf{zxx}}_{k|k-1}, \tag{21}$$

$$c = \tilde{\mathbf{x}}\tilde{\mathbf{z}}^T \mathbf{K}^T \otimes \tilde{\mathbf{x}}^T = \tilde{\mathbf{x}}\tilde{\mathbf{z}}^T \mathbf{K}^T \otimes \tilde{\mathbf{x}}^T \mathbf{I}_{n_x} = (\tilde{\mathbf{x}}\tilde{\mathbf{z}}^T \otimes \tilde{\mathbf{x}}^T)(\mathbf{K}^T \otimes \mathbf{I}_{n_x}), \tag{22}$$

$$\mathsf{E}[c] = \mathbf{M}^{\mathbf{xzx}}_{k|k-1}(\mathbf{K}^T \otimes \mathbf{I}_{n_x}), \tag{23}$$

$$d = \mathbf{K}\tilde{\mathbf{z}}\tilde{\mathbf{z}}^T \mathbf{K}^T \otimes \tilde{\mathbf{x}}^T = \mathbf{K}\tilde{\mathbf{z}}\tilde{\mathbf{z}}^T \mathbf{K}^T \otimes \tilde{\mathbf{x}}^T \mathbf{I}_{n_x} = \mathbf{K}(\tilde{\mathbf{z}}\tilde{\mathbf{z}}^T \otimes \tilde{\mathbf{x}}^T)(\mathbf{K}^T \otimes \mathbf{I}_{n_x}), \tag{24}$$

$$\mathsf{E}[d] = \mathbf{K}\mathbf{M}^{\mathbf{zzx}}_{k|k-1}(\mathbf{K}^T \otimes \mathbf{I}_{n_x}), \tag{25}$$

$$\mathsf{E}[e] = \mathbf{M}^{\mathbf{xxz}}_{k|k-1}\mathbf{K}_k^T, \tag{26}$$

$$\mathsf{E}[f] = \mathbf{K}_k \mathbf{M}^{\mathbf{zxz}}_{k|k-1}\mathbf{K}_k^T, \tag{27}$$

$$g = \tilde{\mathbf{x}}\tilde{\mathbf{z}}^T \mathbf{K}^T \otimes \tilde{\mathbf{z}}^T \mathbf{K}^T = (\tilde{\mathbf{x}}\tilde{\mathbf{z}}^T \otimes \tilde{\mathbf{z}}^T)(\mathbf{K}^T \otimes \mathbf{K}^T), \tag{28}$$

$$\mathsf{E}[g] = \mathbf{M}^{\mathbf{xzz}}_{k|k-1}(\mathbf{K}_k^T \otimes \mathbf{K}_k^T), \tag{29}$$

$$h = \mathbf{K}\tilde{\mathbf{z}}\tilde{\mathbf{z}}^T \mathbf{K}^T \otimes \tilde{\mathbf{z}}^T \mathbf{K}^T = \mathbf{K}(\tilde{\mathbf{z}}\tilde{\mathbf{z}}^T \otimes \tilde{\mathbf{z}}^T)(\mathbf{K}^T \otimes \mathbf{K}^T), \tag{30}$$

$$\mathsf{E}[h] = \mathbf{K}_k \mathbf{M}^{\mathbf{zzz}}_{k|k-1}(\mathbf{K}_k^T \otimes \mathbf{K}_k^T), \tag{31}$$

where $\mathbf{I}_{n_x}$ denotes an $n_x \times n_x$ identity matrix, $\mathbf{M}^{\mathbf{zxx}}_{k|k-1} \triangleq \mathsf{E}[\tilde{\mathbf{z}}_{k|k-1}\tilde{\mathbf{x}}^T_{k|k-1} \otimes \tilde{\mathbf{x}}^T_{k|k-1}]$ and the remaining third moments are defined analogously.

Based on (20)–(31), the final form of the *filtering third moment* (19) is

$$\begin{aligned}\mathbf{M}^{\mathbf{xxx}}_{k|k} = \mathbf{M}^{\mathbf{xxx}}_{k|k-1} - \mathbf{K}_k\mathbf{M}^{\mathbf{zxx}}_{k|k-1} - \mathbf{M}^{\mathbf{xzx}}_{k|k-1}(\mathbf{K}_k^T \otimes \mathbf{I}_{n_x}) + \mathbf{K}_k\mathbf{M}^{\mathbf{zzx}}_{k|k-1}(\mathbf{K}_k^T \otimes \mathbf{I}_{n_x}) - \mathbf{M}^{\mathbf{xxz}}_{k|k-1}\mathbf{K}_k^T \\ + \mathbf{K}_k\mathbf{M}^{\mathbf{zxz}}_{k|k-1}\mathbf{K}_k^T + \mathbf{M}^{\mathbf{xzz}}_{k|k-1}(\mathbf{K}_k^T \otimes \mathbf{K}_k^T) - \mathbf{K}_k\mathbf{M}^{\mathbf{zzz}}_{k|k-1}(\mathbf{K}_k^T \otimes \mathbf{K}_k^T).\end{aligned} \tag{32}$$

The *predictive third moment* of the state is defined as

$$\mathbf{M}^{\mathbf{xxx}}_{k+1|k} \triangleq \mathsf{E}[(\mathbf{x}_{k+1} - \hat{\mathbf{x}}_{k+1|k})(\mathbf{x}_{k+1} - \hat{\mathbf{x}}_{k+1|k})^T \otimes (\mathbf{x}_{k+1} - \hat{\mathbf{x}}_{k+1|k})^T]. \tag{33}$$

Relations (32) and (33) are the recursive relations for the computation of the third conditional moment in the LF framework. The relations are general and in the following part the relations are detailed for the EKF assuming only a scalar case, i.e., $n_x = n_z = 1$ due to space limitations.

3.2. Detailed recursive computation of the third moments for the EKF

For the scalar case (32) reduces to

$$M^{xxx}_{k|k} = M^{xxx}_{k|k-1} - 3K_k M^{xxz}_{k|k-1} + 3(K_k)^2 M^{xzz}_{k|k-1} - (K_k)^3 M^{zzz}_{k|k-1}, \tag{34}$$

where the third moments are defined as

$$M^{xxx}_{k|k-1} \triangleq \mathsf{E}[\tilde{x}^3_{k|k-1}],$$

$$M^{xxz}_{k|k-1} \triangleq \mathsf{E}[\tilde{x}^2_{k|k-1}\tilde{z}_{k|k-1}],$$

$$M^{xzz}_{k|k-1} \triangleq \mathsf{E}[\tilde{x}_{k|k-1}\tilde{z}^2_{k|k-1}],$$

$$M^{zzz}_{k|k-1} \triangleq \mathsf{E}[\tilde{z}^3_{k|k-1}].$$

Following the approximation technique of the EKF given by (11)–(13), where the nonlinear function $h_k(\cdot)$ in measurement equation (2) is approximated using TE1 (10), the required third moments equal to

$$M^{zzz}_{k|k-1} = (H_k)^3(\hat{x}_{k|k-1})M^{xxx}_{k|k-1}, \tag{35}$$

$$M^{xxz}_{k|k-1} = H_k(\hat{x}_{k|k-1})M^{xxx}_{k|k-1}, \tag{36}$$

$$M^{xzz}_{k|k-1} = (H_k)^2(\hat{x}_{k|k-1})M^{xxx}_{k|k-1}. \tag{37}$$

It can be seen that the TE1-based linearization ignores the information about higher order terms caused by the nonlinear transformation. In fact, under assumption of the Gaussian initial condition (imposing $M^{xxx}_{k|k-1} = 0$), the moment $M^{xxx}_{k|k}$ is zero.

Thus, to get the non-zero moment approximation, let the second order Taylor expansion (TE2)

$$z_k = h_k(x_k) + v_k \approx h_k(\hat{x}_{k|k-1}) + H_k(\hat{x}_{k|k-1})\tilde{x}_{k|k-1} + \frac{1}{2}\mathcal{H}_k(\hat{x}_{k|k-1})\tilde{x}^2_{k|k-1} + v_k, \tag{38}$$

be used instead of the TE1 where $\mathcal{H}_k = \mathcal{H}_k(\hat{x}_{k|k-1}) = \frac{d^2h(x_k)}{dx_k^2}|_{x_k=\hat{x}_{k|k-1}}$ is the Hessian. The TE2 is typically used in the second order filter design [26]. Utilization of the TE2 in the higher moments computation requires conditional higher moments of the predictive state estimate, namely $M^{x^4}_{k|k-1}$, $M^{x^5}_{k|k-1}$, and $M^{x^6}_{k|k-1}$, which are not available§. The proposed solution is to approximate those values by the theoretical moments computed under assumption of the Gaussian distribution [27], i.e., as

$$M^{x^4}_{k|k-1} = 3(P^{xx}_{k|k-1})^2, \tag{39}$$

$$M^{x^5}_{k|k-1} = 0, \tag{40}$$

$$M^{x^6}_{k|k-1} = 15(P^{xx}_{k|k-1})^3. \tag{41}$$

After tedious calculations, the required third moments (35)–(37) can be expressed by:

$$M^{zzz}_{k|k-1} = (H_k)^3M^{xxx}_{k|k-1} - \tfrac{3}{2}H_k(\mathcal{H}_k)^2P^{xx}_{k|k-1}M^{xxx}_{k|k-1} + \mathcal{H}^3_k(P^{xx}_{k|k-1})^3 + 3(H_k)^2\mathcal{H}_k(P^{xx}_{k|k-1})^2, \tag{42}$$

$$M^{xxz}_{k|k-1} = H_kM^{xxx}_{k|k-1} + \mathcal{H}_k(P^{xx}_{k|k-1})^2, \tag{43}$$

$$M^{xzz}_{k|k-1} = (H_k)^2M^{xxx}_{k|k-1} - \tfrac{1}{2}(\mathcal{H}_k)^2P^{xx}_{k|k-1}M^{xxx}_{k|k-1} + 2H_k\mathcal{H}_k(P^{xx}_{k|k-1})^2. \tag{44}$$

The predictive moment (33) is given by

$$M^{xxx}_{k+1|k} = (F_k)^3M^{xxx}_{k|k} - \tfrac{3}{2}F_k(\mathcal{F}_k)^2P^{xx}_{k|k}M^{xxx}_{k|k} + \mathcal{F}^3_k(P^{xx}_{k|k})^3 + 3(F_k)^2\mathcal{F}_k(P^{xx}_{k|k})^2, \tag{45}$$

where $\mathbf{F}_k = \mathbf{F}_k(\hat{\mathbf{x}}_{k|k}) = \frac{\partial\mathbf{f}(\mathbf{x}_k)}{\partial\mathbf{x}_k}|_{\mathbf{x}_k=\hat{\mathbf{x}}_{k|k}}$ is the Jacobian of $\mathbf{f}_k(\cdot)$ and $\mathcal{F}_k = \mathcal{F}_k(\hat{x}_{k|k}) = \frac{d^2f(x_k)}{dx_k^2}|_{x_k=\hat{x}_{k|k}}$ is the Hessian of $\mathbf{f}_k(\cdot)$.

3.3. Recursive computation of the third moment for sigma-point based filters

The relations for recursive computation of the third moments by means of the TE2 introduced in the previous section can be readily written for sigma-point based filters such as the UKF, cubature Kalman filter, quadrature Kalman filter, divided difference filter, and stochastic integration filter. For the UKF the relations can be found in [28].

§ This phenomenon is referred in the literature as the moment closure problem [4]

3.4. LF algorithm with NGMs

Having stated the relations necessary for the recursive state higher-order moment computation and the NGMs, the algorithm of the LF with steps for performance diagnosis can be summarized.

Algorithm 2: Local Filter with performance diagnosis

Step 1: **(initialization)** as **Step 1** of Algorithm 1, supplemented with specification of an initial higher moments estimate.

Step 2: **(filtering)** as **Step 2** of Algorithm 1, supplemented with computation of $\mathbf{M}^{\mathbf{xxx}}_{k|k}$ according to (32).

Step 3: **(self-assessment)** If any measure $J_{3,k}(\mathbf{x}_{(n),k})$, $n = 1, 2, \ldots, n_x$, of the filtering estimate is greater than or equal to the respective threshold $J_{3,thr}$, then there is a possibility of a lower estimate quality.

Step 4: **(prediction)** as **Step 3** of Algorithm 1, supplemented with computation of $\mathbf{M}^{\mathbf{xxx}}_{k+1|k}$ according to (33).

Step 5: **(self-assessment)** If any measure $J_{3,k}(\mathbf{x}_{(n),k})$, $n = 1, 2, \ldots, n_x$, of the predictive estimate is greater than or equal to the respective threshold $J_{3,thr}$, then there is a possibility of a lower estimate quality.

Let $k = k + 1$. The algorithm then continues by **Step 2 (filtering)**.

4. Numerical Illustration

Performance diagnosis of the EKF will be illustrated using a model describing ship motion in deep water [29]. The ship motion is described by the following continuous-time model (CTM):

$$\frac{dx}{dt} = K_v v_0 \sin(\psi), \tag{46}$$

$$\frac{dy}{dt} = K_v v_0 \cos(\psi), \tag{47}$$

$$\frac{d\psi}{dt} = K_v \omega, \tag{48}$$

$$\frac{d\omega}{dt} = \frac{v_0^2}{2pL^2}\left(-q\omega\frac{L}{v_0 K_v} - s_{31}\delta\right). \tag{49}$$

The state vector $\mathbf{x} = [x, y, \psi, \omega]^T$ includes the ship coordinates x and y, heading ψ and turn-rate ω. The variable δ denotes the control rudder angle deviation, L is the ship length, $K_v = \left(1 + 1.9\omega^2 L^2 v_0\right)^{-1}$, v_0 is the ship speed at zero turn-rate, $q = q_{21}r_{31} - q_{31}r_{21}$, $p = 0.5(q_{21} + r_{31})$ with q_{21}, r_{21}, q_{31}, r_{31}, and s_{31} being hydrodynamic coefficients depending on the ship geometry and length L. The simulation was performed with the following realistic (for a stable ship) constants $q_{21} = 1.227$, $r_{21} = -0.629$, $q_{31} = -4.64$, $r_{31} = 3.88$, $s_{31} = -1.019$, $v_0 = 30$ [m/s], and $L = 99$ [m] The control rudder angle deviation was simulated as

$$\delta = \begin{cases} 7.5\sin(0.01 \cdot t)\ [^\circ] & \text{if } t \in (250, 450)\ [\text{s}], \\ 0 & \text{if } t \notin (250, 450)\ [\text{s}]. \end{cases} \tag{50}$$

The CTM (46)–(49) was discretized using the Euler method with sampling time $T = 0.5$ [s]. The turn-rate ω_k was supposed to be corrupted by a white zero-mean Gaussian noise w^{ω}_k with its variance depending on the turn-rate as $\Sigma^{\omega}_k = (\gamma_a + \gamma_b\omega_{k-1})^2$, with $\gamma_a = 0.0005$ and $\gamma_b = 0.1$.

The state $\mathbf{x}$ was observed as

$$\mathbf{z}_k = \begin{bmatrix} 1 & 0 & 0 & 0 \\ 0 & 1 & 0 & 0 \end{bmatrix} \mathbf{x}_k + \mathbf{v}_k, \tag{51}$$

which means that the position of the ship was observed directly in Cartesian coordinates. The observation was corrupted by a white zero-mean Gaussian noise $\mathbf{v}_k$ with covariance matrix $\Sigma_k^v = \text{diag}([4, 4])$ [m^2].

The motion was simulated for 900 seconds with initial conditions $\mathbf{x}_0 = [10 \text{ [m]}, 10 \text{ [m]}, 45 \text{ [°]}, 0 \text{ [°/s]}]$. The EKF was initialized with $\hat{\mathbf{x}}_{0|-1} = \bar{\mathbf{x}}_0$ and $\mathbf{P}_0 = \text{diag}([10^4 \text{ [m}^2], 10^4 \text{ [m}^2], (1.5)^2 \text{ [(°)}^2], 1 \text{ [(°/s)}^2]])$. The trajectories of the state are depicted in Figure 1. The estimate error was observed using the MSE for

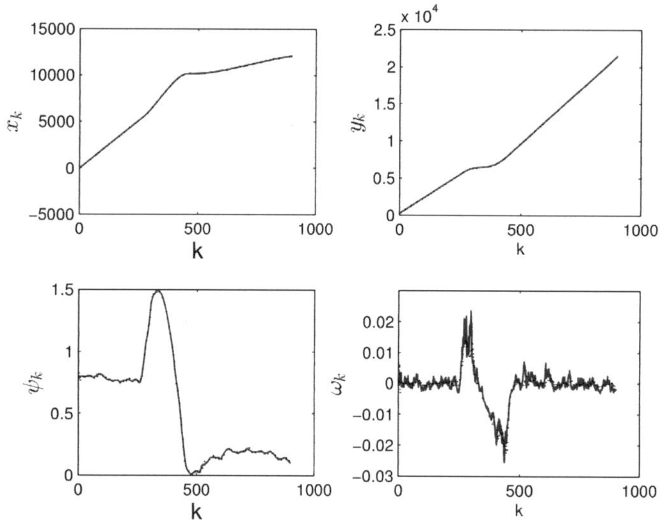

Figure 1. Trajectories of the state variables.

the position in the x direction defined as

$$\text{MSE}_k^x = \frac{1}{N}\sum_{i=1}^{N}(x_{(1),k|k}(i) - x_{(1),k}(i))^2, \tag{52}$$

where $x_{(1),k}(i)$ and $x_{(1),k|k}(i)$ is the first state component at time k and i-th Monte Carlo (MC) simulation and its filtering estimate calculated by the EKF, respectively. The MSE for the position in the y direction MSE_k^y, heading MSE_k^ψ and turn-rate MSE_k^ω are defined analogously. The scenario was simulated $N = 250$ times. The MSEs were compared with the third-order moments M^{xxx}, M^{yyy}, $M^{\psi\psi\psi}$, and $M^{\omega\omega\omega}$, where M^{xxx} is defined as

$$M_k^{xxx} = \frac{1}{N}\sum_{i=1}^{N} M_{k|k}^{xxx}(i)$$

and M^{yyy}, $M^{\psi\psi\psi}$, and $M^{\omega\omega\omega}$ are defined analogously

The resulting MSEs together with the corresponding third-order moments are depicted in Figures 2 and 3. To illustrate instant values, a comparison of the squared error of the state estimate and instant values of the third-order moments are given in Figures 4 and 5. From all Figures 2, 3, 4, and 5 it is clear that time instants with high values of the NGM (given by the third-order moment of the state) correspond to the time instants with deteriorated estimate quality. The deterioration of the estimate quality is caused by a significant change of the turn-rate due to control rudder angle deviation, which although known, shifts the state into a region of severe nonlinearity of the function in the dynamics. The relation between the state errors and the third-order moments confirms our hypothesis that a severe nonlinearity (leading to high values of the NGM) leads to increased estimate error. Therefore, the NGMs can effectively be used for performance diagnosis of nonlinear filters.

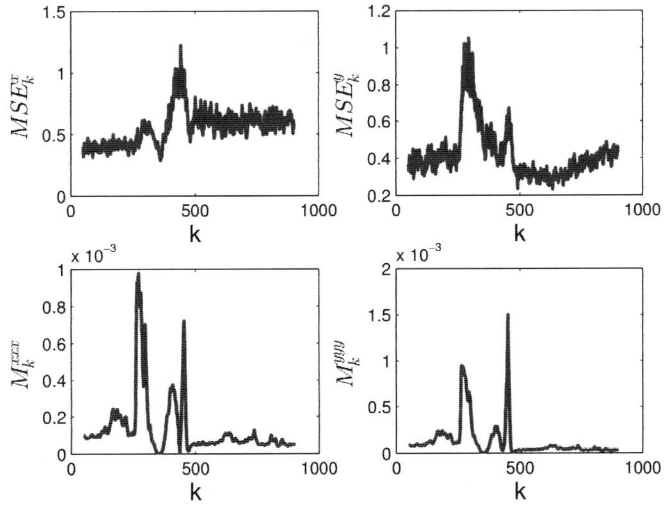

Figure 2. The MSEs MSE_k^x and MSE_k^y and the third moments M_k^{xxx} and M_k^{yyy} for the position components of the state vector.

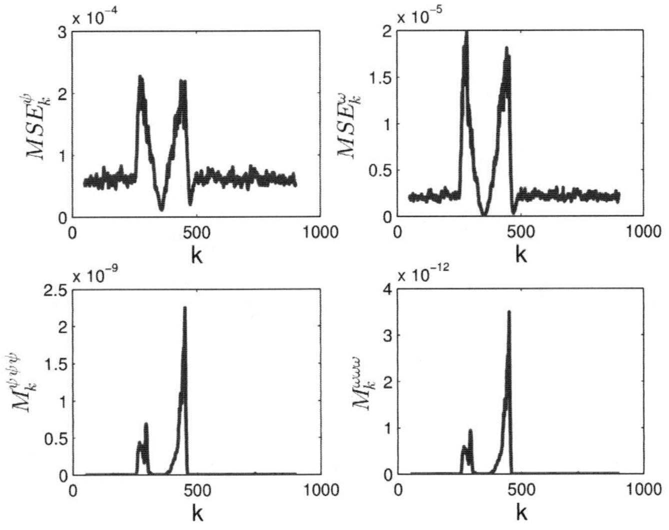

Figure 3. The MSEs MSE_k^ψ and MSE_k^ω and the third moments $M_k^{\psi\psi\psi}$ and $M_k^{\omega\omega\omega}$ for the heading and turn-rate components of the state vector.

5. Concluding Remarks

The paper dealt with state estimation of nonlinear stochastic discrete-time systems. The aim was focused on possibilities of performance diagnosis of nonlinear filters providing point estimates of the state for such systems. Such performance diagnosis would be a convenient tool for the users as it would provide them an additional information about the estimate quality or it may be used for the filter to respond to deteriorating conditions for providing a quality estimate.

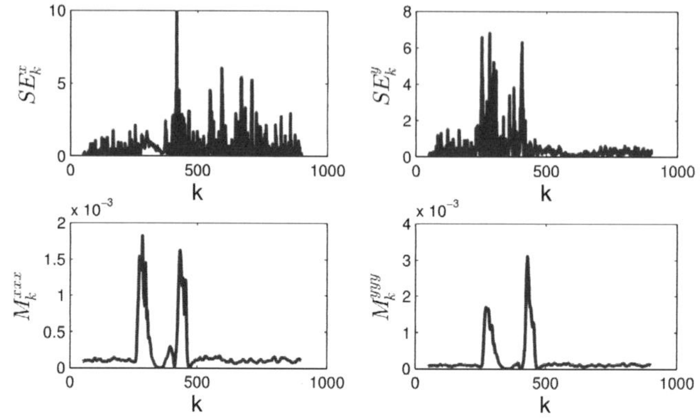

Figure 4. The squared errors SE_k^x and SE_k^y and instant values of the third moments M_k^{xxx} and M_k^{yyy} for the position components of the state vector at a single simulation.

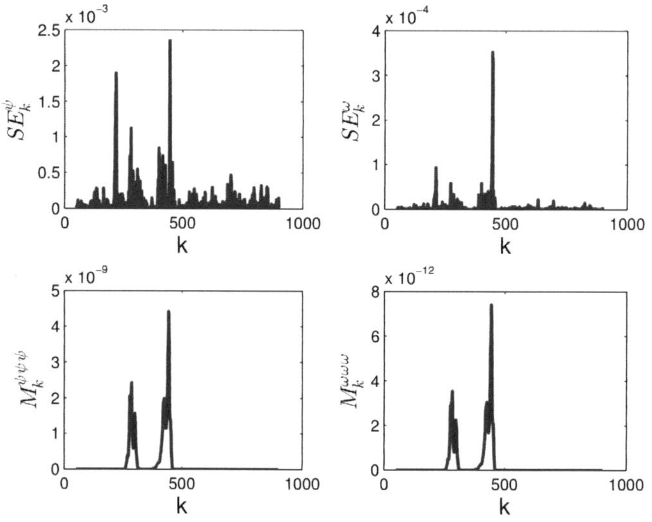

Figure 5. The squared errors SE_k^ψ and SE_k^ω and instant values of the third moments $M_k^{\psi\psi\psi}$ and $M_k^{\omega\omega\omega}$ for the heading and turn-rate components of the state vector at a single simulation.

It was shown that non-Gaussianity measures may be used for this purpose as they can measure an effect of nonlinear transformations which, if they are severe, may lead to a poor estimate quality. Moreover, the NGMs are able to capture cumulating effect of a mildly nonlinear transformation in comparison with the NLMs. The paper focused on the moment-based NGMs, more specifically on the NGM based on the third-order moment. The paper provided general relations for its recursive computation and detailed relations of its computation based on the TE2. Usage of the third-order moment based NGM for performance diagnosis was demonstrated using a numerical example of a ship tracking.

Acknowledgement
This work was supported by the ALFA Program of the Technology Agency of the Czech Republic, project No. TA03030674 (RESPOL).

References

[1] W. Koch. Fixed-interval retrodiction approach to Bayesian IMM-MHT for maneuvering multiple targets. *IEEE Transactions on Aerospace and Electronic Systems*, 36(1):2–14, 2000.
[2] S. Haykin and I. Arasaratnam. *Adaptive Signal Processing*. John Wiley & Sons, Inc., 2010.
[3] Rolf Isermann. *Fault-Diagnosis Systems: An Introduction from Fault Detection to Fault Tolerance*. Springer, Berlin, Germany, 2006.
[4] H. W. Sorenson. On the development of practical nonlinear filters. *Information Sciences*, 7:230–270, 1974.
[5] I. Arasaratnam and S. Haykin. Cubature Kalman filters. *IEEE Transactions on Automatic Control*, 54(6):1254–1269, 2009.
[6] M. Simandl and J. Dunik. Derivative-free estimation methods: New results and performance analysis. *Automatica*, 45(7):1749–1757, 2009.
[7] A. H. Jazwinski. *Stochastic Processes and Filtering Theory*. Academic Press, 1970.
[8] H. W. Sorenson and D. L. Alspach. Recursive Bayesian estimation using Gaussian sums. *Automatica*, 7:465–479, 1971.
[9] F. L. Lewis. *Optimal Estimation*. John Wiley & Sons, New York, 1986. ISBN 0-471-83741-5.
[10] K. Reif, S. Günther, E. Yaz, and R. Unbehauen. Stochastic stability of the discrete-time extended Kalman filter. *IEEE Transactions on Automatic Control*, 44(4):714–728, 1999.
[11] X. R. Li. Measure of nonlinearity for stochastic systems. In *Proceedings of the 15th International Conference on Information Fusion*, Singapore, 2012.
[12] O. Straka, J. Duník, and M. Šimandl. Scaling parameter in unscented transformation: Analysis and specification. In *Proceedings of the 2012 American Control Conference*, Montreal, Canada, June 2012.
[13] E. Jones, M. Scalzo, A. Bubalo, M. Alford, and B. Arthur. Measures of nonlinearity for single target tracking problem. In *Proceedings of the SPIE, Signal Processing, Sensor Fusion, and Target Recognition*, Orlando, Florida, USA, 2011.
[14] J. Duník, O. Straka, and M. Šimandl. Nonlinearity and non-Gaussianity measures for stochastic dynamic systems. In *Proceedings of the 16th International Conference on Information Fusion*, Istanbul, 2013.
[15] B. Jia, M. Xin, and Y. Cheng. Sparse-grid quadrature nonlinear filtering. *Automatica*, 48(2):327–341, 2012.
[16] B. Jia, M. Xin, and Y. Cheng. High-degree cubature Kalman filter. *Automatica*, 49(2):510–518, 2013.
[17] P. X. K. Song. Monte Carlo Kalman filter and smoothing for multivariate discrete state space models. *Canadian Journal of Statistics*, 28(3):641–652, 2000.
[18] J. Duník, O. Straka, and M. Šimandl. Stochastic integration filter. *IEEE Transactions on Automatic Control*, 58(6):1561–1566, 2013.
[19] J. Sarmavuori and S. Särkkä. Fourier-Hermite Kalman filter. *IEEE Transactions on Automatic Control*, 57(6):1511–1515, 2012.
[20] K. Ito and K. Xiong. Gaussian filters for nonlinear filtering problems. *IEEE Transactions on Automatic Control*, 45(5):910–927, 2000.
[21] B. D. O. Anderson and J. B. Moore. *Optimal Filtering*. Englewood Cliffs, New Jersey: Prentice Hall Ins., 1979.
[22] M. Norgaard, N. K. Poulsen, and O. Ravn. New developments in state estimation for nonlinear systems. *Automatica*, 36(11):1627–1638, 2000.
[23] S. J. Julier and J. K. Uhlmann. Unscented filtering and nonlinear estimation. *IEEE Review*, 92(3):401–421, 2004.
[24] J. W. Brewer. Kronecker products and matrix calculus in system theory. *IEEE Transactions on Circuits and Systems*, 25(9):772–781, 1978.
[25] T. Kollo and D. Von Rosen. *Advanced multivariate statistics with matrices, volume 579 of Mathematics and Its Applications*. Springer, 2005.
[26] R. Henriksen. The truncated second-order nonlinear filter revised. *IEEE Transactions On Automatic Control*, 27(1):247–251, 1982.
[27] A. Papoulis and S. U. Pillai. *Probability, Random Variables and Stochastic Processes*. Mc Graw Hill, fourth edition edition, 2002.
[28] O. Straka, J. Duník, and M. Šimandl. Measures of non-Gaussianity in unscented Kaiman filter framework. In *Information Fusion (FUSION), 2014 17th International Conference on*, pages 1–8, 2014.
[29] E.A. Semerdjiev, L.S. Mihaylova, and Tz.A. Semerdjiev. Manoeuvering ship model identification and interacting multiple model tracking algorithm design. In *Proceedings of the 1998 International Conference on Multisource-Multisensor Information Fusion (FUSION'98)*, 1998.

Journal of Physics: Conference Series **570** (2014) 012005 doi:10.1088/1742-6596/570/1/012005

Nonlinear analysis of position estimate in global navigation satellite systems

Ivo Punčochář, Jan Škach, Miroslav Šimandl and Ladislav Král

Department of Cybernetics and NTIS – New Technologies for the Information Society, Faculty of Applied Sciences, University of West Bohemia, Univerzitní 8, 306 14 Plzeň, Czech Republic

E-mail: ivop@kky.zcu.cz, janskach@kky.zcu.cz, simandl@kky.zcu.cz, ladkral@kky.zcu.cz

Abstract. The paper presents a nonlinear analysis of position estimation based on a global navigation satellite system. A classical problem formulation and iterative solution that results in the weighted least squares estimate of the receiver state are assumed. The analysis employs the Taylor's theorem to express the nonlinear measurement model using the first order Taylor polynomial at the state estimate and the Lagrange form of the remainder. A sensitivity analysis of the Jacobian matrix pseudoinverse is performed, and an upper bound on the size of the Lagrange remainder is derived using eigenstructure of the Hessian matrix. The results obtained show that both the sensitivity of the pseudoinverse and the size of the quadratic term are not significant, and thus the linear approximation commonly used to derive stochastic properties of the state estimate is reasonable. Although this result has been experimentally confirmed by numerous successful applications, this analysis can serve as a more rigorous basis when the design procedures for a safety critical system have to be satisfied.

1. Introduction

Global navigation satellite systems (GNSS) have been continuously used in new kinds of applications that require positioning, navigation, or timing services [1, 2]. Besides noncritical (car navigation, geocaching, etc.), there are safety critical applications (aviation navigation, train control systems, etc.) and liability critical applications (road tolling, traffic law enforcement, etc.) that impose increased requirements on accuracy, integrity, continuity, and availability of position, velocity and time information [3, 4]. To make GNSS available for such applications, development has been carried out in several directions. The space and control segments of the fully operational Global Positioning System (GPS) are being gradually modernized, the new European GNSS Galileo is being developed with such applications in mind from the very beginning, satellite-based augmentation systems are being put in use, and GNSS receivers with advanced capabilities such as Receiver Autonomous Integrity Monitoring (RAIM) are being designed.

The development of a GNSS receiver with advanced capabilities to be used in a safety critical application has to comply with international and sector specific standards on the functional safety of systems [5, 6]. Assuming that raw data like pseudoranges and ephemeris are available at sampling steps, the state, i.e., position and clock bias, of the GNSS receiver can be estimated using two different approaches. In the first approach, a static estimator that uses only data from the current sampling step is designed based on a measurement model. In the second approach, global or local nonlinear filtering techniques [7] are employed to design an estimator that uses a dynamical model in addition to the measurement model and data from the preceding sampling steps as well. Although the latter approach usually results into an estimator of a higher quality, it is more challenging to comply with safety standards as the dynamical model is application dependent and undetected faults may propagate into state estimates from the preceding sampling steps. Therefore, a static estimator will be considered in this paper.

The evaluation of state estimate quality and the design of a RAIM algorithm are almost exclusively performed using a linearized measurement model [8, 9]. Nevertheless, a nonlinear analysis of the state estimation error based on two second order Taylor polynomials carried out in [10] showed that unlike the conclusions derived from the linearized model, the state estimate is biased and has a larger variance. It was also noted that the impact on a RAIM algorithm is difficult to evaluate in general because of combined effects. Although approximate expressions for the mean value and covariance matrix of the state estimation error were derived, the nonlinear analysis was not straightforward as it required several approximations to be made. Moreover, the analysis did not provide qualitative evaluation of nonlinear effects.

The aim of this paper is to present a more straightforward nonlinear analysis that utilizes Taylor's theorem with the first order Taylor polynomial and the Lagrange form of the remainder. A sensitivity of the Jacobian matrix pseudoinverse on the linearization point is examined and the eigenvalue analysis is employed to derive the lower and upper bounds on the linearization error. These results are subsequently used to quantify the nonlinear effects on the state estimate and the estimation error.

The paper is structured as follows. The problem of state estimation is presented in Section 2. Section 3 deals with the design of a state estimator that uses a nonlinear model and available measurements to compute the state of the receiver. Properties of the position estimator are analyzed in Section 4. First, the usual results based on linearized model are mentioned and then analysis utilizing the nonlinear model is presented. In Section 5, the results of the nonlinear analysis are demonstrated through a numerical example. Finally, Section 6 provides concluding remarks.

2. Problem formulation

The problem of positioning is considered in a common coordinate system (e.g. Earth-Centered-Earth-Fixed coordinate system) and only data from one time step are used. In this section, a nonlinear measurement model and a static state estimator are presented.

2.1. Nonlinear model of measurement

It is assumed that the measurement of pseudoranges in a GNSS can be described by the following static nonlinear measurement model

$$\mathbf{y} = \mathbf{h}(\mathbf{x}_0) + \mathbf{f} + \mathbf{v}, \tag{1}$$

where $\mathbf{y} \in \mathbb{R}^{n_y}$ is a vector of measured pseudoranges to visible satellites, $\mathbf{h} : \mathbb{R}^{n_x} \mapsto \mathbb{R}^{n_y}$ is a known nonlinear vector-valued function of several variables, $\mathbf{x}_0 \in \mathbb{R}^{n_x}$ is an unknown state of the receiver, $\mathbf{f} \in \mathbb{R}^{n_y}$ represents unknown additive faults in the pseudoranges, and $\mathbf{v} \in \mathbb{R}^{n_y}$ is the measurement noise that has the Gaussian distribution with the zero mean and known positive definite covariance matrix $\boldsymbol{\Sigma}$. The state $\mathbf{x}_0 = [\mathbf{r}_0^{\mathrm{T}},\ b_0]^{\mathrm{T}}$ consists of the unknown receiver position $\mathbf{r}_0 \in \mathbb{R}^3$ and unknown offset $b_0 = c \cdot dt_0 \in \mathbb{R}$ caused by the receiver clock offset $dt_0 \in \mathbb{R}$ with respect to the global GNSS time. Note that c denotes the speed of light as defined in the GNSS. The individual functions of the nonlinear vector function $\mathbf{h}$ have the following form

$$h_j(\mathbf{x}) = \|\mathbf{r} - \mathbf{s}_j\|_2 + b - c \cdot dtsv_j,\ j = 1, 2, \ldots, n_y, \tag{2}$$

where $\|\cdot\|_2$ denotes the $\mathcal{L}^2$-norm of a vector, $\mathbf{x}^{\mathrm{T}} = [\mathbf{r}^{\mathrm{T}}, b] = [r_1,\ r_2,\ r_3,\ b]$ is a point in $\mathbb{R}^{n_x}$, $\mathbf{s}_j = [s_{j,1},\ s_{j,2},\ s_{j,3}]^{\mathrm{T}} \in \mathbb{R}^3$ denotes the known position of the j-th satellite, and $dtsv_j \in \mathbb{R}$ denotes the known j-th satellite clock offset correction term. The measurement noise $\mathbf{v}$ represents a common level of uncertainty in the measurement of the pseudoranges whereas the faults $\mathbf{f}$ represent significant deviations that can be caused by unusual conditions in the ionosphere or troposphere, multipaths, errors in the navigation message, and others. In contrast to the measurement noise, the fault is not a random variable and no detailed information about its behavior is assumed.

2.2. State estimator

The objective is to design a state estimator that has the following form

$$\hat{\mathbf{x}} = \boldsymbol{\sigma}(\mathbf{y}), \tag{3}$$

where $\hat{\mathbf{x}} \in \mathbb{R}^{n_x}$ is an estimate of the unknown state $\mathbf{x}_0$ and $\boldsymbol{\sigma} : \mathbb{R}^{n_y} \mapsto \mathbb{R}^{n_x}$ is a function that represents the state estimator.

2.3. Design criteria

The state estimator (3) is designed using the following criterion. The quality of the state estimator is evaluated through the weighted least squares criterion of the true and predicted measurements. Thus, the state estimate is computed by solving the partial optimization problem

$$\hat{\mathbf{x}} = \boldsymbol{\sigma}(\mathbf{y}) = \underset{\mathbf{x}\in\mathbb{R}^{n_x}}{\arg\min}\, V(\mathbf{x}, \mathbf{y}), \tag{4}$$

where the weighted least squares criterion $V : \mathbb{R}^{n_x+n_y} \mapsto \mathbb{R}$ is a function defined as

$$V(\mathbf{x}, \mathbf{y}) = \mathbf{e}(\mathbf{x}, \mathbf{y})^{\mathrm{T}}\boldsymbol{\Sigma}^{-1}\mathbf{e}(\mathbf{x}, \mathbf{y}) = (\mathbf{y} - \mathbf{h}(\mathbf{x}))^{\mathrm{T}}\boldsymbol{\Sigma}^{-1}(\mathbf{y} - \mathbf{h}(\mathbf{x})) \tag{5}$$

and the equation error is $\mathbf{e}(\mathbf{x}, \mathbf{y}) = \mathbf{y} - \mathbf{h}(\mathbf{x})$.

3. Design of state estimator

The state estimator is obtained by solving the partial optimization problem defined in (4). For a given measurement $\mathbf{y}$, the local minimum of the nonlinear least squares criterion V can occur at a stationary point where any of the first order partial derivatives of V does not exist or the condition

$$\frac{\partial V}{\partial \mathbf{x}}\,|_{\mathbf{x}=\hat{\mathbf{x}}} = \mathbf{0} \tag{6}$$

is satisfied [11]. By taking the first order partial derivatives of the criterion V, the necessary condition (6) shows that the state estimate $\hat{\mathbf{x}}$ has to satisfy the following system of nonlinear equations

$$-2\mathbf{H}\left(\hat{\mathbf{x}}\right)^{\mathrm{T}}\boldsymbol{\Sigma}^{-1}\left(\mathbf{y} - \mathbf{h}\left(\hat{\mathbf{x}}\right)\right) = \mathbf{0}, \tag{7}$$

where $\mathbf{H} : \mathbb{R}^{n_x} \mapsto \mathbb{R}^{n_y \times n_x}$ is the Jacobian matrix of the nonlinear function $\mathbf{h}$ and its j-th row has the following form

$$\left[\mathbf{H}(\mathbf{x})\right]_{(j,\cdot)} = \begin{bmatrix} \frac{(\mathbf{r}-\mathbf{s}_j)^{\mathrm{T}}}{\|\mathbf{r}-\mathbf{s}_j\|_2} & 1 \end{bmatrix}. \tag{8}$$

A geometric interpretation of the necessary condition is that the transformed equation error $\boldsymbol{\Sigma}^{-1}\mathbf{e}(\hat{\mathbf{x}}, \mathbf{y})$ is perpendicular to the tangent hyperplane defined at the stationary point $\hat{\mathbf{x}}$.

Note that the domain of the Jacobian matrix is the whole $\mathbb{R}^{n_x}$ except for the set $\mathcal{X} = \cup_{j=1}^{n_y} \mathcal{X}_j$, where the set $\mathcal{X}_j$ is defined as

$$\mathcal{X}_j = \left\{ \begin{bmatrix} \mathbf{s}_j \\ b \end{bmatrix} : b \in \mathbb{R} \right\}. \tag{9}$$

Since the least squares criterion V is a nonlinear function of $\mathbf{x}$ and the necessary condition for the stationary point is represented by the system of nonlinear equations (7), the state estimator cannot be expressed in the closed form. Instead, the state estimate $\hat{\mathbf{x}}$ is computed using an iterative optimization method for a given measurement $\mathbf{y}$. A frequently used iterative optimization method in GNSS is the Gauss-Newton (GN) method [1]. In its basic version, the GN method starts with an initial estimate $\hat{\mathbf{x}}^{(0)}$ and computes subsequent estimates $\hat{\mathbf{x}}^{(i)}$ using the following update relation

$$\hat{\mathbf{x}}^{(i+1)} = \hat{\mathbf{x}}^{(i)} + \mathbf{H}\left(\hat{\mathbf{x}}^{(i)}\right)^{\dagger}\left(\mathbf{y} - \mathbf{h}\left(\hat{\mathbf{x}}^{(i)}\right)\right), \tag{10}$$

where i is the iteration index and the matrix valued function $\mathbf{H}^{\dagger} : \mathbb{R}^{n_x} \mapsto \mathbb{R}^{n_x \times n_y}$ defined as

$$\mathbf{H}(\mathbf{x})^{\dagger} = \left(\mathbf{H}(\mathbf{x})^{\mathrm{T}} \boldsymbol{\Sigma}^{-1} \mathbf{H}(\mathbf{x})\right)^{-1} \mathbf{H}(\mathbf{x})^{\mathrm{T}} \boldsymbol{\Sigma}^{-1} \tag{11}$$

is the weighted pseudoinverse of the Jacobian matrix $\mathbf{H}(\mathbf{x})$. A state estimate from the previous time step or a fixed nonzero vector can be used as the initial estimate $\hat{\mathbf{x}}^{(0)}$. The usual stopping conditions for the GN method are the maximum number of iterations $i_{\max}$ and a limit ϵ_{GN} on the $\mathcal{L}^2$-norm of the difference between two consecutive state estimates [12]

$$i \leq i_{\max}, \qquad \left\|\hat{\mathbf{x}}^{(i+1)} - \hat{\mathbf{x}}^{(i)}\right\|_2 = \left\|\mathbf{H}\left(\hat{\mathbf{x}}^{(i)}\right)^{\dagger}\left(\mathbf{y} - \mathbf{h}\left(\hat{\mathbf{x}}^{(i)}\right)\right)\right\|_2 \leq \epsilon_{\mathrm{GN}}. \tag{12}$$

In general, the basic version of the GN method may fail to converge. Even if the method converges, the attained minimum can be a local one instead of the global minimum because the nonlinear least squares criterion (5) is not necessarily a convex function. As the iterations of the GN method fail to converge to the point where the nonlinear least squares criterion V is not differentiable, it can be assumed that if the GN method converges, the resulting state estimate $\hat{\mathbf{x}}$ is a stationary point, particularly a local minimum, of the criterion V and the necessary condition (7) is fulfilled up to an additive error that is proportional to ϵ_{GN}. For the GN method to work, it is assumed that the Jacobian matrix has the full column rank at each estimate $\hat{\mathbf{x}}^{(i)}$. Otherwise the GN method is terminated and the state estimate is not computed.

4. Analysis of state estimator properties

The previous section presented the state estimator that was designed by utilizing the nonlinear model. As pointed in [10], the properties of state estimate and a RAIM algorithm are almost exclusively derived from the linearized model despite that the system is nonlinear. To perform a nonlinear analysis, two second order Taylor polynomials of $\mathbf{h}$, one defined at the true receiver state and the other one defined at its estimate, were used in [10]. It was shown that the linearization error can be regarded as an additive noise that makes the state estimate biased and its covariance matrix larger. However, the analysis did not include quantitative evaluation of the nonlinearity. In this section, a nonlinear analysis that uses a slightly different approach and allows the impact of the nonlinearity to be quantified is presented. The Taylor's theorem is employed to obtain the first order Taylor series with the Lagrange form of the remainder suitable for deriving bounds on the quadratic term of the Taylor series.

4.1. Taylor expansion of measurement model

The analysis will be performed using the Taylor's theorem. It is assumed that for the state estimate $\hat{\mathbf{x}}$ there is a nonempty convex neighborhood $\mathcal{O}(\hat{\mathbf{x}})$ where the functions h_j are twice differentiable and that the unknown state $\mathbf{x}_0$ is inside the neighborhood $\mathcal{O}(\hat{\mathbf{x}})$. Then the value of the function h_j at $\mathbf{x}_0$ can be expressed as

$$h_j(\mathbf{x}_0) = h_j(\hat{\mathbf{x}}) + [\mathbf{H}(\hat{\mathbf{x}})]_{(j,\cdot)}(\mathbf{x}_0 - \hat{\mathbf{x}}) + \frac{1}{2}(\mathbf{x}_0 - \hat{\mathbf{x}})^{\mathrm{T}}\mathbf{Q}_j(\boldsymbol{\xi}_j)(\mathbf{x}_0 - \hat{\mathbf{x}}), \tag{13}$$

where $\mathbf{Q}_j : \mathbb{R}^{n_x} \mapsto \mathbb{R}^{n_x \times n_x}$ is the Hessian matrix of the real-valued function h_j, $\boldsymbol{\xi}_j = \hat{\mathbf{x}} + \eta_j(\mathbf{x}_0 - \hat{\mathbf{x}})$ is a point on the line segment between $\hat{\mathbf{x}}$ and $\mathbf{x}_0$, and $\eta_j \in (0,1)$ is an unknown parameter. By computing all second order partial derivatives of function h_j, it can be derived that the Hessian matrix has the following form

$$\mathbf{Q}_j(\mathbf{x}) = \frac{1}{\|\mathbf{r} - \mathbf{s}_j\|_2^3}\left(\begin{bmatrix} \|\mathbf{r} - \mathbf{s}_j\|_2^2 \mathbf{I}_3 & \mathbf{0}_{3\times 1} \\ \mathbf{0}_{1\times 3} & 0 \end{bmatrix} - \begin{bmatrix} (\mathbf{r} - \mathbf{s}_j)(\mathbf{r} - \mathbf{s}_j)^{\mathrm{T}} & \mathbf{0}_{3\times 1} \\ \mathbf{0}_{1\times 3} & 0 \end{bmatrix}\right), \tag{14}$$

where $\mathbf{I}_n$ is the identity matrix of order n and $\mathbf{0}_{m\times n}$ is the zero matrix of size $m \times n$. The neighborhood $\mathcal{O}(\hat{\mathbf{x}})$ can be defined for example as

$$\mathcal{O}(\hat{\mathbf{x}}) = \left\{ \begin{bmatrix} \mathbf{r} \\ b \end{bmatrix} \in \mathbb{R}^{n_x} : b \in \mathbb{R}, \|\mathbf{r} - \hat{\mathbf{r}}\|_2 < R \right\}, \tag{15}$$

where $R = \min_j \|\hat{\mathbf{r}} - \mathbf{s}_j\|_2$. For earth-based positioning applications, the condition $\mathbf{x}_0 \in \mathcal{O}(\hat{\mathbf{x}})$ required for Taylor's theorem to hold, can be considered satisfied because a simple detection technique is usually used to remove implausible position estimates that are too far from the Earth surface. Note that in the case of space-based positioning, a more elaborate analysis would be required.

The nonlinear model (1) can be rewritten using the Taylor's theorem as

$$\mathbf{y} = \mathbf{h}(\hat{\mathbf{x}}) + \mathbf{H}(\hat{\mathbf{x}})(\mathbf{x}_0 - \hat{\mathbf{x}}) + \mathbf{q}(\hat{\mathbf{x}}, \mathbf{x}_0, \boldsymbol{\eta}) + \mathbf{f} + \mathbf{v}, \tag{16}$$

where $\boldsymbol{\eta} = [\eta_1, \ldots, \eta_{n_y}]^{\mathrm{T}}$ are unknown parameters and $\mathbf{q} : \mathbb{R}^{2n_x+n_y} \mapsto \mathbb{R}^{n_y}$ is a function whose j-th element is defined as

$$q_j(\hat{\mathbf{x}}, \mathbf{x}_0, \eta_j) = \frac{1}{2}(\mathbf{x}_0 - \hat{\mathbf{x}})^{\mathrm{T}}\mathbf{Q}_j(\hat{\mathbf{x}} + \eta_j(\mathbf{x}_0 - \hat{\mathbf{x}}))(\mathbf{x}_0 - \hat{\mathbf{x}}). \tag{17}$$

Applying a simple algebraic manipulation, the model (16) can be written as

$$\mathbf{H}(\hat{\mathbf{x}})^{\mathrm{T}}\boldsymbol{\Sigma}^{-1}(\mathbf{y} - \mathbf{h}(\hat{\mathbf{x}})) = \mathbf{H}(\hat{\mathbf{x}})^{\mathrm{T}}\boldsymbol{\Sigma}^{-1}\mathbf{H}(\hat{\mathbf{x}})(\mathbf{x}_0 - \hat{\mathbf{x}}) + \mathbf{H}(\hat{\mathbf{x}})^{\mathrm{T}}\boldsymbol{\Sigma}^{-1}(\mathbf{q}(\hat{\mathbf{x}}, \mathbf{x}_0, \boldsymbol{\eta}) + \mathbf{f} + \mathbf{v}). \tag{18}$$

As the state estimate $\hat{\mathbf{x}}$ is the stationary point of the criterion V and thus meets the condition (7), the left hand side of the relation (18) is zero and the state estimation error $\tilde{\mathbf{x}} = \mathbf{x}_0 - \hat{\mathbf{x}}$ satisfies the relation

$$\tilde{\mathbf{x}} = -\mathbf{H}(\hat{\mathbf{x}})^{\dagger}(\mathbf{q}(\hat{\mathbf{x}}, \mathbf{x}_0, \boldsymbol{\eta}) + \mathbf{f} + \mathbf{v}), \tag{19}$$

where $\mathbf{H}(\hat{\mathbf{x}})^{\dagger} = \left(\mathbf{H}(\hat{\mathbf{x}})^{\mathrm{T}}\boldsymbol{\Sigma}^{-1}\mathbf{H}(\hat{\mathbf{x}})\right)^{-1}\mathbf{H}(\hat{\mathbf{x}})^{\mathrm{T}}\boldsymbol{\Sigma}^{-1}$ is the weighted pseudoinverse of the Jacobian matrix evaluated at the point of the state estimate $\hat{\mathbf{x}}$.

4.2. Linearized model

This subsection presents a common approach for deriving properties of the state estimate. Like the state estimate $\hat{\mathbf{x}}$ the Jacobian matrix and the Hessian matrix are random variables. Since the state estimation error $\tilde{\mathbf{x}}$ appears implicitly also on the right hand side of the relation (19), its exact distribution cannot be derived easily. To circumvent the issue, approximate properties of the state estimation error are usually derived using two following simplifying assumptions.

(i) The Jacobian matrix is evaluated at the computed realization of the state estimate $\hat{\mathbf{x}}_1$ and assumed to be constant when evaluating the mean value and the covariance matrix of the state estimation error $\tilde{\mathbf{x}}$.

(ii) The function $\mathbf{q}$ that represents nonlinear behavior of the measurement model (1) is assumed to be the identically zero function.

Then the state estimation error has the Gaussian distribution with the following mean value and covariance matrix

$$\mathrm{E}\{\tilde{\mathbf{x}}\} = \mathrm{E}\left\{\mathbf{H}(\hat{\mathbf{x}}_1)^{\dagger}(\mathbf{f}+\mathbf{v})\right\} = \mathbf{H}(\hat{\mathbf{x}}_1)^{\dagger}\mathbf{f}, \tag{20}$$

$$\mathrm{cov}\{\tilde{\mathbf{x}}\} = \mathrm{E}\left\{\mathbf{H}(\hat{\mathbf{x}}_1)^{\dagger}\mathbf{v}\left(\mathbf{H}\left(\hat{\mathbf{x}}_1\right)^{\dagger}\mathbf{v}\right)^{\mathrm{T}}\right\} = \left(\mathbf{H}(\hat{\mathbf{x}}_1)^{\mathrm{T}}\boldsymbol{\Sigma}^{-1}\mathbf{H}(\hat{\mathbf{x}}_1)\right)^{-1}. \tag{21}$$

4.3. Nonlinear model

A nonlinear analysis is performed to examine the validity of two above mentioned assumptions used in the previous subsection dealing with the linearized model. First, the sensitivity of the pseudoinverse $\mathbf{H}(\mathbf{x})^{\dagger}$ on $\mathbf{x}$ is examined, and then the worst case size of the quadratic term $\mathbf{q}$ is analyzed. To simplify the analysis, it is assumed that the covariance matrix $\boldsymbol{\Sigma}$ is diagonal

$$\boldsymbol{\Sigma} = \sigma^2\mathbf{I}_{n_y}, \tag{22}$$

where σ^2 is a variance of the measurement noise. Then the pseudoinverse of the Jacobian matrix is $\mathbf{H}(\mathbf{x})^{\dagger} = \left(\mathbf{H}(\mathbf{x})^{\mathrm{T}}\mathbf{H}(\mathbf{x})\right)^{-1}\mathbf{H}(\mathbf{x})^{\mathrm{T}}$. Further, let $\hat{\mathbf{x}}_1 = [\hat{\mathbf{r}}_1^{\mathrm{T}}, \hat{b}_1]^{\mathrm{T}}$ and $\hat{\mathbf{x}}_2 = [\hat{\mathbf{r}}_2^{\mathrm{T}}, \hat{b}_2]^{\mathrm{T}}$ be two state estimates whose position estimates satisfy the inequality

$$\|\hat{\mathbf{r}}_2 - \hat{\mathbf{r}}_1\|_2 = \|\boldsymbol{\Delta}\|_2 \leq M, \tag{23}$$

where M is a given constant that is assumed to be significantly smaller than any distance between the receiver and the satellites.

4.3.1. Sensitivity of Jacobian matrix pseudoinverse In the first part of the analysis, the sensitivity of the Jacobian matrix pseudoinverse to a change in the linearization point is evaluated. Although there are several ways of measuring the sensitivity of a pseudoinverse matrix, the norm $\left\|\mathbf{H}(\hat{\mathbf{x}}_2)^{\dagger} - \mathbf{H}(\hat{\mathbf{x}}_1)^{\dagger}\right\|_2$ is used because the change in the pseudoinverse matrix $\mathbf{H}(\hat{\mathbf{x}})^{\dagger}$ induced by a change in the linearization point $\hat{\mathbf{x}}$ is of interest here. Another point of view is obtained when the expression (19) is used to derive an upper bound on the norm of the difference between two state estimation errors

$$\begin{aligned}\|\tilde{\mathbf{x}}_2 - \tilde{\mathbf{x}}_1\|_2 \leq & \left\|\mathbf{H}(\hat{\mathbf{x}}_2)^{\dagger} - \mathbf{H}(\hat{\mathbf{x}}_1)^{\dagger}\right\|_2 \|\mathbf{q}(\hat{\mathbf{x}}_2, \mathbf{x}_0, \boldsymbol{\eta}_2) + \mathbf{f} + \mathbf{v}\|_2 \\ & + \left\|\mathbf{H}(\hat{\mathbf{x}}_1)^{\dagger}\right\|_2 \|\mathbf{q}(\hat{\mathbf{x}}_2, \mathbf{x}_0, \boldsymbol{\eta}_2) - \mathbf{q}(\hat{\mathbf{x}}_1, \mathbf{x}_0, \boldsymbol{\eta}_1)\|_2 .\end{aligned} \tag{24}$$

Thus the term $\left\|\mathbf{H}(\hat{\mathbf{x}}_2)^{\dagger} - \mathbf{H}(\hat{\mathbf{x}}_1)^{\dagger}\right\|_2$ determines how $\|\mathbf{q}(\hat{\mathbf{x}}_2, \mathbf{x}_0, \boldsymbol{\eta}_2) + \mathbf{f} + \mathbf{v}\|_2$ is amplified when only the difference in the pseudoinverses is considered.

Regardless of the other terms on the right hand side of (24), a general result on pseudoinverse sensitivity [13, 14] leads to the following bound

$$\left\|\mathbf{H}(\hat{\mathbf{x}}_2)^\dagger - \mathbf{H}(\hat{\mathbf{x}}_1)^\dagger\right\|_2 \leq \mu \max\left\{\left\|\mathbf{H}(\hat{\mathbf{x}}_1)^\dagger\right\|_2^2, \left\|\mathbf{H}(\hat{\mathbf{x}}_2)^\dagger\right\|_2^2\right\} \|\mathbf{E}\|_2, \tag{25}$$

where $\mathbf{E} = \mathbf{H}(\hat{\mathbf{x}}_2) - \mathbf{H}(\hat{\mathbf{x}}_1) \in \mathbb{R}^{n_y \times n_x}$ is an additive perturbation matrix and $\mu = \sqrt{2}$ is a constant that corresponds to the $\mathcal{L}^2$-norm. As the structure of the pseudoinverse is the same for both state estimates $\hat{\mathbf{x}}_1$ and $\hat{\mathbf{x}}_2$, upper bounds on $\|\mathbf{H}(\hat{\mathbf{x}}_1)^\dagger\|_2^2$ and $\|\mathbf{E}\|_2$ are derived. The squared norm of the pseudoinverse $\mathbf{H}(\hat{\mathbf{x}}_1)^\dagger$ satisfies

$$\begin{aligned}\left\|\mathbf{H}(\hat{\mathbf{x}}_1)^\dagger\right\|_2^2 &= \sigma_{\max}(\mathbf{H}(\hat{\mathbf{x}}_1)^\dagger)^2 = \lambda_{\max}\left(\left(\mathbf{H}(\hat{\mathbf{x}}_1)^\dagger\right)^{\mathrm{T}} \mathbf{H}(\hat{\mathbf{x}}_1)^\dagger\right) = \lambda_{\max}\left(\mathbf{H}(\hat{\mathbf{x}}_1)^\dagger \left(\mathbf{H}(\hat{\mathbf{x}}_1)^\dagger\right)^{\mathrm{T}}\right) \\ &= \lambda_{\max}\left(\left(\mathbf{H}(\hat{\mathbf{x}}_1)^{\mathrm{T}}\mathbf{H}(\hat{\mathbf{x}}_1)\right)^{-1}\right) = \frac{1}{\lambda_{\min}\left(\mathbf{H}(\hat{\mathbf{x}}_1)^{\mathrm{T}}\mathbf{H}(\hat{\mathbf{x}}_1)\right)},\end{aligned} \tag{26}$$

where the third equality results from $\mathbf{A}^{\mathrm{T}}\mathbf{A}$ and $\mathbf{A}\mathbf{A}^{\mathrm{T}}$ having the same nonzero eigenvalues for any matrix $\mathbf{A}$ [15], $\sigma_{\max}(\mathbf{A})$ denotes the maximum singular value of a matrix $\mathbf{A}$, and $\lambda_{\max}(\mathbf{A})$ and $\lambda_{\min}(\mathbf{A})$ denote the maximum and minimum eigenvalues of a square matrix $\mathbf{A}$, respectively. For certain satellite configurations, the Jacobian matrix $\mathbf{H}(\hat{\mathbf{x}}_1)$ can be almost a rank deficient matrix. In such a case, the minimum eigenvalue of $\mathbf{H}(\hat{\mathbf{x}}_1)^{\mathrm{T}}\mathbf{H}(\hat{\mathbf{x}}_1)$ is arbitrarily close to zero and no meaningful upper bound on the squared norm of the pseudoinverse can be given without any additional assumption. Nevertheless, it is a common practice in GNSS applications to employ so called dilution of precision (DOP) metrics for measuring the quality of the satellite constellation and deciding whether or not the state estimate should be trusted. The geometric dilution of precision (GDOP) metric is defined as [16]

$$\mathrm{GDOP} = \sqrt{\mathrm{tr}\left\{\left(\mathbf{H}(\hat{\mathbf{x}}_1)^{\mathrm{T}}\mathbf{H}(\hat{\mathbf{x}}_1)\right)^{-1}\right\}}, \tag{27}$$

and its relation to the eigenvalues of $\left(\mathbf{H}(\hat{\mathbf{x}}_1)^{\mathrm{T}}\mathbf{H}(\hat{\mathbf{x}}_1)\right)^{-1}$ is given by

$$\mathrm{GDOP}^2 = \mathrm{tr}\left\{\left(\mathbf{H}(\hat{\mathbf{x}}_1)^{\mathrm{T}}\mathbf{H}(\hat{\mathbf{x}}_1)\right)^{-1}\right\} = \sum_{i=1}^{n_x} \lambda_i\left(\left(\mathbf{H}(\hat{\mathbf{x}}_1)^{\mathrm{T}}\mathbf{H}(\hat{\mathbf{x}}_1)\right)^{-1}\right), \tag{28}$$

where $\lambda_i(\mathbf{A})$ is the i-th eigenvalue of a square matrix $\mathbf{A}$ sorted in the descending order. If the maximum admissible geometric dilution of precision for a particular application denoted as $\mathrm{GDOP}_{\max}$ is chosen and considering that all eigenvalues of $(\mathbf{H}(\hat{\mathbf{x}}_1)^{\mathrm{T}}\mathbf{H}(\hat{\mathbf{x}}_1))^{-1}$ are nonnegative, the following upper bound can be written

$$\left\|\mathbf{H}(\hat{\mathbf{x}}_1)^\dagger\right\|_2^2 < \mathrm{GDOP}_{\max}^2. \tag{29}$$

Next, an upper bound on the norm of the perturbation matrix $\mathbf{E}$ is derived. When it is assumed that M is negligible compared to the distance between the receiver and a satellite, the j-th row of the perturbation matrix $\mathbf{E}$ can be approximated as follows

$$[\mathbf{E}]_{(j,\cdot)} = [\mathbf{H}(\hat{\mathbf{x}}_2)]_{(j,\cdot)} - [\mathbf{H}(\hat{\mathbf{x}}_1)]_{(j,\cdot)} = \begin{bmatrix}\frac{(\hat{\mathbf{r}}_2-\mathbf{s}_j)^{\mathrm{T}}}{\|\hat{\mathbf{r}}_2-\mathbf{s}_j\|_2} - \frac{(\hat{\mathbf{r}}_1-\mathbf{s}_j)^{\mathrm{T}}}{\|\hat{\mathbf{r}}_1-\mathbf{s}_j\|_2} & 0\end{bmatrix} \approx \begin{bmatrix}\frac{\boldsymbol{\Delta}^{\mathrm{T}}}{\|\hat{\mathbf{r}}_1-\mathbf{s}_j\|_2} & 0\end{bmatrix}. \tag{30}$$

Now, considering the definition of the $\mathcal{L}^2$ induced norm of a matrix, it can be derived that

$$\begin{aligned}\|\mathbf{E}\|_2 &= \max_{\|\mathbf{z}\|_2=1} \sqrt{\sum_{j=1}^{n_y} \left|[\mathbf{E}]_{(j,\cdot)}\mathbf{z}\right|^2} = \max_{\|\mathbf{z}\|_2=1} \left|[\boldsymbol{\Delta}^{\mathrm{T}}\ 0]\mathbf{z}\right| \sqrt{\sum_{j=1}^{n_y} \frac{1}{\|\hat{\mathbf{r}}_1 - \mathbf{s}_j\|_2^2}} \\ &= \max_{\|\mathbf{z}\|_2=1} \left|\|\boldsymbol{\Delta}\|_2 \|\mathbf{z}\|_2 \cos(\theta)\right| \sqrt{\sum_{j=1}^{n_y} \frac{1}{\|\hat{\mathbf{r}}_1 - \mathbf{s}_j\|_2^2}} \leq \|\boldsymbol{\Delta}\|_2 \sqrt{\sum_{j=1}^{n_y} \frac{1}{\|\hat{\mathbf{r}}_1 - \mathbf{s}_j\|_2^2}},\end{aligned} \tag{31}$$

where $\mathbf{z} \in \mathbb{R}^{n_x}$ is just an auxiliary vector. Using (23), the the $\mathcal{L}^2$ induced norm of the perturbation matrix $\mathbf{E}$ can be written as

$$\|\mathbf{E}\|_2 \leq M \sqrt{\sum_{j=1}^{n_y} \frac{1}{\|\hat{\mathbf{r}}_1 - \mathbf{s}_j\|_2^2}} \leq M \frac{\sqrt{n_y}}{\min_j \|\hat{\mathbf{r}}_1 - \mathbf{s}_j\|_2}. \tag{32}$$

Finally, by using (29) and (32) in (25), an overall bound can be written as

$$\left\|\mathbf{H}(\hat{\mathbf{x}}_2)^\dagger - \mathbf{H}(\hat{\mathbf{x}}_1)^\dagger\right\|_2 < \frac{\sqrt{2n_y}\mathrm{GDOP}_{\max}^2 M}{\min_j \|\hat{\mathbf{r}}_1 - \mathbf{s}_j\|_2} = H_{\max}. \tag{33}$$

Note that the obtained bound depends on M linearly. For example, in the case of a GPS constellation containing $n_y = 8$ visible satellites, maximum admissible geometric dilution of precision $\mathrm{GDOP}_{\max} = 6$, and $M = 100\,\mathrm{m}$, the upper bound can be evaluated to obtain

$$\left\|\mathbf{H}(\hat{\mathbf{x}}_2)^\dagger - \mathbf{H}(\hat{\mathbf{x}}_1)^\dagger\right\|_2 \leq 0.7129 \times 10^{-3}. \tag{34}$$

This value suggests that the sensitivity of the pseudoinverse to the changes in the state estimate is quite small. It confirms that evaluating the mean value (20) and the covariance matrix (21) of the state estimation error by replacing the Jacobian matrix $\mathbf{H}(\hat{\mathbf{x}})$ by a constant matrix $\mathbf{H}(\hat{\mathbf{x}}_1)$ is a reasonable approximation.

4.3.2. Bounds on linearization error In the second part of the analysis, the linearization error represented by the nonlinear function $\mathbf{q}$ is investigated using the eigenstructure of the Hessian matrix $\mathbf{Q}_j$. By computing the eigenvalues λ_i and right eigenvectors $\mathbf{w}_i$ for $i = 1, 2, 3, 4$ from the definition, the following set of eigenpairs is obtained

$$\begin{aligned}\lambda_1 &= \frac{1}{\|\mathbf{r} - \mathbf{s}_j\|_2}, & \mathbf{w}_1^{\mathrm{T}} &= \begin{bmatrix}-(r_3 - s_{j,3}) & 0 & (r_1 - s_{j,1}) & 0\end{bmatrix}, \\ \lambda_2 &= \frac{1}{\|\mathbf{r} - \mathbf{s}_j\|_2}, & \mathbf{w}_2^{\mathrm{T}} &= \begin{bmatrix}-(r_2 - s_{j,2}) & (r_1 - s_{j,1}) & 0 & 0\end{bmatrix}, \\ \lambda_3 &= 0, & \mathbf{w}_3^{\mathrm{T}} &= \begin{bmatrix}\frac{(\mathbf{r}-\mathbf{s}_j)^{\mathrm{T}}}{\|\mathbf{r}-\mathbf{s}_j\|_2} & 0\end{bmatrix}, \\ \lambda_4 &= 0, & \mathbf{w}_4^{\mathrm{T}} &= \begin{bmatrix}0 & 0 & 0 & 1\end{bmatrix}.\end{aligned}$$

Note that the right eigenvectors form a basis of $\mathbb{R}^4$ and the eigenvectors $\mathbf{w}_1$ and $\mathbf{w}_2$ are chosen neither normalized nor mutually orthogonal just to keep the corresponding expressions as simple as possible. Before continuing in the analysis of the linearization error, a geometrical interpretation of the eigenvalues and eigenvectors with respect to the state estimation error $\tilde{\mathbf{x}}$ is provided to give a better insight into the analysis.

The right eigenvector $\mathbf{w}_4$ and zero eigenvalue λ_4 result from the function h_j being linear in its fourth argument. The right eigenvector $\mathbf{w}_3$ and zero eigenvalue λ_3 show that for the position estimation error $\mathbf{r}_0 - \hat{\mathbf{r}}$ lying entirely in the direction $\mathbf{r} - \mathbf{s}_j$, the function h_j behaves like it is linear. The first two eigenvectors $\mathbf{w}_1$ and $\mathbf{w}_2$ have the same nonzero eigenvalue and they are orthogonal to $\mathbf{w}_3$ and $\mathbf{w}_4$. Thus, only the position estimation error that lies in the plane defined by these two eigenvectors makes the function h_j nonlinear. Besides the eigenpair λ_4, $\mathbf{w}_4$ that is common to all functions h_j, the other eigenpairs are different for each function q_j and thus the value of the function $\mathbf{q}$ cannot be zero for nonzero state estimation error $\tilde{\mathbf{x}}$. Finally, as all eigenvalues are nonnegative, the Hessian matrix $\mathbf{Q}_j$ is positive semidefinite and thus the range of the vector-valued function $\mathbf{q}$ is the positive orthant.

To continue in the analysis of the linearization error, the inequality for a quadratic norm with a symmetric matrix $\mathbf{A}$ is used [15]. It holds that

$$\lambda_{\min}(\mathbf{A})\mathbf{z}^{\mathrm{T}}\mathbf{z} \leq \mathbf{z}^{\mathrm{T}}\mathbf{A}\mathbf{z} \leq \lambda_{\max}(\mathbf{A})\mathbf{z}^{\mathrm{T}}\mathbf{z}, \tag{35}$$

where $\mathbf{A} \in \mathbb{R}^{n\times n}$ is a square symmetric matrix and $\mathbf{z} \in \mathbb{R}^n$ is a vector. Using this inequality and substituting $\hat{\mathbf{r}} + \eta_j(\mathbf{r}_0 - \hat{\mathbf{r}}) - \mathbf{s}_j$ for a general point $\mathbf{r}$ in the expression for the eigenvalue λ_1, the function q_j is bounded as follows

$$0 \leq q_j(\hat{\mathbf{x}}, \mathbf{x}_0, \eta_j) \leq \frac{1}{2}\frac{\|\mathbf{r}_0 - \hat{\mathbf{r}}\|_2^2}{\|\hat{\mathbf{r}} + \eta_j(\mathbf{r}_0 - \hat{\mathbf{r}}) - \mathbf{s}_j\|_2}. \tag{36}$$

To eliminate the dependence on the unknown $\eta_j \in (0, 1)$, a lower bound on the denominator is derived. In general, it holds that

$$\begin{aligned}\|\hat{\mathbf{r}} - \mathbf{s}_j\|_2 &= \|\hat{\mathbf{r}} - \mathbf{s}_j + \eta_j(\mathbf{r}_0 - \hat{\mathbf{r}}) - \eta_j(\mathbf{r}_0 - \hat{\mathbf{r}})\|_2 \\ &\leq \|\hat{\mathbf{r}} - \mathbf{s}_j + \eta_j(\mathbf{r}_0 - \hat{\mathbf{r}})\|_2 + \|-\eta_j(\mathbf{r}_0 - \hat{\mathbf{r}})\|_2 \\ &= \|\hat{\mathbf{r}} - \mathbf{s}_j + \eta_j(\mathbf{r}_0 - \hat{\mathbf{r}})\|_2 + \|\eta_j(\mathbf{r}_0 - \hat{\mathbf{r}})\|_2,\end{aligned} \tag{37}$$

and for $\eta_j \in (0, 1)$, the following strict inequality holds

$$\|\eta_j(\mathbf{r}_0 - \hat{\mathbf{r}})\|_2 < \|\mathbf{r}_0 - \hat{\mathbf{r}}\|_2. \tag{38}$$

By combining (37) and (38), it is possible to derive the inequality

$$\|\hat{\mathbf{r}} - \mathbf{s}_j\|_2 - \|\mathbf{r}_0 - \hat{\mathbf{r}}\|_2 < \|\hat{\mathbf{r}} + \eta_j(\mathbf{r}_0 - \hat{\mathbf{r}}) - \mathbf{s}_j\|_2, \tag{39}$$

which allows the right hand side of (36) to be further bounded as

$$\frac{1}{2}\frac{\|\mathbf{r}_0 - \hat{\mathbf{r}}\|_2^2}{\|\hat{\mathbf{r}} + \eta_j(\mathbf{r}_0 - \hat{\mathbf{r}}) - \mathbf{s}_j\|_2} < \frac{1}{2}\frac{\|\mathbf{r}_0 - \hat{\mathbf{r}}\|_2^2}{\|\hat{\mathbf{r}} - \mathbf{s}_j\|_2 - \|\mathbf{r}_0 - \hat{\mathbf{r}}\|_2} = q_{j,\max}(\mathbf{r}_0, \hat{\mathbf{r}}). \tag{40}$$

In the case of the GPS, the upper bound $q_{j,\max}$ can be quantified approximately as follows. For near Earth's surface applications, a typical distance between a receiver and a satellite $\|\mathbf{r}_0 - \mathbf{s}_j\|$ is approximately 20.2×10^6 m. The dependence of the upper bound $q_{j,\max}$ on the $\mathcal{L}^2$-norm of the position estimation error $\|\mathbf{r}_0 - \hat{\mathbf{r}}\|_2$ is depicted in Fig. 1 and a detail for smaller position error estimates is given in Fig. 2. The figures show that the value of the function q_j is lower than 2.5 m even if the $\mathcal{L}^2$-norm of the position estimation error reaches 10 km. It can be concluded that due to the great distance between the Earth's surface and satellites, the linear approximations of the functions h_j are quite accurate even for a large neighborhood of the state estimate. It is also connected with a fast convergence of the GN method which is usually observed in practice.

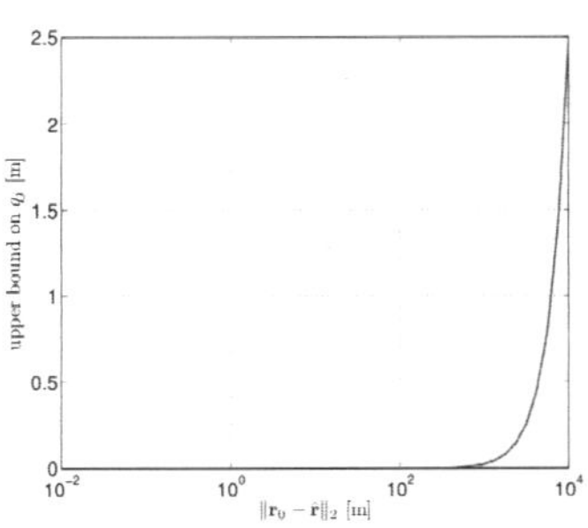

Figure 1. Dependence of the upper bound on the norm of the position estimation error.

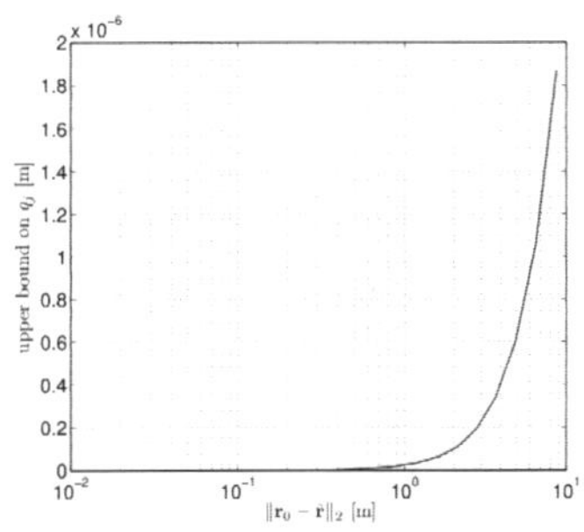

Figure 2. Dependence of the upper bound on the norm of the position estimation error–detail.

Table 1. The geometric dilution of precision *GDOP* and bounds $H_{\max}$, $q_{\max}$ for two different receiver-satellites constellations.

constellation	GDOP	$H_{\max}$	$q_{\max}$
good	1.5269	0.7032×10^{-3}	0.4368×10^{-3}
worse	5.9836	0.5522×10^{-3}	0.4428×10^{-3}

5. Numerical example

The numerical example illustrates the bounds derived in the previous section. One particular state $\mathbf{x}_0$ of the GPS receiver and two different satellite constellations are considered. The first constellation contains 10 visible satellites and is characterized by a good DOP with GDOP $= 1.5269$. The second constellation consists of 6 visible satellites only and has a worse DOP with GDOP $= 5.9836$ which is close to the chosen maximum allowable geometric dilution of precision $\text{GDOP}_{\max} = 6$. Since it is observed in practice that the norm of the position estimation error caused by the measurement noise rarely exceeds 100 m for low GDOP values, it is apriori assumed that $M = 100\,\text{m}$. For both satellite constellations, the following results were obtained using 10000 Monte Carlo simulations where the pseudorange measurements are generated according to the model (1) with $\mathbf{f} = \mathbf{0}$ and $\sigma = 2.7$.

The histogram of the norm of difference between a chosen state estimate realization and other simulated realizations $\|\hat{\mathbf{r}}_l - \hat{\mathbf{r}}_1\|_2$ for the good GDOP is depicted in Figure 3 and for the worse GDOP in Figure 4. It can be seen from these figures, that M is chosen such that the apriori assumption $\|\hat{\mathbf{r}}_l - \hat{\mathbf{r}}_1\|_2 \leq M$ is satisfied. The computed bounds are summarized in Table 1. Note that the worse constellation has a smaller value of $H_{\max}$ mainly because it contains fewer satellites compared to the good constellation. The realization of $\left\|\mathbf{H}(\hat{\mathbf{x}}_l)^\dagger - \mathbf{H}(\hat{\mathbf{x}}_1)^\dagger\right\|_2$ for the good constellation and the worse constellation are depicted in Figure 5 and Figure 6, respectively. By comparing the maximum value of $\left\|\mathbf{H}(\hat{\mathbf{x}}_l)^\dagger - \mathbf{H}(\hat{\mathbf{x}}_1)^\dagger\right\|_2$ with the upper bound $H_{\max}$ in Table 1, it can be observed that the derived upper bound is not particularly tight and true values are much smaller. The realizations of the values of function q_j are depicted in Figures 7 and 8. These histograms show that the nonlinear term is indeed always positive and like in the case of the bound on $\left\|\mathbf{H}(\hat{\mathbf{x}}_2)^\dagger - \mathbf{H}(\hat{\mathbf{x}}_1)^\dagger\right\|_2$, this bound is not tight either.

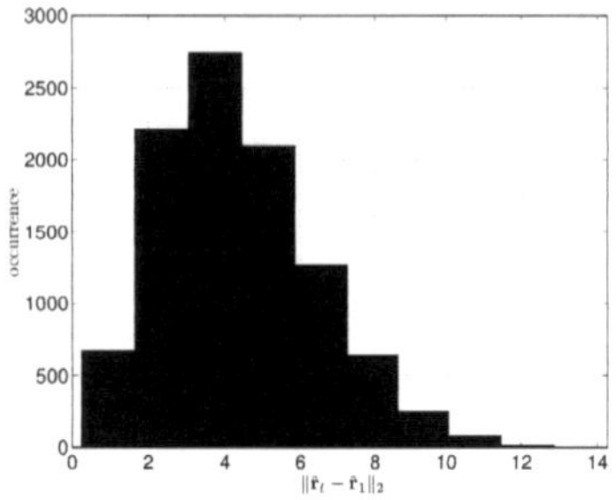

Figure 3. Histogram of $\|\hat{\mathbf{r}}_l - \hat{\mathbf{r}}_1\|_2$ for the good satellite constellation.

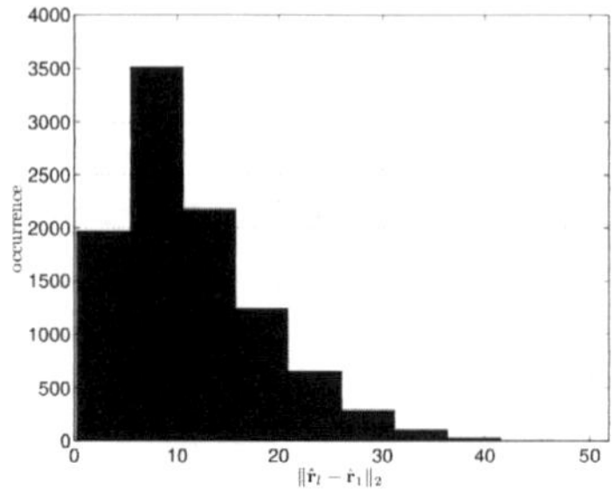

Figure 4. Histogram of $\|\hat{\mathbf{r}}_l - \hat{\mathbf{r}}_1\|_2$ for the worse satellite constellation.

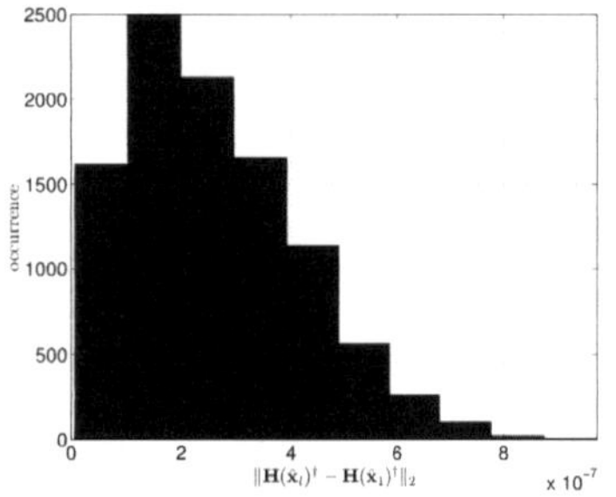

Figure 5. Histogram of $\left\|\mathbf{H}(\hat{\mathbf{x}}_l)^\dagger - \mathbf{H}(\hat{\mathbf{x}}_1)^\dagger\right\|_2$ for the good satellite constellation.

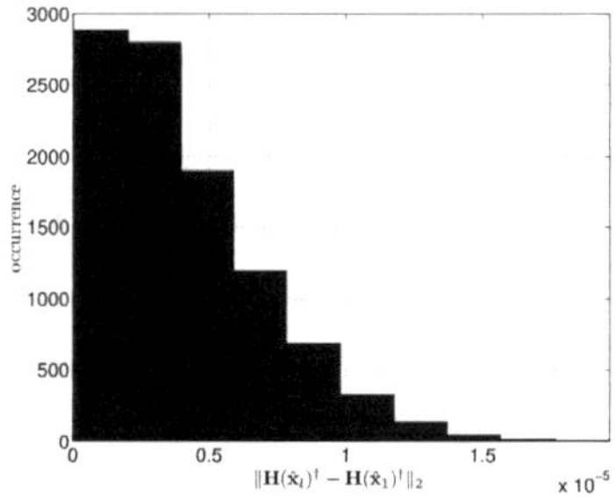

Figure 6. Histogram of $\left\|\mathbf{H}(\hat{\mathbf{x}}_l)^\dagger - \mathbf{H}(\hat{\mathbf{x}}_1)^\dagger\right\|_2$ for the worse satellite constellation.

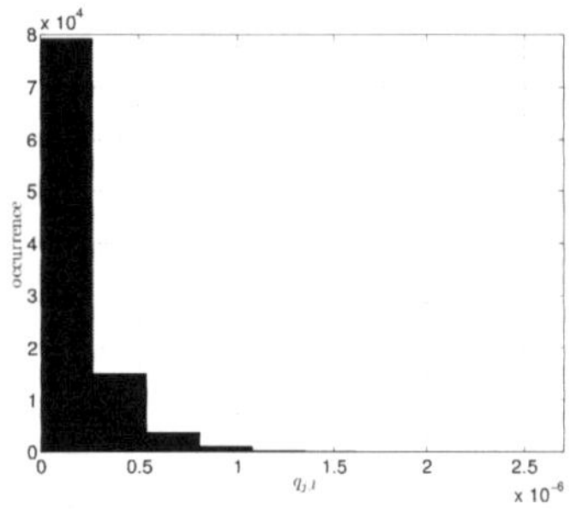

Figure 7. Histogram of $q_{j,l}$ for the good satellite constellation.

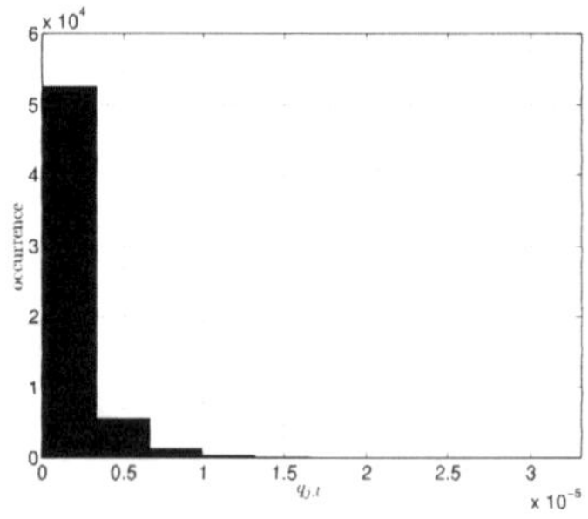

Figure 8. Histogram of $q_{j,l}$ for the worse satellite constellation.

6. Conclusion

The paper dealt with a nonlinear analysis of the state estimate properties in GNSS positioning when a static estimator is used. Basic techniques such as Taylor's theorem and eigenvalue analysis were applied to analyze the state estimate and its error. Also the conditions under which the analysis holds were highlighted. The evaluation of the linearization error and sensitivity of the Jacobian matrix pseudoinverse showed that the linear approximation commonly used when deriving the properties of the state estimate is indeed reasonable. The analysis could be further extended with minor modifications to the case when a dynamical model and a nonlinear filter are used. Another potential extension consists in deriving a robust bound on the estimation error covariance matrix that can be used in a RAIM algorithm. Although the linear approximations have proven to be adequate by their successful use in many noncritical applications so far, the presented nonlinear analysis can serve as a more rigorous basis when GNSS positioning algorithms should be approved for use in safety critical applications.

Acknowledgments

This work was supported by the Technology Agency of the Czech Republic, project No. TE02000202.

References

[1] Kaplan D E and Hegarty J C 2006 *Understanding GPS: principles and applications* (Norwood, MA, USA: Artech House) ISBN 1-58053-894-0

[2] Gleason S and Gebre-Egiabher D 2009 *GNSS Applications and Methods* (Norwood, MA, USA: Artech House) ISBN 978-1596933293

[3] Ochieng W Y, Sauer K, Walsh D, Brodin G, Griffin S and Denney M 2003 *Journal of Navigation* **56** 51–65 ISSN 03734633

[4] Nguyen T P K, Beugin J and Marais J 2014 *Proceedings of the 2nd International Conference on Computing, Management and Telecommunications* (Danang, Vietnam) pp 101–106 ISBN 9781479929030

[5] Redmill F 1998 *Computing & Control Engineering Journal* **9** 205–213

[6] 2006 DO-229D Minimum Operational Performance Standards for Global Positioning System/Wide Area Augmentation System Airborne Equipment

[7] Simon D 2006 *Optimal State Estimation* 1st ed (Hoboken, NJ, USA: Wiley-Interscience) ISBN 9780471708582

[8] Sturza M A 1988 *Navigation: Journal of The Institute of Navigation* **35** 483–502

[9] Wang J and Ober P B 2009 *Journal of Navigation* **62** 251–261 ISSN 0373-4633 URL `http://www.journals.cambridge.org/abstract_S0373463308005158`

[10] Ober P B 1999 *Proceedings of the* 12th *International Technical Meeting of the Satellite Division of The Institute of Navigation* (Nashville, TN, USA) pp 1211–1216

[11] Stewart J 2012 *Calculus* 7th ed (Belmont, CA, USA: Cengage Learning) ISBN 978-0538497817

[12] Fletcher R 1980 *Practical Methods of Optimization: Unconstrained optimization* 1st ed (New York, NY, USA: John Willey & Sons) ISBN 0471277118

[13] Wedin P A k 1973 *BIT Numerical Mathematics* **13** 217–232 ISSN 0006-3835

[14] Meng L and Zheng B 2010 *Linear Algebra and its Applications* **432** 956–963 ISSN 00243795 URL `http://linkinghub.elsevier.com/retrieve/pii/S0024379509005230`

[15] Olver P J and Shakiban C 2005 *Applied Linear Algebra* 1st ed (Upper Saddle River, NJ, USA: Pearson) ISBN 978-0131473829

[16] Langley R B 1999 *GPS World* 52–59

Journal of Physics: Conference Series **570** (2014) 022001 doi:10.1088/1742-6596/570/2/022001

Feedforward Tracking Control of Flat Recurrent Fuzzy Systems

Stefan Gering and Jürgen Adamy

Laboratory of Control Methods and Robotics at Technische Universität Darmstadt, Landgraf-Georg-Str. 4, 64283 Darmstadt, Germany

E-mail: {sgering,adamy}@rmr.tu-darmstadt.de

Abstract. Flatness based feedforward control has proven to be a feasible solution for the problem of tracking control, which may be applied to a broad class of nonlinear systems. If a flat output of the system is known, the control is often based on a feedforward controller generating a nominal input in combination with a linear controller stabilizing the linearized error dynamics around the trajectory. We show in this paper that the very same idea may be incorporated for tracking control of MIMO recurrent fuzzy systems. Their dynamics is given by means of linguistic differential equations but may be converted into a hybrid system representation, which then serves as the basis for controller synthesis.

1. Introduction

The synthesis of controllers is usually based on a known model, which is assumed to sufficiently describe the dynamics of the process that is to be controlled. If only rough expert knowledge or measurement data is available, gray box models based on fuzzy logic are a way to derive an approximate model. Although some methods are known that attempt to control the process without any model (see, e.g., [3, 9, 14]), it is generally preferable to exploit existing rough process knowledge for controller synthesis.

In particular, Recurrent Fuzzy Systems [1, 15] represent the system dynamics approximately by means of linguistic rules, in which the conclusion part consists of the state derivative being described simply as a constant. Thus, Recurrent Fuzzy Systems may also be interpreted as zeroth-order Takagi-Sugeno fuzzy systems. In contrast to the more general Takagi-Sugeno fuzzy systems, the main benefit of Recurrent Fuzzy Systems relies on the possible linguistic interpretability of the rules obtained. Having obtained a recurrent fuzzy system, the question arises on how to control this particular system class. In [11, 12] it was shown how to derive stabilizing feedback for these dynamic fuzzy systems by means of static fuzzy control or switching polynomial controllers for known equilibria.

This paper now addresses the question of tracking control for Recurrent Fuzzy Systems. In the last two decades, the tracking control problem was answered for the broad class of (differentially) flat nonlinear systems [10] in a multitude of works (see, e.g., [4, 8, 13, 22]). See also [18–20] for applications of flatness-based control.

Since it is well known [20], that utilizing an exact linearizing feedback controller may cause severe problems in the presence of model uncertainties, a common approach is to exploit the flatness property to obtain a nominal feedforward control from the inverse system combined with a linear controller for the error system linearized around the trajectory.

In this paper, we show that this approach is general enough to solve the tracking control problem for flat Recurrent Fuzzy Systems as well, whereas for brevity we restrict our attention to square MIMO systems. In a first step, a switched nonlinear system is obtained from the given rule base of the recurrent fuzzy system. By applying a dynamic transformation, this switched nonlinear system is then rendered input-affine, such that feedforward control can be applied locally. By linearizing the error dynamics around the trajectory, a simple PID controller then controls small perturbations around the trajectory as well as model uncertainties, which naturally arise due to the approximate nature of Recurrent Fuzzy Systems.

The benefit of the proposed method clearly is in the analytic derivation of the inverse system, which is then used for generation of the feedforward control signal. This is in contrast to the system class of Piecewise Bilinear Models, where tracking control is usually based on learning techniques for the inverse system [6, 7] and hence, there is no guarantee beforehand that the inverse system exists. On the other hand, the method proposed in this paper can similarly be applied to Piecewise Bilinear Models as well.

The remainder is organized as follows: In Sec. 2, basic definitions for Recurrent Fuzzy Systems are reviewed roughly, and a dynamic transformation for decoupling of the inputs is given. Sec. 3 then reviews well-known results from flat systems in the light of Recurrent Fuzzy Systems. The flatness property is then used in Sec. 4 for trajectory generation and synthesis of the feedforward control, whereas Sec. 5 describes the additional PID control for stabilization of the error dynamics. A numerical example is given in Sec. 6 and concluding remarks in Sec. 7.

2. Preliminaries

2.1. Recurrent Fuzzy Systems

This section briefly revisits the definition of Recurrent Fuzzy Systems (RFS) as detailed in [15] and [1]. Its block diagram is depicted in Fig. 1, with dynamic function $\mathbf{f}(\mathbf{x}, \mathbf{u})$ and output function $\mathbf{g}(\mathbf{x})$ being static fuzzy systems defined in the input-state space $\mathcal{Z} = \mathcal{X} \times \mathcal{U}$, $\mathcal{X} = \mathbb{R}^n$, $\mathcal{U} \subset \mathbb{R}^m$ with $\mathbf{u} \in [\mathbf{u}_{\min}, \mathbf{u}_{\max}]$. The output function $\mathbf{g}(\mathbf{x})$ is assumed not to contain feedthrough and is thus defined over the state space only. The output values are bounded to $[\mathbf{y}_{\min}, \mathbf{y}_{\max}]$.

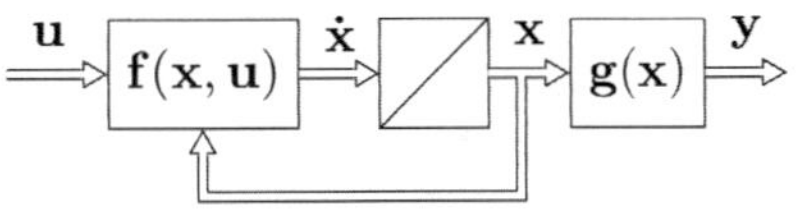

Figure 1: Recurrent Fuzzy System

Each linguistic rule of the complete rule base describing the system dynamics is of the form

$$\begin{aligned} &\text{If } \mathbf{x} = \mathbf{L}_{\mathbf{j}}^{\mathbf{x}} \text{ and } \mathbf{u} = \mathbf{L}_{\mathbf{q}}^{\mathbf{u}}, \\ &\text{then } \dot{\mathbf{x}} = \mathbf{L}_{\mathbf{w}(\mathbf{j},\mathbf{q})}^{\dot{\mathbf{x}}}, \end{aligned} \tag{1}$$

with $\mathbf{L}_{\mathbf{j}}^{\mathbf{x}}$, $\mathbf{L}_{\mathbf{q}}^{\mathbf{u}}$ denoting vectors of linguistic values in the state and input space. Similarly, $\mathbf{L}_{\mathbf{w}(\mathbf{j},\mathbf{q})}^{\dot{\mathbf{x}}}$ denotes vectors of linguistic values describing gradients quantitatively. For the output function, similar rules of the form

$$\begin{aligned} &\text{If } \mathbf{x} = \mathbf{L}_{\mathbf{j}}^{\mathbf{x}}, \\ &\text{then } \mathbf{y} = \mathbf{L}_{\mathbf{j}}^{\mathbf{y}} \end{aligned} \tag{2}$$

are obtained.

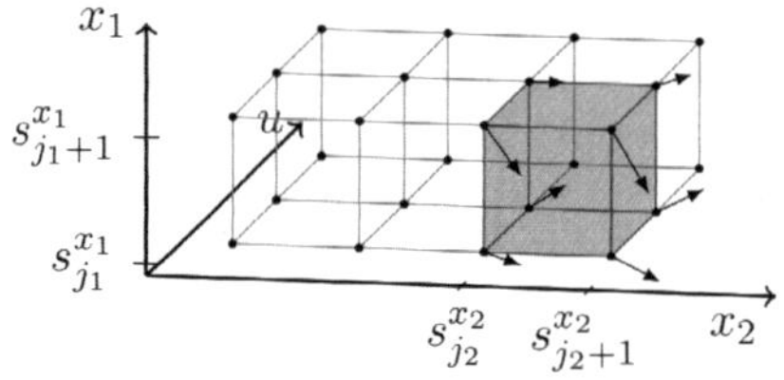

Figure 2: Hypersquare decomposition of RFS and examples of core position gradients defining the system dynamics within the colored hypersquare.

In order to transfer the linguistic rules into a numerical representation, core position vectors $\mathbf{s}^{\mathbf{x}}_{\mathbf{j}} \in [\mathbf{x}_{\min}, \mathbf{x}_{\max}]$, $\mathbf{s}^{\mathbf{u}}_{\mathbf{q}} \in [\mathbf{u}_{\min}, \mathbf{u}_{\max}]$, $\mathbf{s}^{\mathbf{y}}_{\mathbf{j}} \in [\mathbf{y}_{\min}, \mathbf{y}_{\max}]$ and core position gradients $\mathbf{s}^{\dot{\mathbf{x}}}_{\mathbf{w}} \in \mathbb{R}^n$ are associated with the linguistic vectors. For simplicity, the sets of core position vectors $\mathbf{s}^{\mathbf{x}}_{\mathbf{j}}$ of the dynamic function $\mathbf{f}(\mathbf{x}, \mathbf{u})$ and those of the output function $\mathbf{g}(\mathbf{x})$ are assumed to be identical. This assumption is rather mild, since an RFS with different sets can always be converted into one with identical sets by an augmentation of the respective grids.

Clearly, the rule base defines gradients at discrete grid points $(\mathbf{s}^{\mathbf{x}}_{\mathbf{j}}, \mathbf{s}^{\mathbf{u}}_{\mathbf{q}})$, which is also visualized in Fig. 2. It becomes obvious, that the core position vectors induce a rectangular grid in $\mathcal{Z}$. The interpolation region of core position vectors being adjacent to a vector $(\mathbf{x}, \mathbf{u}) \in \mathcal{Z}$ is termed *hypersquare*, denoted $H_{\mathbf{l}}$. Therein, $\mathbf{l} = (\mathbf{j}, \mathbf{q})$ consist of the indices of the lower core positions. Note, that $\mathbf{l}$ might therefore index a single core position $\mathbf{s}_{\mathbf{l}}$ or a hypersquare $H_{\mathbf{l}}$.

In order to obtain real-valued dynamic and output functions in the entire input-state space, membership functions $\mu^{x_i}_{j_i}(x_i)$, $\mu^{u_p}_{q_p}(u_p)$ are introduced, assigning to states and inputs a degree of membership to linguistic values with associated core positions $s^{x_i}_{j_i}$, $s^{u_p}_{q_p}$. Having thereby fuzzified states and inputs, the algebraic product is used for aggregation and implication, the simple sum for accumulation of the single rules, and the center of singleton method for defuzzification [2]. Thus, the representation

$$\dot{\mathbf{x}} = \sum_{\mathbf{j},\mathbf{q}} \mathbf{s}^{\dot{\mathbf{x}}}_{\mathbf{w}(\mathbf{j},\mathbf{q})} \cdot \prod_{i=1}^{n} \mu^{x_i}_{j_i}(x_i) \cdot \prod_{p=1}^{m} \mu^{u_p}_{q_p}(u_p) \tag{3a}$$

$$\mathbf{y} = \sum_{\mathbf{j}} \mathbf{s_j}^{\mathbf{y}} \cdot \prod_{i=1}^{n} \mu^{x_i}_{j_i}(x_i) \tag{3b}$$

is obtained for the state derivative (and output), which is a weighted summation over all core position gradients (or core position outputs, respectively). The weighting by the premise is the product of memberships in every dimension. If for membership functions $\mu^{x_i}_{j_i}, \mu^{u_p}_{q_p}$ of states and inputs, triangular and ramp shaped functions are assumed, i.e.

$$\mu^{x_i}_{j_i}(x_i) = \begin{cases} \frac{x_i - s^{x_i}_{j_{i-1}}}{s^{x_i}_{j_i} - s^{x_i}_{j_{i-1}}}, & x_{\mathrm{i,min}} \le s^{x_i}_{j_{i-1}} \le x_i < s^{x_i}_{j_i} \le x_{\mathrm{i,max}} \\ \frac{s^{x_i}_{j_{i+1}} - x_i}{s^{x_i}_{j_{i+1}} - s^{x_i}_{j_i}}, & x_{\mathrm{i,min}} \le s^{x_i}_{j_i} \le x_i < s^{x_i}_{j_{i+1}} \le x_{\mathrm{i,max}} \\ 1, & x_{\mathrm{i,min}} \ge x_i \vee x_i \ge x_{\mathrm{i,max}} \\ 0, & \text{else,} \end{cases} \tag{4}$$

then the dynamics (3a) evaluates to the product of piecewise affine functions, which can be further simplified to

$$\begin{aligned}\dot{\mathbf{x}} &= \mathbf{a}_0^{\mathbf{l}} + \sum_{i=1}^{n} \mathbf{a}_{x_i}^{\mathbf{l}} x_i + \sum_{p=1}^{m} \mathbf{a}_{u_p}^{\mathbf{l}} u_p + \sum_{j=2}^{n} \sum_{i=1}^{j-1} \mathbf{a}_{x_{ij}}^{\mathbf{l}} x_i x_j \\ &+ \sum_{p=1}^{m} u_p \sum_{j=2}^{N} \mathbf{a}_{x_i u_p}^{\mathbf{l}} x_i + \cdots + \mathbf{a}_{x_{1\ldots n} u_{1\ldots m}}^{\mathbf{l}} x_1 \ldots x_n u_1 \ldots u_m\end{aligned} \tag{5}$$

for every $H_{\mathbf{l}}$. By rearranging terms,

$$\dot{\mathbf{x}} = \mathbf{a}_{\mathbf{l}}(\mathbf{x}) + \mathbf{B}_{\mathbf{l}}(\mathbf{x})\mathbf{Z}(\mathbf{u}) \tag{6}$$

is obtained, with the vector of monomials in the inputs u_p being denoted as $\mathbf{Z}(\mathbf{u}) = \begin{bmatrix} u_1, & u_2, & u_1u_2, & \ldots & u_m, & u_1u_m, & \ldots & u_1 \cdots u_m \end{bmatrix}^T \in \mathbb{R}^{2^m-1}$. Obviously, the system dynamics of an RFS with $m > 1$ are not input affine, which is disadvantageous for the controller synthesis. Nevertheless, with the following dynamic transformation, an equivalent input-affine dynamics can be achieved:

First, the system input is augmented by an integrator, such that inputs $\mathbf{v} \in \mathbb{R}^m$, $\dot{\mathbf{u}} = \mathbf{v}$ are obtained. Introducing new state variables $\mathbf{w} = \mathbf{Z}(\mathbf{u})$, the differential equations of these new states then read

$$\dot{\mathbf{w}} = \begin{bmatrix} \dot{u}_1 \\ \dot{u}_2 \\ \dot{u}_1 u_2 + u_1 \dot{u}_2 \\ \vdots \end{bmatrix} = \begin{bmatrix} v_1 \\ v_2 \\ v_1 u_2 + u_1 v_2 \\ \vdots \end{bmatrix} = \frac{\partial \mathbf{Z}(\mathbf{u})}{\partial \mathbf{u}} \mathbf{v}. \tag{7}$$

Thus, the state space equations of the prolonged system read

$$\begin{bmatrix} \dot{\mathbf{x}} \\ \dot{\mathbf{w}} \end{bmatrix} = \begin{bmatrix} \mathbf{a}_{\mathbf{l}}(\mathbf{x}) + \mathbf{B}_{\mathbf{l}}(\mathbf{x})\mathbf{w} \\ \mathbf{0} \end{bmatrix} + \begin{bmatrix} \mathbf{0} \\ \frac{\partial \mathbf{Z}(\mathbf{u})}{\partial \mathbf{u}} \end{bmatrix} \mathbf{v}. \tag{8}$$

Therefore, the following considers without loss of generality the input-affine dynamics

$$\dot{\mathbf{x}} = \mathbf{a}_{\mathbf{l}}(\mathbf{x}) + \sum_{p=1}^{m} \mathbf{B}_{\mathbf{l},p}(\mathbf{x}) \cdot u_p, \tag{9a}$$

$$\mathbf{y} = \mathbf{g}_{\mathbf{j}}(\mathbf{x}) \tag{9b}$$

for each interpolation region $H_{\mathbf{l}}$.

2.2. Further notations

Throughout the paper, we use the standard notation rank $\{\mathbf{M}\}$ for the rank of a matrix and $\mathbf{I}_n$ for the $n \times n$-identity matrix. By $L_{\mathbf{a}}\mathbf{f}(\mathbf{x}) = \sum_{i=1}^{n} \frac{\partial \mathbf{f}(\mathbf{x})}{\partial x_i} \mathbf{a}_i$, a Lie derivative of a function $\mathbf{f}$ regarding $\mathbf{a}$ is denoted. Eigenvalues are denoted by $\lambda \in \mathbb{C}$. Usually, hypersquares $H_{\mathbf{l}}$ and core position vectors $\mathbf{s}_{\mathbf{j}}^{\mathbf{x}}$, $\mathbf{s}_{\mathbf{q}}^{\mathbf{u}}$ are indexed by an integer vector $\mathbf{l}$ corresponding to a grid position. If ambiguity is excluded, sometimes a scalar index notation H_i, H_j is used for enumeration of different hypersquares and core positions.

Journal of Physics: Conference Series **570** (2014) 022001 doi:10.1088/1742-6596/570/2/022001

3. Flatness Property of RFS

In the following, we briefly review the property of (differential) flatness, which is detailed in length, e.g., in [10, 13]. Nevertheless, the description is carried out from the perspective of RFS.

Definition 1 *An RFS* (9) *is* locally (differentially) flat *on a hypersquare* $H_{\mathfrak{l}}$, *iff a (virtual) flat output* $\mathbf{z} \in \mathbb{R}^m$ *with*

$$\mathbf{z} = \boldsymbol{\varphi}_{\mathfrak{l}}(\mathbf{x}, \mathbf{u}, \dot{\mathbf{u}}, \ldots, \mathbf{u}^{(\alpha)}), \tag{10}$$

exists, such that

$$\mathbf{x} = \boldsymbol{\psi}_{\mathfrak{l}}^{x}(z_1, \dot{z}_1, \ldots, z_1^{(\beta_1 - 1)}, \ldots, z_m, \ldots, z_m^{(\beta_m - 1)}), \tag{11a}$$

$$\mathbf{u} = \boldsymbol{\psi}_{\mathfrak{l}}^{u}(z_1, \dot{z}_1, \ldots, z_1^{(\beta_1)}, \ldots, z_m, \ldots, z_m^{(\beta_m)}). \tag{11b}$$

The RFS is globally (differentially) flat, *iff it is locally flat for every* $H_{\mathfrak{l}} \in \mathcal{H}$.

Often, the flat outputs $\mathbf{z}$ are incident with the measured outputs $\mathbf{y}$. For the order of the derivatives β_p in (11), it holds in general that $\sum_{p=1}^{m} \beta_p \geq n$. As noted in [21], for $\sum_{p=1}^{m} \beta_p = n$ it is always possible to transform the system dynamics (locally) into a linear form by means of static state feedback. Furthermore, in this case $\mathbf{z} = \boldsymbol{\varphi}_{\mathfrak{l}}(\mathbf{x})$, i.e., the flat outputs are not dependent on the input or its derivatives. Also, flatness is then directly related to observability of the system's states via the flat outputs z_i and in contrast to the case $\sum_{p=1}^{m} \beta_p > n$, observability is not dependent on the input signal.

For simplicity, the following therefore considers the case $\sum_{p=1}^{m} \beta_p = n$. Then, by local change of coordinates

$$\begin{aligned} \boldsymbol{\xi} &= \left[\xi_{1,1}, \xi_{1,2}, \ldots, \xi_{1,\delta_{\mathfrak{l},1}}, \ldots, \xi_{m,1}, \ldots, \xi_{m,\delta_{\mathfrak{l},m}}\right]^T \\ &= \left[z_1, \dot{z}_1, \ldots, z_1^{(\delta_{\mathfrak{l},1}-1)}, \ldots, z_m, \ldots, z_m^{(\delta_{\mathfrak{l},m}-1)}\right]^T, \end{aligned} \tag{12}$$

(6) is diffeomorphic to the controller normal form

$$\begin{aligned} \dot{\xi}_{i,j} &= \xi_{i,j+1}, \quad j = 1, \ldots, \delta_{\mathfrak{l},i} - 1 \\ \dot{\xi}_{i,\delta_{\mathfrak{l},i}} &= \alpha_{\mathfrak{l},i}(\boldsymbol{\xi}, \mathbf{u}). \end{aligned} \tag{13}$$

Therein, $\delta_{\mathfrak{l},p}$ denotes the local vector relative degree regarding the p-th output, defined by

$$L_{\mathbf{b}_{\mathfrak{l},p}} L_{\mathbf{a}_{\mathfrak{l}}}^{(i)} \mathbf{g}_{\mathbf{j},p}(\mathbf{x}) = 0, \quad i = 1, \ldots, \delta_{\mathfrak{l},p} - 2, \tag{14a}$$

$$L_{\mathbf{b}_{\mathfrak{l},p}} L_{\mathbf{a}_{\mathfrak{l}}}^{(\delta_{\mathfrak{l},p}-1)} \mathbf{g}_{\mathbf{j},p}(\mathbf{x}) \neq 0, \tag{14b}$$

which has to hold for all $\mathbf{x} \in H_{\mathfrak{l}}$. The local vector relative degree $\boldsymbol{\delta}_{\mathfrak{l}} = [\delta_1, \ldots, \delta_m]^T$ then combines all m relative degrees. By letting $\alpha_{\mathfrak{l},i}(\boldsymbol{\xi}, \mathbf{u}) = v_i$, (6) is equivalent to a system in MIMO Brunovský form

$$\begin{aligned} \dot{\xi}_{i,j} &= \xi_{i,j+1}, \quad j = 1, \ldots, \delta_{\mathfrak{l},i} - 1, \\ \dot{\xi}_{i,\delta_{\mathfrak{l},i}} &= v_i, \end{aligned} \tag{15}$$

as shown in [13]. Furthermore and according to [17], if rank $\{\mathbf{A}_{\mathfrak{l}}\} = m$ holds for the matrix

$$\mathbf{A}_{\mathfrak{l}} = \begin{bmatrix} L_{\mathbf{b}_{\mathfrak{l},1}} L_{\mathbf{a}_{\mathfrak{l}}}^{(\delta_{\mathfrak{l},1}-1)} \mathbf{g}_{\mathbf{j},1}(\mathbf{x}) & \ldots & L_{\mathbf{b}_{\mathfrak{l},m}} L_{\mathbf{a}_{\mathfrak{l}}}^{(\delta_{\mathfrak{l},1}-1)} \mathbf{g}_{\mathbf{j},1}(\mathbf{x}) \\ \vdots & & \vdots \\ L_{\mathbf{b}_{\mathfrak{l},1}} L_{\mathbf{a}_{\mathfrak{l}}}^{(\delta_{\mathfrak{l},m}-1)} \mathbf{g}_{\mathbf{j},m}(\mathbf{x}) & \ldots & L_{\mathbf{b}_{\mathfrak{l},m}} L_{\mathbf{a}_{\mathfrak{l}}}^{(\delta_{\mathfrak{l},m}-1)} \mathbf{g}_{\mathbf{j},m}(\mathbf{x}) \end{bmatrix}, \tag{16}$$

then a static state feedback may be applied to (6), such that it is input-output decoupled.

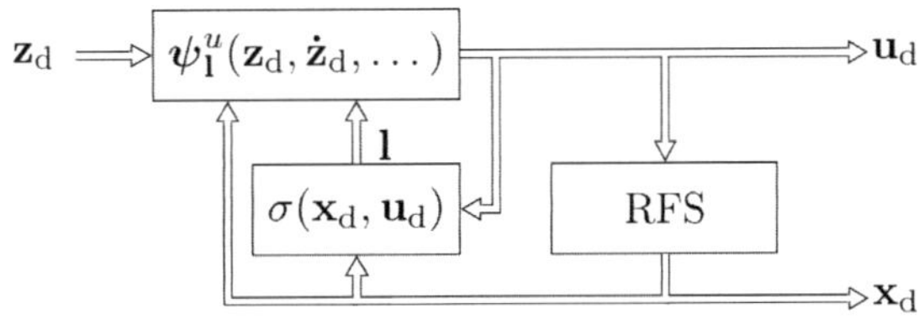

Figure 3: Inverse System

4. Inverse System and Feedforward Control

According to (11), state and input can be reconstructed from the flat outputs and a finite number of their derivatives. Under the assumption that initial conditions are consistent with the system, e.g.,

$$\mathbf{x}_0 = \boldsymbol{\psi}_\mathbf{l}^x(z_1(0), \ldots, z_1^{(\beta_1 - 1)}(0), \ldots, z_m(0), \ldots, z_m^{(\beta_m - 1)}(0)), \tag{17}$$

the nominal control $\mathbf{u}_\mathrm{d}(t)$ may be computed from (11b) given a sufficiently smooth trajectory $\mathbf{z}_\mathrm{d}(t)$.

Due to the piecewise definition of the system dynamics (9), direct computation of (11b) is not possible, since $\boldsymbol{\psi}_\mathbf{l}^u$ is implicit due to the dependence of $\mathbf{l}$ on $\mathbf{x}$ and $\mathbf{u}$. To circumvent this problem, the inverse system as shown in Fig. 3 continuously reconstructs the current hypersquare $H_\mathbf{l}$ by means of $\mathbf{l} = \boldsymbol{\sigma}(\mathbf{x}_\mathrm{d}, \mathbf{u}_\mathrm{d})$. Within the closed-loop, the nominal input $\mathbf{u}_\mathrm{d}$ is then reconstructed from the nominal trajectory $\mathbf{z}_\mathrm{d}$ and the states $\mathbf{x}_\mathrm{d}$. An alternative would be to obtain necessary derivatives of $\mathbf{z}_\mathrm{d}$ from the trajectory generator, or calculation of the derivatives, which is always difficult in the presence of noise.

In addition, the system depicted in Fig. 3 is diffeomorphic to a linear system, thus its eigenvalues can be chosen arbitrarily stable. To see this, consider from (13) the linear dynamics in transformed coordinates

$$\begin{bmatrix} \dot{\xi}_{1,\delta_{\mathbf{l},1}} \\ \vdots \\ \dot{\xi}_{m,\delta_{\mathbf{l},m}} \end{bmatrix} = \begin{bmatrix} \alpha_{\mathbf{l},1}(\boldsymbol{\xi}, \mathbf{u}) \\ \vdots \\ \alpha_{\mathbf{l},m}(\boldsymbol{\xi}, \mathbf{u}) \end{bmatrix} = \begin{bmatrix} L_{\mathbf{a_l}}^{(\delta_{\mathbf{l},1})} \mathbf{g}_{\mathbf{j},1}(\mathbf{x}) \\ \vdots \\ L_{\mathbf{a_l}}^{(\delta_{\mathbf{l},m})} \mathbf{g}_{\mathbf{j},m}(\mathbf{x}) \end{bmatrix} + \mathbf{A}_\mathbf{l}(\mathbf{x}) \cdot \mathbf{u}. \tag{18}$$

If for every $H_\mathbf{l}$, (16) is invertible, then global input-output decoupling and tracking is achieved by means of the control law

$$\mathbf{u} = \mathbf{A}_\mathbf{l}(\mathbf{x})^{-1} \left(- \begin{bmatrix} L_{\mathbf{a_l}}^{(\delta_{\mathbf{l},1})} \mathbf{g}_{\mathbf{j},1}(\mathbf{x}) \\ \vdots \\ L_{\mathbf{a_l}}^{(\delta_{\mathbf{l},m})} \mathbf{g}_{\mathbf{j},m}(\mathbf{x}) \end{bmatrix} - \mathbf{K}_\mathbf{l} \cdot \boldsymbol{\xi} + \mathbf{v} \right). \tag{19}$$

Therein, either $v_i = z_i^{(\delta_{\mathbf{l},i})}$ and $\mathbf{K} = \mathbf{0}$ is chosen such that in (15), $\dot{\xi}_{i,\delta_{\mathbf{l},i}} = z_i^{(\delta_{\mathbf{l},i})}$, or $\mathbf{K}$ is chosen such that the closed-loop linear dynamics

$$\begin{bmatrix} \dot{\xi}_{i,1} \\ \vdots \\ \dot{\xi}_{i,\delta_{\mathbf{l},i}} \end{bmatrix} = \begin{bmatrix} 0 & 1 & & \\ & \ddots & \ddots & \\ & & 0 & 1 \\ -k_{i,1} & -k_{i,2} & \cdots & k_{i,\delta_{\mathbf{l},i}} \end{bmatrix} \begin{bmatrix} \xi_{i,1} \\ \vdots \\ \xi_{i,\delta_{\mathbf{l},i}} \end{bmatrix} + \begin{bmatrix} 0 \\ \vdots \\ 0 \\ v_i \end{bmatrix} \tag{20}$$

are stable and $\mathbf{v}$ is set such that $\mathbf{g}_{\mathrm{l}}(\mathbf{x}_{\mathrm{d}}) = \mathbf{z}_{\mathrm{d}}$. While the first choice allows in theory for the exact tracking of $\mathbf{z}_{\mathrm{d}}$, fast changes of the nominal trajectory may cause a high amplitude of $\mathbf{u}_{\mathrm{d}}$. In the second case, poles of the closed-loop linear system can be placed such that the transient behavior of (20) will not cause saturation effects due to the usual bounds on the control input. Obviously, the following holds:

Proposition 1 *If for each hypersquare* H_{l}, $\alpha_{\mathrm{l},i}(\boldsymbol{\xi}, \mathbf{u})$ *are such that the linear closed-loop local dynamics* (20) *have identical eigenvalues, the controlled RFS has a global linear input-output behavior.*

It follows immediately, that if the desired trajectory $\mathbf{z}_{\mathrm{d}}$ is sufficiently smooth, so will be the output $\mathbf{z}$ of the RFS. On the other hand, this smoothness property does in general not hold for the nominal input $\mathbf{u}_{\mathrm{d}}$. To see this, consider the dynamics of the RFS (3a), which is globally Lipschitz by definition. Due to the choice of triangular membership functions (4), the piecewise defined function ψ_{l}^{u} and more specifically, (19) is dependent on Lie derivatives of the output function. Continuity of $\mathbf{u}$ then implies, e.g.,

$$\lim_{x_i \nearrow s_{j_i}^{x_i}} \sum_{i=1}^{n} \frac{\partial \mathbf{g}_{\mathrm{l}}(\mathbf{x})}{\partial x_i} \mathbf{a}_{\mathrm{l},i}(\mathbf{x}) \stackrel{!}{=} \lim_{x_i \searrow s_{j_i}^{x_i}} \sum_{i=1}^{n} \frac{\partial \mathbf{g}_{\mathrm{l}}(\mathbf{x})}{\partial x_i} \mathbf{a}_{\mathrm{l},i}(\mathbf{x}), \tag{21}$$

which does not hold in general. Thus, $\mathbf{u}_{\mathrm{d}}$ may be discontinuous, even if $\mathbf{z}_{\mathrm{d}}$ is smooth.

5. Extended PID control of Error Dynamics

In addition to the flatness-based feedforward control derived from the RFS model of the plant, an additional feedback controller is necessary in order to stabilize the error dynamics around the trajectory, as described in numerous works (see, e.g., [5, 13, 20, 22]). The necessity of the feedback controller is caused by minor inconsistencies of the system's initial state with the reference trajectory as well as noise and disturbances. In the light of RFS, the main reason for the additional feedback is due to model uncertainties arising from the approximate gray box modeling.

Several synthesis concepts are proposed in the literature, which are mostly based on linearization of the error dynamics and application of linear controllers. Here, the approach of an extended PID controller as proposed in [13] is applied due to its simplicity.

Let the error state of the i-th linear subsystem be denoted

$$\mathbf{e}_i = \boldsymbol{\xi}_i - \boldsymbol{\xi}_{i,\mathrm{d}} \tag{22}$$

with $\boldsymbol{\xi}_i^T = \left[\xi_{i,1}, \ldots, \xi_{i,\delta_{\mathrm{l},i}}\right]$, $i = 1, \ldots, m$. Extending this error state by the integrated difference between the measured output and the reference leads to

$$\bar{\mathbf{e}}_i^T = \begin{bmatrix} e_{i,0} & \mathbf{e}_i^T \end{bmatrix}, \tag{23a}$$

$$e_{i,0} = \int_0^t e_{i,1}(\tau) d\tau. \tag{23b}$$

Thus, the nonlinear error dynamics read

$$\dot{\bar{e}}_{i,j} = \bar{e}_{i,j+1}, \quad j = 0, \ldots, \delta_{\mathrm{l},i} - 1, \tag{24a}$$

$$\begin{aligned} \dot{\bar{e}}_{i,\delta_{\mathrm{l},i}} &= \alpha_{\mathrm{l},i}(\boldsymbol{\xi}, \mathbf{u}) - \boldsymbol{\xi}_i^{(\delta_{\mathrm{l},i})} \\ &= \alpha_{\mathrm{l},i}(\boldsymbol{\xi}_{\mathrm{d}} + \mathbf{e}, \mathbf{u}_{\mathrm{d}} + \Delta\mathbf{u}) - \boldsymbol{\xi}_i^{(\delta_{\mathrm{l},i})}. \end{aligned} \tag{24b}$$

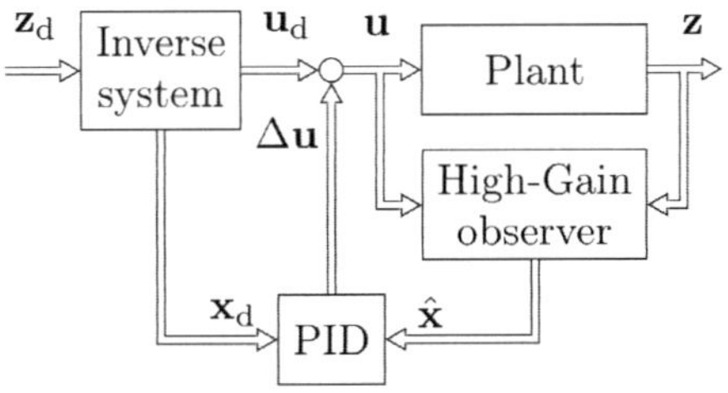

Figure 4: Complete controlled system with flatness-based feedforward control derived from RFS, high-gain observer and extended PID control

By means of an online linearization,

$$\Delta\dot{\bar{e}}_{i,j} = \Delta\bar{e}_{i,j+1} \quad j = 0, \dots, \delta_{1,i} - 1 \tag{25a}$$

$$\begin{aligned} \Delta\dot{\bar{e}}_{i,\delta_{1,i}} &= \left.\frac{\alpha_{1,i}(\boldsymbol{\xi}_\mathrm{d} + \mathbf{e}, \mathbf{u}_\mathrm{d} + \Delta\mathbf{u})}{\partial\mathbf{e}}\right|_{\mathbf{e}^*, \Delta\mathbf{u}^*} \cdot \Delta\mathbf{e} \\ &+ \left.\frac{\alpha_{1,i}(\boldsymbol{\xi}_\mathrm{d} + \mathbf{e}, \mathbf{u}_\mathrm{d} + \Delta\mathbf{u})}{\partial\Delta\mathbf{u}}\right|_{\mathbf{e}^*, \Delta\mathbf{u}^*} \cdot \Delta\mathbf{u}' \end{aligned} \tag{25b}$$

is then obtained, for which a stabilizing controller

$$\Delta\mathbf{u}' = \mathbf{k}_i^T \Delta\bar{\mathbf{e}}_i \tag{26}$$

can be determined. Overall, the resulting control law combining the feedforward part as well as the extended PID control reads

$$\mathbf{u} = \mathbf{u}_\mathrm{d} + \Delta\mathbf{u}. \tag{27}$$

Since (22) is dependent on the full state $\boldsymbol{\xi}$, a high-gain observer [16] is utilized to reconstruct non-measurable states. The complete control structure is depicted in Fig. 4.

6. Simulation Results

In the following, the aforementioned tracking control is applied to the numerical example of the nonlinear dynamics

$$\begin{bmatrix} \dot{x}_1 \\ \dot{x}_2 \\ \dot{x}_3 \end{bmatrix} = \begin{bmatrix} x_2 \\ -x_1 x_2 + x_3 + x_2^3 + u_1 \\ x_2(1 - x_3) - x_2^2 + u_2 \end{bmatrix} \tag{28a}$$

$$\begin{bmatrix} y_1 \\ y_2 \end{bmatrix} = \begin{bmatrix} x_1 \\ x_3 \end{bmatrix}, \tag{28b}$$

which will serve as ground truth reference model. Usually, when utilizing RFS, the *exact* plant dynamics are unknown and the fuzzy system is obtained by expert knowledge or measured data. Here, an RFS is derived from (28) by sampling the nonlinear dynamics at core positions $s_{j_i}^{x_i} \in \{-2, -1, 0, 1, 2\}$ and $s_{q_p}^{u_p} \in \{-10, 0, 10\}$ with associated linguistic values $L_{j_i}^{x_i} \in$ {neg. big, neg. small, zero, pos. small, pos. big} and $L_{q_p}^{u_p} \in$ {negative, zero, positive}. The complete rule base defines a partition of the input-state space with $(5-1)^3 \cdot (3-1)^2 = 256$

Journal of Physics: Conference Series **570** (2014) 022001 doi:10.1088/1742-6596/570/2/022001

hypersquares $H_\mathbf{l}$, whereas the output model is defined over a grid with $(5-1)^3 = 64$ hypersquares $H_\mathbf{j}^\mathbf{x}$. Each local dynamics is then defined by (6). E.g., for the hypersquare H_1 bounded by $-2 \leq x_j \leq -1, j \in \{1,2,3\}$, and $-10 \leq u_q \leq 0, q \in \{1,2\}$, the local system

$$\dot{\mathbf{x}} = \begin{bmatrix} 0 & 0 & 1 & 0 & 0 & 0 & 0 & 0 \\ 6 & 0 & 7 & -1 & 1 & 0 & 0 & 0 \\ 2 & 0 & 4 & 0 & 0 & 0 & -1 & 0 \end{bmatrix} \begin{bmatrix} 1 \\ x_1 \\ x_2 \\ x_1x_2 \\ x_3 \\ x_1x_3 \\ x_2x_3 \\ x_1x_2x_3 \end{bmatrix} + \begin{bmatrix} 0 & 0 & 0 \\ 1 & 0 & 0 \\ 0 & 1 & 0 \end{bmatrix} \begin{bmatrix} u_1 \\ u_2 \\ u_1u_2 \end{bmatrix}, \tag{29a}$$

$$\mathbf{y} = \begin{bmatrix} 0 & 1 & 0 & 0 \\ 0 & 0 & 1 & 0 \end{bmatrix} \begin{bmatrix} 1 \\ x_1 \\ x_3 \\ x_1x_3 \end{bmatrix} \tag{29b}$$

is obtained. For each hypersquare $H_\mathbf{l}$, the local relative degrees

$$\boldsymbol{\delta}_i^T = \begin{bmatrix} 2 & 1 \end{bmatrix}, \quad i = 1, \ldots, 256 \tag{30}$$

are then computed. Due to $\sum_{j=1}^{n} \delta_{i,j} = n$, the system has no internal dynamics and can be stabilized via static state feedback. Thus, $\mathbf{y} = \mathbf{z}$ is a flat output. Furthermore, for each hypersquare, the matrix $\mathbf{A}_\mathbf{l}$ in (16) is invertible, e.g., for (29) it is computed to $\mathbf{A}_1 = \mathbf{I}_m$. Therefore, a local controller can be found for each hypersquare achieving global input-output decoupling.

In this example, the functions $\alpha_{\mathbf{l},i}$ were chosen such that the local linear dynamics (20) have stable eigenvalues at $\lambda_i = -10$. Since all $\alpha_{\mathbf{l},i}$ can be pre-computed offline, fast online determination of the nominal control $\mathbf{u}_\mathrm{d}$ is possible by means of a look-up table and the evaluation of a function. Furthermore, the extended PID control is computed online as linear quadratic regulator. The online linearization of the error dynamics and computation of the feedback are computationally more complex than the feedforward control, and thus become the bottleneck regarding real time implementation.

In order to test the controller, the signals

$$y_{1,\mathrm{d}} = -1.5 \sin \frac{\pi}{8} t, \tag{31a}$$

$$y_{2,\mathrm{d}} = 0.75 \cos \frac{\pi}{6} t \tag{31b}$$

are considered, which are to be tracked by the flat outputs of the system. The nominal trajectories (31) are applied to the controlled ground truth (28), having an initial state of $\mathbf{x}_0 = [0.05, \ -0.5, \ 0.5]^T$, whereas the initial state of the inverse system as well as the high-gain observer is $\hat{\mathbf{x}}_0 = [0, \ 0, \ 0]^T$. As becomes apparent from the simulation results shown in Fig. 5, the flat outputs follow the nominal trajectory very well. Note that tracking is achieved even in case of initial states not being consistent with the trajectory. Furthermore, Fig. 6 depicts the development of the control input $\mathbf{u} = \mathbf{u}_\mathrm{d} + \Delta\mathbf{u}$. Apparently, the bound on the control input $|u_q| \leq 10$ is not exceeded, even during the initial transient behavior.

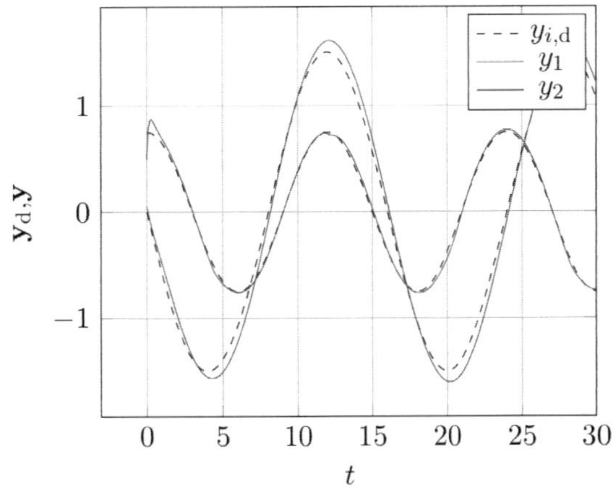

Figure 5: Simulation results: Reference trajectory $\mathbf{y}_{\mathrm{d}}$ and development of flat outputs $\mathbf{y}$.

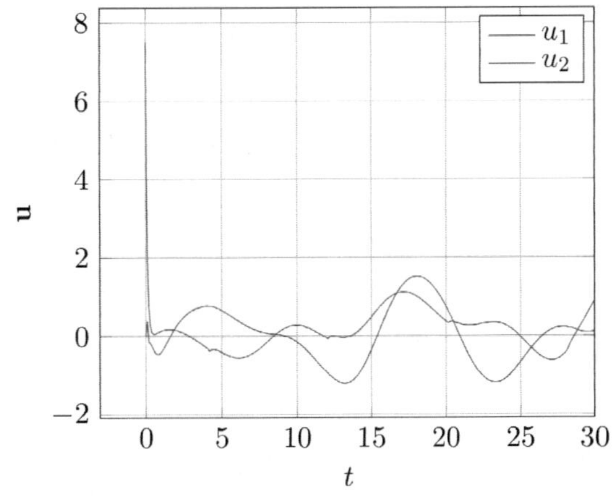

Figure 6: Control input during simulation of controlled plant

From Fig. 5, it can also be seen that the plant output does not converge asymptotically to the nominal trajectory, since small deviations of $\mathbf{y}$ from $\mathbf{y}_{\mathrm{d}}$ remain. The reason for this effect is mainly due to the fact that the RFS, by which the nominal control law $\mathbf{u}_{\mathrm{d}}$ is computed, is only an approximation of the ground truth model, to which the controller is applied to. In addition, the inverse system (19) does not follow the reference trajecotry arbitrarily fast. By assigning faster poles to the linear subsystems (20), this effect can be mildered, as shown in Fig. 7. Therein, different simulation runs are shown, each with unique set of equal eigenvalues of (20). As can be seen, neglecting the initial transient behavior, the error $\mathbf{e} = \mathbf{y} - \mathbf{y}_{\mathrm{d}}$ becomes smaller with faster eigenvalues. At the same time, the magnitude of $\mathbf{u}_{\mathrm{d}}$ increases with faster eigenvalues, such that especially during the initial displacement of the outputs from the reference trajectory, the control input exceeds its bound. Thus, a trade-off between fast decay of the transients and meeting the control input bounds has to be made.

7. Discussion

In this paper, an approach for flatness-based feedforward control of Recurrent Fuzzy Systems was presented. It was shown that existing techniques for tracking control of flat nonlinear systems

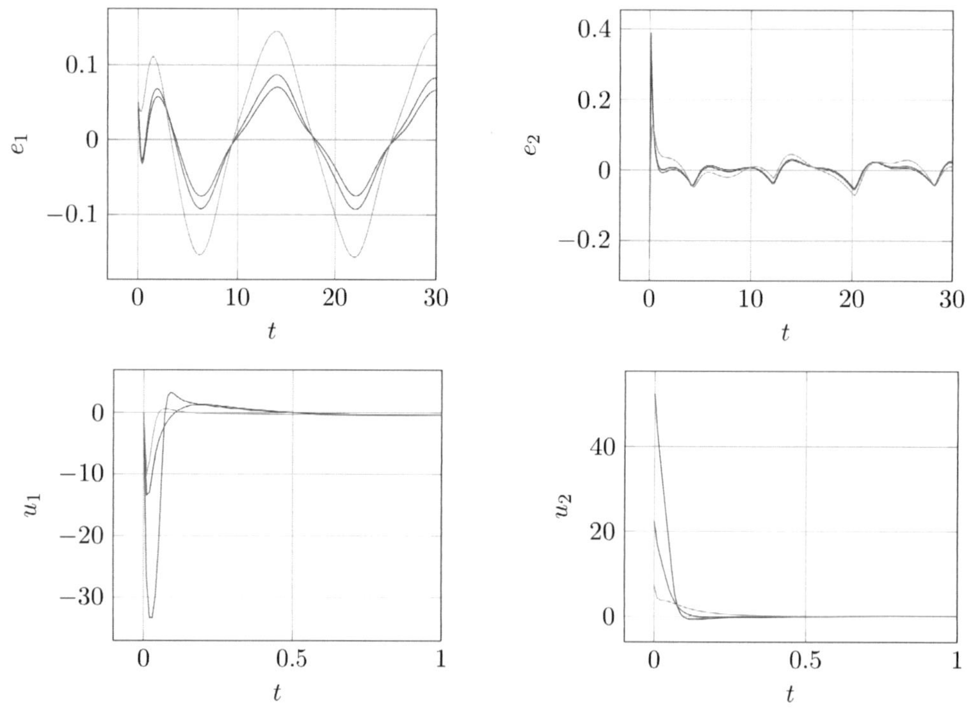

Figure 7: Comparison of control error and control input for different eigenvalues of the linearized dynamics: (—) $\lambda_i = -10$, (—) $\lambda_i = -30$, and (—) $\lambda_i = -70$.

may also be applied to Recurrent Fuzzy Systems. Since the dynamics of a recurrent fuzzy system is defined by means of a linguistic rule base, it was first shown how to derive local real valued system dynamics under mild assumptions. Thus, a partitioning of the input-state space was obtained, whereas the system dynamics within each partition was described by a vector polynomial being multi-affine in the inputs and states. By means of a dynamic transformation, these local dynamics were then shown to be always equivalent to a prolonged input-affine system.

Due to space limitations, several simplifying assumptions were made throughout the paper, which do not lessen the underlying idea, but should be addressed in future work: First, only Recurrent Fuzzy Systems without internal dynamics were considered, whereas the application to those with stable internal dynamics (i.e., *minimum phase* Recurrent Fuzzy Systems) is straightforward. Second, only flat Recurrent Fuzzy Systems stabilizable with static state feedback were addressed. A similar discussion for the more general case of dynamic feedback remains an open topic, but may also be tackled by means of the flatness approach. In addition, only square systems with global decoupling of each input-output pair were discussed, in contrast to the more general form of Recurrent Fuzzy Systems with a number of inputs different from the number of outputs.

Finally, for feedback control of the error system, an extended PID controller was utilized, stabilizing the linearized error dynamics around the reference trajectory. Here, the PID controller was mainly chosen due to its simple implementation. Nevertheless, as mentioned in [13], any kind of linear controller can be applied. In particular, a comparison especially with robust controllers, such as H_∞ controllers, would be interesting in the light of the approximate nature of the system model.

Acknowledgments
This work was supported by the Deutsche Telekom Stiftung (www.telekom-stiftung.de) and the Studienstiftung des Deutschen Volkes (www.studienstiftung.de).

References
[1] J. Adamy and A. Flemming. Equilibria of continuous-time recurrent fuzzy systems. *Fuzzy Sets and Systems*, 157(22):2913–2933, 2006.

[2] J. Adamy and R. Kempf. Regularity and chaos in recurrent fuzzy systems. *Fuzzy Sets and Systems*, 140(2): 259–284, 2003.

[3] F. Allgöwer, J. Ashman, and A. Ilchmann. High-gain adaptive λ-tracking for nonlinear systems. *Automatica*, 33(5):881–888, 1997.

[4] F. Antritter. On the relations between different flatness based design methods for tracking controllers. In *American Control Conference, 2008*, pages 1740–1745. IEEE, 2008.

[5] S. Devasia, D. Chen, and B. Paden. Nonlinear inversion-based output tracking. *Automatic Control, IEEE Transactions on*, 41(7):930–942, 1996.

[6] L. Eciolaza, T. Taniguchi, M. Sugeno, D. Filev, and Y. Wang. Piecewise bilinear models for feedback error learning: On-line feedforward controller design. In *IEEE International Conference on Fuzzy Systems*, pages 1–8, 2013.

[7] L. Eciolaza, T. Taniguchi, and M. Sugeno. Brain style control scheme: Simultaneous forward and inverse model identification and controller design. In *IEEE International Conference on Fuzzy Systems (FUZZ-IEEE)*, pages 2457–2464, July 2014. doi: 10.1109/FUZZ-IEEE.2014.6891893.

[8] N. Faiz, S. K. Agrawal, and R. M. Murray. Trajectory planning of differentially flat systems with dynamics and inequalities. *Journal of Guidance, Control, and Dynamics*, 24(2):219–227, 2001.

[9] M. Fliess and C. Join. Model-free control. *International Journal of Control*, 86(12):2228–2252, 2013.

[10] M. Fliess, J. Lévine, P. Martin, and P. Rouchon. Flatness and defect of non-linear systems: introductory theory and examples. *International journal of control*, 61(6):1327–1361, 1995.

[11] S. Gering and J. Adamy. Synthesis of local state feedback for continuous-time recurrent fuzzy systems. In *Proceedings of the IEEE International Conference on Fuzzy Systems (FUZZ-IEEE)*, Hyderabad, India, 2013.

[12] S. Gering, A. Schwung, T. Gußner, and J. Adamy. Sum of squares approaches for control of continuous-time recurrent fuzzy systems. In *Proc. of the 21st IEEE Mediterranean Conference on Control and Automation*, pages 271–277, Chania, Greece, June 2013.

[13] V. Hagenmeyer and E. Delaleau. Exact feedforward linearization based on differential flatness. *International Journal of Control*, 76(6):537–556, 2003.

[14] A. Ilchmann and E. P. Ryan. High-gain control without identification: a survey. *GAMM-Mitteilungen*, 31 (1):115–125, 2008.

[15] R. Kempf and J. Adamy. Equilibria of recurrent fuzzy systems. *Fuzzy Sets and Systems*, 140(2):231–257, 2003.

[16] H. K. Khalil. High-gain observers in nonlinear feedback control. In *New directions in nonlinear observer design*, pages 249–268. Springer, 1999.

[17] H. Nijmeijer and A. van der Schaft. *Nonlinear Dynamical Control Systems*. Springer, 1990.

[18] M. Pahlevaninezhad, P. Das, J. Drobnik, P. Jain, and A. Bakhshai. A new control approach based on the differential flatness theory for an ac/dc converter used in electric vehicles. *IEEE Transactions on Power Electronics*, 27(4):2085–2103, April 2012. ISSN 0885-8993. doi: 10.1109/TPEL.2011.2170098.

[19] P. Thounthong, S. Pierfederici, J.-P. Martin, M. Hinaje, and B. Davat. Modeling and control of fuel cell/supercapacitor hybrid source based on differential flatness control. *IEEE Transactions on Vehicular Technology*, 59(6):2700–2710, 2010.

[20] M. J. Van Nieuwstadt and R. M. Murray. Real time trajectory generation for differentially flat systems. 1997.

[21] S. Waldherr and M. Zeitz. Flat inputs in the mimo case. In *IFAC Symposium on Nonlinear Control Systems*, 2010.

[22] M. Zerar, F. Cazaurang, and A. Zolghadri. Robust tracking of nonlinear mimo uncertain flat systems. In *IEEE International Conference on Systems, Man and Cybernetics*, volume 1, pages 536–541. IEEE, 2004.

The modified Poincare-Dulac method in analysis of auto-oscillations of nonlinear mechanical systems.

Gennady I. Melnikov

Prof., Department of Theoretical and Applied Mechanics, University ITMO, St.Petersburg, Russia
E-mail: melnikov.ifmo@yandex.ru

Sergei E. Ivanov

Ph.D., Department of Information Systems, University ITMO, St.Petersburg, Russia
E-mail: serg ie@mail.ru

Vitaly G. Melnikov

Head of Department of Theoretical and Applied Mechanics, University ITMO, St.Petersburg, Russia
E-mail: melnikov@mail.ifmo.ru

Abstract. A dynamic equation of a mechanical system with one degree of freedom with functionally odd nonlinear recover forces and small dissipative forces is considered. An improved method of transformation and integration of the equation, based on a method of normalization of Poincare-Dulac is developed. The improvement of the method is in application of the Chebyshev's economization to high-order nonlinear terms.

1. Introduction.

In this paper we consider the Poincare-Dulac method [6], [8], [7] in a polynomial form. We apply Chebyshev economization method [2], [10] to approximate high-order polynomials using low-order polynomials.

2. Polynomial transformation and Chebyshev economization.

Let us consider a homogeneous cubic differential equation of a motion of a nonlinear autonomous mechanical system

$$\ddot{q}+2\alpha\dot{q}+\gamma q=\varepsilon P(\dot{q},q);P(\dot{q},q)=\sum_{\nu=0}^{3}p_{\nu}\dot{q}^{\nu}q^{3-\nu}\ , \tag{1}$$

where α,ε - are small constants.

We consider this equation in the dimensionless form in a square field of the phase plane. We use the mapping of the right-angled field to the square field using the linear scale transformation of the variable.

$$\bar{G}=\{(\dot{q},q):|\dot{q}|\le r,|q|\le r\};r\le 1 \tag{2}$$

Suppose that the roots of the linear part of system are complex and have small real parts.

$$\lambda=-\alpha+i\beta,\bar{\lambda}=-\alpha-i\beta;\beta=\sqrt{\gamma-\alpha^{2}} \tag{3}$$

Using the differentiation operator $D=d/dr$ we get two forms of the left-hand part.

$$(D-\lambda)(D-\bar{\lambda})q=(D-\lambda)x=(D-\bar{\lambda})\bar{x}\ ,$$

where we introduce the complex variables as follows:

$$x = \dot{q} - \bar{\lambda} q, \bar{x} = \dot{q} - \lambda q \ , \tag{4}$$

or

$$\dot{q} = -i\mu(\lambda x - \bar{\lambda}\bar{x}), q = -i\mu(x - \bar{x}); \mu = 1/2\beta \tag{5}$$

The new variables fulfill differential equations with equal nonlinear polynomials.

$$\dot{x} = \lambda x + \varepsilon f(x, \bar{x}), \dot{\bar{x}} = \bar{\lambda}\bar{x} + \varepsilon f(x, \bar{x}) \tag{6}$$

Note that the function

$$Q = x\bar{x} = \dot{q}^2 + 2\alpha\dot{q}q + \gamma q^2$$

satisfies to a differential equation

$$\dot{Q} + 2\alpha Q = 2\varepsilon(\dot{q} + \alpha q)P(\dot{q}, q); \dot{Q} + 2\alpha Q = 2\varepsilon \sum_{\nu=0}^{4} (p_{\nu-1} + \alpha p_\nu)\dot{q}^\nu q^{4-\nu} \tag{7}$$

Using the Poincare-Dulac method [1, 2, 3] for vibrating systems and for non-vibrating systems [4, 5] we change the variables in the equation (1), i.e.the nonlinear cubic variables

$$y = x + \varepsilon \sum_{\nu=0}^{3} a'_\nu x^\nu \bar{x}^{3-\nu}, \bar{y} = \bar{x} + \varepsilon \sum_{\nu=0}^{3} \bar{a}'_\nu x^{3-\nu} \bar{x}^\nu \ , \tag{8}$$

have the coefficients a'_ν . We replace these nonlinear variables so that the converted equation will have only one unremovable cubic monomial and a small error δ' .

$$\dot{y} = \lambda y + \varepsilon a' y^2 \bar{y} + \delta', \dot{\bar{y}} = \bar{\lambda}\bar{y} + \varepsilon \bar{a}' y \bar{y}^2 + \bar{\delta}' \ , \tag{9}$$

and in the exponential form

$$y = \rho \exp(i\theta), u = \rho^2 \tag{10}$$

$$\dot{y}e^{-i\theta} = \dot{\rho} + \rho i\theta = (-\alpha + i\beta)\rho - \varepsilon(a^* + ib^*)\rho^3$$

$$\dot{\rho} = -\alpha\rho - \varepsilon a^* \rho^3, \dot{\theta} = \beta - \varepsilon b^* \rho^2$$

$$\dot{u} = -2\alpha u - 2\varepsilon a u^2 \tag{11}$$

$$\dot{\theta} = \beta - \varepsilon b u \tag{12}$$

Let us get the equation (11) within the limits of transformation of real variables, but of structure (11), (12).Let us introduce a new real variable of the form

$$u = Q + \varepsilon V; V = \sum_{\nu=0}^{4} a_\nu \dot{q}^\nu q^{4-\nu} \ , \tag{13}$$

with undetermined coefficients. For brevity we will not write further summation signs. Firstly, we will reduce terms in expression by means of a shift in a reference of indexes.

$$\dot{V} = (4 - \nu')a_{\nu'}\dot{q}^{\nu'+1} q^{3-\nu'} - 2\alpha\nu a_\nu \dot{q}^\nu q^{4-\nu} - \gamma\nu' a_{\nu'} \dot{q}^{\nu'-1} q^{5-\nu'} + \varepsilon R$$

$$\dot{V} = U + \varepsilon R; U = [(5-\nu)a_{\nu-1} - 2\alpha\nu a_\nu - \gamma(\nu+1)a_{\nu+1}]\dot{q}^\nu q^{4-\nu} \tag{14}$$

$$R = \nu a_\nu \dot{q}^{\nu-1} q^{4-\nu} P = \nu a_\nu p_{\nu'} \dot{q}^{\nu'+\nu-1} q^{7-\nu'-\nu} = \nu a_\nu p_{k+1-\nu} \dot{q}^k q^{6-k},$$

$$R = \sum_{k=0}^{6} R_k \dot{q}^k q^{6-k}; R_k = a_1 p_k + 2a_2 p_{k-1} + 3a_3 p_{k-2} + 4a_4 p_{k-3} \ , \tag{15}$$

and also in the expression

$$QV = a_\nu \dot{q}^{\nu+2} q^{4-\nu} + 2\alpha a_\nu \dot{q}^{\nu+1} q^{5-\nu} + \gamma a_\nu \dot{q}^\nu q^{6-\nu} = (a_{k-2} + 2\alpha a_{k-1} + \gamma a_k)\dot{q}^k q^{6-k}$$

$$QV = \sum_{k=0}^{6} R'_k \dot{q}^k q^{6-k}; R'_k = a_{k-2} + 2\alpha a_{k-1} + \gamma a_k \tag{16}$$

Let us find the expression for the error using (6),(15),(16) neglecting the small terms $O(\varepsilon^3)$

$$U = Q + \varepsilon V \tag{17}$$

$$\dot{u} \approx -2\alpha u - 2\varepsilon a u^2, \quad \delta = \dot{u} + 2\alpha u + 2\varepsilon a u^2 \tag{18}$$

Let us substitute expressions (17) into the formula (18)

$$\delta = \dot{Q} + 2\alpha Q + \varepsilon[\dot{V} + 2\alpha V + 2aQ^2] + 4\varepsilon^2 aQV + 2\varepsilon^3 aV^2 \tag{19}$$

We select items with a factor ε^2

$$\delta = \varepsilon[2(p_{\nu-1} + 2p_\nu)\dot{q}^\nu q^{4-\nu} + \nu a_\nu(-2\alpha\dot{q} - \gamma q)\dot{q}^{\nu-1}q^{4-\nu} + 2\alpha V + 2qQ^2] + \varepsilon^2 W_6 \tag{20}$$

$$W_6 = 4aQV + \nu a_\nu \dot{q}^{\nu-1}q^{4-\nu}p = 4a(\dot{q}^2 + 2\alpha\dot{q}q + \gamma q^2)a_{\nu'}\dot{q}^{\nu'}q^{4-\nu'} + \nu a_\nu p\dot{q}^{\nu'+\nu-1}q^{7-\nu'-\nu}$$

$$W_6 = 4a(a_{k-2} + 2\alpha a_{k-1} + \gamma a_k)\dot{q}^k q^{6-k} + \nu a_\nu p_{k+1-\nu}\dot{q}^k q^{6-k} \tag{21}$$

Let us present equation (21) as

$$W_6 = \sum_{k=0}^{6} A_k \dot{q}^k q^{6-k}, \quad A_k = \sum_{\nu=0}^{3} \nu a_\nu p_{k+1-\nu} + 4a(\gamma a_k + 2\alpha a_{k-1} + a_{k-2})$$

We will use Chebyshev economization in the domain $h = \frac{3}{4}r^2$ for the forms of the degree $\varepsilon^2 W_6$.

$$\dot{q}^3 \approx h\dot{q}, q^3 \approx hq; (\dot{q}, q) \in \bar{G}$$

Let us approximate the form W_6

$$W_6 \approx W_4 = h\left(A_0 q^4 + A_1\dot{q}q^3 + A_2\dot{q}^2q^2 + \frac{1}{2}A_3(\dot{q}^3 q + \dot{q}q^3) + A_4\dot{q}^2q^2 + A_5\dot{q}^3 q + A_6\dot{q}^4\right)$$

$$W_6 \approx W_4 = \sum_{\nu=0}^{4} B_\nu \dot{q}^\nu q^{4-\nu}; B_0 = A_0; B_1 = A_1 + \frac{1}{2}A_3; B_2 = A_2 + A_4; B_3 = A_5 + \frac{1}{2}A_3; B_4 = A_6 \tag{22}$$

Let us substitute expression (22) in (20)

$$\delta = \varepsilon\{[2(\alpha p_\nu + p_{\nu-1}) - 2\alpha\nu a_\nu - \gamma(\nu+1)a_{\nu+1} + 2\alpha a_\nu + \varepsilon h B_\nu]\dot{q}^\nu q^{4-\nu} + 2aQ^2\} \tag{23}$$

$$Q^2 \equiv \sum_{\nu=0}^{4} c_\nu \dot{q}^\nu q^{4-\nu}; c_0 = \gamma^2; c_1 = 4\alpha\gamma; c_2 = 4\alpha^2 + 2\gamma; c_3 = 4\alpha; c_4 = 1$$

Let us equate expression in the formula (23) to null, according to Poincare-Dulac method.

$$2(\alpha p_\nu + p_{\nu-1}) - \gamma(\nu+1)a_{\nu+1} - 2\alpha(\nu-1)a_\nu + 2ac_\nu + \varepsilon h B_\nu = 0$$
$$(\nu = 0, \ldots, 4) \tag{24}$$

Let us find the solution of a set of equations (24). Suppose $\varepsilon \approx 0, \alpha \approx 0$

$$a_0 = 0; a = -p_3; a_1 = -2\gamma p_3; a_2 = \frac{p_0}{\gamma}; a_3 = \frac{2(p_1 - 2\gamma p_3)}{3\gamma}; a_4 = \frac{p_2}{2\gamma}$$

Let us obtain an accurate formula for the coefficient a at $\varepsilon \neq 0, \alpha \neq 0$

$$a = -(24\alpha^3\varepsilon h p_0^2 + 3\alpha\varepsilon h p_1^2\left(4\alpha^2 + 3\gamma^2\right) +$$
$$3\alpha p_0\left(\varepsilon h p_2\left(4\alpha^2 + 3\gamma^2\right) + 2\gamma\left(4\alpha^2 + 4\alpha(4\gamma+1)\varepsilon h p_3 + 3\gamma\varepsilon h p_4\right) - 16\alpha\gamma\varepsilon h p_1\right) -$$
$$2\gamma p_1\left(\alpha\varepsilon h p_3\left(4\alpha^2(3\gamma+5) + 3\gamma(5\gamma+1)\right) + 12\alpha^2\varepsilon h p_2 + 6\left(-2\alpha^4 + 2\alpha^2\gamma + \gamma^2\varepsilon h p_4\right)\right) +$$
$$\gamma^2(6\gamma p_3\left(3\alpha^2 - 2\gamma + 4\varepsilon h p_4\left(\alpha^2 + \gamma\right)\right) + \alpha\varepsilon h p_3^2\left(\alpha^2(96\gamma + 68) + 3\gamma(17\gamma + 4)\right)) +$$
$$\gamma^2(6p_2\left(-4\alpha^3 + 3\alpha\gamma + 2\varepsilon h p_3\left(2\alpha^2(\gamma+3) + \gamma\right)\right) - 12\alpha\gamma^2 p_4)) / 12\gamma^3\left(2\alpha^2 - \gamma\right)$$

As an example, let us consider the Van der Pol equation, written in the form (1):

$$\ddot{q} + 2\alpha\dot{q} + \gamma q = \varepsilon\dot{q}q^2,$$

where $(q, \dot{q})$ are the phase coordinates.

For the Van der Pol equation we obtaine the unknown coefficients.

$$a = \alpha\left(-8\alpha^3\gamma + 8\alpha\gamma^2 + 4\alpha^2\varepsilon h + 3\gamma^2\varepsilon h\right)/4\gamma^3\left(2\alpha^2 - \gamma\right),$$

$$a_1 = \alpha\left(-8\alpha^3\gamma + 8\alpha\gamma^2 + 4\alpha^2\varepsilon h + 3\gamma^2\varepsilon h\right)/2\gamma^2\left(2\alpha^2 - \gamma\right),$$

$$a_2 = (-16\alpha^5\gamma + 20\alpha^3\gamma^2 - 2\alpha\gamma^3 + 8\alpha^4\varepsilon h + 2\alpha^2(3\gamma - 1)\gamma\varepsilon h + \gamma^2\varepsilon h)/2\gamma^3\left(\gamma - 2\alpha^2\right),$$

$$a_3 = (-12\alpha^4\gamma + 14\alpha^2\gamma^2 - 2\gamma^3 + 6\alpha^3\varepsilon h - \alpha\gamma(1 - 3\gamma)\varepsilon h)/3\gamma^3\left(2\alpha^2 - \gamma\right),$$

$$a_4 = (8\alpha^3\gamma - 8\alpha\gamma^2 - 4\alpha^2\varepsilon h - 3\gamma^2\varepsilon h)/(24\alpha^2\gamma^3 - 12\gamma^4)$$

Let $\varepsilon = 0.6, \alpha = -0.3, h = 2$ then the unknown coefficients

$$a = -0.169024, a_1 = -0.338049, a_2 = 1.10283, a_3 = -0.712033, a_4 = 0.187805$$

The graph of vibration amplitude is presented in figure 1.

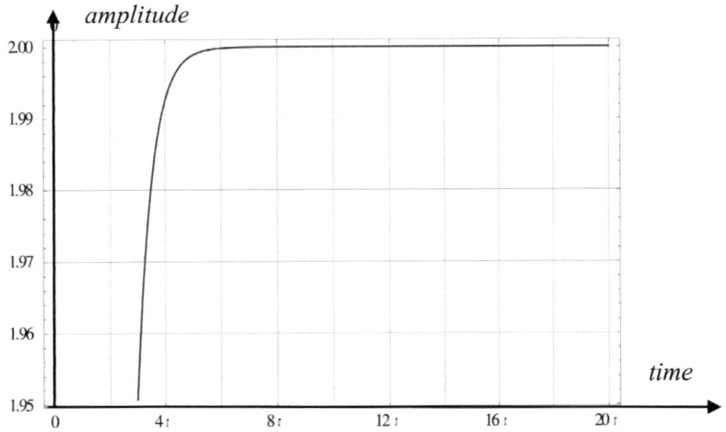

Figure 1. The graph of a vibration amplitude.

3. Conclusion

In this paper we considered the motion of a nonlinear autonomous oscillatory system with one degree of freedom. The differential equation of motion contains a non-linear function in the cubic form from the phase coordinates. The conversion is based on Poincare-Dulac method and Chebyshev's economization. The polynomial transformation was performed without transition to integrated state variables. The example shows a significant increase in the accuracy of the obtained solutions. As a result, the error of the transformed equations is reduced by an order.

3.1. Acknowledgments

The research was supported by the State Financial Support of the Leading National Universities, no. 074-U01, the Ministry of Education and Science of the Russian Federation, project no.14.Z50.31.0031, and by Committee of Science and Higher School of St. Petersburg Government.

References

[1] H. Dulac, "Solution dun systeme dequations differentielles dans le voisinage des valeurs singulieres," Bull.Soc. Math., vol. 40, pp. 324-383, 1912.

[2] T. J. Rivlin, Chebyshev Polynomials: From Approximation Theory to Algebra and Numbers Theory. New York: Wiley, 1990.

[3] R. W. Hamming, Numerical methods for scientists and engineers. Dover, 1986.

[4] V. G. Melnikov, "Chebyshev economization in transformations of nonlinear systems with polynomial structure," in 14th WSEAS international conference on systems, vol. 1: Latest trends on systems, Corfu, 2010, pp. 301-303.

[5] V. G. Melnikov, "Chebyshev economization in poincare-dulac transformations of nonlinear systems," Nonlinear Analysis: Theory, Methods & Applications, vol. 63, no. 5-7, pp. 1351-1355, 2005.

[6] V.I. Arnold and Yu.S. Ilyashenko, Ordinary diferential equations; in: Encyclopaedia of Mathematical Sciences, vol. 1 Dynamical Systems I, (D.V. Anosov and V.I. Arnold eds.), pp. 1-148, Springer, Berlin 1988.

[7] G.I. Melnikov, Dynamics of nonlinear mechanical and electromechanical systems, Mashinostroenie, Moskow, 1975.

[8] A.D. Bruno, Power geometry in algebraic and diferential equations, North- Holland, Amsterdam, 2000.

[9] S.E.Ivanov., G.I. Melnikov, "Method of autonomation of the nonlinear dynamic systems." Scientific and Technical Journal of Information Technologies, Mechanics and Optics, №1 (89), pp. 151-156, 2014.

[10] G. Arfken, Mathematical Methods for Physicists, 2nd ed. New York: Academic Press, p. 631, 1970.

Advanced embedded nonlinear observer design and HIL validation using a Takagi-Sugeno approach with unmeasurable premise variables

S C Olteanu[1], **A Aitouche**[2], **L Belkoura**[3]
LAGIS UMR CNRS 8219
[1] University of Lille 1, France; E-mail: sc.olteanu@ed.univ-lille1.fr
[2] HEI Lille, France; E-mail: abdel.aitouche@hei.fr
[3] University of Lille 1, France; E-mail: lotfi.belkoura@univ-lille1.fr

Abstract. The article's goals are to illustrate the feasibility of implementing a Takagi Sugeno state observer on an embedded microcontroller based platform and secondly to present a methodology for validating a physical embedded system using a Hardware In The Loop architecture, where a simulation software replaces the process. As an application, a three water tank system was chosen. For the validation part, LMS AMESim software is employed to reproduce the process behaviour. The interface to the embedded platform is assured by Simulink on a Windows operating system, chosen as it is the most commonly used operating system. The lack of real time behaviour of the operating system is compensated by a real time kernel that manages to offer deterministic response times. The Takagi-Sugeno observer in the case of this process has the complex form that considers the premise variables to be unmeasurable. The embedded system consists of two Arduino boards connected in parallel, thus offering distributed resources.

1. Introduction

A consistent increase has been seen in the last decade in the usage of small scaled embedded systems in control, diagnostics and signal manipulation, as these solution become cheaper and more powerful [1]. Also, as the requirements of more complex algorithms augment, the study of the possibilities and the feasibility of implementing complex algorithms on these types of platforms becomes more and more useful. As such, we have implemented a nonlinear state observer using a Takagi-Sugeno (T-S) [2].

Because it is preferred to test an embedded system prior to its implementation on a real world process in order to avoid unforeseen errors in modeling or communication, it is a useful practice to test the system in a hardware in the loop configuration, where the process itself is replaced by a simulation software, as the AMESim software [3], [4]. For the embedded part, based on versatility and size/price ratio, a microcontroller platform (on an Arduino board) is employed.

The T-S methodology, that can be seen as a polytopic linear differential inclusion, has seen a growing interest in the last decade as it has a simple implementation mechanism, as well as a simple structure. A common solution for obtaining a TS model from a general nonlinear one, is to apply a sector nonlinearity transformation [5] to the state space form, thus giving an exact approximation. This model consists of a blend of linear systems that are coupled together by means of weighting functions.

As an application, a nonlinear analysis of a three tank system has been selected as it represents a basic example that brings some interesting complexity when looked upon from a nonlinear perspective. After the construction of the state space model, the T-S transformation is applied, and for the newly built representation, the observer is developed. A similar approach, also with unmeasurable premise variables can be seen in [6]. Others have used the application in nonlinear studies, even fault diagnosis like in [7].

The article begins in section 2 with a general description of the T-S representation and the construction of the observer, followed in section 3 by the embedded system description as well as the HIL validation technique. Section 4 deals with the application process, meaning the three tank system's modelling and its TS representation. Section 5 presents the implementation of the observer for the system both in simulation and on the physical process. The paper ends with a conclusions section.

2. Takagi-Sugeno Observer

2.1. Takagi-Sugeno Representation

A general TS model presents itself as a sum of linear systems as in (1), where each system contributes in a percentage given by the weighting function w_i. Here we consider B and C as non dependent on states and we consider i as the number of a fuzzy rule (this representation may also be seen as a fuzzy system).

$$\begin{cases} \dot{x} = \sum_{i=1}^{8} w_i(z) A_i x + Bu \\ y = Cx \\ z = \text{nonlinear terms(premise variables)} \end{cases} \tag{1}$$

It becomes obvious that this way of writing the system moves the nonlinearities from inside the system matrices into the weighting functions, adding the fact that they respect a convex sum property:

$$\sum_i w_i = 1 \tag{2}$$

To arrive from the general nonlinear form to a TS representation the most common procedure is to apply sector nonlinearity technique. But first of all, the system is rewritten in form (3):

$$\begin{cases} \dot{x} = A(z(x)) \cdot x + B \cdot u; \\ y = C \cdot x \end{cases} \tag{3}$$

The z variables, that are dependent on states, have the name of premise variables, and they contain the nonlinear information for the system. Now the sector nonlinearity method [8] can be applied by determining the minimum and maximum of each z_i. If we consider a general triangular membership function one arrives at (4) where we design MFs as membership functions:

$$\begin{cases} i = 1..\text{n} \\ MF_{min}(z_i) = \dfrac{z_{i,Max} - z_i}{z_{i,Max} - z_{i,\min}} \\ MF_{Max}(z_i) = \dfrac{z_i - z_{i,\min}}{z_{i,Max} - z_{i,\min}} \\ w_i = MF_{Max/Min}(z_1) \cdot \ldots \cdot MF_{Max/Min}(z_n) \end{cases} \tag{4}$$

2.2. Takagi-Sugeno Observer

Using the T-S representation, the observer itself can be constructed. The observer design will materialize as an optimization problem, that will be resolved by means of linear matrix inequalities (LMIs). The observer chosen here has a classical Luenberger form adapted to TS systems as in (5).

$$\begin{cases} \dot{\hat{x}} = \sum_i w_i(\hat{z})\left(A_i\hat{x} + B_i u + L_i(\hat{y} - y)\right) \\ \hat{y} = \sum_i w_i(\hat{z}) C_i \hat{x} \end{cases} \tag{5}$$

As it can be seen, the weighting functions depend on estimated premise variables z, therefore the computations are more complex. In the literature, there are many cases where premise variables are directly measurable as in [5], a supposition that indeed simplifies the stability study ($w_i(\hat{z})=w_i(z)$). This is not the case for the current work.

In order to construct the observer (6), we need to calculate the observer gains. This will be done by employing Matlab's implicit LMI solver or YALMIP toolbox to resolve the LMIs presented in (6).

$$\begin{cases} \text{For } i = 1:8 \\ \begin{pmatrix} A_i^T P - C^T Q_i + P A_i - Q_i C + I + 2\alpha P & P \\ P & -\lambda \end{pmatrix} < 0 \\ size(P) = size(A) \\ \mathrm{P} > 0 \\ P = P^T \\ Q_i = P \cdot L_i \\ size(Q_i) = size(A_i.C) \\ \alpha > 0 \text{ ; } scalar \end{cases} \tag{6}$$

Proof:

As the goal of any observer is to reach stability in 0 of the estimation error $e = x - \hat{x}$, the methodology adopted is Lyapunov stability analysis. This dictates to impose condition (8) upon a well chosen Lyapunov function. We have chosen a quadratic one as in (7).

$$V = e^T P e \text{ ; } P = P^T > 0 \tag{7}$$

Its derivative being:

$$\dot{V} = \dot{e}^T P e + e^T P \dot{e} < 0 \tag{8}$$

By developing the expression of the estimation error's derivative as in (9), one can rearrange it to get expression (10).

$$\dot{e} = \sum_i w_i(\hat{z})\left(A_i x + B_i u\right) + \sum_i \left(w_i(z) - w_i(\hat{z})\right)\left(A_i x + B_i u\right) - \sum_i w_i(\hat{z})\left(A_i\hat{x} + B_i u + L_i(\hat{y} - y)\right); \tag{9}$$

$$\begin{cases} \dot{e} = \sum_i w_i(\hat{z})\left(A_i \cdot e + B_i \cdot u - L_i \cdot C \cdot e\right) + \Delta; \\ \Delta \overset{not}{=} \sum_i \left(w_i(z) - w_i(\hat{z})\right)\left(A_i \cdot x + B_i \cdot u\right) \end{cases} \tag{10}$$

We have separated the Δ term as it contains a difference between measured weighting function and estimated one, this expression posing a problem in the resolution of the stability requirement. A solution seen generally is to use Lipchitz constants, yet this solution is quite restrictive for systems where the constant is big. Therefore we adopted another method, which is to consider Δ as a perturbation, a value that converges to 0 when t→inf. Therefore the stability problem passes from an asymptotic one to a bounded stability, where the bound is as small as possible. Similar approach has been treated in [9].

From equation (8), making use of (10), one obtains:

$$\dot{V} = e^T \left(\sum_i w_i(\hat{z})\left(A_i - L_i C\right)^T P + \sum_i w_i(\hat{z}) P \left(A_i - L_i C\right) \right) e + \Delta^T P e + e^T P \Delta \tag{11}$$

Moreover, imposing stability with rejection of the virtual perturbation Δ, (12) is reached.

$$\dot{V} + e^T Re - \xi^2 \Delta^T \Delta < 0 \tag{12}$$

Where R>0, usually taken as the identity matrix and ξ is an small as possible positive coefficient. This leading to:

$$e^T \left(\sum_i w_i(\hat{z}) \left[(A_i - L_i C)^T P + P(A_i - L_i C) \right] \right) e + \Delta^T Pe + e^T P\Delta + e^T Re - \xi^2 \Delta^T \Delta < 0; \tag{13}$$

These inequalities still contain variable terms multiplied with other variable terms, so it still has nonlinearities in the last part of the sum. In order to eliminate this, the following known inequality is employed, which is available for all positive λ:

$$\forall\ \lambda > 0, X, Y : X^T Y + Y^T X \le \lambda X^T X + \lambda^{-1} Y^T Y \tag{14}$$

Which in our context means:

$$\Delta^T Pe + e^T P\Delta + e^T Re - \xi^2 \Delta^T \Delta \le \lambda \Delta^T \Delta + \lambda^{-1} e^T PPe + e^T Re - \xi^2 \Delta^T \Delta; \tag{15}$$

Applying this to (13) will get us to:

$$e^T \left(\sum_i w_i(\hat{z}) \left[(A_i - L_i C)^T P + P(A_i - L_i C) \right] + I \right) e + \lambda \Delta^T \Delta + \lambda^{-1} e^T PPe - \xi^2 \Delta^T \Delta < 0; \tag{16}$$

As λ >0 can be any scalar, then one can choose $\lambda = \xi^2$, therefore considering also the convex sum property, the inequality becomes:

$$\sum_i w_i(\hat{z}) \left[(A_i - L_i C)^T P + P(A_i - L_i C) + I + \lambda^{-1} PP \right] < 0; \tag{17}$$

To finalize and to obtain a set of LMI, we use a restrictive hypothesis that each term in the sum is negative, and as w_i are all positive, then (18) is reached.

$$\begin{cases} \text{For } i = 1:8 \\ (A_i - L_i C)^T P + P(A_i - L_i C) + I + \lambda^{-1} PP < 0; \end{cases} \tag{18}$$

Using the notation $Q_i = PL_i$ and making use of the Schur transformation, we thus avoid a Bilinear Matrix problem (BMI).

Also, optionally we can improve the performance by imposing exponential stability, by replacing (12) with (19), where α is the exponential decay rate.

$$\dot{V} + e^T Re - \xi^2 \Delta^T \Delta + 2\alpha V < 0 \tag{19}$$

Of course this would also generate a BMI problem, but this can be easily resolved by choosing manually by trial, a suitable α. So from (18). Adding the exponential stability, we have proven the LMIs (6).

3. The Embedded system and HIL validation platform

The embedded platform is based on two Arduino Mega 2560, as the number of input/outputs of the ATMEGA 2560 is bigger than the standard UNO version. The two microcontrollers are connected in between themselves as in figure 1. The connection can be done both by serial transfer or parallel one (by digital pins). In the application chosen in the current article both configurations give acceptable results, yet it has to be taken into account that floating point data is transferred (stored on 16 bits), so the serial communication has a slower transfer time.

The implementation of the observer on the microcontroller and possible unpredicted errors can be tested by a hardware in the loop (HIL) architecture, where the microcontrollers based platform is connected to a computer that hosts the simulation software (AMESim). The communication protocol between the computer hosting the software and the embedded system has been chosen to be a general one, adaptable for any situation, therefore an Ethernet UDP connection fits the requirement. For this, a 5 ms sampling time can be reached at the 16Mhz of the microcontroller. An important inconvenient of microcontroller systems is that the floating point operations are done at software level, increasing therefore the processing power.

Figure 1. Hardware implementation

The basic functional schematic at the base of the implementation of the embedded HIL validation structure is shown in Figure 2. Although the operating system is Windows, which by default does not assure real time operation, by using the Matlab Real Time Kernel, only the processing threads of interest to the task at hand are kept at a top priority level being overpassed just by core root threads like system inputs and operating system messaging interrupts.

As AMESim needs Simulink for operating a real time data exchange, the creation of a co-simulation AMESim/Simulink is required. As the co-simulation is built, the Real Time Windows Target toolbox in Simulink manages the compilation of the program that will be executed in the Real Time Windows Kernel of Matlab.

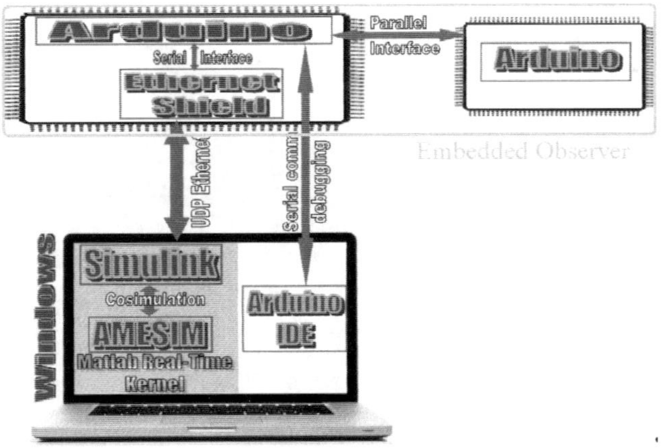

Figure 2. Functional block of the real time system

The coding done in the Arduino IDE is embedded C language, more robust for larger problems than assembly language. A variable discretization is used for the observer, meaning it is recalculated each sampling time inside each microcontroller, this helping in talking the problem of lost samples

The time lost because of the computations done in floating point (which is done at software level) and also due to the fact that TS model is represented as a sum, gave us the possibility to implement a distributed solution, where 2 processors divides the workload at each sampling time. The logical diagram is presented in Figure 3.

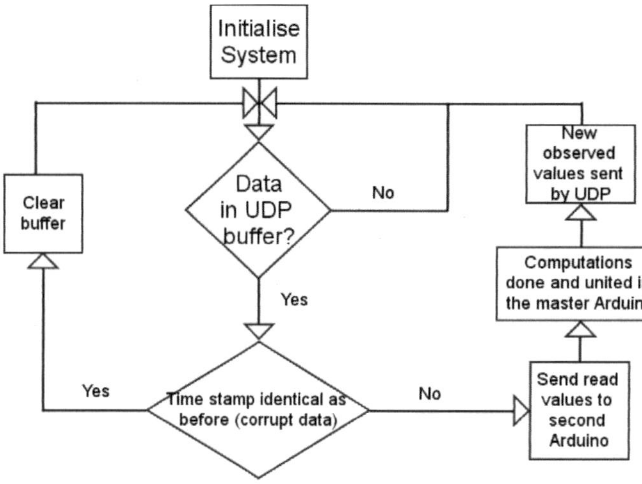

Figure 3. Logical diagram of the Arduino code

4. The application

4.1. System description and modeling

In order to show the functioning principles, an application example has been chosen, specifically a system of three interconnected water tanks. The system is analyzed from a nonlinear perspective to show the potential of using T-S techniques on an embedded platform. As it can be seen in Figure 4, the height of the first tank and the third tank has been considered measurable while the second tank's height is the estimated parameter.

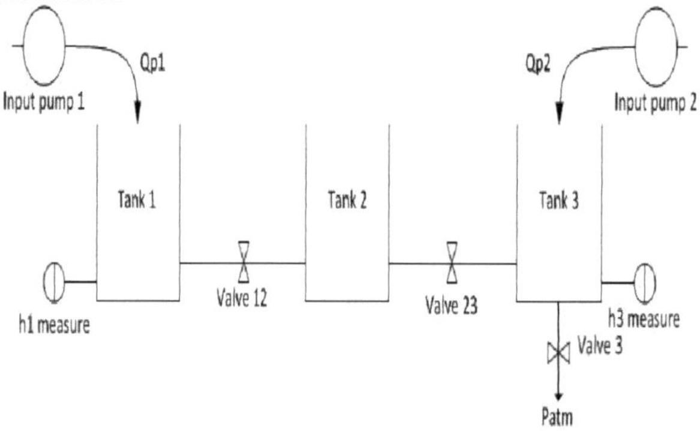

Figure 4. Three tanks system representation

In this system, we have an input pump supplying the first tank Q_{p1} as well as another pump for the third tank Q_{p2}. The last tank presents a valve evacuating water outwards, where atmospheric pressure P_{atm} is encountered. Also, a valve between it and the middle tank is present. The second tank has only the two interconnection valves while the first tank presents only one valve supplying the second tank. A supposition is made, that the flow of the third tank pump is smaller than the first one, thus the levels in tanks will always remain in a descending step, the first the highest level, then the second and the smallest water level in the third at all times.

$$\begin{cases} \frac{d(V_1)}{dt} = Q_{P_1} - Q_{12} \\ \frac{d(V_2)}{dt} = Q_{12} - Q_{23} \\ \frac{d(V_3)}{dt} = Q_{P_2} + Q_{23} - Q_3 \\ V_1 = S_1 h_1;\ V_2 = S_2 h_2;\ V_3 = S_3 h_3 \\ Q_3 = C_{P_3}\sqrt{P_3} \\ Q_{12} = C_{P_{12}} \operatorname{sgn}(P_1 - P_2)\sqrt{|P_1 - P_2|} \\ Q_{23} = C_{P_{23}} \operatorname{sgn}(P_2 - P_3)\sqrt{|P_2 - P_3|} \\ P_1 = \rho g h_1;\ P_2 = \rho g h_2;\ P_3 = \rho g h_3 \end{cases} \tag{20}$$

We will use the notations h_i as heights of liquid in each tank and Cp the flow coefficient of each valve, all valves being considered identical. By neglecting the atmospheric pressure, equations (20) are found. Developing equation (20), the system equations become (21).

$$\begin{cases} \frac{dh_1}{dt} = \frac{1}{S_1}\left(Q_{P_1} - C_{P_{12}} \operatorname{sgn}(h_1 - h_2)\sqrt{\rho g}\sqrt{|h_1 - h_2|}\right) \\ \frac{dh_2}{dt} = \frac{1}{S_2}\left(C_{P_{12}} \operatorname{sgn}(h_1 - h_2)\sqrt{\rho g}\sqrt{|h_1 - h_2|} - C_{P_{23}} \operatorname{sgn}(h_2 - h_3)\sqrt{\rho g}\sqrt{|h_2 - h_3|}\right) \\ \frac{dh_3}{dt} = \frac{1}{S_3}\left(Q_{P_2} + C_{P_{23}} \operatorname{sgn}(h_2 - h_3)\sqrt{\rho g}\sqrt{|h_2 - h_3|} - C_{P_3}\sqrt{\rho g}\sqrt{h_3}\right) \end{cases} \tag{21}$$

Now, the equivalent state-space form can be built, considering $S_1 = S_2 = S_3$ and $CP_{12} = CP_{23} = CP_3$. As it can be seen, the inputs are the pump flows, and the states are the three heights and finally the output is the first and last tanks' height:

$$\begin{cases} u = \begin{bmatrix} Q_{P_1} \\ Q_{P_2} \end{bmatrix} \\ x = \begin{bmatrix} h_1 \\ h_2 \\ h_3 \end{bmatrix} \\ y = \begin{bmatrix} h_1 \\ h_3 \end{bmatrix} = \begin{bmatrix} x_1 \\ x_3 \end{bmatrix} = h(x) \end{cases} \tag{22}$$

$$\begin{cases} \dot{x} = \frac{C_P\sqrt{\rho g}}{S}\begin{bmatrix} \operatorname{sgn}(h_1 - h_2)\sqrt{|h_1 - h_2|} \\ -\operatorname{sgn}(h_1 - h_2)\sqrt{|h_1 - h_2|} + \operatorname{sgn}(h_2 - h_3)\sqrt{|h_2 - h_3|} \\ -\operatorname{sgn}(h_2 - h_3)\sqrt{|h_2 - h_3|} + \sqrt{h_3} \end{bmatrix} + \begin{bmatrix} \frac{1}{S} & 0 \\ 0 & 0 \\ 0 & \frac{1}{S} \end{bmatrix} u \\ y = \begin{bmatrix} x_1 \\ x_3 \end{bmatrix} = \begin{bmatrix} 1 & 0 & 0 \\ 0 & 0 & 1 \end{bmatrix} x \end{cases} \tag{23}$$

As it can be seen, the nonlinearities are present as square roots, that have no differentiability in 0, yet for the regime in which we work, we will not reach this point.

4.2. Takagi-Sugeno transformation

In order to obtain the state observer, it is first needed to modify the state space form, so that it can be used to write the Takagi-Sugeno (TS) representation. So equation (23) is written as (24):

$$\begin{cases} \dot{x} = -\frac{C_P\sqrt{\rho g}}{S}\begin{bmatrix} z_1 & -z_1 & 0 \\ -z_1 & z_1+z_2 & -z_2 \\ 0 & -z_2 & z_2+z_3 \end{bmatrix} x + \frac{1}{S}\begin{bmatrix} 1 & 0 \\ 0 & 0 \\ 0 & 1 \end{bmatrix} u \\ y = x_1 = \begin{bmatrix} 1 & 0 & 0 \\ 0 & 0 & 1 \end{bmatrix} x \end{cases} \tag{24}$$

Where:

$$z = \begin{bmatrix} z_1 \\ z_2 \\ z_3 \end{bmatrix} = \begin{bmatrix} \operatorname{sgn}(x_1 - x_2)\frac{1}{\sqrt{|x_1 - x_2|}} \\ \operatorname{sgn}(x_2 - x_3)\frac{1}{\sqrt{|x_2 - x_3|}} \\ \frac{1}{\sqrt{x_3}} \end{bmatrix} \tag{25}$$

For the sector nonlinearity transformation we have to establish the conditions for the heights, where some additional constraints are added so that singular values are eliminated:

$$\begin{cases} 0,05 \le h_1 \le 0,6m \\ 0,05 \le h_2 \le 0,6m \\ 0,05 \le h_3 \le 0,6m \end{cases} \tag{26}$$

The three valves have similar characteristics, and the flow on the left pump is bigger, therefore the only possibility when the water levels would be equal is when the pump starts or when the tanks are empty, in which case the observer is not useful. Therefore, it is impossible that the difference between levels of the 3 tanks will change their sign. Even if this had happened, one could resolve this problem, by separating the system into multiple systems that switch according to the value of the difference between heights. Once all the observers are built, the one that will be active is chosen in a switching manner. Of course the main regime is the case when all the square roots are different from 0.

The parameters for the system are:

$$\begin{aligned} &Cp = 1.57 \cdot 10^{-6} m^3.s^{-1}.Pa^{-1/2} \\ &\rho = 1000 kg.m^{-3} \\ &g = 9.81 kg.m.s^{-2} \\ &S = 0.0164 m^2 \\ &Qp = 10^{-4} m^3.s^{-1} \end{aligned} \tag{27}$$

$$\begin{aligned} &z_{1_{\min}} = 0,5 \ ; \ z_{1_{\max}} = 2,5 \ ; \ z_{2_{\min}} = 2 \ ; \\ &z_{2_{\max}} = 10 \ ; \ z_{3_{\min}} = 1,6 \ ; \ z_{3_{\max}} = 2,6 \end{aligned} \tag{28}$$

5. Observer Implementation

5.1. The AMESim System

For the validation part, one has to first construct the AMESim model, by adding all the elements, configuring their parameters, and then the co-simulation block with Simulink needs to be added in Amesim as well as in figure 6. For the Simulink part (Figure 5), Matlab 2012a have been used. The Amesim version is R12.

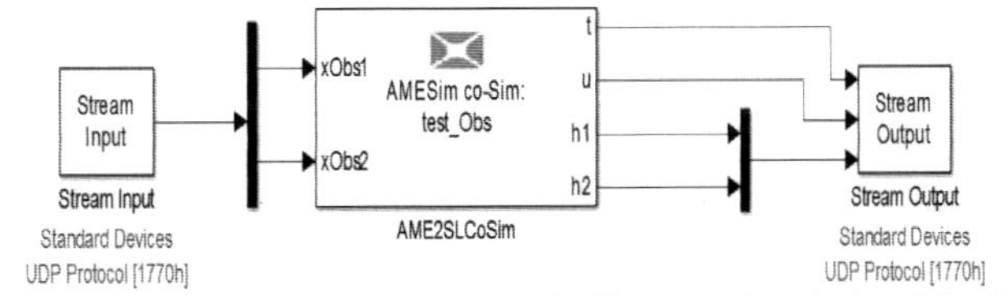

Figure 5. Simulink diagram including co-simulation block

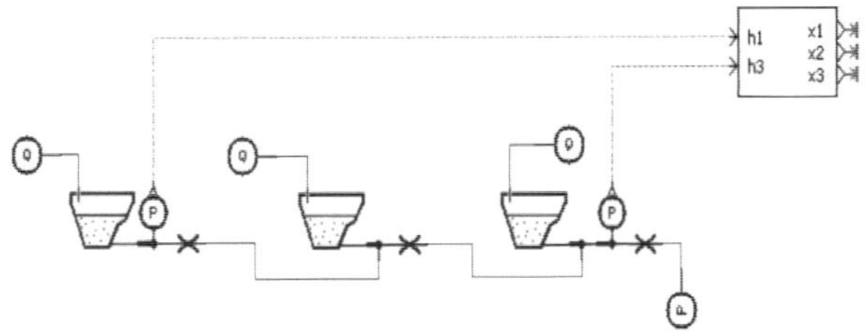

Figure 6. AMESim implementation of the three tank system

As the connection between Amesim and Simulink is done, by using Matlab real time workshop we add the input/output blocks that will connect via Ethernet to the microcontroller. The real time compilation is then done, and the whole system is run from Simulink.

5.2. The real time simulation

By means of the experimental setup described previously, a performance analysis is done in real time, and the parameters are configured for optimal results.

As it can be observed in figures 7 and 8, the estimation of the water level in the second tank stabilizes in about 10 seconds, which is an adequate time response. Also the overshoot of the estimation is in an acceptable interval.

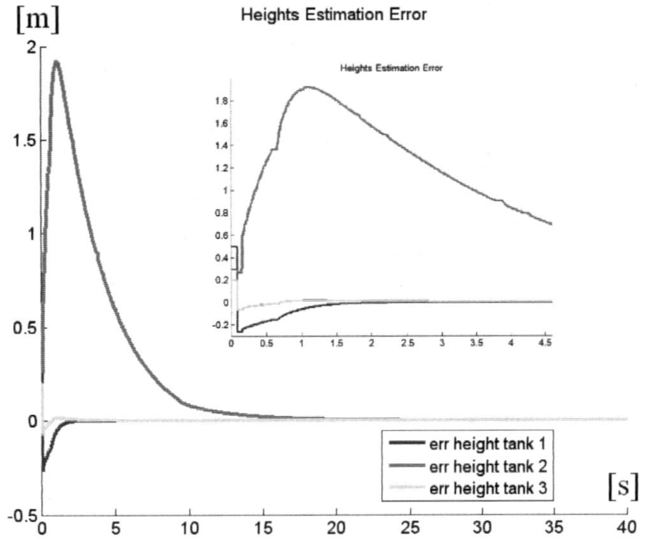

Figure 7. Estimation error dynamics of the 3 heights (tank 1, 2 and 3)

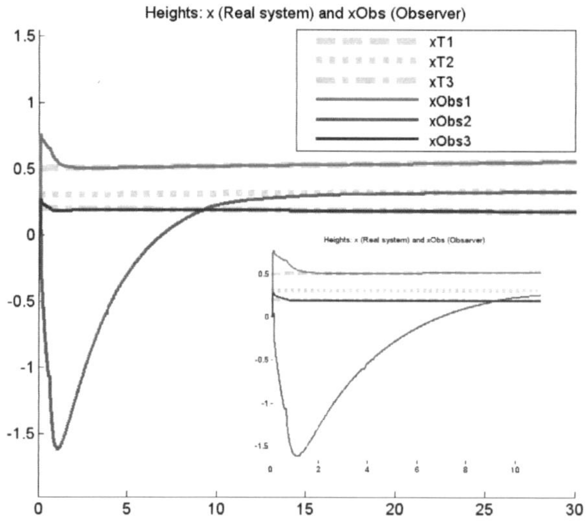

Figure 8. Real/Estimated water levels evolution

It is important to notice that perturbation added to the communication in a forcedly manner, can be seen to affect the system like in figure 7 at second 0.7 (in the zoomed pane), but performance remains good overall, thus proving its robustness to data loss.

5.3. Validation on a real system

Once the tested had been finished the embedded observer has been installed on a real 3 tank platform AMIRA DTS200 (Figure 9).

Figure 9. AMIRA DTS 200

After the implementation, also adding some filters on the inputs and outputs, the estimation error stabilizes in 8 seconds as can be seen in the following figure 10:

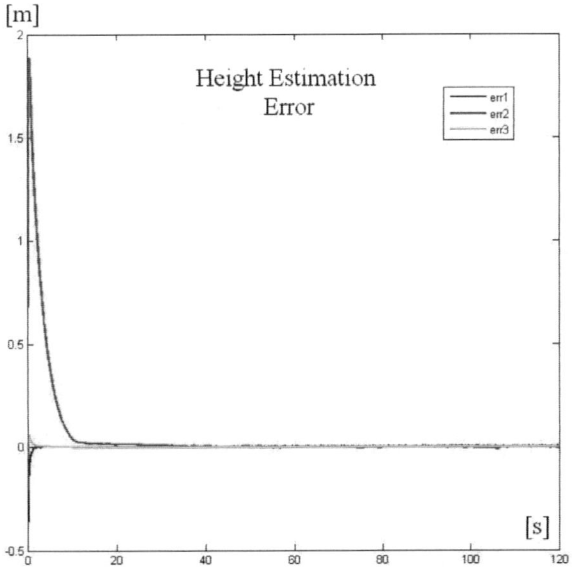

Figure 10. AMIRA DTS 200 estimation error dynamics in time

By introducing short perturbations, into the system, by means of purge valves openings, the observer still manages to regain stability (as in figure 11):

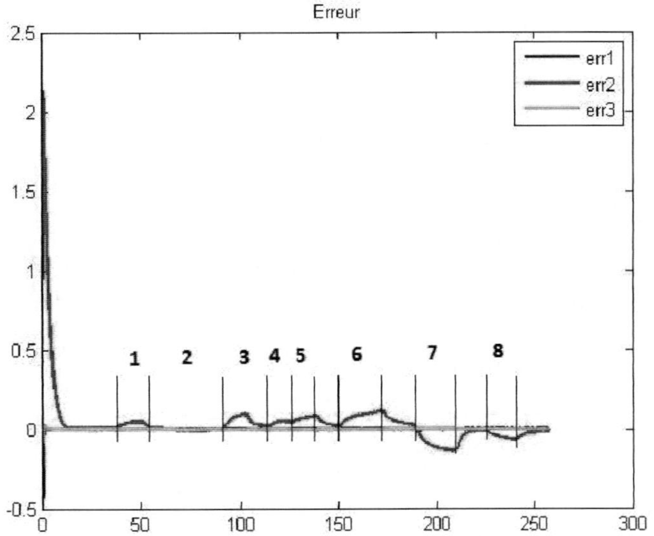

Figure 11. AMIRA DTS 200

Where the numbered areas are:

1) Leakage in tank 1 (fault 1)
2) Leakage in tank 2 (fault 2)
3) Leakage in tank 3 (fault 3)
4) Pump of tank 1 does not work (no input flow) (fault 4)
5) Pump of tank 3 does not work (no input flow) (fault 5)
6) Valve between tank 2 and tank 3 is blocked (fault 6)
7) Valve between tank 3 and atmosphere is blocked (fault 7)
8) Valve between tank 1 and tank 2 is blocked (fault 8).

6. Conclusions and Perspectives

An embedded nonlinear observer was developed using the Takagi Sugeno approach and applied to a three tank system with a prior real time HIL validation. As previously seen, the observer's performances on the HIL testing bench reproduce the behaviour of the real system thus reducing implementation time and also avoiding potential dangerous faults on the real system. The testing also takes into consideration communication losses and memory and computation timing errors. Of course the testing itself can be reproduced for any process that is supported by the simulation platform. The developed observer was applied to a real three tank system to confirm the results. Also another objective of the paper was to show the capability of a T-S system to be represented on small scaled microcontroller platforms. By doing the parallel calculus in between two microcontrollers, the complexity of the system can be augmented even further.

7. Acknowledgments

This work was carried out in the framework of PN-7-022-BE i-MOCCA (Interregional Mobility and Competence Centres in Automation), an European territorial cooperation project part-funded by the European Regional Development Fund (ERDF) through the INTERREG IV A 2 Seas Programme and the Ministry of Education and Research, France.

8. References

[1] Chen M, Edwards D, Boehmer E, Eller N, Slack J, Speck C, Brown S, Williams H, Wilson S, Gillum C, Lewin G, Sherriff M, Garner G, Designing a spatially aware and autonomous quadcopter, *Systems and Information Engineering Design Symposium (SIEDS), 2013 IEEE* , pp.213,218, 26-26 April 2013

[2] Tanaka K and Wang H, Fuzzy Control Systems Design and Analysis: A Linear Matrix Inequality Approach, *John Wiley and Sons*, 2001.

[3] Bourdon T, Saussol L and Varoquié B, Integration of Physical AMESim® Engine Model in Hardware in the Loop Environment, Dedicated to Engine Control Unit Testing, *SAE Technical Paper,* January 2007, doi:10.4271/2007-01-1300.

[4] Feingesicht M, Olteanu S, Aitouche A and Belkoura L, HIL validation of an embedded system acting as a nonlinear Takagi-Sugeno state observer on an Arduino board , P*roceeding of the 6th European Embedded Design in Education and Research,* pp. 250-254, Milano, 11-12 September, 2014.

[5] Lendek Z, Guerra T, Babuska R and de Schutter B, Stability Analysis and Nonlinear Observer Design Using Takagi-Sugeno Fuzzy Models, *Springer*, 2010.

[6] Ghorbel H, El Hajjaji A, Souissi M and Chaabane M, Robust Tracking Control for Takagi-Sugeno Fuzzy Systems With Unmeasurable Premise Variables: Application to Tank System, *Journal of Dynamic Systems, Measurement and Control*, vol. 136, July. 2014.

[7] Rincon-Pasaye J, R Martinez-Guerra and A Soria-Lopez, Fault diagnosis in nonlinear systems: An application to a three-tank system, *American Control Conference ,* pp. 2136 – 2141, Seattle, June 2008.

[8] Herrera-Lopez E, Castillo-Toledo B, Ramirez-Cordova J and Ferreira E, Takagi-Sugeno Fuzzy Observer for a Switching Bioprocess: Sector Nonlinearity Approach, *New Developments in Robotics Automation and Control*, InTech, chap. 8, pp. 155, 2008.

[9] Yacine Z, Ichalal D, Oufroukh N, Mammar S, Vehicle nonlinear observer for state and tire-road friction estimation, *(ITSC), 2013 16th International IEEE Conference on Intelligent Transportation Systems* , pp.2181,2186, 6-9 Oct. 2013

A conformal mapping approach to a root-clustering problem

Gennady I. Melnikov

Prof., Department of Theoretical and Applied Mechanics, University ITMO, St.Petersburg, Russia

E-mail: melnikov.ifmo@yandex.ru

Nataly A. Dudarenko

Ph.D, Department of Management Systems and Informatics, University ITMO, St.Petersburg, Russia

E-mail: dudarenko@yandex.ru

Vitaly G. Melnikov

Head of Department of Theoretical and Applied Mechanics, University ITMO, St.Petersburg, Russia

E-mail: melnikov@mail.ifmo.ru

Abstract. This paper presents a new approach for matrix root-clustering in sophisticated and multiply-connected regions of the complex plane. The parametric sweeping method and a concept of the closed forbidden region covered by a set of modified three-parametrical Cassini regions are used. A conformal mapping approach was applied to formulate the main results of the paper. An application of the developed method to the problem of matrix root-clustering in a multiply connected region is shown for illustration.

1. Introduction

The properties of control systems depend on the location of roots of a system matrix on the complex plane. In this connection the problem of root-clustering is topical for control systems analysis. Some control techniques like state-space control [1–5] require the roots of a linear part of the system to be localised in a defined regions of the complex plane. The most-used approach this problem is based on the generalized Lyapunov matrix inequality. The necessary and sufficient conditions for matrix root-clustering in Γ-transformable and Ω- transformable algebraic regions obtained in the works of Gutman and colleagues [6] made the fundamental contributions in this direction. The problem of root clustering for robust systems using the generalized Lyapunov matrix inequality approach was investigated in many works, the recent of them are: [7–16].

This paper is devoted to a problem of matrix root-clustering in the sophisticated regions given as an intersection of several algebraic regions of the fourth and smaller order. Such regions can have one or several boundaries. The boundaries may be smooth or piecewise smooth, convex or nonconvex; the boundaries may be formed by segments of algebraic curves or by curves enveloping a continual set of curves. The region may be multiply- or simply- connected, nonconvex or convex. We consider more general statement of the problem than in the literature. The main attention is focused on a complement of the root-clustering region to $\mathbb{C}$; these regions are called the *closed forbidden region.* The forbidden region we approximate (sweep) by a set of algebraic subregions; we offer to use a set of simply-connected and multiply-connected three-parametric regions of the fourth order called the *modified Cassini oval regions* as these algebraic subregions.

In the introduced terms the root-clustering region is an intersection of the complements of a set of the forbidden subregions. Necessary and sufficient conditions of matrix spectrum localization in the specified region are obtained as a union of a set of conditions of spectrum location in outside of each forbidden subregion. In this paper we use the conformal mapping approach instead of generalized Lyapunov matrix inequality one to obtain the results.

The paper is laid out as follows. First, general information about Cassini ovals is given and some modifications of Cassini ovals are suggested. Thereafter, a sweeping technique in a root-clustering problem is presented and a conformal mapping of Cassini regions and matrix root-clustering approach are described. The paper is finished with some concluding remarks and acknowledgments.

2. Cassini ovals

In this section we recall some general information about Cassini ovals and introduce some modifications. We present the regions bounded by the modified Cassini ovals as the convenient subregions for covering (sweeping) of forbidden regions.

As it is known, the equation of Cassini ovals includes two parameters (a, c) determining the shape of the ovals.

$$(x^2+y^2)^2 - 2c^2(x^2-y^2) = a^4 - c^4 \tag{1}$$

For example, Fig. 1 illustrates the Cassini ovals at $a = [0.6c, 0.8c, c, 1.2c,\ 1.4c, 1.6c, ...]$. In [17] one can find the recent example of application of Cassini ovals in the control theory (for the multi-agent systems).

Let us introduce the modified Cassini ovals, tree-parametric ovals defined by the equation

$$\Gamma(a,c,\mu) = \{s = x + \mathrm{j}y : ((x+\mu)^2+y^2)^2 + + 2c((x+\mu)^2-y^2) + c^2 - a^2 = 0\},$$

where $a > 0, c \in [-a/2, 0] \cup [a, \infty)$. These ovals are oriented along an imaginary axis at $c > 0$ or along a real axis at $c < 0$. A third parameter μ provides the parallel shift of the ovals along the real axis. Note that we get the parallel shift to the left-hand side at $\mu > 0$ and the parallel shift to the right-hand side at $\mu < 0$.

For numerical calculations (2) can be rewritten as

$$\Gamma = \{x, y : y = (\pm)\sqrt{c - (x+\mu)^2 \pm \sqrt{a^2 - 4c(x+\mu)^2}}\}.$$

Note that it is necessary to consider four different combinations of values of the double marks.

The restriction $c \in [-a/2, 0] \cup [a, \infty)$ means that among different forms of Cassini lines we shall consider only ovals.

Equation (2) in terms of a complex variable s has the form

$$\Gamma = \{s \in \mathbb{C} : ((\bar{s}+\mu)^2 + c)((s+\mu)^2 + c) - a^2 = 0\}, \tag{2}$$

where $a > 0, c \in [-a/2, 0] \cup [a, \infty)$.

At $c = 0$ we have the unit circle. At $c < 0$ we obtain the oval with semi-axis $(\sqrt{a-c})$ and $(\sqrt{a}+c)$, extended along the real axis. At $c = -a/2$ the curvature of the oval in the top and

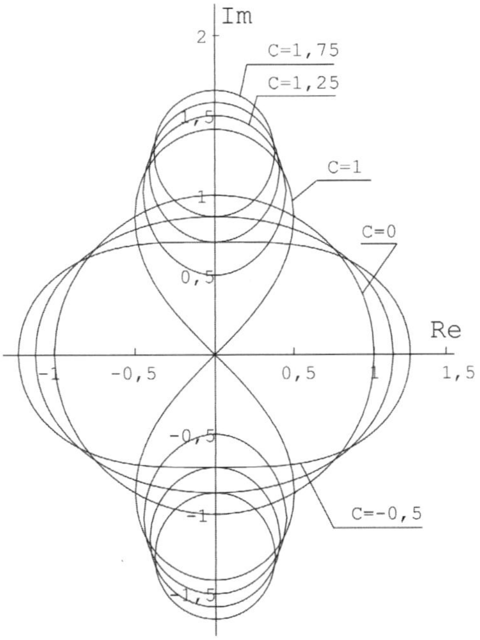

Figure 1. Cassini ovals at $a = [0.6c, 0.8c, c, 1.2c, 1.4c, 1.6c, ...]$

in the bottom points are equal to zero. At $c = a$ we have the lemniscate. Finally, at any values $c > a$ we have the pair of ovals symmetrically located at both sides of the real axis having the height of $(\sqrt{c+a} - \sqrt{c-a})$ and having the maximal width of $(a/\sqrt{c})$.

In this subsection we give some definitions.

Suppose the closed region H satisfies the inequality

$$\bar{H}(a, c, \mu) = \{s \in \mathbb{C} : \quad |(s + \mu)^2 + c|^2 - a^2 \leqslant 0\}, \tag{3}$$

where $a > 0, c \in [-a/2, 0] \cup [a, \infty)$. Then the region $\bar{H}(a, c, \mu)$ is called the *modified three-parametric Cassini oval region*. It is clear that curve (2) is the boundary of $\bar{H}$.

Further, the complement of H to $\mathbb{C}$ is called the *external oval region* and denoted by G. The region G satisfies the inequality

$$G(a, c, \mu) = \{s \in \mathbb{C} : f(s) = a^2 - |(s + \mu)^2 + c|^2 < 0\}, \tag{4}$$

where $a > 0, c \in [-a/2, 0] \cup [a, \infty)$.

3. A sweeping technique in a root-clustering problem

In the literature a problem of matrix root - clustering is usually considered as a problem of a membership of all roots to a specified open region D of the complex plane $\mathbb{C}$. Let us consider another treatment of this problem.

3.1. A Modified treatment of a matrix root - clustering problem

Let the closed region $\bar{S}$ be the complement to $\mathbb{C}$ of D, i.e., $\bar{S} = \mathbb{C} \backslash D$; then the region $\bar{S}$ is called *the forbidden region.*

Let us consider 2 cases.

Journal of Physics: Conference Series **570** (2014) 022004 doi:10.1088/1742-6596/570/2/022004

Case 1 Suppose the region $\bar{S}$ is defined as below:

$$\bar{S} = \cup \bar{S}_\alpha \quad \forall \quad \alpha = \overline{1, \nu}, \tag{5}$$

where $\{\bar{S}_\alpha, \quad \forall \alpha = \overline{1, \nu}\}$ is a *final* set of ν intersecting or nonintersecting subregions; the envelope of this set forms the border of $\bar{S}$. Then by definition put

$$\{D_\alpha : \quad D_\alpha = \mathbb{C} \backslash \bar{S}_\alpha \supseteq D\} \quad \forall \quad \alpha = \overline{1, \nu}, \tag{6}$$

where the regions D_α are called *the expanded root-clustering regions*. Note that $D_\alpha \supseteq D$; this follows from $\bar{S}_\alpha \subseteq \bar{S} \Rightarrow D_\alpha = \mathbb{C} \backslash \bar{S}_\alpha \supseteq \mathbb{C} \backslash \bar{S} = D$.

Case 2 Let $\{\bar{\Omega}(\mu), \mu \in [0, \tau]\}$ be a *continual* set of the subregions covering the region $\bar{S}$

$$\bar{S} = \cup \bar{\Omega}(\mu) \quad \forall \quad \mu \in [0, \tau] \tag{7}$$

Then by definition, put

$$\{D(\mu) : \quad D(\mu) = \mathbb{C} \backslash \bar{\Omega}(\mu)\} \quad \forall \quad \mu \in [0, \tau], \tag{8}$$

where $D(\mu) \supseteq D \quad \forall \mu \in [0, \tau]$ is the continual parametric set of the expanded root-clustering regions.

Let the forbidden region $\bar{S}$ for the matrix spectrum $s(A)$ be a union of the regions (5); then the root-clustering region $D = \mathbb{C} \backslash \bar{S}$ satisfies the equality

$$D = D^*, \, D^* = \cap D_\alpha = \cap (\mathbb{C} \backslash \bar{S}_\alpha), \, \alpha = \overline{1, \nu} \tag{9}$$

i.e., D is the intersection of the set of the expanded root - clustering regions (6).

Let the forbidden region S be a union of the parametric continual set (7); then the root-clustering region D satisfies the equality

$$D = D^*, \, D^* = \cap D(\mu), \, D(\mu) = \mathbb{C} \backslash \bar{\Omega}(\mu). \tag{10}$$

3.2. Conformal mapping of Cassini regions

The important property of the modified Cassini ovals making them suitable for application in root-clustering problems is in the fact that the function of a complex variable

$$z = a[(s + \mu)^2 + c]^{-1}, \quad c > a > 0, \quad \mu \geqslant 0, \tag{11}$$

conformally maps the closed Cassini oval region $\bar{H}(a, c, \mu)$ onto the complement of the closed unit central disk $\{z : |z| \geqslant 1\}$. The reader will easily check that substituting (11) for (3), (4). Note also that function (11) maps the complementary region $G(a, c, \mu) = \mathbb{C} \backslash \bar{H}(a, c, \mu)$ into the open unit disk $\{z : |z| < 1\}$. The bilinear function

$$w = (z + 1)(z - 1)^{-1}, \tag{12}$$

conformally maps the unit central disk $\{z : |z| < 1\}$ and the exterior disk $\{z : |z| \geqslant 1\}$ respectively into the left half-plane and onto the right half-plane. Substituting (12) for (11), we get

$$w(a, c, \mu) = (a + c + (s + \mu)^2)(a - c - (s + \mu)^2)^{-1}. \tag{13}$$

This function conformally maps the open region $G(a, c, \mu)$ and the closed region $\bar{H}(a, c, \mu)$ respectively into the open left half-pane and onto the right half plane.

3.3. Matrix root-clustering

Suppose the forbidden region $\bar{S}$ is covered by the final set of regions (3), i.e.,

$$\bar{S} = \cup \bar{H}(a_\alpha, c_\alpha, \mu_\alpha), \quad \forall \alpha = \overline{1, \nu}, \tag{14}$$

where

$$\bar{H}(a_\alpha, c_\alpha, \mu_\alpha) = \{s \in \mathbb{C} : |(s + \mu_\alpha)^2 + c_\alpha|^2 - a_\alpha^2 \leqslant 0\} \tag{15}$$

The following functionally transformed matrices correspond to functions (11) and (13):

$$Z_\alpha = a_\alpha((A + \mu_\alpha E)^2 + c_\alpha E)^{-1} \forall \alpha \in [1, \nu] \tag{16}$$

$$W_\alpha = ((a_\alpha + c_\alpha)E + (A + \mu_\alpha E)^2)((a_\alpha - c_\alpha)E - (A + \mu_\alpha E)^2)^{-1} \tag{17}$$

Using the property of matrix eigenvalues we get the following conditions. All roots of a real matrix A are clustered in $D = \mathbb{C} \backslash \bar{S}$, where $\bar{S}$ is the forbidden region (14), (15), if and only if all eigenvalues of all functionally transformed matrices (16) are clustered in the unit central disk, or if all eigenvalues of all matrices (17) are clustered in the left half-plane.

This leads to the Hurwitz conditions for matrices (17) or Lyapunov inequality for them; or this leads to the conditions of Schur-Cohn for matrices (16).

To illustrate the method we consider an example adopted from a band-filtration problem. Let a band filter has two suppression frequency bands (low and high). In the presented above terms these bands form the multiply connected forbidden region. Now we will obtain the algebraic conditions for matrix root-clustering in outside of this forbidden region. First we approximate the forbidden bands by a set of modified three-parametric Cassini oval regions. We approximate the low frequency band by a set of the simply-connected oval regions and approximate the high band by another set of multiply-connected oval regions As a whole we approximate the forbidden region by two sets of simply-connected and conjugate oval regions with corresponding values of the parameters $\{a, c, \mu\}$. Secondly in the light of the obtained conditions all roots of matrix A are clustered in outside of this forbidden region, if and only if the functionally transformed matrix (17) is Hurwitz for all values of the parameters $a_\alpha, c_\alpha, \mu_\alpha$ corresponding to the defined sets.

4. Conclusion

This paper presents a new conditions for matrix root-clustering in sophisticated and multiply-connected regions of the complex plane using a conformal mapping approach and a parametric sweeping method by a set of modified three-parametrical Cassini regions. An application of the developed method to the problem of matrix root-clustering in a multiply connected region is shown for illustration.

4.1. Acknowledgments

The research was supported by the State Financial Support of the Leading National Universities, no. 074-U01, the Ministry of Education and Science of the Russian Federation, project no. 14.Z50.31.0031, and by Committee of Science and Higher School of St. Petersburg Government.

References

[1] S. Tararykin and V. Tyutikov, "A robust model control for dynamic systems," *Automation and Remote Control*, vol. 63, no. 5, pp. 730–742, 2002.

[2] H. Schulte, "Approximate modeling of a class of nonlinear mechanical oscillators using fuzzy systems and its application to control design," *IEEE/ASME International Conference on Advanced Intelligent Mechatronics, AIM*, vol. 2, pp. 1058–1064, 2005.

[3] H. Schulte and K. Guelton, "Modelling and simulation of two-link robot manipulators based on takagi sugeno fuzzy descriptor systems," *Proceedings of the IEEE International Conference on Industrial Technology*, pp. 2692–2697, 2006.

[4] G. Leonov, "On the stability in the large of nonlinear systems in the critical case of two zero roots," *Journal of Applied Mathematics and Mechanics*, vol. 45, no. 4, pp. 557–560, 1981.

[5] N. Dudarenko, O. Slita, and A. Ushakov, "Algebraic conditions of generalized modal control," *IFAC Proceedings Volumes*, vol. 7, no. 1, pp. 150–155, 2012.

[6] S. Gutman and E. I. Jury, "A general theory for root-clustering in subregions of the complex plane," *IEEE Trans. Automat. Contr.*, vol. 26, pp. 853–863, 1981.

[7] O. Bachelier and D. Mehdi, "Robust matrix du-stability analysis," *Int. J. of Robust and Nonlinear Control*, vol. 13, pp. 533–558, 2003.

[8] J. Bosche, O. Bachelier, and D. Mehdi, "An approach for robust matrix root-clustering analysis in a union of regions," *IMA Journal of Mathematical Control and Information*, vol. 22, pp. 227–239, 2005.

[9] O. Rejichi, O. Bachelier, M. Chaabane, and D. Mehdi, "Robust root-clustering analysis in a union of subregions for descriptor systems," *IET Control Theory Appl.*, vol. 2(7), pp. 615–624, 2008.

[10] Y. Sun, R. Gau, and J. Hsieh, "Simple criteria for sector root clustering of uncertain systems with multiple time delays," *Chaos, Solitons and Fractals*, vol. 39 (1), pp. 65–71, 2009.

[11] V. Spitsa, A. Alexandrovitz, and E. Zeheb, "Robust approach to a small-signal stability analysis of power systems," *WSEAS Transactions on Circuits and Systems*, vol. 4 (11), pp. 1449–1456, 2005.

[12] O. Bachelier and D. Mehdi, "Robust matrix root-clustering analysis through extended kyp lemma," *SIAM Journal on Control and Optimization*, vol. 45 (1), pp. 368–381, 2006.

[13] V. G. Melnikov, "A sweeping method for matrix root clustering," in *Prepr. of the 18th International Federation of Automatic Control World Congress*, IFAC. Milan: Elsevier, 2011, pp. 168–171.

[14] V. Melnikov and N. Dudarenko, "Method of forbidden regions in the dynamic system matrices root clustering problem," *Automation and Remote Control*, vol. 75, no. 8, pp. 1393–1401, 2014.

[15] V. G. Melnikov, "About root-clustering in sophisticated regions," in *14th WSEAS international conference on systems*, vol. 1: Latest trends on systems, Corfu, 2010, pp. 297–300.

[16] S. Vassilyev and A. Kosov, "Analysis of hybrid systems' dynamics using the common lyapunov functions and multiple homomorphisms," *Automation and Remote Control*, vol. 72, no. 6, pp. 1163–1183, 2011.

[17] B. Sharma, S. Ramakrishnan, and M. Kumar, "Cyclic pursuit in a multi-agent robotic system with double-integrator dynamics under linear interactions," *Robotica*, vol. 31, no. 7, pp. 1037–1050, 2013.

State control of discrete-time linear systems to be bound in state variables by equality constraints

Anna Filasová, Dušan Krokavec and Vladimír Serbák

Department of Cybernetics and Artificial Intelligence, Faculty of Electrical Engineering and Informatics, Technical University of Košice, Letná 9, 042 00 Košice, Slovakia

E-mail: `anna.filasova@tuke.sk; dusan.krokavec@tuke.sk; vladimir.serbak@tuke.sk`

Abstract. The paper is concerned with the problem of designing the discrete-time equivalent PI controller to control the discrete-time linear systems in such a way that the closed-loop state variables satisfy the prescribed equality constraints. Since the problem is generally singular, using standard form of the Lyapunov function and a symmetric positive definite slack matrix, the design conditions are proposed in the form of the enhanced Lyapunov inequality. The results, offering the conditions of the control existence and the optimal performance with respect to the prescribed equality constraints for square discrete-time linear systems, are illustrated with the numerical example to note effectiveness and applicability of the considered approach.

1. Introduction

The control of the plant operates in such a way that some variables reflect certain physical limits in the prescribed demarcation occurs in practice very often. The underlying stimulus relies on the fact that the industrial systems may operate at acting conditions which are defined not only by given amplitude limitations but also by the ratios of certain process variables. With regard to this fact, the constraint control conceptions in such a way generalized have widely theoretical significance and practical sense.

Since it is possible to design the controllers that stabilize systems and concurrently force the closed-loops some kind of bounds [1], [13], to cope this problem efficiently exploiting the technique based on linear matrix inequalities (LMI), the control design task has to be formulated using the same approaches dealing with the state variable ratios as well as with the prescribed system constraints [15].

One way to implement above given task is to reformulate the control action manner as the stabilized control which also ensures that the closed-loop system state variables will satisfy the prescribed equality constraints in all time instants except the initial one. Because systems with the state equality constraints generally do not be deemed as descriptor systems [7], the design task has to be interpreted as a singular problem and the associated methods need to be developed to compute the controller parameters. Following the idea of the linear quadratic control, the technique which makes use of the equality constraints formulation for the discrete-time linear systems has been introduced in [8], and was extensively applied in the reconfigurable control design tasks [6], [9]. The reflection of this technique to the stochastic systems constrained in the state variables can be found in [10], and the design conditions based on the system H_∞ norm properties were presented in [5], [11].

Because the above mentioned control algorithms as a rule use the static decoupling principle in setting the system working points, the prescribed values of equality constraints in the enforced mode are slithered in an additive offset [5]. To become values of the equality constraints linearly dependent on the desired system output values, the subject of the paper is to design the discrete version of PI controller in the structure which ensures the system stability and sets up the required linear dependency of the equality constraints.

Reflecting the fact that the design task is singular (i.e., the direct extension of the Lyapunov principle could give very conservative solutions since resulting LMIs are ill conditioned), the enhanced form of Lyapunov inequality [4] is adapted to obtain the regular set of LMIs. Using the parametrization approach which allows to convert the PI controller synthesis into the set of LMIs, the controller structure takes the optimal performance with respect to the prescribed equality constraints. It should be noted that the control law parameters could be altered while the desired control input values are changed.

The outline of this paper is as follows. Ensuing the introduction given in Sec. I, the system model and the control law structures are briefly described in Sec. II, while Sec. III places the suggested approaches within the context of the existing design requests and the basic preliminaries. In Sec. IV the proposed control design conditions are derived and, subsequently, these results are illustrated in Sec. V by a numerical example. Finally, Sec. VI presents some concluding remarks.

Throughout the paper the notations is narrowly standard in such way that $\boldsymbol{x}^T$, $\boldsymbol{X}^T$ denotes the transpose of the vector $\boldsymbol{x}$ and the matrix $\boldsymbol{X}$, diag[$\cdot$] enters up a block diagonal matrix, rank($\cdot$) remits the rank of a matrix, $\boldsymbol{X}^{\ominus 1}$ is Moore-Penrose inverse of the non-square matrix $\boldsymbol{X}$, for a square matrix $\boldsymbol{X} < 0$ means that $\boldsymbol{X}$ is a symmetric negative definite matrix, the symbol $\boldsymbol{I}_n$ indicates the n-th order unit matrix, Z_+ is the set of all positive integers, $I\!R$ notes the set of real numbers, and $I\!R^n$, $I\!R^{n\times r}$ refer to the set of all n-dimensional real vectors and $n \times r$ real matrices, respectively.

2. Problem formulation

Through this paper the task is concerned with design of the state feedback constrained in the state variables and controlling the square discrete-time linear dynamic systems given by the set of state equations

$$\boldsymbol{q}(i+1) = \boldsymbol{F}\boldsymbol{q}(i) + \boldsymbol{G}\boldsymbol{u}(i), \tag{1}$$

$$\boldsymbol{y}(i) = \boldsymbol{C}\boldsymbol{q}(i), \tag{2}$$

where $\boldsymbol{q}(i) \in I\!R^n$, $\boldsymbol{u}(i) \in I\!R^r$, $\boldsymbol{y}(i) \in I\!R^m$ are vectors of the state, input and output variables, respectively, $\boldsymbol{F} \in I\!R^{n\times n}$, $\boldsymbol{G} \in I\!R^{n\times r}$, $\boldsymbol{C} \in I\!R^{m\times n}$ are real matrices and $i \in Z_+$. It is considered that the pair $(\boldsymbol{F}, \boldsymbol{G})$ is controllable.

In practice [2], [3], ratio control can be used to maintain the relationship between two closed-loop state variables, defined as

$$\frac{q_h(i+1)}{q_k(i+1)} = a_h \Rightarrow q_h(i+1) - a_h q_k(i+1) = 0 \tag{3}$$

for all $i \in Z$, or more compactly as

$$\boldsymbol{e}_{hk}^T \boldsymbol{q}(i+1) = 0, \tag{4}$$

$$\boldsymbol{e}_{hk}^T = \begin{bmatrix} 0_1 & \cdots & 0_{h-1} & 1_h & 0_{h+1} & \cdots & 0_{k-1} & -a_k & 0_{k+1} & \cdots & 0_n \end{bmatrix}. \tag{5}$$

Generalization of the above task formulation means the design of the stable closed-loop system with the linear state feedback control of the form

$$\boldsymbol{u}(i) = -\boldsymbol{K}\boldsymbol{q}(i), \tag{6}$$

where $\boldsymbol{K} \in \mathbb{R}^{r \times n}$ is the controller feedback gain matrix and the design constraint is given in the matrix equality form

$$\boldsymbol{E}\boldsymbol{q}(i+1) = \boldsymbol{E}(\boldsymbol{F} - \boldsymbol{G}\boldsymbol{K})\boldsymbol{q}(i) = \boldsymbol{E}\boldsymbol{F}_c\boldsymbol{q}(i) = 0 \tag{7}$$

with $\boldsymbol{E} \in \mathbb{R}^{k \times n}$, rank$\boldsymbol{E} = k \le r$, and

$$\boldsymbol{F}_c = \boldsymbol{F} - \boldsymbol{G}\boldsymbol{K}. \tag{8}$$

In that sense, $\boldsymbol{E}$ reflects prescribed fixed ratios of two or more state variables.

Considering the square system, i.e., $r = m$, the control law can be formulated as [14]

$$\boldsymbol{u}(i) = -(\boldsymbol{K}_p\boldsymbol{q}(i) + \boldsymbol{K}_s\sum_{j=1}^{i-1}\boldsymbol{e}(j)), \tag{9}$$

$$\boldsymbol{e}(i) = \boldsymbol{w}(i) - \boldsymbol{C}\boldsymbol{q}(i), \tag{10}$$

where $\boldsymbol{K}_p \in \mathbb{R}^{r \times n}$, $\boldsymbol{K}_s \in \mathbb{R}^{m \times m}$, are the control gain matrices and $\boldsymbol{w}(i) \in \mathbb{R}^m$ is the reference output variable vector.

With the definition of the summation state variable

$$\boldsymbol{z}(i) = \sum_{j=0}^{i-1}\boldsymbol{e}(j) \tag{11}$$

it yields

$$\boldsymbol{z}(i+1) = \sum_{j=0}^{i}\boldsymbol{e}(j) = \boldsymbol{e}(i) + \sum_{j=0}^{i-1}\boldsymbol{e}(j), \tag{12}$$

$$\boldsymbol{z}(i+1) = \boldsymbol{w}(i) - \boldsymbol{C}\boldsymbol{q}(i) + \boldsymbol{z}(i), \tag{13}$$

respectively.

To describe the closed-loop system structure with such controller, the extended form of the system description can be constructed as follows

$$\begin{bmatrix} \boldsymbol{q}(i+1) \\ \boldsymbol{z}(i+1) \end{bmatrix} = \begin{bmatrix} \boldsymbol{F} & \boldsymbol{0} \\ -\boldsymbol{C} & \boldsymbol{I}_m \end{bmatrix}\begin{bmatrix} \boldsymbol{q}(i) \\ \boldsymbol{z}(i) \end{bmatrix} + \begin{bmatrix} \boldsymbol{G} \\ \boldsymbol{0} \end{bmatrix}\boldsymbol{u}(i) + \begin{bmatrix} \boldsymbol{0} \\ \boldsymbol{I}_m \end{bmatrix}\boldsymbol{w}(i), \tag{14}$$

$$\boldsymbol{y}(i) = \begin{bmatrix} \boldsymbol{C} & \boldsymbol{0} \end{bmatrix}\begin{bmatrix} \boldsymbol{q}(i) \\ \boldsymbol{z}(i) \end{bmatrix} \tag{15}$$

and using the notations

$$\boldsymbol{q}^{\circ T}(i) = \begin{bmatrix} \boldsymbol{q}^T(i) & \boldsymbol{z}^T(i) \end{bmatrix}, \tag{16}$$

$$\boldsymbol{F}^\circ = \begin{bmatrix} \boldsymbol{F} & \boldsymbol{0} \\ -\boldsymbol{C} & \boldsymbol{I}_m \end{bmatrix}, \quad \boldsymbol{G}^\circ = \begin{bmatrix} \boldsymbol{G} \\ \boldsymbol{0} \end{bmatrix}, \quad \boldsymbol{W}^\circ = \begin{bmatrix} \boldsymbol{0} \\ \boldsymbol{I}_m \end{bmatrix}, \tag{17}$$

$$\boldsymbol{K}^\circ = \begin{bmatrix} \boldsymbol{K}_p & \boldsymbol{K}_s \end{bmatrix}, \quad \boldsymbol{C}^\circ = \begin{bmatrix} \boldsymbol{C} & \boldsymbol{0} \end{bmatrix}, \tag{18}$$

$\boldsymbol{F}^\circ \in \mathbb{R}^{(n+m)\times(n+m)}$, $\boldsymbol{G}^\circ$, $\boldsymbol{W}^\circ \in \mathbb{R}^{(n+m)\times m}$, $\boldsymbol{K}^\circ$, $\boldsymbol{C}^\circ \in \mathbb{R}^{m\times(n+m)}$, then the closed-loop system state-space equations can be written as follows

$$\boldsymbol{q}^\circ(i+1) = \boldsymbol{F}_c^\circ\boldsymbol{q}^\circ(i) + \boldsymbol{W}^\circ\boldsymbol{w}(i), \tag{19}$$

$$\boldsymbol{y}(i) = \boldsymbol{C}^\circ\boldsymbol{q}^\circ(i), \tag{20}$$

Journal of Physics: Conference Series **570** (2014) 022005 doi:10.1088/1742-6596/570/2/022005

where

$$\boldsymbol{F}_c^\circ = \boldsymbol{F}^\circ - \boldsymbol{G}^\circ \boldsymbol{K}^\circ. \tag{21}$$

Thus, analogously, the design constraint can be defined as follows

$$\boldsymbol{E}^\circ \boldsymbol{q}^\circ(i+1) = \boldsymbol{E}^\circ \boldsymbol{F}_c^\circ \boldsymbol{q}^\circ(i) = 0, \tag{22}$$

where $\boldsymbol{E}^\circ \in I\!R^{k\times(n+m)}$, $\text{rank}\boldsymbol{E}^\circ = k \leq r$

The task is to design the feedback control gain matrices $\boldsymbol{K}$, $\boldsymbol{K}^\circ$ in such a way that the closed-loop system is stable and the constraints (7), (22), respectively, are satisfied. Note, such formulated design problem is singular [6].

3. Basic preliminaries

Proposition 1 *[12] If $\boldsymbol{\Pi}$ is a matrix variable and $\boldsymbol{A}$, $\boldsymbol{B}$, $\boldsymbol{\Lambda}$ are known non-square matrices of the appropriate dimensions such that the equality*

$$\boldsymbol{A}\boldsymbol{\Pi}\boldsymbol{B} = \boldsymbol{\Lambda} \tag{23}$$

can be set, then all solution to $\boldsymbol{\Pi}$ means

$$\boldsymbol{\Pi} = \boldsymbol{A}^{\ominus 1}\boldsymbol{\Lambda}\boldsymbol{B}^{\ominus 1} + \boldsymbol{\Pi}^\bullet - \boldsymbol{A}^{\ominus 1}\boldsymbol{A}\boldsymbol{\Pi}^\bullet\boldsymbol{B}\boldsymbol{B}^{\ominus 1}, \tag{24}$$

where $\boldsymbol{A}^{\ominus 1}$ and $\boldsymbol{B}^{\ominus 1}$ is Moore-Penrose inverse of $\boldsymbol{A}$, $\boldsymbol{B}$, respectively, and $\boldsymbol{\Pi}^\bullet$ is an arbitrary matrix of appropriate dimension. □

Proposition 2 *[5] If $\boldsymbol{H} \in I\!R^{n\times n}$ is a real square matrix with non-repeated eigenvalues, satisfying the constraint*

$$\mathbf{e}^T\boldsymbol{H} = \mathbf{0}, \tag{25}$$

then one from its eigenvalues is zero, and (normalized) $\mathbf{e}^T$ is the left raw eigenvector of $\boldsymbol{H}$ associated with the zero eigenvalue. □

Since the formulated design task is singular, the design conditions can be stated using the following lemma which is a derivative to the theorem presented in [4].

Lemma 1 *The static feedback control (6) to the system (1), (2) exists if there exist symmetric positive definite matrices $\boldsymbol{R}$, $\boldsymbol{T} \in I\!R^{n\times n}$ and a matrix $\boldsymbol{Y} \in I\!R^{r\times m}$ such that*

$$\boldsymbol{R} = \boldsymbol{R}^T > 0, \quad \boldsymbol{T} = \boldsymbol{T}^T > 0, \tag{26}$$

$$\begin{bmatrix} -\boldsymbol{T} & \boldsymbol{R}\boldsymbol{F}^T - \boldsymbol{Y}^T\boldsymbol{G}^T \\ \boldsymbol{F}\boldsymbol{R} - \boldsymbol{G}\boldsymbol{Y} & \boldsymbol{T} - 2\boldsymbol{R} \end{bmatrix} < 0. \tag{27}$$

The control law gain matrix is given by the equation

$$\boldsymbol{K} = \boldsymbol{Y}\boldsymbol{R}^{-1} \tag{28}$$

when the above conditions are affirmative. □

The inequality (27) is an enhanced representation of the Lyapunov matrix inequality for the linear discrete-time systems. It is linear with respect to the system variables but does not involve any product of the Lyapunov matrix $\boldsymbol{T}$ and the system matrices $\boldsymbol{F}$, $\boldsymbol{G}$. By introducing a new variable $\boldsymbol{R}$, the matrix products are relaxed to the discrepant products $\boldsymbol{F}\boldsymbol{R}$ and $\boldsymbol{G}\boldsymbol{K}\boldsymbol{R}$ which can be singular. This offers the possibility of this LMI structure to be applied in a singular task analysis. It is obvious that if $\boldsymbol{T} = \boldsymbol{R}$ the standard form of the Lyapunov inequality implies from (27).

4. Constrained control design

4.1. Design task formulation

Prescribed by a matrix $\boldsymbol{E}^\circ \in \mathbb{R}^{k\times 2n}$, rank$\boldsymbol{E}^\circ = k \le r$, it is considered that the design constraint (22) will be satisfied for all nonzero natural numbers i. Moreover, it is supposed that $\boldsymbol{E}^\circ$ is specified in such a way that the equalities

$$\boldsymbol{E}^\circ(\boldsymbol{F}^\circ - \boldsymbol{G}^\circ\boldsymbol{K}^\circ) = \boldsymbol{0}, \tag{29}$$

$$\boldsymbol{E}^\circ\boldsymbol{F}^\circ = \boldsymbol{E}^\circ\boldsymbol{G}^\circ\boldsymbol{K}^\circ, \tag{30}$$

respectively, can be solved as well as that the closed-loop system matrix $\boldsymbol{F}_c^\circ$ for designed $\boldsymbol{K}^\circ$ is stable (all eigenvalues of the closed-loop system matrix $\boldsymbol{F}_c^\circ$ site in the unit circle in the complex plane $\mathcal{Z}$).

Solving (30) with respect to $\boldsymbol{K}^\circ$ then (24) implies all solutions of $\boldsymbol{K}^\circ$ as follows

$$\boldsymbol{K}^\circ = (\boldsymbol{E}^\circ\boldsymbol{G}^\circ)^{\ominus 1}\boldsymbol{E}^\circ\boldsymbol{F}^\circ + (\boldsymbol{I}_m - (\boldsymbol{E}^\circ\boldsymbol{G}^\circ)^{\ominus 1}\boldsymbol{E}^\circ\boldsymbol{G}^\circ)\boldsymbol{K}^\bullet, \tag{31}$$

where $\boldsymbol{K}^\bullet \in \mathbb{R}^{m\times(n+m)}$ is an arbitrary matrix and

$$(\boldsymbol{E}^\circ\boldsymbol{G}^\circ)^{\ominus 1} = (\boldsymbol{E}^\circ\boldsymbol{G}^\circ)^T(\boldsymbol{E}^\circ\boldsymbol{G}^\circ(\boldsymbol{E}^\circ\boldsymbol{G}^\circ)^T)^{-1}, \tag{32}$$

where $(\boldsymbol{E}^\circ\boldsymbol{G}^\circ)^{\ominus 1}$ is Moore-Penrose inverse of $\boldsymbol{E}^\circ\boldsymbol{G}^\circ$.

Thus, it is possible to express (31) as follows

$$\boldsymbol{K}^\circ = \boldsymbol{J}^\circ + \boldsymbol{L}^\circ\boldsymbol{K}^\bullet, \tag{33}$$

where

$$\boldsymbol{J}^\circ = (\boldsymbol{E}^\circ\boldsymbol{G}^\circ)^{\ominus 1}\boldsymbol{E}^\circ\boldsymbol{F}^\circ \tag{34}$$

and

$$\boldsymbol{L}^\circ = \boldsymbol{I}_m - (\boldsymbol{E}^\circ\boldsymbol{G}^\circ)^T(\boldsymbol{E}^\circ\boldsymbol{G}^\circ(\boldsymbol{E}^\circ\boldsymbol{G}^\circ)^T)^{-1}\boldsymbol{E}^\circ\boldsymbol{G}^\circ \tag{35}$$

is the projection matrix (the orthogonal projector onto the null space $\mathcal{N}_{\boldsymbol{E}^\circ\boldsymbol{G}^\circ}$ of $\boldsymbol{E}^\circ\boldsymbol{G}^\circ$ [12]).

4.2. Control Law Parameter Design

Theorem 1 *The unforced closed-loop system (19), (20) satisfying the constraint (30) is stable if there exist positive definite symmetric matrices $\boldsymbol{R}^\bullet$, $\boldsymbol{T}^\bullet \in \mathbb{R}^{(n+m)\times(n+m)}$ and a matrix $\boldsymbol{Y}^\bullet \in \mathbb{R}^{m\times(n+m)}$ such that*

$$\boldsymbol{R}^\bullet = \boldsymbol{R}^{\bullet T} > 0, \quad \boldsymbol{T}^\bullet = \boldsymbol{T}^{\bullet T} > 0, \tag{36}$$

$$\begin{bmatrix} -\boldsymbol{T}^\bullet & * \\ \boldsymbol{F}^\bullet\boldsymbol{R}^\bullet - \boldsymbol{G}^\bullet\boldsymbol{Y}^\bullet & \boldsymbol{T}^\bullet - 2\boldsymbol{R}^\bullet \end{bmatrix} < 0, \tag{37}$$

where

$$\boldsymbol{F}^\bullet = \boldsymbol{F}^\circ - \boldsymbol{G}^\circ\boldsymbol{J}^\circ, \quad \boldsymbol{G}^\bullet = \boldsymbol{G}^\circ\boldsymbol{L}^\circ. \tag{38}$$

When the above conditions hold, the control law gain matrices are given as follows

$$\boldsymbol{K}^\bullet = \boldsymbol{Y}^\bullet(\boldsymbol{R}^\bullet)^{-1}, \tag{39}$$

$$\boldsymbol{K}^\circ = \boldsymbol{J}^\circ + \boldsymbol{L}^\circ\boldsymbol{K}^\bullet, \tag{40}$$

respectively, and it yields

$$\boldsymbol{F}_c^\circ = \boldsymbol{F}^\circ - \boldsymbol{G}^\circ\boldsymbol{K}^\circ = \boldsymbol{F}^\bullet - \boldsymbol{G}^\bullet\boldsymbol{K}^\bullet. \tag{41}$$

Here and hereafter, $*$ denotes the symmetric item in a symmetric matrix.

Proof: Substituting (21) into (19) gives

$$\boldsymbol{q}^{\circ}(i+1) = (\boldsymbol{F}^{\circ} - \boldsymbol{G}^{\circ}\boldsymbol{K}^{\circ})\boldsymbol{q}^{\circ}(i) + \boldsymbol{W}^{\circ}\boldsymbol{w}(i) \tag{42}$$

and using (33) then (42) can be rearranged as

$$\boldsymbol{q}^{\circ}(i+1) = (\boldsymbol{F}^{\circ} - \boldsymbol{G}^{\circ}\boldsymbol{J}^{\circ} - \boldsymbol{G}^{\circ}\boldsymbol{L}^{\circ}\boldsymbol{K}^{\bullet})\boldsymbol{q}^{\circ}(i) + \boldsymbol{W}^{\circ}\boldsymbol{w}(i), \tag{43}$$

$$\boldsymbol{q}^{\circ}(i+1) = (\boldsymbol{F}^{\bullet} - \boldsymbol{G}^{\bullet}\boldsymbol{K}^{\bullet})\boldsymbol{q}^{\circ}(i) + \boldsymbol{W}^{\circ}\boldsymbol{w}(i), \tag{44}$$

respectively, where the notations (38) were used. It is evident that (42) and (44) implies (41).

Considering the unforced closed-loop system, (44) can be rewritten as

$$\boldsymbol{F}_c^{\circ}\boldsymbol{q}^{\circ}(t) - \boldsymbol{q}^{\circ}(i+1) = \boldsymbol{0} \tag{45}$$

and for an arbitrary symmetric positive definite matrix $\boldsymbol{Q}^{\circ} \in I\!R^{(n+m)\times(n+m)}$ it yields

$$\boldsymbol{q}^{\circ T}(i+1)\boldsymbol{Q}^{\circ}(\boldsymbol{F}_c^{\circ}\boldsymbol{q}^{\circ}(t) - \boldsymbol{q}^{\circ}(i+1)) = \boldsymbol{0}. \tag{46}$$

Defining the Lyapunov function candidate of the form

$$v(\boldsymbol{q}^{\circ}(i)) = \boldsymbol{q}^{T}(i)\boldsymbol{P}^{\circ}\boldsymbol{q}^{\circ}(i) > 0, \tag{47}$$

where $\boldsymbol{P}^{\circ} \in I\!R^{(n+m)\times(n+m)}$ is a symmetric positive definite matrix, then the difference of the Lyapunov function along a trajectory of the unforced closed-loop system is

$$\Delta v(\boldsymbol{q}^{\circ}(i)) = \boldsymbol{q}^{\circ T}(i+1)\boldsymbol{P}^{\circ}\boldsymbol{q}^{\circ}(i+1) - \boldsymbol{q}^{\circ T}(i)\boldsymbol{P}^{\circ}\boldsymbol{q}^{\circ}(i) < 0 \tag{48}$$

and adding (46) as well as its transposition to (48) it yields

$$\begin{aligned}\Delta v(\boldsymbol{q}^{\circ}(i)) \; &= \boldsymbol{q}^{\circ T}(i+1)\boldsymbol{P}^{\circ}\boldsymbol{q}^{\circ}(i+1) - \boldsymbol{q}^{\circ T}(i)\boldsymbol{P}^{\circ}\boldsymbol{q}^{\circ}(i) \\ &+ \boldsymbol{q}^{\circ T}(i+1)\boldsymbol{Q}^{\circ}(\boldsymbol{F}_c^{\circ}\boldsymbol{q}^{\circ}(t) - \boldsymbol{q}^{\circ}(i+1)) \\ &+ (\boldsymbol{F}_c^{\circ}\boldsymbol{q}^{\circ}(t) - \boldsymbol{q}^{\circ}(i+1))^T\boldsymbol{Q}^{\circ}\boldsymbol{q}^{\circ}(i+1) < 0.\end{aligned} \tag{49}$$

Thus, introducing the composed vector

$$\boldsymbol{q}_c^{\circ T}(i) = \begin{bmatrix} \boldsymbol{q}^{\circ T}(i) & \boldsymbol{q}^{\circ T}(i+1) \end{bmatrix} \tag{50}$$

and using the notation (41), then (49) can be rewritten as follows

$$\Delta v(\boldsymbol{q}^{\circ}(i)) = \boldsymbol{q}_c^{\circ T}(i)\boldsymbol{P}_c^{\circ}\boldsymbol{q}_c^{\circ}(i) < 0, \tag{51}$$

where

$$\boldsymbol{P}_c^{\circ} = \begin{bmatrix} -\boldsymbol{P}^{\circ} & (\boldsymbol{F}^{\bullet} - \boldsymbol{G}^{\bullet}\boldsymbol{K}^{\bullet})^T\boldsymbol{Q}^{\circ} \\ \boldsymbol{Q}^{\circ}(\boldsymbol{F}^{\bullet} - \boldsymbol{G}^{\bullet}\boldsymbol{K}^{\bullet}) & \boldsymbol{P}^{\circ} - 2\boldsymbol{Q}^{\circ} \end{bmatrix} < 0. \tag{52}$$

Since it is supposed that $\boldsymbol{Q}^{\circ}$ is positive definite, defining the transform matrix $\boldsymbol{W}^{\circ} \in I\!R^{2(n+m)\times 2(n+m)}$ as follows

$$\boldsymbol{W}^{\circ} = \mathrm{diag}\begin{bmatrix} \boldsymbol{R}^{\bullet} & \boldsymbol{R}^{\bullet} \end{bmatrix}, \quad \boldsymbol{R}^{\bullet} = (\boldsymbol{Q}^{\circ})^{-1}, \tag{53}$$

pre-multiplying the both side of (49) by $\boldsymbol{W}^{\circ}$ leads to the inequality

$$\begin{bmatrix} -\boldsymbol{R}^{\bullet}\boldsymbol{P}^{\circ}\boldsymbol{R}^{\bullet} & \boldsymbol{R}^{\bullet}\boldsymbol{F}^{\bullet T} - \boldsymbol{R}^{\bullet}\boldsymbol{K}^{\bullet T}\boldsymbol{G}^{\bullet T} \\ \boldsymbol{F}^{\bullet}\boldsymbol{R}^{\bullet} - \boldsymbol{G}^{\bullet}\boldsymbol{K}^{\bullet}\boldsymbol{R}^{\bullet} & \boldsymbol{R}^{\bullet}\boldsymbol{P}^{\circ}\boldsymbol{R}^{\bullet} - 2\boldsymbol{R}^{\bullet} \end{bmatrix} < 0. \tag{54}$$

Finally, using the notations

$$\boldsymbol{T}^{\bullet} = \boldsymbol{R}^{\bullet}\boldsymbol{P}^{\circ}\boldsymbol{R}^{\bullet}, \quad \boldsymbol{Y}^{\bullet} = \boldsymbol{K}^{\bullet}\boldsymbol{R}^{\bullet}, \tag{55}$$

then (52) implies (37). This concludes the proof. ■

4.3. Static decoupling principle

Because the integral feedback control increases the order of the closed-loop system dynamics, in such cases where this control structure is not appropriate, the principle of static decoupling can be used in the forced mode to set-point the desired values of the closed-loop output variables.

The state control in a forced mode for a square linear system is defined by the control policy

$$\boldsymbol{u}(i) = -\boldsymbol{K}\boldsymbol{q}(i) + \boldsymbol{W}_w\boldsymbol{w}(i), \tag{56}$$

where $\boldsymbol{w}(i) \in I\!R^m$ is a desired output vector signal, and $\boldsymbol{W}_w \in I\!R^{m\times m}$ is the signal gain matrix. Subsequently, the closed-loop system is described as

$$\boldsymbol{q}(i+1) = (\boldsymbol{F} - \boldsymbol{G}\boldsymbol{K})\boldsymbol{q}(i) + \boldsymbol{W}_w\boldsymbol{w}(i). \tag{57}$$

If the system (1), (2) satisfy the following condition [17]

$$\operatorname{rank}\begin{bmatrix} \boldsymbol{F} & \boldsymbol{G} \\ \boldsymbol{C} & \boldsymbol{0} \end{bmatrix} = n + m, \tag{58}$$

$\boldsymbol{W}_w$ can be designed as

$$\boldsymbol{W}_w = (\boldsymbol{C}(\boldsymbol{I}_n - (\boldsymbol{F} - \boldsymbol{G}\boldsymbol{K}))^{-1}\boldsymbol{G})^{-1}. \tag{59}$$

The constraint (7) is now defined by a matrix $\boldsymbol{E} \in I\!R^{k\times n}$, rank$\boldsymbol{E} = k \leq r$, specified in such a way that the equality

$$\boldsymbol{E}(\boldsymbol{F} - \boldsymbol{G}\boldsymbol{K}) = \boldsymbol{0} \tag{60}$$

can be solved to obtain the stable matrix $\boldsymbol{F}_c$ of the form (8).

Resolving (60) then (24) implies all solutions of $\boldsymbol{K}$ as follows

$$\boldsymbol{K} = (\boldsymbol{E}\boldsymbol{G})^{\ominus 1}\boldsymbol{E}\boldsymbol{F} + (\boldsymbol{I}_m - (\boldsymbol{E}\boldsymbol{G})^{\ominus 1}\boldsymbol{E}\boldsymbol{G})\boldsymbol{K}^{\diamond}, \tag{61}$$

where $\boldsymbol{K}^{\diamond} \in I\!R^{m\times n}$ is an arbitrary matrix and

$$(\boldsymbol{E}\boldsymbol{G})^{\ominus 1} = (\boldsymbol{E}\boldsymbol{G})^T(\boldsymbol{E}\boldsymbol{G}(\boldsymbol{E}\boldsymbol{G})^T)^{-1}, \tag{62}$$

where $(\boldsymbol{E}\boldsymbol{G})^{\ominus 1}$ is Moore-Penrose inverse of $\boldsymbol{E}\boldsymbol{G}$.

Expressing (61) as follows

$$\boldsymbol{K} = \boldsymbol{J} + \boldsymbol{L}\boldsymbol{K}^{\diamond}, \tag{63}$$

where

$$\boldsymbol{J} = (\boldsymbol{E}\boldsymbol{G})^{\ominus 1}\boldsymbol{E}\boldsymbol{F}, \tag{64}$$

$$\boldsymbol{L} = \boldsymbol{I}_m - (\boldsymbol{E}\boldsymbol{G})^T(\boldsymbol{E}\boldsymbol{G}(\boldsymbol{E}\boldsymbol{G})^T)^{-1}\boldsymbol{E}\boldsymbol{G}, \tag{65}$$

then it yields

$$\boldsymbol{F}_c = \boldsymbol{F} - \boldsymbol{G}\boldsymbol{K} = \boldsymbol{F} - \boldsymbol{G}\boldsymbol{J} - \boldsymbol{G}\boldsymbol{L}\boldsymbol{K}^{\diamond} = \boldsymbol{F}^{\diamond} - \boldsymbol{G}^{\diamond}\boldsymbol{K}^{\diamond}, \tag{66}$$

where

$$\boldsymbol{F}^{\diamond} = \boldsymbol{F} - \boldsymbol{G}\boldsymbol{J}, \quad \boldsymbol{G}^{\diamond} = \boldsymbol{G}\boldsymbol{L}. \tag{67}$$

Supposing that the realization of $(\boldsymbol{F}^{\diamond}, \boldsymbol{G}^{\diamond})$ is controllable, analogously as above this implies the following theorem which links the static decoupling principle to the solutions of feasible LMIs for the discrete equivalent of PI control, both with respect to the in state constrained closed-loop system.

Journal of Physics: Conference Series **570** (2014) 022005 doi:10.1088/1742-6596/570/2/022005

Theorem 2 *The unforced closed-loop system (1), (2) satisfying the constraint (60) is stable if there exist positive definite symmetric matrices $\boldsymbol{R}^{\diamond}$, $\boldsymbol{T}^{\diamond} \in I\!R^{n\times n}$ and a matrix $\boldsymbol{Y}^{\diamond} \in I\!R^{m\times n}$ such that*

$$\boldsymbol{R}^{\diamond} = \boldsymbol{R}^{\diamond T} > 0, \quad \boldsymbol{T}^{\diamond} = \boldsymbol{T}^{\diamond T} > 0, \tag{68}$$

$$\begin{bmatrix} -\boldsymbol{T}^{\diamond} & * \\ \boldsymbol{F}^{\diamond}\boldsymbol{R}^{\diamond} - \boldsymbol{G}^{\diamond}\boldsymbol{Y}^{\diamond} & \boldsymbol{T}^{\diamond} - 2\boldsymbol{R}^{\diamond} \end{bmatrix} < 0, \tag{69}$$

where $\boldsymbol{F}^{\diamond}$, $\boldsymbol{G}^{\diamond}$ are defined in (67).

When the above conditions hold, the control law gain matrices are given as follows

$$\boldsymbol{K}^{\diamond} = \boldsymbol{Y}^{\diamond}(\boldsymbol{R}^{\diamond})^{-1}, \tag{70}$$

$$\boldsymbol{K} = \boldsymbol{J} + \boldsymbol{L}\boldsymbol{K}^{\circ}, \tag{71}$$

where $\boldsymbol{J}$, $\boldsymbol{L}$ are defined in (64), (65), respectively.

The proof follows analogously the same steps as the proof of Theorem 1 and therefore is omitted.

4.4. Constrain projection in forced modes

Theorem 3 *If the closed-loop system state variables satisfy the state constraint (29) where $\boldsymbol{E}^{\circ}$ takes the form*

$$\boldsymbol{E}^{\circ} = \begin{bmatrix} \boldsymbol{E} & \boldsymbol{X} \end{bmatrix}, \tag{72}$$

$\boldsymbol{w}(i) = \boldsymbol{w}_s$ for all i and $m = r \geq 2$, then the common state variable vector $\boldsymbol{q}_d(i) = \boldsymbol{E}^{\circ}\boldsymbol{q}^{\circ}(i)$, $\boldsymbol{q}_d(i) \in I\!R^{k}$, of the system (16) in the forced mode attains the zero value vector at all time instants $i \in Z_+$ if there exist a nonzero matrix $\boldsymbol{X} \in I\!R^{k\times m}$ such that

$$\boldsymbol{X}\boldsymbol{w}_s = \boldsymbol{0}. \tag{73}$$

Proof: Applying the constraint matrix $\boldsymbol{E}^{\circ}$ on the closed-loop system in the forced mode then (19) implies

$$\boldsymbol{E}^{\circ}\boldsymbol{q}^{\circ}(i+1) = \boldsymbol{E}^{\circ}(\boldsymbol{F}^{\circ} - \boldsymbol{G}^{\circ}\boldsymbol{K}^{\circ})\boldsymbol{q}^{\circ}(i) + \boldsymbol{E}^{\circ}\boldsymbol{W}^{\circ}\boldsymbol{w}(i). \tag{74}$$

When $\boldsymbol{E}^{\circ}$ takes the form (72) then the equation (74) conditioned by (29) gives

$$\boldsymbol{E}^{\circ}\boldsymbol{q}^{\circ}(i+1) = \boldsymbol{E}^{\circ}\boldsymbol{W}^{\circ}\boldsymbol{w}(i) = \begin{bmatrix} \boldsymbol{E} & \boldsymbol{X} \end{bmatrix}\begin{bmatrix} \boldsymbol{0} \\ \boldsymbol{I}_m \end{bmatrix}\boldsymbol{w}(i) = \boldsymbol{X}\boldsymbol{w}(i). \tag{75}$$

It is evident that the common state variable vector $q_d(i)$ of the closed-loop system in the forced mode with $\boldsymbol{w}(i) = \boldsymbol{w}_s$ is zero vector if $\boldsymbol{X}\boldsymbol{w}_s = \boldsymbol{0}$ for $\boldsymbol{X} \neq \boldsymbol{0}$. This concludes the proof. ■

Note, considering $m = k = 1$, a solution of this problem exists with the prescribed precision $\varepsilon > 0$ if $\boldsymbol{X} = \varepsilon$, $\varepsilon \in I\!R$.

Theorem 4 *If the closed-loop system state variables satisfy the state constraint (60) and $\boldsymbol{w}(i) = \boldsymbol{w}_s$ for all $i \in Z_+$, then the common state variable vector $\boldsymbol{q}_d(i) = \boldsymbol{E}\boldsymbol{q}(i)$, $\boldsymbol{q}_d(i) \in I\!R^{k}$, of the system (57) in the forced mode attains the value*

$$\boldsymbol{q}_d = \boldsymbol{E}\boldsymbol{G}\boldsymbol{W}_w\boldsymbol{w}_s. \tag{76}$$

Proof: Using the control policy (56) then

$$\boldsymbol{E}\boldsymbol{q}(i+1) = \boldsymbol{E}(\boldsymbol{F} - \boldsymbol{G}\boldsymbol{K})\boldsymbol{q}(i) + \boldsymbol{E}\boldsymbol{G}\boldsymbol{W}_w\boldsymbol{w}(i). \tag{77}$$

Since (77) conditioned by (60) implies

$$\boldsymbol{E}\boldsymbol{q}(i+1) = \boldsymbol{E}\boldsymbol{G}\boldsymbol{W}_w\boldsymbol{w}(i) = \boldsymbol{q}_d(i), \tag{78}$$

it is evident that the common state variable $q_d(i)$ of the closed-loop system in a steady state is linear dependent on the steady value $\boldsymbol{w}_s$. This concludes the proof. ■

5. Illustrative example

To demonstrate the properties of the proposed approach, the system with two-inputs and two-outputs is used in the example. The matrix parameters of the system are

$$\boldsymbol{F} = \begin{bmatrix} 0.9993 & 0.0987 & 0.0042 \\ -0.0212 & 0.9612 & 0.0775 \\ -0.3875 & -0.7187 & 0.5737 \end{bmatrix}, \; \boldsymbol{G} = \begin{bmatrix} 0.0010 & 0.0010 \\ 0.0206 & 0.0197 \\ 0.0077 & -0.0078 \end{bmatrix}, \; \boldsymbol{C} = \begin{bmatrix} 1 & 2 & -2 \\ 1 & -1 & 0 \end{bmatrix},$$

respectively, for sampling period $\Delta t = 0.1$ s. The state constraint was specified as

$$\frac{q_1(t) - 0.4\, q_3(t)}{q_2(t)} = 0.1,$$

which implies

$$\boldsymbol{E} = \begin{bmatrix} 1 & -0.1 & -0.4 \end{bmatrix}.$$

Prescribing the desired value of the closed-loop system in the forced mode as follows

$$\boldsymbol{w}_s = \begin{bmatrix} 1 & -0.5 \end{bmatrix}^T,$$

then it can be chosen

$$\boldsymbol{X} = \begin{bmatrix} 5 & 10 \end{bmatrix},$$

$$\boldsymbol{E}^{\circ} = \begin{bmatrix} \boldsymbol{E} & \boldsymbol{X} \end{bmatrix} = \begin{bmatrix} 1 & -0.1 & -0.4 & 5 & 10 \end{bmatrix},$$

Solving (36), (37) with respect to the LMI matrix variables $\boldsymbol{R}^{\circ}$, $\boldsymbol{T}^{\circ}$, $\boldsymbol{Y}^{\circ}$ using Self-Dual-Minimization (SeDuMi) package [16] for Matlab, the design conditions imply the control gain matrix parameters

$$\boldsymbol{K}_p = 10^3 \begin{bmatrix} 5.4867 & -0.0335 & -3.8042 \\ 4.1263 & 0.0704 & -2.7825 \end{bmatrix},$$

$$\boldsymbol{K}_s = 10^3 \begin{bmatrix} -1.9209 & -3.8417 \\ -1.3733 & -2.7464 \end{bmatrix},$$

respectively, and the eigenvalues spectrum of the extended closed-loop loop system is

$$\rho(\boldsymbol{A}_c^{\circ}) = \{\; 0.0000 \quad 0.1389 \quad 0.2104 \quad 0.5153 \quad 0.9483 \;\}.$$

Solving (68), (69) the solution of the control design conditions gives the following control gain matrix

$$\boldsymbol{K} = \begin{bmatrix} -181.4457 & -39.5609 & 40.2261 \\ 188.4813 & 58.7340 & -30.9274 \end{bmatrix}$$

and the stable closed-loop eigenvalues spectrum is

$$\rho(\boldsymbol{A}_c) = \{\; 0.0000 \quad 0.8170 \pm 0.0445\,\mathrm{i} \;\}.$$

The previously described control design methods were applied to the simulation benchmark. Since the same input variables in the forced modes were utilized to assess the each controller ability response and to demonstrated performance with respect to asymptotic properties, the results of the proposed design methods can be compared. Moreover, the initial conditions for the state vector $\boldsymbol{q}(i)$ in both cases were simply chosen to be $\boldsymbol{q}(0) = \boldsymbol{0}$ and, moreover, in the system with the summation part was set $\boldsymbol{z}(0) = \boldsymbol{0}$. To complete the forced mode parameters, the signal gain matrix $\boldsymbol{W}_w$ was computed using (59) as follows

$$\boldsymbol{W}_w = \begin{bmatrix} -32.3526 & -124.5375 \\ 30.0938 & 134.0236 \end{bmatrix}.$$

Journal of Physics: Conference Series **570** (2014) 022005 doi:10.1088/1742-6596/570/2/022005

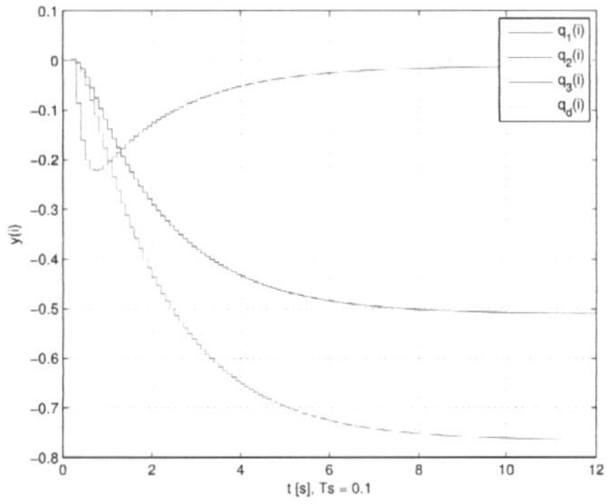

Figure 1. State responses for the control designed pursuant to Theorem 1.

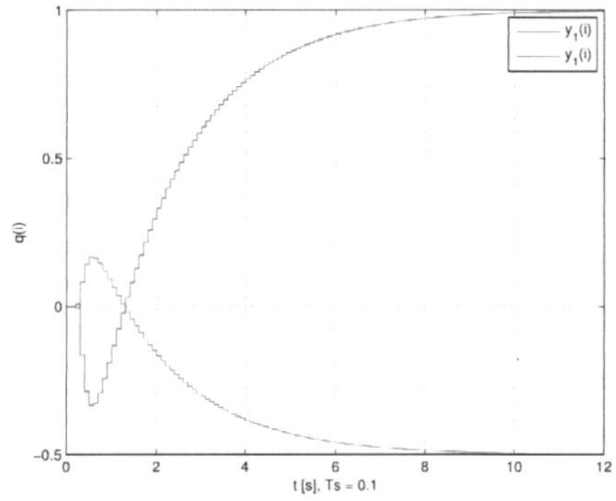

Figure 2. Output responses for the control designed pursuant to Theorem 1.

It was verified, that the condition (62) is satisfied at all time instant except initial time in such a way that

$$\boldsymbol{q}_d = \boldsymbol{E}\boldsymbol{W}^{\circ}\boldsymbol{w}_s = 0, \quad \boldsymbol{q}_d = \boldsymbol{E}\boldsymbol{G}\boldsymbol{W}_w\boldsymbol{w}_s = -0.2032.$$

(see $q_d(i)$ in Fig. 1, Fig. 3, respectively).

The closed-loop responses for the equally prescribed desired system outputs are shown in Fig. 1 to Fig. 4. From these figures it is obvious that the both control laws, which parameters were obtained by the solutions of the LMI problems specified by Theorem 1 and Theorem 2, respectively, can successfully provide for the closed-loop system steady-state properties and asymptotic dynamics. Due to the relatively large eigenvalues that causes the summation part of the control law (9), the response of the system under control designed using Theorem 1 is considerably slower then is the response of the system with the control law computed pursuant to Theorem 2. On the other hand, this disadvantage is compensated by zero value of the variable $\boldsymbol{q}_d(i)$. It should be noted that the $\boldsymbol{K}^{\circ} = [\boldsymbol{K}_p,\ \boldsymbol{K}_s]$ can be altered while the desired control input $\boldsymbol{w}_s$ is changed.

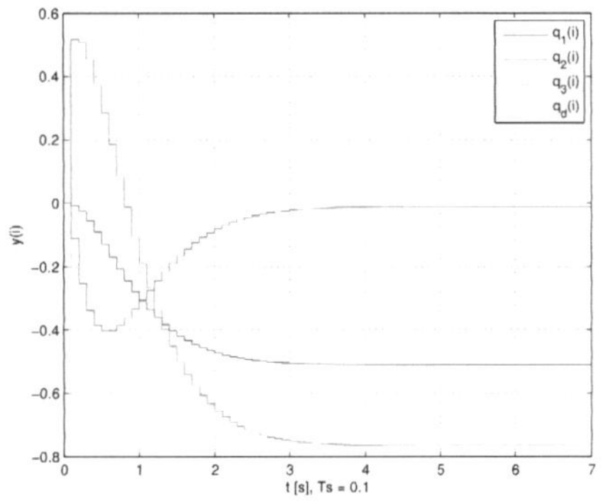

Figure 3. State responses for the control designed pursuant to Theorem 2.

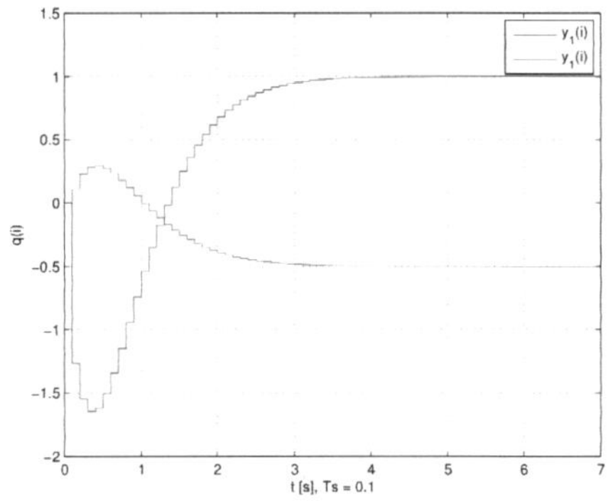

Figure 4. Output responses for the control designed pursuant to Theorem 2.

6. Concluding remarks

The paper presents the novel approach for synthesis the discrete-time equivalent PI controller which is destined to control the discrete-time linear MIMO systems in such a way that the closed-loop state variables satisfy the prescribed equality constraints. To circumvent the singular design task, the resulting LMIs do not involve any product of the Lyapunov matrix and the system matrices, which provides a suitable way for the control law determination, conditioned by the singular closed-loop system matrix. Design conditions are established in the terms of LMI as a feasible problem, accomplishing the manipulation in the manner giving guaranty the asymptotic stability of the closed-loop system and satisfying the prescribed equality constraints in the forced regime. Comparing the proposed PI state-space design with the principle based on the static decoupling, the presented illustrative example also confirms the effectiveness of the proposed design method.

Acknowledgment

The work presented in this paper was supported by VEGA, the Grant Agency of the Ministry of Education and the Academy of Science of Slovak Republic under Grant No. 1/0348/14. This support is very gratefully acknowledged. The authors would also like to thank the anonymous reviewers for their detailed comments and suggestions, which helped to improve the paper.

References

[1] Benzaouia A and Gurgat C 1988 Regulator problem for linear discrete-time systems with nonsymmetrical constrained control *Int. J. Control* **48** 2441-51
[2] Cakmakci M and Ulsoy A G 2009 Modular discrete optimal MIMO controller for a VCT engine *2009 American Control Conference* St. Louis USA 1359-64
[3] Debiane L Ivorra B Mohammadi B Nicoud F Ernz A Poinsot T and Pitsch H 2004 Temperature and pollution control in flames *Proc. Summer Program 2004* University of Montpellier France 1-9
[4] de Oliveira M C Bernussou J and Geromel J C 1999 A new discrete-time robust stability condition *Systems & Control Letters* **37** 261-5
[5] Filasová A and Krokavec D 2012 H_∞ control of discrete-time linear systems constrained in state by equality constraints *Int. J. Applied Mathematics and Computer Science*, **22** 551-60
[6] Filasová A and Krokavec D 2012 An active reconfiguration scheme for systems with single sensor faults *Prepr. 8th IFAC Symp. Fault Detection, Supervision and Safety of Technical Processes SAFEPROCESS 2012* Mexico City Mexico 624-9
[7] Hahn H 1992 Linear systems controlled by stabilized constraint relations *Proc. 31st IEEE Conf. Decision and Control* Tucson USA 840-8
[8] Ko S and Bitmead R R 2007 Optimal control for linear systems with state equality constraints *Automatica* **43** 1573-82
[9] Krokavec D and Filasová A 2009 Control reconfiguration based on the constrained LQ control algorithms *Prepr. 7th IFAC Symp. Fault Detection, Supervision and Safety of Technical Processes SAFEPROCESS 2009* Barcelona Spain 686-91
[10] Krokavec D and Filasová A 2011 Control of discrete-time stochastic systems with state equality constraints *Prepr. 18th IFAC World Congress* Milano Italy 3210-16
[11] Krokavec D and Filasová A 2014 H_∞ control of discrete-time stochastic state-multiplicative systems constrained in state by equality constraints *Prepr. 19th IFAC World Congress* Cape Town South Africa 8699-704
[12] Laub A J 2005 *Matrix Analysis for Scientists and Engineers* (Philadelphia: SIAM)
[13] Mason O and Shorten R 2004 On common quadratic Lyapunov functions for stable discrete-time LTI systems *IMA J. Applied Mathematics* **69** 271-83
[14] Mukaidani H Umeda T Tanaka Y and Tsuji T 2009 Guaranteed cost PI control for uncertain discrete-time systems with additive gain *Proc. European Control Conference 2009* Budapest Hungary 2319-24
[15] Nguyen H N 2014 *Constrained Control of Uncertain, Time-Varying, Discrete-Time Systems. An Interpolation-Based Approach* (Cham: Springer)
[16] Peaucelle D Henrion D Labit Y and Taitz K 2002 *User's Guide for SeDuMi Interface 1.04* (Toulouse: LAAS-CNRS)
[17] Wang Q G 2003 *Decoupling Control* (Berlin Heidelberg: Springer)

A receding horizon scheme for discrete-time polytopic linear parameter varying systems in networked architectures

Giuseppe Franzè, Walter Lucia and Francesco Tedesco
Università della Calabria, DIMES, Via Pietro Bucci, Cubo 42-C, Rende (CS), 87036, ITALY
E-mail: {franze,wlucia,tedesco}@dimes.unical.it

Abstract. This paper proposes a Model Predictive Control (MPC) strategy to address regulation problems for constrained polytopic Linear Parameter Varying (LPV) systems subject to input and state constraints in which both plant measurements and command signals in the loop are sent through communication channels subject to time-varying delays (Networked Control System (NCS)). The results here proposed represent a significant extension to the LPV framework of a recent Receding Horizon Control (RHC) scheme developed for the so-called robust case. By exploiting the parameter availability, the pre-computed sequences of one-step controllable sets inner approximations are less conservative than the robust counterpart. The resulting framework guarantees asymptotic stability and constraints fulfilment regardless of plant uncertainties and time-delay occurrences. Finally, experimental results on a laboratory two-tank test-bed show the effectiveness of the proposed approach.

1. Introduction

The term Networked Control System [2, 16] describes a complex process: the plant to be regulated, actuators, sensors and the controller, all exchanging data via a communication channel. Relevant contributions on this subject ranges from Lyapunov strategies for nonlinear MIMO systems [25] to state/observer based methodologies for linear continuous time NCSs [20].

Here we point out our attention to the class of constrained RHC strategies which are an extremely appealing methodology for dealing with the NCS stabilization problem thanking to its intrinsic capability to generate, at each time instant, a sequence of virtual inputs which can be transmitted within a single data-packet [9, 23, 21]. In [9], the authors consider a receding horizon strategy for nonlinear systems under wireless and asynchronous measurements. In order to regulate the state towards an equilibrium point while minimizing a given performance index, a Lyapunov-based model predictive controller is designed by explicitly taking into account data losses both in the optimization problem formulation and in the controller implementation. This proposed scheme allows an explicit characterization of the stability region and guarantees that such a set is invariant for the closed-loop system under data-losses. Gupta and Quevedo [23] extend a control scheme for nonlinear plants, popular in real-time systems to tolerate the presence of time-varying processing resources (such as variable delays, packet losses/drops etc.) and known as *anytime* algorithm. In [21], by following similar ideas as in [9] a nonlinear RHC scheme exploiting a Network Delay Compensation strategy is proposed to efficiently manage the simultaneous presence of constraints, model uncertainties, time-varying transmission delays and data-packet losses. The main aim consists in overcoming the inherent difficulties related to a MPC design based on linear process models and the joint presence of constraints, nonlinearities and NCS features. Moving from these considerations in [10] a discrete time RHC strategy for NCSs which are described by means of uncertain polytopic linear plants, subject to time-varying delays and data-losses, is developed. There, the framework prescribes that measurements and control commands need to be sent over communications links which bring to varying transmission delays and packet dropouts between plant-controller and *vice-versa*. The key idea is the following: pre-computed inner approximations sequences of the one-step ahead state prediction sets are on-line exploited as target sets for the actual state prediction vector to compute the commands to be applied to the plant in a receding horizon fashion and the time-varying delays are taken into account by resorting to both Independent-of-Delay (**IOD**) and Delay Dependent (**DD**) stability concepts [13]. Following the same lines of [10], a more general framework based on polynomial approach has been presented in [11, 12].

The class of LPV systems has attracted a significant attention within the time-delay systems research field: delay-independent analysis and synthesis [26], delay-dependent stability conditions [24], H_∞ controller design problems for NCSs with packet dropouts [15], continuous-time networked control systems with random sensor-delays [18] and references therein. However, from these contributions it is noted that very few efforts have been paid on the study of efficient strategies for dealing with constrained NCS-LPV systems subject to time-varying delays on the communication medium, which mainly motivates this present study.

In this paper, the main aim is to extend the networked control scheme of [10] to the LPV framework. As it is well-known, the LPV system paradigm provides an interesting modelling framework which is capable to describe special classes of non-linear and/or time-varying dynamics. Because of its relevance in practical applications, several Model Predictive Control (MPC) schemes have been recently developed for the LPV framework. For a comprehensive literature on the matter see e.g. [5, 22, 8]. One difficulty dealing with LPV plants governed by linear scheduled control laws is that the one-step ahead state prediction set is non-convex. Here, a tight outer convex approximation proposed in [7] will be exploited for control design purposes. Then, the networked control strategy is adapted to the class of constrained polytopic LPV systems by exploiting the outer convex approximations of the one-step state prediction set. Specifically, one-step controllable set sequences are formally determined used for managing time-delay occurrences on the signals sent through transmission channels.

Effectiveness and applicability of the proposed control framework are shown via simulation studies on a cascaded two-tank process.

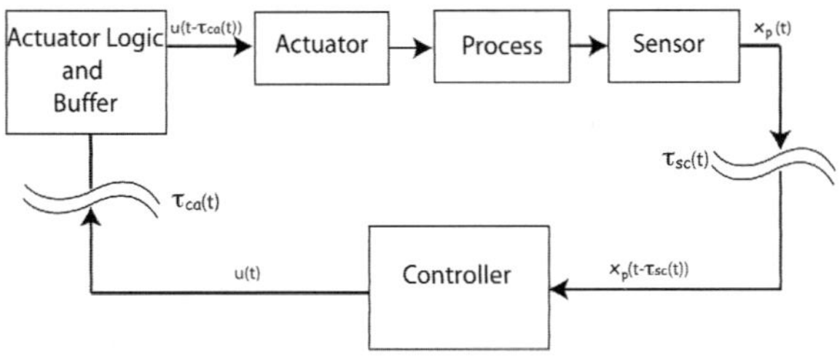

Figure 1. Networked control system structure

2. Problem formulation

In the sequel, we shall consider the networked control systems depicted in Fig. 1 where both the communication channels are subject to time-varying induced delays.

A discrete-time polytopic LPV system is used to describe the process dynamics

$$x_p(t+1) = \Phi(\alpha(t))x_p(t) + G(\alpha(t))u(t) \tag{1}$$

where $t \in \mathbb{Z}_+ := \{0, 1, ...\}$, $x_p(t) \in \mathbf{R}^n$ denotes the state plant and $u(t) \in \mathbf{R}^m$ the control input. The measurable time-varying parameter vector $\alpha(t) \in \mathbf{R}^l$ belongs to the unit simplex

$$\mathcal{P}_l := \left\{ \alpha \in \mathbf{R}^l : \sum_{i=1}^{l} \alpha_i = 1,\ \alpha_i \geq 0 \right\} \tag{2}$$

while the system matrices $\Phi(\alpha)$ and $G(\alpha)$ to

$$\Sigma(\mathcal{P}_l) := \left\{ (\Phi(\alpha), G(\alpha)) = \sum_{i=1}^{l} \alpha_i (\Phi_i, G_i), \alpha \in \mathcal{P}_l \right\} \tag{3}$$

with the pairs (Φ_i, G_i) denoting the polytope vertices of $\Sigma(\mathcal{P}_l)$, viz. $(\Phi_i, G_i) \in \text{vert}\{\Sigma(\mathcal{P}_l)\}, i = 1, \ldots, l$. Moreover, the following constraints are prescribed

$$u(t) \in \mathcal{U},\ \forall t \geq 0,\ \ \mathcal{U} := \{u \in \mathbb{R}^m | u^T u \leq \bar{u}\}, \tag{4}$$

$$x_p(t) \in \mathcal{X},\ \forall t \geq 0,\ \ \mathcal{X} := \{x_p \in \mathbb{R}^n | x_p^T x_p \leq \bar{x}\}, \tag{5}$$

with $\bar{u} > 0, \bar{x} > 0$, and $\mathcal{U}$, $\mathcal{X}$ compact subsets of $\mathbb{R}^m$ and $\mathbb{R}^n$, respectively.
The actuator buffer is used to store the last applied command, hereafter named u^R_{-1}, whereas at each time instant t the logic is instructed to apply the current command $u(t)$ if more recent than u^R_{-1}.
Assumption 1 *Let us assume that:*

- $\tau_{ca}(t) : \mathbb{Z}_+ \to \mathbb{Z}_+$ *such that* $\tau_{ca}(t) \leq \bar{\tau}_c, \forall t \geq 0, \bar{\tau}_c \in \mathbb{Z}_+$;
- $\tau_{sc}(t) : \mathbb{Z}_+ \to \mathbb{Z}_+$ *such that* $\tau_{sc}(t) \leq \bar{\tau}_s, \forall t \geq 0, \bar{\tau}_s \in \mathbb{Z}_+$.

By using the following definitions:

Definition 1 *At each time instant t, the age of the state measurement used by the controller to compute the command input is defined as*

$$\tau_m(t) \leq \tau_{ca}(t)$$

Definition 2 *At each time instant t, the age of the command used by the actuator is defines as*

$$\tau_c(t) \leq \tau_{sc}(t)$$

we have that the following age cumulative induced delay should be considered:

$$\tau_{NL}(t) = \tau_m(t) + \tau_c(t - \tau_m(t))$$

in order to compute the control action $u(t)$ on the plant-controller link.

Then to properly comply with time-delay occurrences, it is mandatory to equip the NCS with a time-stamp mechanism and a clock synchronization procedure capable to provide at each time instant t sensor and actuator latencies and the actuator latency $\tau_m(t)$ and $\tau_c(t)$. As it is well-known, the round-trip delay $\tau_c(t - \tau_m(t))$ cannot be available at the controller side, therefore in the sequel we shall consider the upper bound $\bar{\tau}_c$ on the controller-actuator link for design purposes and the following network induced delay is exploited:

$$\tau(t) = \tau_m(t) + \bar{\tau}_c,\ \forall t \tag{6}$$

This paper focuses on the problem of designing a RHC strategy that is capable via the control action

$$u(\cdot) = g(x_p(\cdot), \alpha(\cdot)) \tag{7}$$

to asymptotically stabilizes the closed-loop system under the following time-delay occurrences:

- **Sensor-to-Controller link -** each time-delay occurrence is bounded: $\tau(t) < \bar{\tau}$, with the upper bound $\bar{\tau}$ the maximum admissible time interval (MATI) [25];
- **Controller-to-Actuator link -** at each time instant t the actuator receives a control action $u(t)$.

From now on we refer to the above statement as the **Network Constrained Stabilization (NCS)** problem. It will be addressed by exploiting the basic dual-mode predictive scheme based on the set-theoretic approach, see [1], [7] and references therein. Essentially, the controlscheme consists of the following phases:

- *Off-line* - Stabilizing state-feedback control laws for (1) subject to (4), (5) and (6) are first derived by resorting to **DD** and **IOD** stability concepts. Then, the corresponding working regions are enlarged by deriving sets of states that can be steered into **DD** and **IOD** terminal ellipsoids in a finite number of steps.
- *On-line* - At each time t, the "smallest" ellipsoidal set (**DD** or **IOD** regions) compatible with the current cumulative delay instance $\tau(t)$ is first determined. Then, the command action is computed by minimizing a running cost under the condition that the one-step state prediction belongs to the successor set [3].

3. Outer approximation of the one-step ahead state prediction set under scheduled control laws

In this section, the availability at each time instant of the system parameter $\alpha(t)$ is used to characterize the one-step ahead state sets, see [7] for details.

To this end, we shall consider the unconstrained evolutions of the polytopic system (1) and shall define the one-step ahead state prediction set as

$$\mathcal{X}^+ := \{x_p^+ \mid x_p^+ = \Phi(\alpha)x_p + G(\alpha)u, \ \forall\alpha\in\mathcal{P}_l\} \tag{8}$$

where $x_p \in \mathbb{R}^n$ is a given initial state and $u \in \mathbb{R}^m$ the command input. By using the following class of scheduled control laws

$$u(t) = \sum_{j=1}^{l} \alpha_j(t)\,u_j(t) \tag{9}$$

with $u_j(t) \in \mathbb{R}^m$, each element $x^+ \in \mathcal{X}^+$ can be rewritten as follows

$$x_p^+ = \sum_{i=1}^{l}\sum_{j=1}^{l} \alpha_i\alpha_j\Phi_i x_p + \sum_{i=1}^{l}\sum_{j=1}^{l} \alpha_i\alpha_j G_i u_j, \forall\alpha\in\mathcal{P}_l \tag{10}$$

Then, it is straightforwardly to show that expression (10) can be rewritten as

$$x_p^+ = \left(\left(\mathbf{1}_l^T\otimes I_{n\times n}\right)\left(\alpha\alpha^T\otimes I_{n\times n}\right)\begin{bmatrix}\Phi_1\\ \vdots\\ \Phi_l\end{bmatrix}\right)x + \left(\begin{bmatrix}G_1 & \dots & G_l\end{bmatrix}\left(\alpha\alpha^T\otimes I_{m\times m}\right)\right)\bar{u}, \forall\alpha\in\mathcal{P}_l \tag{11}$$

where $\otimes$ denotes the Kronecker product and $\mathbf{1}_l \triangleq [1\ \dots\ 1]^T \in \mathbb{R}^l$ and $\bar{u}^T = \left[u_1^T \dots u_l^T\right]^T \in \mathbb{R}^{lm}$. Since the above expression shows a quadratic dependence of x^+ from the parameter vector $\alpha\in\mathcal{P}_l$, the dyadic rank-one matrix $\alpha\alpha^T$, the region $\mathcal{X}^+$ is not a convex set. With this in mind, a standard way to proceed is to search for an appropriate convexification of $\mathcal{X}^+$ by solving the following problem:

Outer Approximation Problem (OAP) - *Determine a collection of matrices* $\Pi_i \in \mathbb{R}^{l\times l}, i = 1,\dots,l_c$ *such that*

$$\left\{\alpha\alpha^T \mid \alpha\in\mathcal{P}_l\right\} \subseteq \Pi \tag{12}$$

where $\Pi := conv\{\Pi_1,\dots,\Pi_{l_c}\}$ *is the set of all convex combinations of* $\{\Pi_1,\dots,\Pi_{l_c}\}$. □

Note that the integer l_c represent the minimum number of vertices of the outer convex convexification complying with (12). Then, the one-step ahead state prediction set (8) can be embedded into

$$\mathcal{X}_{l_c}^+ := \{x_p^+ \in\mathbb{R}^n \mid x_p^+ = \bar{A}(\bar{\alpha})x_p + \bar{B}(\bar{\alpha})\bar{u}, \ \forall\bar{\alpha}\in\mathcal{P}_{l_c}\} \tag{13}$$

that is generated by means of the following multi-model linear system

$$x_p(t+1) = \bar{\Phi}(\bar{\alpha}(t))x_p(t) + \bar{G}(\bar{\alpha}(t))\bar{u}(t),\ \bar{\alpha}(t)\in\mathcal{P}_{l_c} \tag{14}$$

with

$$\bar{\Phi}(\bar{\alpha}) = \sum_{i=1}^{l_c} \bar{\alpha}_i \bar{\Phi}_i, \ \bar{\Phi}_i = \left(\left(1_l^T \otimes I_{n\times n}\right)\left(\Pi_i \otimes I_{n\times n}\right) \begin{bmatrix} \Phi_1 \\ \vdots \\ \Phi_l \end{bmatrix} \right) \tag{15}$$

$$\bar{G}(\bar{\alpha}) = \sum_{i=1}^{l_c} \bar{\alpha}_i \bar{G}_i, \ \bar{G}_i = \left(\begin{bmatrix} G_1 & \cdots & G_l \end{bmatrix} \left(\Pi_i \otimes I_{m\times m}\right) \right) \tag{16}$$

The starting point to achieve an adequate solution to the **OAP** approximation consists in rewriting the matrix dyadic $\alpha\alpha^T = \sum_{i=1}^{l}\sum_{j=1}^{l} \alpha_i\alpha_j e_i e_j^T$, $e_i \in \mathbb{R}^l$ being the i-th vector of the canonical basis of $\mathbb{R}^l$, as

$$\alpha\alpha^T = \sum_{i=1}^{l} \alpha_i^2 \left[e_i e_i^T\right] + \sum_{i=1}^{l}\sum_{j=i+1}^{l} 2\alpha_i\alpha_j \left[\frac{e_i e_j^T + e_j e_i^T}{2}\right] \tag{17}$$

where the property $\alpha_i\alpha_j = \alpha_j\alpha_i$, $\forall i \neq j$ has been exploited. Then it is straightforward to show that the polytopic set

$$\Omega_{l(l+1)/2} = \mathcal{P}_{l(l+1)/2} \tag{18}$$

encloses vectors of the form $\left[\alpha_1^2,\ldots,\alpha_l^2, 2\alpha_1\alpha_2,\ldots\ 2\alpha_{l-1}\alpha_l\right]^T$, $\forall\alpha \in \mathcal{P}_l$ and the following family of vertices for Π is obtained

$$vert\{\Pi\} \ = \ \left\{ \left\{e_i e_i^T\right\}_{i=1}^{l}, \left\{\tfrac{1}{2}\left(e_i e_j^T + e_j e_i^T\right)\right\}_{\substack{i=1\\ j=i+1}}^{l} \right\} \tag{19}$$

In [7] it has been proved that a polytope, which embeds the vector $[\alpha_1^2,\ldots,\alpha_l^2, 2\alpha_1\alpha_2,\ldots,2\alpha_{l-1}\alpha_l]^T$ into a more refined region than (18), can be derived by resorting to the following result:

$$2\alpha_i\alpha_j \leq \frac{1}{2},\ i,j = 1,\ldots,l, \forall\alpha \in \mathcal{P}_l, \tag{20}$$

i.e. the intersection of the unit simplex $\mathcal{P}_{l(l+1)/2}$ with the family of half-spaces (20):

$$\Omega_{l(l+1)/2} := \left(\mathcal{P}_{l(l+1)/2}\right) \cap \left(\bigcap_{i=l+1}^{l(l+1)/2} \left\{ \alpha \in \mathbb{R}^{l(l+1)/2} \middle| \alpha_i \leq 1/2 \right\} \right) \tag{21}$$

whose $l_c = \frac{l^4+2l^3-5l^2+10l}{8}$ vertices are given by

$$vert\{\Omega\} = \left\{ \{e_i\}_{i=1,\ldots,l}, \left\{\frac{(e_i+e_j)}{2}\right\}_{\substack{i=l+1,\ldots,l(l+1)/2,\\ j=1,\ldots,l}}, \left\{\frac{(e_i+e_j)}{2}\right\}_{\substack{i=l+1,\ldots,l(l+1)/2,\\ j=i+1,\ldots,l(l+1)/2}} \right\} \tag{22}$$

Finally, based on (22), the following more tight polytopic outer approximation of $\alpha\alpha^T, \forall\alpha \in \mathcal{P}_l$, results

$$vert\{\Pi\} = \left\{ \left\{e_i e_i^T\right\}_{i=1}^{l}, \left\{\frac{1}{4}\left(e_k e_s^T + e_s e_k^T\right) + \frac{1}{2} e_i e_i^T\right\}_{\substack{i=1\\ k=1\\ s=k+1}}^{l}, \left\{\frac{1}{4}\left(e_i e_j^T + e_j e_i^T\right) + \frac{1}{4}\left(e_i e_s^T + e_s e_i^T\right)\right\}_{\substack{i=1\\ j=i+1\\ s=j+1}}^{l}, \right.$$
$$\left. \left\{\frac{1}{4}\left(e_i e_j^T + e_j e_i^T\right) + \frac{1}{4}\left(e_k e_s^T + e_s e_k^T\right)\right\}_{\substack{i=1\\ j=i+1\\ k=i+1\\ s=k+1}}^{l} \right\} \tag{23}$$

4. Networked LPV MPC strategy

In this section, the availability of $\alpha(t)$ will be taken into account by referring to the multi-model state space description (14)-(16) with the uncertainty characterized by the polytope vertices (23). This means that the **NCS** problem is solved by resorting to an uncertain (robust) paradigm where the hypothesised parameter knowledge is used to develop a more insight state space description.

4.1. *Off-line Phase*

In the first part of this section, for the sake of notational simplicity and without loss of generality the **DD** and **IOD** acronyms will be dropped out.

Recall that given the plant (1) under a time-delay free scenario the sets of states i-step controllable to $\mathcal{T}$ are determined via the following recursion:

$$\begin{array}{rl} \mathcal{T}_0 := & \mathcal{T} \subset \mathcal{X} \\ \mathcal{T}_i := & \{x_p \in \mathcal{X} : \exists \bar{u} \text{ such that } u_j \in \mathcal{U}, j=1,\dots,l,: \forall \alpha \in \mathcal{P}_{l_c}, \bar{\Phi}(\bar{\alpha})x_p + \bar{G}(\bar{\alpha})\bar{u} \in \mathcal{T}_{i-1}\} \end{array} \tag{24}$$

where $\mathcal{T}_i$ is the set of states that can be steered into $\mathcal{T}_{i-1}$ using a single control move [3]. To extend such a concept to the proposed framework, it is important to notice that the one-step state prediction needs to be evaluated along the model

$$x(t+1) = \bar{\Phi}(\bar{\alpha}(t))x(t) + \bar{G}(\bar{\alpha}(t))\bar{u}(t) \tag{25}$$

with $\bar{u}(t)$ chosen according to a delayed state plant information, i.e. $x(t) = x_p(t-\tau(t))$. Therefore, at each time instant t there exists a difference between the process measurement $x_p(t)$ and the model state $x(t)$ that must be adequately addressed. By following the developments of [10], we will resort to the following transition map

$$x(t+1) = \bar{\Phi}(\bar{\alpha}(t))x(t-\tau(t)) + \bar{G}(\bar{\alpha}(t))\bar{u}(t), \tau(t) \in [0, \bar{\tau}] \tag{26}$$

in order to compute the sequence of sets $\mathcal{T}_i$.

By considering the auxiliary state $y(t)$ the following descriptor form results

$$\begin{bmatrix} x(t+1) \\ 0 \end{bmatrix} = \begin{bmatrix} y(t)+x(t) \\ -y(t)+\bar{\Phi}(\bar{\alpha}(t))x(t-\tau(t))+\bar{G}(\bar{\alpha}(t))\bar{u}(t)-x(t) \end{bmatrix} \tag{27}$$

Hence, by noticing that $x(t-\tau(t)) = x(t) - \sum_{j=t-\tau(t)}^{t-1} y(j)$ and by imposing the worst-case occurrence on the time-varying delay $\tau(t)$, i.e. $\tau(t) = \bar{\tau}$, $\forall t$, we have that

$$\hat{E}\hat{x}(t+1) = \hat{\Phi}(\bar{\alpha}(t))\hat{x}(t) + \hat{G}(\alpha)\bar{u}(t) - \hat{G}_y(\bar{\alpha}(t)) \sum_{j=t-\bar{\tau}}^{t-1} y(j) \tag{28}$$

where $\hat{x}(t) = [x(t)^T \; y(t)^T]^T$, $\hat{E} = \text{diag}\{I, 0\}$,

$$\hat{\Phi}(\bar{\alpha}(t)) = \begin{bmatrix} I & I \\ \bar{\Phi}(\bar{\alpha}(t)) - I & -I \end{bmatrix}, \hat{G}(\bar{\alpha}(t)) = \begin{bmatrix} 0 \\ \bar{G}(\bar{\alpha}(t)) \end{bmatrix}, \hat{G}_y(\bar{\alpha}(t)) = \begin{bmatrix} 0 \\ \bar{\Phi}(\bar{\alpha}(t)) \end{bmatrix}$$

The following proposition provides inner approximations of the exact one-step controllable sets related to the description (26).

Proposition 1 *Let $\mathcal{T}_0 \neq \emptyset \subset \mathcal{X}$ be a given robustly invariant ellipsoidal region complying with the input constraints and $\hat{x}_{aug} = \left[x^T \; y^T \; z_1^T \; z_2^T\right]^T$ the augmented state describing the dynamics (28) with $z_1, z_2 \in \mathbf{R}^n$ accounting for all the cumulative sum vectors $y(j)$, $j = t-\bar{\tau}, \dots, t-1$. Then, the ellipsoidal sets sequence*

$$\begin{array}{rcl} \mathcal{E}_0 & = & \mathcal{T}_0 \\ \mathcal{E}_i & = & Proj_x\{In[\hat{x}_{aug} \in \mathbf{R}^{4n} \, with \, x \in \mathcal{X}, z_1, z_2 \in \mathcal{E}_{i-1} : \exists \bar{u} \text{ such that } u_j \in \mathcal{U}, j=1,\dots,l, \forall \alpha \in \mathcal{P}_l, \\ & & Proj_x\{\hat{\Phi}(\bar{\alpha})_{aug}\hat{x}_{aug} + \hat{G}(\bar{\alpha})_{aug}\bar{u}\} \in \mathcal{E}_{i-1}]\}, \end{array}$$

$$\hat{\Phi}(\bar{\alpha})_{aug} = \begin{bmatrix} I & I & 0 & 0 \\ \bar{\Phi}(\bar{\alpha}) - I & -I & -\bar{\tau}\bar{\Phi}(\bar{\alpha}) & \bar{\tau}\bar{\Phi}(\bar{\alpha}) \\ 0 & 0 & 0 & I \\ 0 & 0 & 0 & \bar{\Phi}(\bar{\alpha}) \end{bmatrix}, \hat{G}(\bar{\alpha})_{aug} = \begin{bmatrix} 0 & \bar{G}(\bar{\alpha}) & 0 & 0 \end{bmatrix}^T \tag{29}$$

if non-empty, satisfies $\mathcal{E}_i \subset \mathcal{T}_i$.

Proof - The proof follows *mutatis mutandis* the same arguments of [10]. □

From now on we will assume that for (14) there exist

(i) **IOD** and **DD** stabilizing feedback control laws

$$\bar{u}(t) = K_{IOD}x_p(t-\tau(t)), \forall \tau(t) \leq \bar{\tau},$$
$$\bar{u}(t) = K_{DD}x_p(t-\tau(t)), \forall \tau(t) \leq \tau_{max} \leq \bar{\tau}$$

(ii) **IOD** and **DD** non-empty robust positively invariant regions for the closed-loop state evolutions

$$\mathcal{E}_{IOD} \triangleq \{x_p \in \mathbb{R}^n \,|\, x_p^T Q_{IOD} x_p \leq 1\} \subset \mathbb{R}^n,$$
$$\mathcal{E}_{DD} \triangleq \{x_p \in \mathbb{R}^n \,|\, x_p^T Q_{DD} x_p \leq 1\} \subset \mathbb{R}^n$$

complying with input and state constraints (4)-(5), see [13, 10] for technical details.

Finally, the same reasoning on the time-delay occurrence management outlined in [10] can be here applied: two nested one-step controllable ellipsoidal sequences with $N+1$ elements ($N>0$) $\{\mathcal{T}_i^{DD}\}_{i=0}^N$ and $\{\mathcal{T}_i^{IOD}\}_{i=0}^N$, have to be computed by using recursions (29) under the additional condition

$$\mathcal{T}_0^{DD} \subseteq \mathcal{T}_N^{IOD} \tag{30}$$

4.2. On-line phase

We will resort to a time-stamp mechanism in order to characterize the data-packet dispatch along the sensor-to-controller and controller-to-actuator channels. Specifically, the event (x_p, t_x) denotes the state measurement x_p sent at the time instant t_x, whereas (u, t_u) has the following meaning: u is the command sent by the controller and marked by the stamp t_u. Moreover, the stored command u_{-1}^R is time stamped with $t_{u_{-1}^R}$.

On the controller side, at each instant t the computation of the control action $u(t)$ is performed by using the following arguments:

- *Consider the most recent data-packet* (x_p, t_x);
- *Determine the set containing* x_p : *if* $x_p \in \mathcal{T}_i^{IOD}$ *then the set* $\mathcal{T}_i^{IOD}$ *is selected, otherwise find the smallest index i such that* $x_p \in \mathcal{T}_i^{DD}$;
- *Solve the following optimization:*
  - *if* $x_p(t-\tau(t)) \in \mathcal{T}_i^{IOD}$:

$$\bar{u}(t) = \arg\min J_{i(t)}(x_p(t-\tau(t)), \bar{u}) \quad \textit{subject to} \tag{31}$$

$$\begin{gathered}\bar{\Phi}(\bar{\alpha})x_p(t-\tau(t))\,(t) + \bar{G}(\bar{\alpha})\bar{u} \in \mathcal{T}_{i(t)-1}^{IOD}, \\ \bar{\alpha} \in \mathcal{P}_{l_c}, u_j \in \mathcal{U}, j=1,\ldots,l,\end{gathered} \tag{32}$$

  - *else* $x_p(t-\tau(t)) \in \mathcal{T}_i^{DD}$:

$$\bar{u}(t) = \arg\min J_{i(t)}(x_p(t-\tau(t)), \bar{u}) \quad \textit{subject to} \tag{33}$$

$$\begin{gathered}\bar{\Phi}(\bar{\alpha})x_p(t-\tau(t))\,(t) + \bar{G}(\bar{\alpha})\bar{u} \in \mathcal{T}_{i(t)-1}^{DD}, \\ \bar{\alpha} \in \mathcal{P}_{l_c}, u_j \in \mathcal{U}, j=1,\ldots,l,\end{gathered} \tag{34}$$

where

$$J_{i(t)}(x(t), \bar{u}) = \max_j \|\bar{\Phi}_j x_p(t-\tau(t))\,(t) + \bar{G}_j \bar{u}\|^2_{P^{(\cdot)}_{i(t)-1}} + \|\bar{u}\|^2_{R_u} \tag{35}$$

with $R_u = R_u^T > 0$, *and* $P_{i(t)-1}^{(\cdot)} > 0$ *the shaping matrix of the one-step controllable set* $\mathcal{T}_{i(t)-1}^{(\cdot)}$. *The apex notation* $(\cdot)$ *refers to both* **IOD** *and* **DD** *acronyms.*

Journal of Physics: Conference Series **570** (2014) 032001 doi:10.1088/1742-6596/570/3/032001

Notice that when the delayed state belongs to some $\mathcal{T}_i^{DD}$, the command $u(t)$ can be computed using the optimization (31)-(32) with $\mathcal{T}_i^{DD}$ in place of $\mathcal{T}_i^{IOD}$ and the feasibility is retained thanking to (30).

All the above developments allows one to write down a computable MPC scheme:

NCS-LPV-MPC-Algorithm

Initialization

0.1 Given the scalars $\bar{\tau}_c$, τ_{max} and $\bar{\tau}$, compute the non-empty robust invariant ellipsoidal regions $\mathcal{T}_0^{DD} \subset \mathbb{R}^n$, $\mathcal{T}_0^{IOD} \subset \mathbb{R}^n$ and the stabilizing state feedback gains K_{DD}, K_{IOD} satisfying (4)-(5) $\forall \tau(t) \leq \bar{\tau}$;

0.2 Generate and store the sequences of N one-step controllable sets $\mathcal{T}_i^{DD}$ and $\mathcal{T}_i^{IOD}$ complying with (30).

On-line phase

SENSOR SIDE - *for all $t \in \mathbb{Z}_+$*

1.1 send the packet (x_p, t_{x_p}) with $t_{x_p} = t$ the time-stamp;

CONTROLLER SIDE - *for all $t \in \mathbb{Z}_+$ and for all arrived packets (x_p, t_{x_p})*

1.1 compute $\tau(t) = \tau_m(t) + \bar{\tau}_c$ with $\tau_m(t) = t - t_{x_p}$;

1.1.1 If there exists $i(t) := \min\{i : x_p \in \mathcal{T}_i^{IOD}\}$ then
- *a- If $i(t) = 0$ then $\bar{u}(t) = K_{IOD} x_p$*
- *b- else solve (31)-(32);*

1.1.2 else find $i(t) := \min\{i : x_p \in \mathcal{T}_i^{DD}\}$
- *a- If $i(t) = 0$ then $\bar{u}(t) = K_{DD} x_p$*
- *b- else solve (33)-(34);*

1.2 Send the packet $(\bar{u}, t_{\bar{u}})$.

ACTUATOR SIDE - *for all $t \in \mathbb{Z}_+$ and for all arrived packets $(\bar{u}, t_{\bar{u}})$*

1.1 If $t_{\bar{u}} > t_{u_{-1}^R}$ then $u_{-1}^R := \bar{u}$, $t_{u_{-1}^R} := t_{\bar{u}}$ and apply $\bar{u}$;

1.2 else apply u_{-1}^R.

The next proposition proves feasibility retention and closed-loop stability of the proposed **NCS-LPV-MPC** Algorithm.

Proposition 2 *Let the set sequences $\mathcal{T}_i^{DD}$ and $\mathcal{T}_i^{IOD}$ be non-empty and $x_p(0) \in \mathcal{T}_N^{DD} \cup \mathcal{T}_N^{IOD}$. Then, the* **NCS-LPV-MPC** *algorithm always satisfies the constraints and ensures asymptotic stability for all bounded time-delay occurrences.*

Proof - On the controller side, existence of solutions at time t implies existence of solutions at time $t+1$, because the optimization problems in steps 1.1.1 and 1.1.2 are always feasible and under the boundedness of $\tau_c(t)$ the timely arrival feature of the computed command $\bar{u}(t)$ is guaranteed. Let us consider without loss of generality that at the generic instant t the state $x_p(t - \tau(t)) \in \mathcal{T}_{i(t)}^{DD}$. First note that by construction there exists an input vector $\bar{u}$ satisfying (4)-(5) such that the one-step state evolution $\bar{\Phi}(\bar{\alpha})x_p(t-\tau(t))\,(t) + \bar{G}(\bar{\alpha})\bar{u}$ belongs to $\mathcal{T}_{i(t)-1}^{DD}$, $\forall \tau(t)$ with $\bar{\tau}$. Then, thanking to the requirement (30) and to recursions (29), at the next time instant $t+1$ $u(t)$ drives $\bar{\Phi}(\bar{\alpha})x_p(t-\tau(t))\,(t) + \bar{G}(\bar{\alpha})\bar{u}$ into $\mathcal{T}_{i(t)-1}^{DD}$. Therefore, the existence of a solution $\bar{u}(t+1)$ for steps 1.1.1 and 1.1.2 is ensured.

On the actuator side the feasibility is guaranteed by using the properties of the ellipsoidal sequences $\mathcal{T}_i^{DD}$ and $\mathcal{T}_i^{IOD}$. In fact, at each time instant the actuator logic is able to apply an admissible control input because the last stored command u_{-1}^R is updated at most after $\bar{\tau}$ time instants. The asymptotic stability follows from the same arguments. □

Tank 1	Value	Tank 2	Value
S_1	2500 cm^2	S_2	1600 cm^2
A_1	9 cm^2	A_2	4 cm^2
$\bar{h}_1$	35 cm	$\bar{h}_2$	200 cm
$\underline{h}_1$	1cm	$\underline{h}_2$	10 cm

Table 1. Tanks and constraints values

Parameters	Value
g	980 $cm/(sec^2)$
ρ	0.001 $Kg/(cm^3)$
V_{max}	4
T_c	0.5 sec

Table 2. Parameter values

5. A cascaded two-tank system

A two-tank process (see Figure 2) is used to evaluate the performance of the proposed **NCS-LPV-MPC** algorithm. A schematic diagram of the process is shown in Figure 2.b and the goal is to regulate the water level $h_2(t)$ (plant output) at a given set-point by acting on the incoming water flow via the supply pump voltage $V(t)$ (plant input).

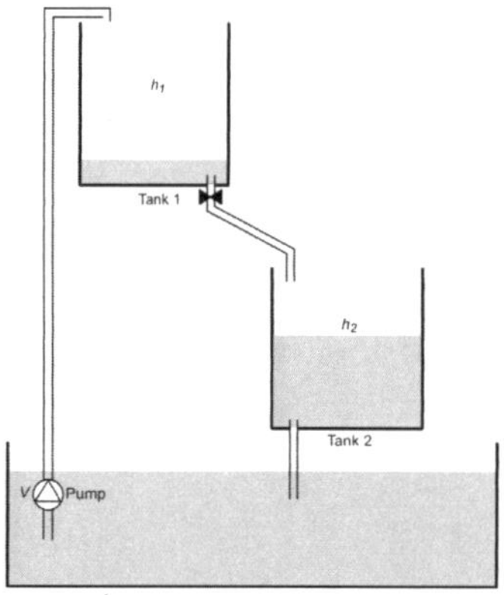

Figure 2. Plant description

The cascaded two-tank model is described by the following non-linear differential equations

$$\begin{cases} \rho S_1 \dot{h}_1 &= -\rho A_1 \sqrt{2gh_1} + u \\ \rho S_2 \dot{h}_2 &= \rho A_1 \sqrt{2gh_1} - \rho A_2 \sqrt{2gh_2} \end{cases} \tag{36}$$

where u is the water flow supplied by the pump whose command is the voltage V, S_i, $i = 1,2$, are the tank cross sections, h_i the water levels in the tanks, A_i the pipe sections connecting the tanks, g the gravity constant, and ρ the water density. A simple static equation is used to model the relationship between the input voltage $V(t)$ and the incoming water flow $u(t)$

$$u(t) = \begin{cases} V(t) & \text{if} \quad V(t) \geq 0 \\ 0 & \text{if} \quad V(t) < 0 \end{cases} \tag{37}$$

Moreover, the following constraints are enforced (see Tables I-II)

$$\underline{h}_1 \leq h_1 \leq \bar{h}_1,\ \underline{h}_2 \leq h_2 \leq \bar{h}_2,\ 0 \leq V \leq V_{max} \tag{38}$$

An LPV multi-model description of (36) has been exploited by following the lines indicated in [6]. Hence, the plant has been discretized with a sampling time $T_c = 0.5$ sec. and it is required that h_2 tracks a set-point r whose value changes from 105 to 107 at $t = 100$ sec. (see Fig. 8 - dashed line). It is supposed that the initial water levels h_1 and h_2 are equal to $[11\ 112]^T$.

In order to implement the **NCS-LPV-MPC** strategy, a test rig has been built up: it includes two computers and the two tank process (see Fig. 3). The notebook marked as A in Fig. 3 is a Toshiba

Satellite A665-S6058 Intel©Core^{TM} i5-450M Processor (3M cache, 2.40 GHz 4gb RAM) and acts as the software interface of both the actuator and sensor units within the control scheme of Fig. 1. The second PC (an Alienware M15x with processor Intel©Core^{TM} i7 940XM (8M Cache, 2.13 GHz 4GB RAM)) has the role of the controller and runs Matlab 2012b©. The signals are physically transmitted between two Intranet IP addresses 160.97.27.112 and 160.97.27.54, which are both located on the department WiFi Network DEIS42c3 physically implemented on Ubiquiti©PicoStation^{TM} M2 HP (see Fig. 5) and the TCP protocol is used. Moreover to simulate the network communication latency, delays have been artificially added into the Intranet environment (see Fig. 4) so that the network MATI is $\bar{\tau} = 1.8\,sec.$ (i.e. 4 time units in the proposed framework), while $\tau_{max} = 1\,sec.$ (2 time units) has been estimated by applying the heuristics described in [10].

Figure 3. Networked control test rig, where computer A is the sensor and the actuator and computer B is the controller

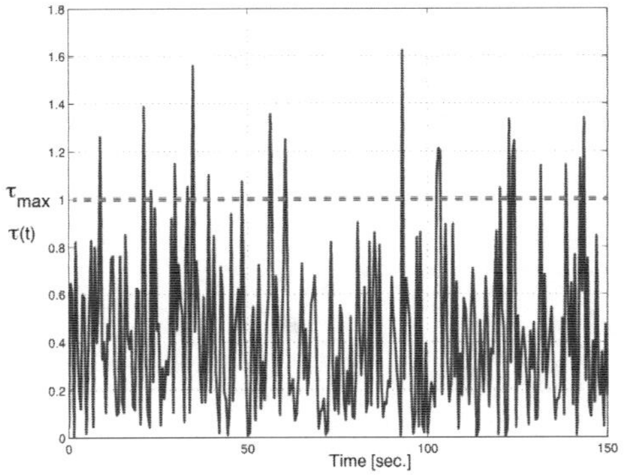

Figure 4. Network Latency

Finally, in order to characterize the set of admissible disturbances/measurement errors acting on the tanks, the following region

$$\mathcal{D} = \{d \in \mathbb{R}^2 \,|\, d^T d \leq 0.5\}$$

has been considered and used in the proposed RHC setting as suggested in [7]. For this simulation, we have used the scenario of Fig. 4 which depicts the cumulative time-delay occurring at each time instant on the network. A family of 50 regions $\mathcal{T}_i$ centered at $[20.74\ 105]^T$ has been off-line computed by using recursions (29), see Fig. 5.

All the relevant results achieved under the action of the proposed **NCS-LPV-MPC** control strategy are summarized in Figs. 6-8. First, it is worth to remark that both input and state constraints are always fulfilled. Then, Fig. 8 shows a good capability of the proposed scheme to reach the prescribed set-points regardless of any time-delay occurrence on the packet transmission. Finally, it is important to underline that the on-line burdens, evaluated by computing the average CPU time (seconds per step), confirms a modest computational load ($0.0283\,sec.$) when compared to the sampling time T_c (about one order less). This is relevant because it can be viewed as a tangible clue that the proposed approach can be affordable in practical applications

6. Conclusions

In this paper, a receding horizon strategy for LPV polytopic networked systems subject to input state and communication constraints has been proposed. First, by exploiting ellipsoidal calculus and viability theory, a novel fast ellipsoidal MPC scheme capable to properly manage time-delay occurrences on the network channels is proposed. Then, it is worth to underline that the requested computational effort and resources requirements appear modest so rendering more simple the applicability of such an approach.

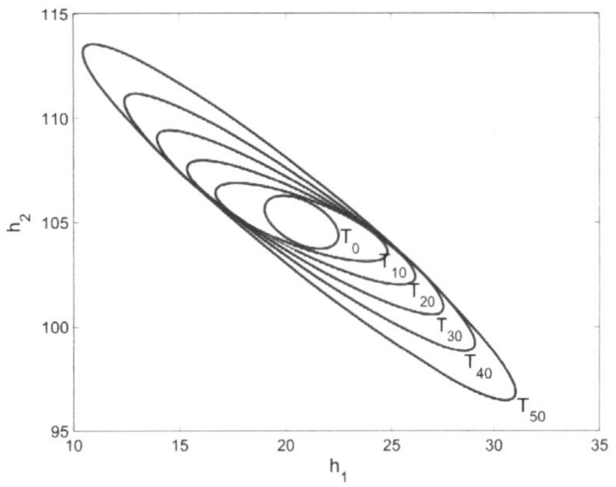

Figure 5. Controllable sets sequence computation for the **NCS-LPV-MPC** algorithm

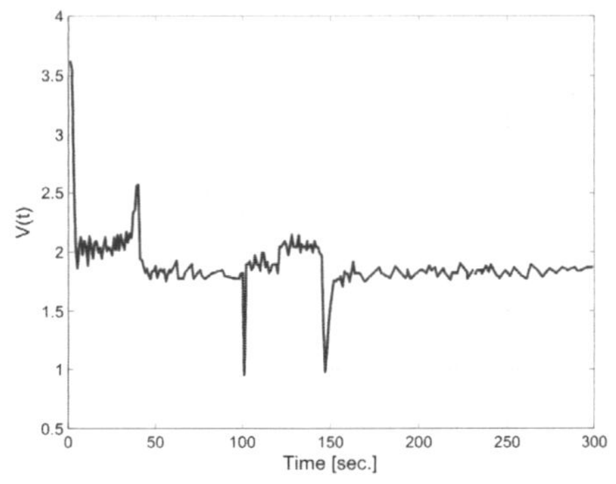

Figure 6. Regulated Plant input: $0 \leq u \leq 4$

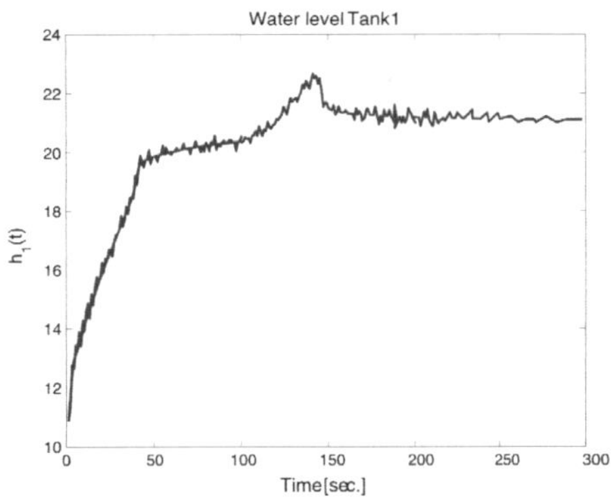

Figure 7. Regulated Water level Tank 1: $1 \leq h_1 \leq 35$

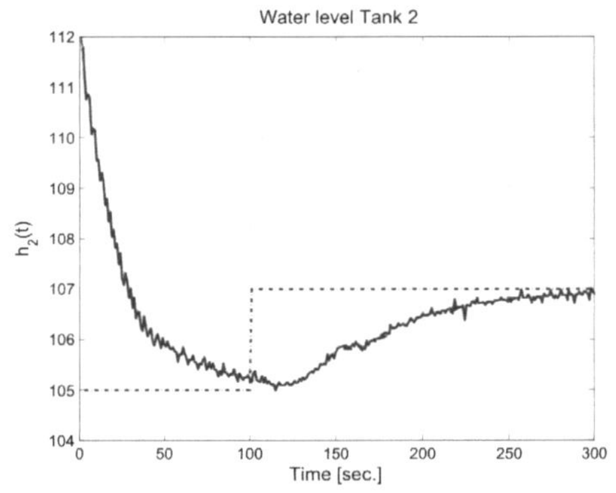

Figure 8. Regulated Water level Tank 2: $10 \leq h_2 \leq 200$

The effectiveness of the proposed scheme has been shown by means of an illustrative example that has put in evidence the **NCS-LPV-MPC** capability to on-line take care of time-varying delays on both sensor-to-controller and controller-to-actuator links.

Acknowledgments
This work has been partially supported by the European Commission, the European Social Fund and the Calabria Region. The authors are solely responsible for the content of this paper and the European Commission and Calabria Region disclaim any responsibility for the use that may be made of the information contained therein.

7. References

[1] D. Angeli, A. Casavola, G. Franzè and E. Mosca, "An Ellipsoidal Off-line MPC Scheme for Uncertain Polytopic Discrete-time Systems", *Automatica*, Vol. 44, pp. 3113–3119, 2008.
[2] J. Ballieul and P.J. Antsaklis "Control and Communication Challenges in Networked Real-Time Systems", *Proceedings of the IEEE*, Vol. **95**, No. 1, pp. 9–28, 2007.
[3] F. Blanchini and S. Miani. "Set-Theoretic Methods in Control", *Birkäuser*, Boston, 2008.

[4] S. Boyd, L. El Ghaoui, E. Feron and V. Balakrishnan, "Linear Matrix Inequalities in System and Control Theory", *SIAM Studies in Applied Mathematics*, SIAM, London.
[5] A. Casavola, D. Famularo and G. Franzè, "A feedback minmax predictive control algorithm for LPV systems subject to bounded rates of change of parameters", *IEEE Transactions on Automatic Control*, Vol. 47, pp. 1147-1153, 2002.
[6] A. Casavola, D. Famularo and G. Franzè, "Predictive control of constrained nonlinear systems via LPV linear embeddings", *International Journal of Robust and Nonlinear Control*, Vol. 13, No. 3-4, pp. 281-294, 2003.
[7] A. Casavola, D. Famularo, G. Franzè and E. Garone, "A fast ellipsoidal MPC scheme for discrete-time polytopic linear parameter varying systems", *Automatica*, Vol. 48, No. 10, pp. 2620–2626, 2012.
[8] G. Franzè, E. Garone, D. Famularo and A. Casavola, "A dilated MPC control strategy for LPV systems with a time-varying terminal set", *IET Control Theory & Applications*, Vol. 3, No. 1, pp. 110-120, 2009.
[9] D. M. de la Peña and P. D. Christofides, "Lyapunov-Based Model Predictive Control of Nonlinear Systems Subject to Data Losses", *IEEE Trans. Aut. Contr.*, Vol. 53, pp. 2076-2089, 2008.
[10] G. Franzè, D. Famularo and F. Tedesco, "Receding horizon control for constrained networked systems subject to data-losses", *Proc. of the 50th CDC-ECC*, Orlando, Florida, USA, 2011.
[11] D. Famularo, G. Franzè, and W. Lucia. "Networked control systems with state, input and communication constraints: A nonlinear approach". *Proc. of the 51th IEEE Conference on Decision and Control*, Maui, HI, 2012.
[12] G. Franzè and F. Tedesco, "Networked control systems: a polynomial receding horizon approach", *IEEE Transactions on Control of Network Systems*, doi: 10.1109/TCNS.2014.2357502, September 2014.
[13] E. Fridman and U. Shaked. "Delay-dependent H-infinity control of uncertain discrete delay systems", *European Journal of Control* , Vol. 11, pp. 29–37, 2005.
[14] H. Gao and T. Chen. "New Results on Stability of Discrete-Time Systems With Time-Varying State Delay", *IEEE Trans. Aut. Contr.*, Vol. 52, No. 2, pp. 328-334, 2007.
[15] W. W. Che, J. L. Wang, G. H. Yang, "Observer-based H_{infty} control in multiple channel networked control systems with random packet dropouts", *Journal of Control Theory and Applications*, Vol. 8, pp. 359-367, 2010.
[16] J. Hespanha, P. Naghshtabrizi and Y. Xu, "A survey of recent results in networked control systems," *Proc. of the IEEE*, Vol. 95, pp. 138-162, 2007.
[17] Y. Jianyong, Y. Shimin, and W. Haiqing, "Survey on the performance analysis of networked control systems," *Proceedings of IEEE Int. Conf. on System, Man and Cyber.*, The Hague, Netherlands, pp. 5068-5073, 2004.
[18] C. Lin, Z. D. Wang, F. W. Yang, "Observer-based networked control for continuous-time systems with random sensor delays", *Automatica*, Vol. 45, pp. 578-584, 2009.
[19] L. A. Montestruque and P. J. Antsaklis, "On the model-based control of networked systems," *Automatica*, Vol. 39, pp. 1837-1843, 2003.
[20] L. A. Montestruque and P. J. Antsaklis, "Stability of Model-Based Networked Control Systems With Time-Varying Transmission Times," *IEEE Trans. Aut. Contr.*, Vol. 49, pp. 1562-1572, 2004.
[21] G. Pin and T. Parisini, "Networked Predictive Control of Uncertain Constrained Nonlinear Systems: Recursive Feasibility and Input-to-State Stability Analysis", *IEEE Trans. Aut. Contr.*, Vol. 56, No.1, pp. 72-87, 2011.
[22] B. Pluymers, J. Rossiter, J.A.K. Suykens and B. De Moor, "Interpolation based MPC for LPV systems using polyhedral invariant sets" *American Control Conference 2005*, pp. 810-815, 2005.
[23] D. E. Quevedo and V. Gupta, "Sequence-Based Anytime Control", *IEEE Trans. Aut. Contr.*, Vol. 58, pp. 377-389, 2013.
[24] M. Sun, Y. Jia, J. Du and S. Yuan, "Delay-dependent H_{infty} control for LPV systems with time delays", *IEEE Conference on Decision and Control*, New Orleans, LA, pp. 2089-2093, 2007
[25] G.Walsh, O. Beldiman, and L. Bushnell, "Asymptotic behavior of nonlinear networked control systems", *IEEE Trans. Automat. Control.*, Vol. 46, No. 7, pp. 1093–1097, Jul. 2001.
[26] F. Wu and K. M. Grigoriadis, "LPV systems with parameter-varing time delays: analysis and control", *Automatica*, Vol. 37, pp. 221-229, 2001.

Robust MPC for a non-linear system - a neural network approach

Marcel Luzar, Marcin Witczak

Institute of Control and Computation Engineering, University of Zielona Góra, ul. Podgórna 50, 65–246 Zielona Góra, Poland

E-mail: {m.luzar,m.witczak}@issi.uz.zgora.pl

Abstract.
The aim of the paper is to design a robust actuator fault-tolerant control for a non-linear discrete-time system. Considered system is described by the Linear Parameter-Varying (LPV) model obtained with recurrent neural network. The proposed solution starts with a discrete-time quasi-LPV system identification using artificial neural network. Subsequently, the robust controller is proposed, which does not take into account actuator saturation level and deals with the previously estimated faults. To check if the compensation problem is feasible, the robust invariant set is employed, which takes into account actuator saturation level. When the current state does not belong to the set, then a predictive control is performed in order to make such set larger. This makes it possible to increase the domain of attraction, which makes the proposed methodology an efficient solution for the fault-tolerant control. The last part of the paper presents an experimental results regarding wind turbines.

1. Introduction

A continuous growth of the complexity, efficiency, and reliability of modern industrial systems necessitates a continuous development in control and fault diagnosis. Combinations of these two paradigms are intensively studied under the name of Fault-Tolerant Control (FTC). The Fault-Tolerant Control systems are classified into two distinct classes: passive and active. In the first one, controllers are designed to deal with a set of predefined faults by being robust. Therefore there is no necessity for fault diagnosis, but this kind of design usually degrades the overall performance. Contrary, the active FTC schemes [7] react to faults by reconfiguring or redesigning control scheme, simultaneously preserving system stability and acceptable performance. To achieve that, the control system relies on the Fault Detection and Isolation (FDI) [6, 18, 20] as well as an accommodation technique [5]. Most of the past and current works treat the FDI and FTC problems individually. Since the perfect FDI and fault identification are impossible, there always exists related inaccuracy. Thus, there is a need for integrated FDI and FTC schemes for linear as well as non-linear systems.

The approach proposed in this paper provides an elegant way of fault diagnosis (particularly fault identification) employed in fault-tolerant control framework. The proposed method is based on a triple stage procedure that starts from fault estimation, subsequently the fault is compensated with a robust controller. The robust controller is designed without considering input constraints related with the actuator saturation. Thus, to check the compensation feasibility, the robust invariant set is introduced. It takes into account the input constraints.

If the current state does not belong to the robust invariant set, a predictive control scheme enhances the invariant set. This appealing phenomenon makes possible to increase the domain of attraction, making the proposed framework an efficient solution for the FTC. The presented solution can be perceived as an extension of the recent developments in this area [14, 15, 21], which shows a fault estimation and compensation strategy for non-linear systems. The novelty of the scheme holds:

- robustness to exogenous disturbances, through the $\mathcal{H}_\infty$ approach,
- triple stage procedure: fault estimation, fault compensation with robust controller, and predictive control enhancing the applicability of the approach,
- extension of the work of [12] to the case with exogenous disturbances,
- introduction of robust invariant set extending the usual framework proposed by [12].

The paper is organised as follows. Section 2 presents a method for transforming a neural state-space model into a discrete-time polytopic quasi-LPV model. Section 3 presents preliminaries regarding the problem being undertaken. Robust fault estimation and control approach is proposed in Section 4. Subsequently, Section 5 presents the development of a robust invariant set while Section 6 presents an efficient robust predictive fault-tolerant control strategy, which enhances the performance of the overall scheme. The final part of the paper contains a numerical example, which shows the performance of the proposed approach.

2. Neural network into quasi-LPV model: transformation method

A non-linear dynamic system can be described in a relatively simple way by a quasi-LPV model. According to this model, a non-linear dynamic systems can be linearised around a number of operating points. Each of these linear models represents the local system behaviour around the operating point. Let us consider the following discrete-time non-linear model:

$$\boldsymbol{x}_{k+1} = \boldsymbol{g}\left(\boldsymbol{x}_k, \boldsymbol{u}_k\right), \tag{1}$$

$$\boldsymbol{y}_k = \boldsymbol{C}(\boldsymbol{x}_k). \tag{2}$$

The goal of this section is to represent this model in the form of a discrete-time polytopic quasi-LPV model:

$$\boldsymbol{x}_{k+1} = \boldsymbol{A}(h_k)\boldsymbol{x}_k + \boldsymbol{B}(h_k)\boldsymbol{u}_k, \tag{3}$$

$$\boldsymbol{y}_k = \boldsymbol{C}(h_k)\boldsymbol{x}_k, \tag{4}$$

where $\boldsymbol{A}(h_k)$, $\boldsymbol{B}(h_k)$, $\boldsymbol{C}(h_k)$ are state-space matrices and $h_k \in \mathbb{R}^l$ is a time-varying parameter vector which ranges over a fixed polytope. The dependence of the $\boldsymbol{A}$,$\boldsymbol{B}$,$\boldsymbol{C}$ on h_k represents a general quasi-LPV model. The model considered in experiments has matrices $\boldsymbol{A}$ and $\boldsymbol{B}$ parameter dependent. Bearing this in mind, it is possible to simplify (3)-(4) into following form:

$$\boldsymbol{x}_{k+1} = \boldsymbol{A}(h_k)\boldsymbol{x}_k + \boldsymbol{B}(h_k)\boldsymbol{u}_k, \tag{5}$$

$$\boldsymbol{y}_k = \boldsymbol{C}\boldsymbol{x}_k, \tag{6}$$

where matrix $\boldsymbol{C}$ is known. Vector h_k is assumed to depend on vector of measurable signals $\boldsymbol{\rho}_k \in \mathbb{R}^r$ referred to scheduling signals, according to:

$$h_k = \boldsymbol{s}(\boldsymbol{\rho}_k) \tag{7}$$

where $\boldsymbol{s} \in \mathbb{R}^r \rightarrow \mathbb{R}^l$ is continuous mappings. Note that in the quasi-LPV systems, input and output determines all or some scheduling parameters. A polytope can be presented as a matrix and is defined as a convex hull of a finite number of matrices $\boldsymbol{N}_i$ with the same dimension:

$$\mathrm{Co}\{\boldsymbol{N}_i,\, i = 1, \ldots, l\} := \left\{ \sum_{i=1}^{l} \alpha_i \boldsymbol{N}_i,\, \sum_{i=1}^{l} \alpha_i = 1,\, \alpha_i \geq 0 \right\}. \tag{8}$$

The time varying parameter h_k varies in a polytope $\boldsymbol{\Theta}$, which is assumed to be a compact set, with vertices $v_1, v_2, \ldots, v_r$, that is:

$$h_k \in \boldsymbol{\Theta} := \mathrm{Co}\{v_1, v_2, \ldots, v_r\}. \tag{9}$$

Thus, the problem considered is to transform the neural state-space model (1)-(2) into a polytopic quasi-LPV model (3)-(4) which has the above properties with

$$[\boldsymbol{A}(h_k)\, \boldsymbol{B}(h_k)] \in \tilde{\boldsymbol{P}}_h := \mathrm{Co}\{[\boldsymbol{A}_i\, \boldsymbol{B}_i], \quad i = 1, \ldots, l\} \tag{10}$$

where $\tilde{\boldsymbol{P}}_h \subset \mathbb{R}^l$. To construct the above quasi-LPV model, the Multi-Layer Perceptron (MLP) feedforward neural network is proposed. From neural network theory it is known that it is possible to estimate the states to a desired accuracy with a single hidden layer with l neurons (assuming l is chosen large enough) [8]. Thus, the neural network structure is similar to that proposed in [13] and presented in Fig. 1. Moreover, the modified method described in [1] is used. Let us assume, that the non-linear system dynamics are represented in the following neural

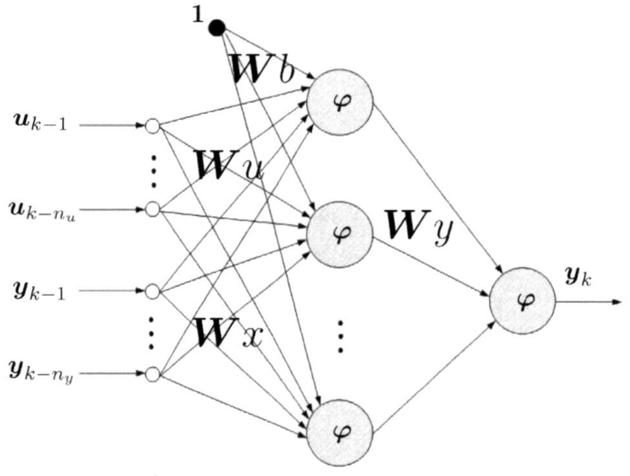

Figure 1. The structure of dynamic neural model

state-space form, which has zero bias in the output layer:

$$\bar{\boldsymbol{x}}_{k+1} = \boldsymbol{W}_y\, tanh\, (\boldsymbol{W}_x \bar{\boldsymbol{x}}_k + \boldsymbol{W}_u \bar{\boldsymbol{u}}_k + \boldsymbol{W}_b) \tag{11}$$

$$\bar{\boldsymbol{y}}_k = \boldsymbol{C} \bar{\boldsymbol{x}}_k \tag{12}$$

where vector $\bar{\boldsymbol{x}}_{k+1} \in \mathbb{R}^p$ is the state, $\bar{\boldsymbol{u}}_k \in \mathbb{R}^m$ is a control signal and $\boldsymbol{W}_y \in \mathbb{R}^{p \times l}$, $\boldsymbol{W}_x \in \mathbb{R}^{l \times n}$, $\boldsymbol{W}_u \in \mathbb{R}^{l \times m}$ contain the output and hidden layer weights, respectively. $\boldsymbol{W}_b \in \mathbb{R}^l$ contains a set of biases in the hidden layer. During the neural network training, a satisfactory estimation values of the weight matrices are obtained. For a training process, the Levenberg-Marquardt algorithm is used. Note that the choice of the number of hidden neurons l will determine the number of scheduling parameters of the quasi-LPV model.

In order to remove the bias $\boldsymbol{W}_b$ from (11), the method first discussed in [2] is used, but for the sake of completeness it is reiterated in the following.

It is assumed that there exist an equilibrium point $(\bar{\boldsymbol{x}}_k, \bar{\boldsymbol{u}}_k) = (\bar{\boldsymbol{x}}_k^o, \bar{\boldsymbol{u}}_k^o)$ such that

$$0 = \boldsymbol{W}_y \, tanh\,(\boldsymbol{W}_x\bar{\boldsymbol{x}}_k^o + \boldsymbol{W}_u\bar{\boldsymbol{u}}_k^o + \boldsymbol{W}_b) \tag{13}$$

This equilibrium can be determined as

$$\begin{bmatrix} \bar{\boldsymbol{x}}_k^o \\ \bar{\boldsymbol{u}}_k^o \end{bmatrix} = -\bar{\boldsymbol{W}}^+\boldsymbol{W}_b \tag{14}$$

where $\bar{\boldsymbol{W}} = [\boldsymbol{W}_x \, \boldsymbol{W}_u]$ is assumed to have full column rank and $\bar{\boldsymbol{W}}^+$ is a right inverse. Then the network coordinates can be changed such that the new coordinates are

$$\boldsymbol{x}_k = \bar{\boldsymbol{x}}_k - \bar{\boldsymbol{x}}_k^o, \quad \boldsymbol{u}_k = \bar{\boldsymbol{u}}_k - \bar{\boldsymbol{u}}_k^o. \tag{15}$$

As a result, (11) can be written as

$$\boldsymbol{x}_{k+1} = \boldsymbol{W}_y \, tanh\,(\boldsymbol{W}_x\boldsymbol{x}_k + \boldsymbol{W}_u\boldsymbol{u}_k) \tag{16}$$

Now, define the following time varying parameter

$$h_k^j = \begin{cases} \frac{tanh(\boldsymbol{W}_x^j\boldsymbol{x}_k+\boldsymbol{W}_u^j\boldsymbol{u}_k)}{(\boldsymbol{W}_x^j\boldsymbol{x}_k+\boldsymbol{W}_u^j\boldsymbol{u}_k)} & , \quad (\boldsymbol{W}_x^j\boldsymbol{x}_k + \boldsymbol{W}_u^j\boldsymbol{u}_k) \neq 0 \\ 1 & , \quad (\boldsymbol{W}_x^j\boldsymbol{x}_k + \boldsymbol{W}_u^j\boldsymbol{u}_k) = 0, \end{cases} \tag{17}$$

for $1 \leq j \leq l$, where $\boldsymbol{W}_x^j$, $\boldsymbol{W}_u^j$ denote the jth rows in the respective hidden layer weight matrices. Then (16) can be rewritten as

$$\boldsymbol{x}_{k+1}^i = \boldsymbol{W}_y^i\boldsymbol{\Theta_d}\,(\boldsymbol{W}_x\boldsymbol{x}_k + \boldsymbol{W}_u\boldsymbol{u}_k) \tag{18}$$

for $1 \leq i \leq p$, where $\boldsymbol{\Theta_d} \in \mathbb{R}^{l \times l}$ is a diagonal matrix

$$\boldsymbol{\Theta_d} = \begin{bmatrix} h_k^1 & 0 & \dots & 0 \\ 0 & h_k^2 & \dots & 0 \\ \vdots & \vdots & \ddots & \vdots \\ 0 & 0 & \dots & h_k^l \end{bmatrix} \tag{19}$$

and $\boldsymbol{W}_y^i$ denotes the ith row in the outer layer weight matrix. In this way, the neural network model is transformed into a quasi-LPV model (5) where

$$\boldsymbol{A}(h_k) = \begin{bmatrix} \boldsymbol{A}_{11}(h_k) & \boldsymbol{A}_{12}(h_k) & \dots & \boldsymbol{A}_{1n}(h_k) \\ \boldsymbol{A}_{21}(h_k) & \boldsymbol{A}_{22}(h_k) & \dots & \boldsymbol{A}_{2n}(h_k) \\ \vdots & \vdots & \ddots & \vdots \\ \boldsymbol{A}_{p1}(h_k) & \boldsymbol{A}_{p2}(h_k) & \dots & \boldsymbol{A}_{pn}(h_k) \end{bmatrix} \tag{20}$$

with $\boldsymbol{A}_{ri}(h_k) = \sum_{j=1}^{l} w_y^{rj} h_k^j w_x^{ji}$, where w_y^{rj} and w_x^{ji} are the elements of $\boldsymbol{W}_y^j$ and $\boldsymbol{W}_x^j$, respectively,

$$\boldsymbol{B}(h_k) = \begin{bmatrix} \boldsymbol{B}_{11}(h_k) & \boldsymbol{B}_{12}(h_k) & \dots & \boldsymbol{B}_{1(m\times n_u)}(h_k) \\ \boldsymbol{B}_{21}(h_k) & \boldsymbol{B}_{22}(h_k) & \dots & \boldsymbol{B}_{2(m\times n_u)}(h_k) \\ \vdots & \vdots & \ddots & \vdots \\ \boldsymbol{B}_{p1}(h_k) & \boldsymbol{B}_{p2}(h_k) & \dots & \boldsymbol{B}_{p(m\times n_u)}(h_k) \end{bmatrix} \tag{21}$$

with $\boldsymbol{B}_{ri}(h_k) = \sum_{j=1}^{l} w_o^{rj} h_k^j w_u^{ij}$.

This completes the transformation of the neural state-space model (11)-(12) into a quasi-LPV model in the polytopic representation (3) with the condition (10) where $\boldsymbol{A}(h_k)$ is given by (20), $\boldsymbol{B}(h_k)$ is given by (21), $\boldsymbol{C}$ are known and depend on the system considered. The time varying parameter h_k is defined by (17).

3. A general description of the fault-tolerant scheme

Let us consider the non-linear discrete-time quasi-LPV model, where the state equation is given as:

$$\boldsymbol{x}_{f,k+1} = \boldsymbol{A}(h_k)\boldsymbol{x}_{f,k} + \boldsymbol{B}(h_k)\boldsymbol{u}_{f,k} + \boldsymbol{B}(h_k)\boldsymbol{f}_k + \boldsymbol{W}\boldsymbol{w}_k, \tag{22}$$

where $\boldsymbol{A}(h_k)$ and $\boldsymbol{B}(h_k)$ are known matrices with apropriate dimentions and h_k is a known time-varying parameter (with a slight abuse of notation, furthermore h should be interpreted as h_k), $\boldsymbol{x}_{f,k} \in \mathbb{R}^n$ stands for the state, $\boldsymbol{u}_{f,k} \in \mathbb{R}^r$ denotes the nominal control input, $\boldsymbol{f}_k \in \mathbb{R}^r$ describes the actuator fault and $\boldsymbol{w}_k \in l_2$ is an exogenous disturbance vector, while:

$$l_2 = \{\mathbf{w} \in \mathbb{R}^n | \ \|\mathbf{w}\|_{l_2} < +\infty\}, \quad \|\mathbf{w}\|_{l_2} = \left(\sum_{k=0}^{\infty} \|\boldsymbol{w}_k\|^2\right)^{\frac{1}{2}}. \tag{23}$$

Moreover, the control limits are given as follows:

$$-\bar{\boldsymbol{u}}_i \leq \boldsymbol{u}_{i,k} \leq \bar{\boldsymbol{u}}_i, \quad i = 1, \dots, r. \tag{24}$$

where $\bar{\boldsymbol{u}}_i > 0$, $i = 1, \dots, r$ are given control limits. In the sake of the simplicity, assumed limits are symmetrical around zero.

The subsequent part of the paper deals with the design the control strategy in such a way that the system (22) will converge to the origin irrespective of the presence of the fault $\boldsymbol{f}_k$. The proposed control scheme is:

$$\boldsymbol{u}_{f,j} = \begin{cases} -\boldsymbol{K}\boldsymbol{x}_j - \hat{\boldsymbol{f}}_{k-1} + \boldsymbol{c}_j, & j = k, \dots, k + n_c - 1, \\ -\boldsymbol{K}\boldsymbol{x}_j - \hat{\boldsymbol{f}}_{k-1}, & j \geq k + n_c. \end{cases} \tag{25}$$

where:

- n_c is the prediction horizon,
- $\boldsymbol{K}$ is the $\mathcal{H}_\infty$ robust controller (with respect to exogenous disturbances $\boldsymbol{w}_k$),
- $\hat{\boldsymbol{f}}_{k-1}$ is the fault estimate
- $\boldsymbol{c}_j$ is a vector introducing additional design freedom, which should be exploited when the fault compensation does not provide the expected results due to the actuator saturation.

Note that beyond the control horizon n_c, $\boldsymbol{c}_j$ is set to zero, which denotes the feasibility of the $\boldsymbol{H}_\infty$ control. Thus, the design of the proposed control strategy boils down to solving a set of problems:

- design a robust controller $\boldsymbol{K}$ that guarantees its convergence to the origin, with prescribed disturbance attenuation level in respect of $\boldsymbol{x}_{f,k}$
- fault $\boldsymbol{f}_k$ estimation,
- determine a set of states for which the robust controller along with the fault compensation (under the control constraints) is feasible,
- determine $\boldsymbol{c}_j$ in such a way as to enhance a set of states and, hence make the control problem feasible.

Since the general scheme is given, the remaining part of the paper is devoted to solving the above-mentioned design problems.

4. Fault estimation and robust control

In this section, the fault estimation technique will be proposed, which will be used to compensate the effect of a fault and feed the system to converge the system to the origin (along with the robust controller $\boldsymbol{K}$). Note that the designs of the fault estimator and the robust controller are realised for the unconstrained case. Moreover, the free control parameter $\boldsymbol{c}_j$ (cf. (25)) is set to zero.

Let us also assume that the system is controllable and $\boldsymbol{B}(h_k)$ is a full rank one. Thus, following [10], it is possible to compute $\boldsymbol{H}(h_k) = \boldsymbol{B}(h_k)^+$. Subsequently, multiplying (22) by $\boldsymbol{H}(h_k)$ and then extracting $\boldsymbol{f}_k$ gives:

$$\boldsymbol{f}_k = \boldsymbol{H}(h_k)\boldsymbol{A}(h_k)\boldsymbol{x}_{f,k+1} - \boldsymbol{u}_{f,k} - \boldsymbol{H}(h_k)\boldsymbol{W}\boldsymbol{w}_k, \tag{26}$$

while its estimate can be given as:

$$\hat{\boldsymbol{f}}_k = \boldsymbol{H}(h_k)\boldsymbol{A}(h_k)\boldsymbol{x}_{f,k+1} - \boldsymbol{u}_{f,k}, \tag{27}$$

with the associated estimation error

$$\varepsilon_{f,k} = \boldsymbol{f}_k - \hat{\boldsymbol{f}}_k = -\boldsymbol{H}(h_k)\boldsymbol{W}\boldsymbol{w}_k. \tag{28}$$

In order to obtain $\hat{\boldsymbol{f}}_k$ it is necessary to have $\boldsymbol{x}_{f,k+1}$. Thus, the only choice to compensate $\boldsymbol{f}_k$ in (22) is to use $\hat{\boldsymbol{f}}_{k-1}$. This determines the above-proposed control strategy

$$\boldsymbol{u}_{f,k} = -\hat{\boldsymbol{f}}_{k-1} - \boldsymbol{K}\boldsymbol{x}_{f,k}. \tag{29}$$

Bearing in mind that in any physical system $\boldsymbol{f}_k$ is bounded, without a loss of generality, it is possible to write

$$\hat{\boldsymbol{f}}_k = \hat{\boldsymbol{f}}_{k-1} + \boldsymbol{v}_k, \quad \boldsymbol{v}_k \in l_2, \tag{30}$$

Thus, (29) can be written equivalently as

$$\boldsymbol{u}_{f,k} = -\hat{\boldsymbol{f}}_k + \boldsymbol{v}_k - \boldsymbol{K}\boldsymbol{x}_{f,k}, \tag{31}$$

which will be used for further deliberations. Substituting (31) into (22) gives

$$\boldsymbol{x}_{f,k+1} = \boldsymbol{A}_1(h_k)\boldsymbol{x}_{f,k} + \left[\boldsymbol{I} - \boldsymbol{B}(h_k)\boldsymbol{H}(h_k)\right]\boldsymbol{W}\boldsymbol{w}_k + \boldsymbol{B}(h_k)\boldsymbol{v}_k \tag{32}$$

with $\boldsymbol{A}_1(h_k) = \boldsymbol{A}(h_k) - \boldsymbol{B}(h_k)\boldsymbol{K}$ defining an quasi-LPV polytopic system [19] with

$$\begin{aligned}\mathbb{A} &= \left\{\boldsymbol{A}(h_k): \quad \boldsymbol{A}(h_k) = \sum_{i=1}^{N} h_i\boldsymbol{A}_i, \sum_{i=1}^{N} h_i = 1,\, h_i \geq 0\right\},\\ \mathbb{B} &= \left\{\boldsymbol{B}(h_k): \quad \boldsymbol{B}(h_k) = \sum_{j=1}^{N} h_j\boldsymbol{B}_j, \sum_{j=1}^{N} h_j = 1,\, h_j \geq 0\right\}\end{aligned} \tag{33}$$

The equation (32) can be equivalently written as:

$$\boldsymbol{x}_{f,k+1} = \boldsymbol{A}_1(h_k)\boldsymbol{x}_{f,k} + \bar{\boldsymbol{W}}(h_k)\bar{\boldsymbol{w}}_k \tag{34}$$

with

$$\bar{\boldsymbol{W}}(h_k) = [[\boldsymbol{I} - \boldsymbol{B}(h_k)\boldsymbol{H}(h_k)]\,\boldsymbol{W} \quad \boldsymbol{B}(h_k)], \quad \bar{\boldsymbol{w}}_k = \begin{bmatrix}\boldsymbol{w}_k^T, & \boldsymbol{v}_k^T\end{bmatrix}^T \in l_2.$$

Before providing the control design procedure, let us consider the following lemma, which can be perceived as the generalisation of the one presented in [19] for quasi-LPV systems.

Lemma 1. *The following statements are equivalent*

(i) There exists $\boldsymbol{X}(h_k) \succ \boldsymbol{0}$ *such that*

$$\boldsymbol{V}(h_k)^T\boldsymbol{X}(h_k)\boldsymbol{V}(h_k) - \boldsymbol{W}(h_k) \prec \boldsymbol{0}, \tag{35}$$

(ii) There exists $\boldsymbol{X}(h_k) \succ \boldsymbol{0}$ *such that*

$$\begin{bmatrix} -\boldsymbol{W}(h_k) & \boldsymbol{V}(h_k)^T\boldsymbol{U}^T \\ \boldsymbol{U}\boldsymbol{V}(h_k) & \boldsymbol{X}(h_k) - \boldsymbol{U} - \boldsymbol{U}^T \end{bmatrix} \prec \boldsymbol{0}, \tag{36}$$

where

$$\boldsymbol{W}(h_k) = \sum_{i=1}^{N} h_i\boldsymbol{W}^i, \quad \boldsymbol{V}(h_k) = \sum_{i=1}^{N} h_i\boldsymbol{V}^i, \quad \boldsymbol{X}(h_k) = \sum_{i=1}^{N} h_i\boldsymbol{X}^i.$$

Proof 1. *For proof, see [16].*

It is easy to show that (36) is satisfied when there exist matrices $\boldsymbol{X}^i \succ \boldsymbol{0}$ such that

$$\begin{bmatrix} -\boldsymbol{W}^{(i)} & \boldsymbol{V}^{(i)^T}\boldsymbol{U}^T \\ \boldsymbol{U}\boldsymbol{V}^{(i)} & \boldsymbol{X}^{(i)} - \boldsymbol{U} - \boldsymbol{U}^T \end{bmatrix} \prec \boldsymbol{0}, \quad i = 1, \ldots, N. \tag{37}$$

The following theorem constitutes the main result of this section.

Theorem 1. *For a prescribed disturbance attenuation level* $\mu > 0$ *for the* $\boldsymbol{x}_{f,k}$*, the* $\mathcal{H}_\infty$ *controller design problem for the system (22) is solvable if there exist* $\boldsymbol{U}$*,* $\boldsymbol{N}$ *and* $\boldsymbol{P} \succ \boldsymbol{0}$ *such that the following LMIs are satisfied:*

$$\begin{bmatrix} \boldsymbol{I} - \boldsymbol{P}^{(ij)} & \boldsymbol{0} & \boldsymbol{A}^{(i)}\boldsymbol{U} - \boldsymbol{B}^{(j)}\boldsymbol{N} \\ \boldsymbol{0} & -\mu^2\boldsymbol{I} & \bar{\boldsymbol{W}}^{(j)T}\boldsymbol{U}^T \\ \boldsymbol{U}^T\boldsymbol{A}^{(i)^T} - \boldsymbol{N}^T\boldsymbol{B}^{(j)^T} & \boldsymbol{U}\bar{\boldsymbol{W}}^{(j)} & \boldsymbol{P}^{(ij)} - \boldsymbol{U} - \boldsymbol{U}^T \end{bmatrix} \prec \boldsymbol{0}, \tag{38}$$

with $i = 1, \ldots, N$*,* $j = 1, \ldots, N$*.*

Proof 2. *For proof, see [16].*

Finally, the design procedure boils down to solving (38) with respect to $\boldsymbol{U}$, $\boldsymbol{N}$ and $\boldsymbol{P}$, and then calculating

$$\boldsymbol{K} = \boldsymbol{N}\boldsymbol{U}^{-1}. \tag{39}$$

The objective of this section was to provide a fault estimation and compensation scheme excluding the control limits. In the presence of faults, disturbances and control limits the set of the states that can be reached from $\boldsymbol{x}_{f,k}$ is significantly smaller that the one obtained without these unappealing phenomena. Thus, the objective of the subsequent section is to provide a useful description of such a set, while the Section 6 presents an on-line optimisation strategy that can be used for enlarging this set.

5. Derivation of a robust invariant set

As it was mentioned in the previous section, in order to maintain a desired system behaviour, the idea of an invariant set of state variables is to be employed [4, 11].

In this section the ellipsoidal bounding will be used for describing the robust invariant set for

$$\boldsymbol{x}_{f,k+1} = \boldsymbol{A}_1(h_k)\boldsymbol{x}_{f,k} + \bar{\boldsymbol{W}}(h_k)\bar{\boldsymbol{w}}_k \tag{40}$$

with an additional assumption that:

$$\bar{\boldsymbol{w}}_k^T\boldsymbol{Q}^{-1}\bar{\boldsymbol{w}}_k \leq 1, \quad \boldsymbol{Q} \succ \boldsymbol{0}. \tag{41}$$

The proposed ellipsoidal bounding strategy can be perceived as an inner approximation of the exact invariant set [9]. The clear drawback of the proposed approach is, that the obtained set is smaller than the exact one. However, the simplicity of the ellipsoidal description will make it possible to use it for on-line optimisation, described in Section 6. In particular, $\boldsymbol{E}_{\boldsymbol{x}_f}$ is a robust invariant set for (40) if

$$\boldsymbol{x}_k \in \boldsymbol{E}_{\boldsymbol{x}_f} \Longrightarrow \boldsymbol{x}_{k+1} \in \boldsymbol{E}_{\boldsymbol{x}_f}. \tag{42}$$

Thus, the ellipsoidal robust invariant set is given by

$$\boldsymbol{E}_{\boldsymbol{x}_f} = \{\boldsymbol{x}_f | \boldsymbol{x}_f^T\boldsymbol{P}(h_k)\boldsymbol{x}_f \leq 1\}, \quad \boldsymbol{P}(h_k) \succ \boldsymbol{0}. \tag{43}$$

The above definition implies the following constraints:

$$\begin{aligned} &\boldsymbol{x}_{f,k}^T\boldsymbol{P}(h_k)\boldsymbol{x}_{f,k} \leq 1 \\ &\boldsymbol{x}_{f,k+1}^T\boldsymbol{P}(h_k)\boldsymbol{x}_{f,k+1} \leq 1 \end{aligned} \tag{44}$$

which, by applying the S-procedure along with (41), yields the following coupled constraint

$$\begin{aligned} &(\boldsymbol{x}_{f,k}^T\boldsymbol{A}_1(h_k)^T + \bar{\boldsymbol{w}}_k^T\bar{\boldsymbol{W}}(h_k)^T)\boldsymbol{P}(h_k)(\boldsymbol{A}_1(h_k)\boldsymbol{x}_{f,k} + \bar{\boldsymbol{W}}(h_k)\bar{\boldsymbol{w}}_k) - 1 + \\ &\gamma(1 - \boldsymbol{x}_{f,k}^T\boldsymbol{P}(h_k)\boldsymbol{x}_{f,k}) + \beta(1 - \bar{\boldsymbol{w}}_k^T\boldsymbol{Q}^{-1}\bar{\boldsymbol{w}}_k) = \\ &\boldsymbol{x}_{f,k}^T\boldsymbol{A}_1(h_k)^T\boldsymbol{P}(h_k)^{-1}\boldsymbol{A}_1(h_k)\boldsymbol{x}_{f,k} + \boldsymbol{x}_{f,k}^T\boldsymbol{A}_1(h_k)^T\boldsymbol{P}(h_k)\bar{\boldsymbol{W}}(h_k)\bar{\boldsymbol{w}}_k + \\ &\bar{\boldsymbol{w}}_k^T\bar{\boldsymbol{W}}(h_k)^T\boldsymbol{P}(h_k)\boldsymbol{A}_1(h_k)\boldsymbol{x}_{f,k} + \bar{\boldsymbol{w}}_k^T\bar{\boldsymbol{W}}(h_k)^T\boldsymbol{P}(h_k)\bar{\boldsymbol{W}}(h_k)\bar{\boldsymbol{w}}_k - 1 + \gamma \\ &- \gamma\boldsymbol{x}_{f,k}^T\boldsymbol{P}(h_k)\boldsymbol{x}_{f,k} + \beta - \beta\bar{\boldsymbol{w}}_k^T\boldsymbol{Q}^{-1}\bar{\boldsymbol{w}}_k \leq 0 \end{aligned}$$

with $\gamma > 0$ and $\beta > 0$, written in a more compact form:

$$\begin{aligned} &\boldsymbol{x}_{f,k}^T\left[\boldsymbol{A}_1(h_k)^T\boldsymbol{P}(h_k)\boldsymbol{A}_1 - \gamma\boldsymbol{P}(h_k)\right]\boldsymbol{x}_{f,k} + \boldsymbol{x}_{f,k}^T\boldsymbol{A}_1(h_k)^T\boldsymbol{P}(h_k)\bar{\boldsymbol{W}}(h_k)\bar{\boldsymbol{w}}_k + \\ &\bar{\boldsymbol{w}}_k^T\bar{\boldsymbol{W}}(h_k)^T\boldsymbol{P}(h_k)\boldsymbol{A}(h_k)_1\boldsymbol{x}_{f,k} + \bar{\boldsymbol{w}}_k^T\left[\bar{\boldsymbol{W}}(h_k)^T\boldsymbol{P}(h_k)\bar{\boldsymbol{W}}(h_k) - \beta\boldsymbol{Q}^{-1}\right]\bar{\boldsymbol{w}}_k - 1 + \gamma + \beta \leq 0, \end{aligned} \tag{45}$$

or described in a matrix form:

$$\begin{bmatrix} \boldsymbol{A}_1(h_k)^T\boldsymbol{P}(h_k)\boldsymbol{A}_1(h_k) - \gamma\boldsymbol{P}(h_k) & \boldsymbol{A}_1(h_k)^T\boldsymbol{P}(h_k)\bar{\boldsymbol{W}}(h_k) & 0 \\ \bar{\boldsymbol{W}}(h_k)^T\boldsymbol{P}(h_k)\boldsymbol{A}_1(h_k) & \bar{\boldsymbol{W}}(h_k)^T\boldsymbol{P}(h_k)\bar{\boldsymbol{W}}(h_k) - \beta\boldsymbol{Q}^{-1} & 0 \\ 0 & 0 & \gamma + \beta - 1 \end{bmatrix} \preceq \boldsymbol{0} \tag{46}$$

Studying (46) leads directly to:

$$\gamma + \beta \leq 1 \Rightarrow \beta = 1 - \gamma \Rightarrow 0 \leq \gamma \leq 1. \tag{47}$$

and makes (48)

$$\begin{bmatrix} \boldsymbol{A}_1(h_k)^T \\ \bar{\boldsymbol{W}}(h_k)^T \end{bmatrix} \boldsymbol{P}(h_k)\begin{bmatrix} \boldsymbol{A}_1(h_k) & \bar{\boldsymbol{W}}(h_k) \end{bmatrix} + \begin{bmatrix} -\gamma\boldsymbol{P}(h_k) & 0 \\ 0 & -(1-\gamma)\boldsymbol{Q}^{-1} \end{bmatrix} \preceq \boldsymbol{0}, \quad 0 \leq \gamma \leq 1, \tag{48}$$

a perfect form for applying Lemma 1, which subsequently gives:

$$\begin{bmatrix} -\gamma \boldsymbol{P}^{(ij)} & \boldsymbol{0} & {\boldsymbol{A}_1^{(i)}}^T \boldsymbol{U}^T \\ \boldsymbol{0} & -(1-\gamma)\boldsymbol{Q}^{-1} & {\bar{\boldsymbol{W}}^{(j)}}^T \boldsymbol{U}^T \\ \boldsymbol{U}\boldsymbol{A}_1^{(i)} & \boldsymbol{U}\bar{\boldsymbol{W}}^{(j)} & \boldsymbol{P}^{(ij)} - \boldsymbol{U} - \boldsymbol{U}^T \end{bmatrix} \preceq \boldsymbol{0}, \tag{49}$$

$i = 1, \ldots, N; j = 1, \ldots, N, \quad 0 \le \gamma \le 1.$
Note that for a fixed $0 \le \gamma \le 1$, the inequality (49) becomes the usual LMI. Another strategy is to formulate (49) as a generalised eigenvalue optimisation problem. Both of them can be efficiently solved with the numerical packages like Matlab.

6. Efficient predictive FTC

The robust fault-tolerant control presented in Sec. 4 is based on the idea of estimating the fault, and further compensation it with a suitable increasement or decreasement of the control feeding the faulty actuator. It is easy to see its main drawback, - insensivity to any saturation rules. Thus, the idea behind the approach presented in this section is to employ controler reconfiguration strategy to overcome above drawback. This should ensure overall control problem feasibility, by increasing the robust invariant set. The subsequent part of this section is devoted to the implementation of such a strategy.

Thus, subsequent part of this section deals with a suitable control strategy that takes into account the actuator saturation. For this purpose, the efficient predictive control scheme introduced by [12] is utilised. In particular, the proposed scheme is suitably extended to cope with the external disturbances, and hence, achieving robustness.

It can be easily shown that the input constraints (24) for the ith input are (cf. (25))

$$-\bar{\boldsymbol{u}}_i \le \begin{bmatrix} -\boldsymbol{K}_i^T & 1 \end{bmatrix} \begin{bmatrix} \boldsymbol{x}_{f,k} \\ \boldsymbol{c}_{i,k} - \hat{\boldsymbol{f}}_{i,k-1} \end{bmatrix} \le \bar{\boldsymbol{u}}_i, i = 1, \ldots, r \tag{50}$$

where $\boldsymbol{K}_i$ stands for the ith row of $\boldsymbol{K}$. Thus, predictions at time k are generated as follows [12]:

$$\boldsymbol{z}_{k+1} = \boldsymbol{Z}(h_k)\boldsymbol{z}_k + \tilde{\boldsymbol{W}}(h_k)\bar{\boldsymbol{w}}_k. \tag{51}$$

where

$$\tilde{\boldsymbol{W}}(h_k) = \begin{bmatrix} \bar{\boldsymbol{W}}(h_k) \\ \boldsymbol{0} \end{bmatrix}, \quad \boldsymbol{Z}(h_k) = \begin{bmatrix} \boldsymbol{A}(h_k) - \boldsymbol{B}(h_k)\boldsymbol{K} & \boldsymbol{B}(h_k)\boldsymbol{T} \\ \boldsymbol{0} & \boldsymbol{M} \end{bmatrix}, \quad \boldsymbol{M} = \begin{bmatrix} \boldsymbol{0}_{(n_c-1)r\times r} & \boldsymbol{I} \\ \boldsymbol{0}_{r\times r} & \boldsymbol{0}_{r\times(n_c-1)r} \end{bmatrix},$$

$$\boldsymbol{z}_k = \begin{bmatrix} \boldsymbol{x}_{f,k} \\ \boldsymbol{\omega}_k \end{bmatrix}, \quad \boldsymbol{\omega}_k = \begin{bmatrix} \boldsymbol{c}_k \\ \boldsymbol{c}_{k+1} \\ \cdots \\ \boldsymbol{c}_{k+n_c-1} \end{bmatrix}, \quad \boldsymbol{T} = \begin{bmatrix} \boldsymbol{I}_{r\times r} & \boldsymbol{0} & \ldots & \boldsymbol{0} \end{bmatrix}.$$

Note that the stability (in the $\boldsymbol{H}_\infty$ sense) of the autonomous system (51) is guaranteed by the stability of $\boldsymbol{A}(h_k) - \boldsymbol{B}(h_k)\boldsymbol{K}$. Following [12], it can be pointed out that if there exists robust invariant set $\boldsymbol{E}_{\boldsymbol{x}_f}$ (cf. (43)) for (40), then there must exist at least one robust invariant set $\boldsymbol{E}_{\boldsymbol{z}}$ for (51). Thus, (49) can be easily adapted for (51), which gives the robust invariant set for the proposed fault-tolerant predictive scheme:

$$\begin{bmatrix} -\gamma \boldsymbol{P}^{(ij)} & \boldsymbol{0} & {\boldsymbol{Z}^{(ij)}}^T \boldsymbol{U}^T \\ \boldsymbol{0} & -(1-\gamma)\boldsymbol{Q}^{-1} & {\tilde{\boldsymbol{W}}^{(j)}}^T \boldsymbol{U}^T \\ \boldsymbol{U}\boldsymbol{Z}^{(ij)} & \boldsymbol{U}\tilde{\boldsymbol{W}}^{(j)} & \boldsymbol{P}^{(ij)} - \boldsymbol{U} - \boldsymbol{U}^T \end{bmatrix} \preceq \boldsymbol{0}, \quad 0 \le \gamma \le 1, \tag{52}$$

Since the robust invariant set for (51) is given then it is possible to introduce the input constraints (50). Such a procedure is described in [17] in details.

7. Illustrative example

The objective of this section is to provide an illustrative example regarding the proposed robust fault-tolerant model predictive control. The presented example is based on the wind turbine description provided in [3] with suitable modifications. Let us assume that the wind turbine can be described as:

$$\boldsymbol{x}_{k+1} = \boldsymbol{A}(\alpha)\boldsymbol{x}_k + \boldsymbol{B}(\alpha)\boldsymbol{u}_k + \boldsymbol{W}\boldsymbol{w}_k, \tag{53}$$

with

$$\boldsymbol{A}(\boldsymbol{\alpha}) = \boldsymbol{I} - h\begin{bmatrix} -\frac{1}{\tau} & 0 & 0 & 0 \\ \frac{nlt_\beta(\alpha)}{J_r} & -\frac{nlt_{\omega_r}(\alpha)+B_s}{J_r} & \frac{B_s}{J_r} & -\frac{K_s}{J_r} \\ 0 & \frac{B_s}{J_g} & -\frac{B_s+B_g}{J_g} & \frac{K_s}{J_g} \\ 0 & 1 & -1 & 0 \end{bmatrix}, \boldsymbol{B}(\boldsymbol{\alpha}) = h\begin{bmatrix} \frac{1}{\tau} & 0 \\ 0 & nlt_b(\boldsymbol{\alpha}) \\ 0 & 0 \\ 0 & 0 \end{bmatrix}, \boldsymbol{x}_k = \begin{bmatrix} \beta \\ \omega_r \\ \omega_g \\ \theta_\Delta \end{bmatrix}, \boldsymbol{u}_k = \begin{bmatrix} \beta_r \\ v_w \end{bmatrix}$$

while $h = 0.01s$ is a sampling time and $\alpha = [\bar{\beta}, \bar{\omega_r}, \bar{v}]$. The remaining parameters are defined as follows:

$$nlt_\beta(\boldsymbol{\alpha}) = \frac{T_a}{C_q}\frac{\partial C_q}{d\beta}, \quad nlt_{\omega_r}(\boldsymbol{\alpha}) = \frac{T_a}{\omega}\frac{\partial C_q}{d\lambda}\frac{\lambda}{C_q}, \quad nlt_b(\boldsymbol{\alpha}) = \frac{T_a}{v_w}\left(2 - \frac{\partial C_q}{d\lambda}\frac{\lambda}{C_q}\right)$$

where $T_a = \frac{1}{2}\rho R^3 C_q(\lambda, \beta)v_w^2$ is aerodynamic torque, $\dot{\theta}_\Delta$ is shaft angular velocity difference between generator and rotor tips, while ω_r stands for the rotor shaft speed, v_r wind speed, β pitch angle of blades, λ tip-speed ratio, J_r and J_g stands for rotor and generator shafts inertias respectively as well as B_s and B_g denotes appropriate dampings of transmission and generator respectively. Figs. 2–5 presents the results of the neural network modelling. The modelling error is low which proves, that neural model excellent reflects the wind turbine behaviour. As a fault, the constant bias +0.5 was introduced in $300 < k < 1800$ for the ω_g, which estimate is presented in Fig. 6. From Fig. 6 is clear, that robust predictive controller compensate the fault effectively.

8. Conclusions

The main contribution of the paper was to propose a robust predictive fault-tolerant control scheme for both linear and non-linear systems. Indeed, the contribution can be divided into a few important points: extension of the efficient predictive control to the robust case with exogenous external disturbances, development of robust fault estimation and compensation scheme an integration of the developed schemes within a unified robust predictive fault-tolerant control framework. It is worth to note that the framework was also suitably extended to the linear parameter-varying systems. All the proposed approaches can be efficiently implemented, i.e., the off-line computations boils down to solving a number of linear matrix inequalities while the on-line computation reduces to the application of the Newton-Raphson method.

8.1. Acknowledgments

The work was supported by the National Science Centre of Poland under grant: 2014-2017 (UMO-2013/11/B/ST7/01110).

References

[1] H. Abbas and H. Werner. Polytopic quasi-LPV models based on neural state-space models and application to air charge control of a si engine. In *Proc. of the 17th IFAC World Congress, Seoul, Korea*, pages 6466–6471, 2008.

[2] J. Bendtsen and K. Trangbæk. Transformation of neural state space models into LFT models for robust control design. In *Proc. of the 6th International Conference on Control, Automation, Robotics and Vision*, 2000.

[3] F.D. Bianchi, H. De Battista, and R.J. Mantz. *Wind turbine control systems: principles, modelling and gain scheduling design*. Advances in industrial control. Springer, 2007.

[4] F. Blanchini. Set invariance in control. In *Automatica*, number 35, pages 1747–1767. 1999.

[5] M. Blanke, M. Kinnaert, J. Lunze, and M. Staroswiecki. *Diagnosis and Fault-Tolerant Control*. Springer-Verlag, New York, 2003.

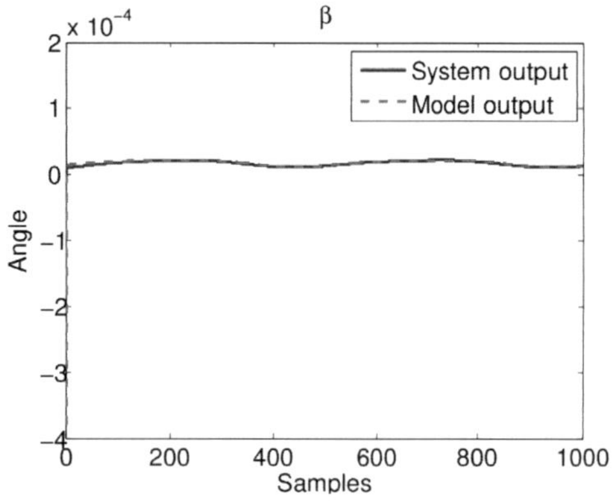

Figure 2. Results of the neural modelling for β

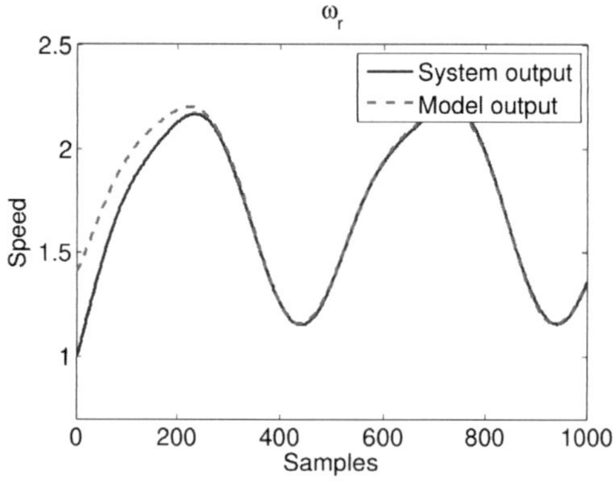

Figure 3. Results of the neural modelling for ω_r

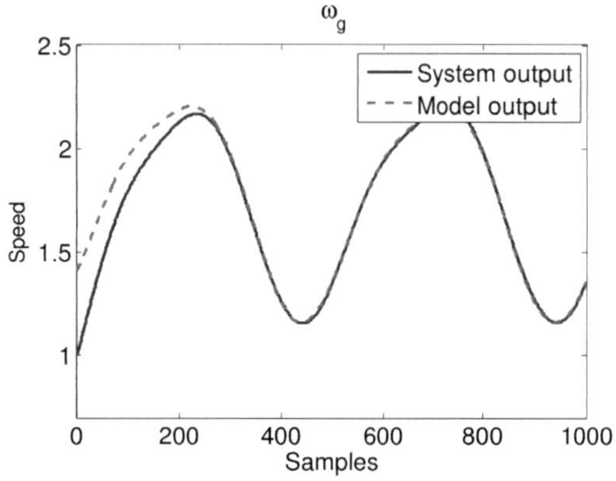

Figure 4. Results of the neural modelling for ω_g

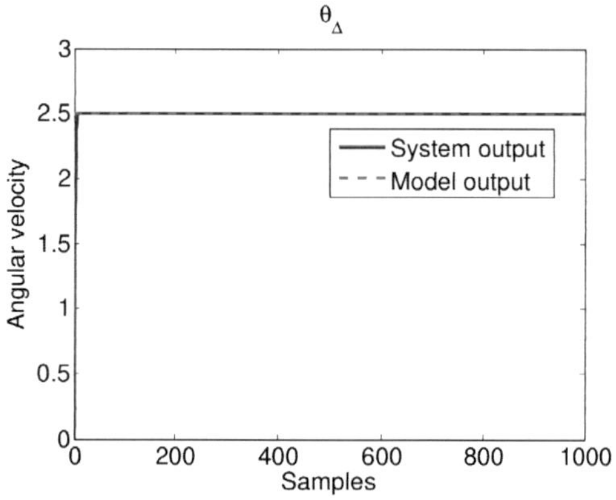

Figure 5. Results of the neural modelling for θ_Δ

[6] W. Chen, A.Q. Khan, M. Abid, and S.X. Ding. Integrated design of observer based fault detection for a class of uncertain nonlinear systems. *Int. J. Applied Mathematics and Computer Science*, 21(3):423–430, 2011.

[7] S. De Oca, V. Puig, M. Witczak, and L. Dziekan. Fault-tolerant control strategy for actuator faults using lpv techniques: application to a two degree of freedom helicopter. *International Journal of Applied Mathematics and Computer Science*, 22(1):161–171, 2012.

[8] A.I. Galushkin. *Neural network theory*. Springer-Verlag, Berlin, 2007.

[9] E.G. Gilbert and K. T. Tan. Linear systems with state and control constraints: The theory and application of maximal output admissible sets. *IEEE Transactions on Automatic Control*, 36(9):1008–1020, 1991.

[10] S. Gillijns and B. De Moor. Unbiased minimum-variance input and state estimation for linear discrete-time systems. *Automatica*, 43:111–116, 2007.

[11] B. Kouvaritakis, Y.I. Lee, and M. Cannon. Extended invariance and its use in model predictive control. *Automatica*, 41, 2005.

[12] B. Kouvaritakis, J. A. Rossiter, and Schuurmans J. Efficient robust predictive control. *IEEE Transactions On Automatic Control*, 45(8):1545–1549, August 2000.

[13] M. Ławryńczuk. *Computationally Efficient Model Predictive Control Algorithms: A Neural Network Approach*. Springer, 2014.

[14] M. Luzar, M. Witczak, M. Buciakowski, and V. Puig. Robust $\mathcal{H}_\infty$ actuator fault diagnosis and fault-tolerant control for a multi-tank system. In *Proceedings of the 2nd Conference on Control and Fault-Tolerant Systems (SysTol), Nice, France*, pages 758–763, 2013.

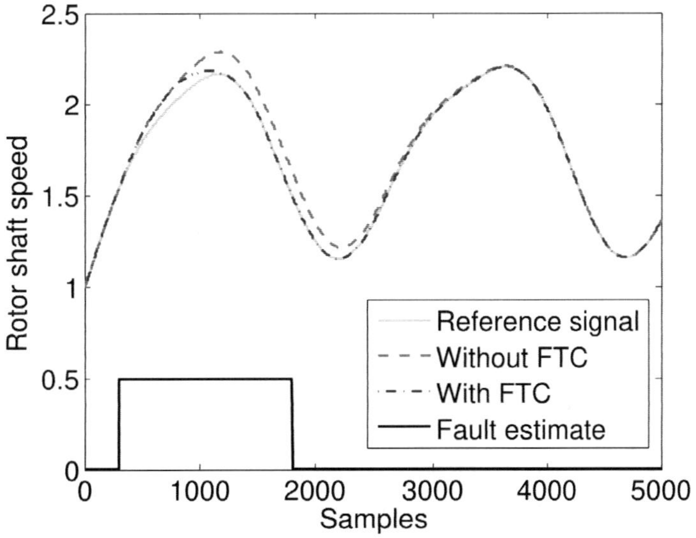

Figure 6. Simulation of results for FTC (blue) and non-FTC (red) for ω_g

[15] M. Luzar, M. Witczak, J. Korbicz, and P. Witczak. Neural-network based robust FTC: Application to wind turbines. In *Lecture Notes in Artificial Intelligence*, volume 8467, 2014.

[16] M. Luzar, M. Witczak, and P. Witczak. Robust $\mathcal{H}_\infty$ actuator fault diagnosis with neural network. In *Methods and Models in Automation and Robotics - MMAR 2013 : Proceedings of the 18th IEEE International Conference*, pages 200–205, 2013.

[17] M. Luzar, M. Witczak, P. Witczak, and J. Korbicz. Neural-network based robust predictive fault-tolerant control for multi-tank system. In *Proceedings of the 13th European Control Conference, Strasbourg, France*, 2014.

[18] M. Mrugalski. An unscented kalman filter in designing dynamic gmdh neural networks for robust fault detection. *International Journal of Applied Mathematics and Computer Science*, 23(1):157–169, 2013.

[19] M.C. Oliveira, J. Bernussou, and J.C. Geromel. A new discrete-time robust stability condition. *System and Control Letters*, 37(4):261–265, 1999.

[20] M. Witczak, A. Obuchowicz, and J. Korbicz. Genetic programming based approaches to identification and fault diagnosis of non-linear dynamic systems. *International Journal of Control*, 75(13):1012–1031, 2002.

[21] M. Witczak, V. Puig, and S. De Oca. A fault-tolerant control strategy for non-linear discrete-time systems: application to the twin-rotor system. *International Journal of Control*, 86(10):1788–1799, 2013.

Journal of Physics: Conference Series **570** (2014) 032003 doi:10.1088/1742-6596/570/3/032003

Implementation of PFC (Predictive Functional Control) in a PLC (Programmable Logic Controller) for a HVAC (Heating, Ventilation and Air Conditioning) system

M. Kreutz
A. Nattermann & Cie GmbH (A Sanofi Company), 50829 Köln, Nattermannallee 1, Germany,
(e-mail: mirco.kreutz@sanofi.com)
J. Richalet
Consultant, Louveciennes, France,
(e-mail: jacques.richalet@wanadoo.fr)
K. Mocha, R. Haber
Institute of Process Engineering and Plant Design, Laboratory of Process Automation, Cologne University of Applied Sciences, D-50679 Köln, Betzdorfer Str. 2, Germany,
(e-mail: karl.mocha@fh-koeln.de, robert.haber@ fh-koeln.de)

Abstract. HVAC systems of industrial buildings consume a lot of energy. Therefore it is important to know the performance of these systems and strategies to optimize the hardware and the control. Tackling the temperature control of the HVAC system promises quick savings by tuning the control within specified tolerance limits, which mostly can be done by low investment. This paper mainly deals with the implementation strategy of a new controller in a PLC using the predictive functional control for temperature control. The different stages of the implementation from the simulation over the SCL code till to the real-time operation are presented. A bumpless switch between the PI(D) and the PFC control was realized, as well.

1. Introduction

When looking at the main consumer of an industrial site mostly the HVAC system could be addressed. The integration of an energy management to industrial management structures will be more and more interesting to handle all processes which are related to the energy consumption and costs. Beside the acquisition of significant energy data, methods have to be defined which can identify wasteful processes and present solutions which can optimize the performance of the related equipment to reduce the energy consumption, therefore costs and the CO_2 emissions. When tackling HVAC systems energy savings can be achieved either by implementing new modern HVAC devices, like a new high efficient ventilator, or to apply a better control of the actual HVAC devices. Latter task is solved with a lower investment. The premise is that the new control algorithm is more effective than the actual used one. Temperature records of the actual HVAC systems have shown that the existing PI(D) control sometimes is not optimal, because the prescribed temperature limits were often violated. Another problem was observed during the crossover from heating to cooling and vice versa, as often both heating and cooling was active for a short transition period. This phenomenon caused additional energy costs. For this defective control a strategy was wanted, which keeps the temperature within the allowed limits and prevents a simultaneous working of heating and cooling.

We had two choices, either tuning PI(D) control better and using a dead zone or to apply a whole new control strategy which considers the physical processes of heating and cooling. It is known from the literature that predictive control is a model based control and therefore it is superior over PI(D) control. MPC (Model-based Predictive Control) has been implemented in hundreds of big process engineering plants like refineries and lead to cost savings and safety improvement in opposite to coupled PI(D) control. In such plants usually several variables and constraints are to be controlled to be handled by expensive MPC packages. In a HVAC system however there are only few controlled variables (temperature and humidity) and usually only the manipulated and controlled variables are limited. PFC (Predictive Functional Control) is optimal for this case. The big advantage against big multivariable MPC packages lies in the simplicity of the PFC algorithm which allows a direct implementation in a PLC (Programmable Logical Controller) or a DCS (Decentralized Control System).

The modernization of the HVAC control in our case is divided into two stages:
1. model building, design and simulation of PFC control on the model of the HVAC plant,
2. real-time implementation of PFC control (with backup PI(D) control).

Because of the technological aims and possibilities only temperature control has been designed and elaborated so far. The first realization was tested on a HVAC system with constant recirculation. The operator could select the controllers (PI(D) or PFC) and could change between the control modes. Different interfaces were defined to ensure the data transfer between the two control types. In this way a bumpless switch between the two controllers is possible. The configuration, parameterization and visualization of the new controller are described in the sequel.

Simulations have shown the advantage of variable recirculation flap control; see (Kreutz, et al., 2013). This work is in the simulation stage and the real-time realization has been started.

2. HVAC SYSTEM

The selected HVAC system consists of a mixing chamber in which the fresh air is mixed with recirculated air from the conditioned room. Following the sequence in Fig. 1 the air is heated or cooled depending on the difference of the mixed air temperature and the desired set-point. The conditioned air is directed into the room by a ventilator, a part is recirculated and another part is discharged into the atmosphere.

Fig. 1 shows a detailed PI (Piping and Instrumentation) diagram of the HVAC system.

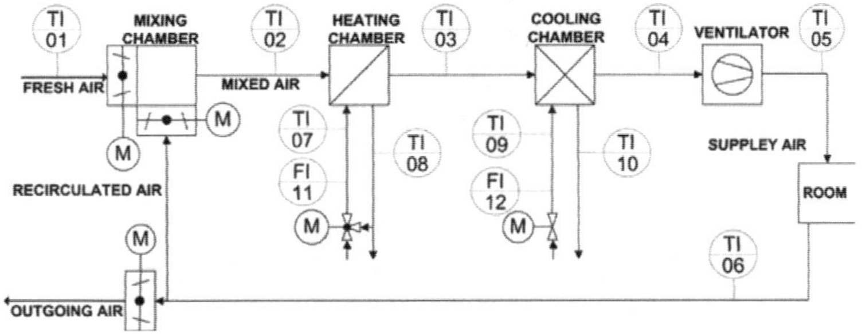

Fig. 1. PI(D) diagram of the HVAC system

Journal of Physics: Conference Series **570** (2014) 032003 doi:10.1088/1742-6596/570/3/032003

The measuring points are as follows. 01: ambient air temperature, 02: mixed air temperature, 03: air temperature after heater, 04: air temperature after cooler, 05: air temperature after ventilator, 06: room temperature, 07: heater fluid inlet temperature, 08: heater fluid outlet temperature, 09: cooler fluid inlet temperature, 10: cooler fluid outlet temperature, 11: mass flow heater fluid, 12: mass flow cooler fluid.

Fig. 2 and Fig. 3 show the flow, inlet and outlet temperature sensors placed on the heat exchangers of the HVAC system to balance the energy over the following enthalpy equation:

$$P_{HE} = \dot{Q}_{\mathrm{HE}} = \dot{m}_{\mathrm{fluid}} \cdot c_{\mathrm{p,fluid}} \cdot (\vartheta_{\mathrm{fluid\_in}} - \vartheta_{\mathrm{fluid\_out}}) \tag{1}$$

where

$P_{HE} = \dot{Q}_{\mathrm{HE}}$: heating power

$\dot{m}_{\mathrm{fluid}}$: mass flow of the heat exchanger fluid

$c_{\mathrm{p,fluid}}$: specific heat capacity of the fluid

$\vartheta_{\mathrm{fluid\_out}}$: outlet temperature of the heat exchanger fluid

$\vartheta_{\mathrm{fluid\_in}}$: inlet temperature of the heat exchanger fluid

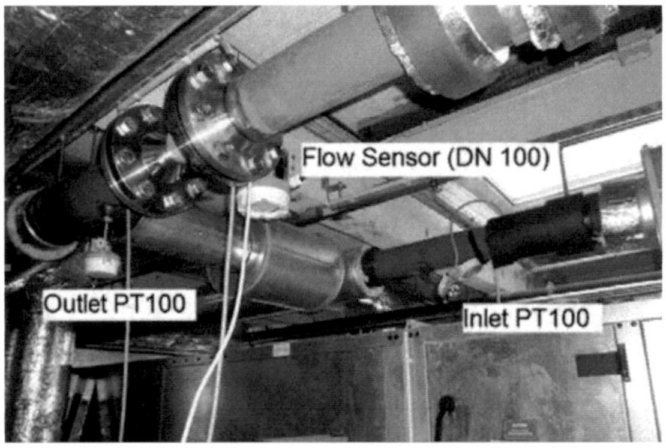

Fig. 2. Flow, inlet and outlet temperature sensors placed on the cooler

Fig. 3 shows the flow, inlet and outlet temperature sensors placed on the heater of the HVAC system.

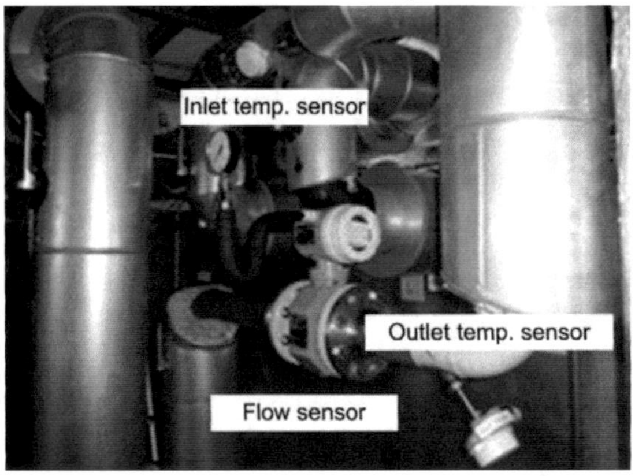

Fig. 3. Flow, inlet and outlet temperature sensors placed on the heater

Journal of Physics: Conference Series **570** (2014) 032003 doi:10.1088/1742-6596/570/3/032003

3. HVAC CONTROL

Fig. 4 shows the plot of a defective control behavior of a PI(D) controlled air handling unit coming from a sporadic data recording. This alternating heating and cooling result in a serious energy waste. The reason is not the sampling time (100 ms) but the bad tuning and switching strategy applied.

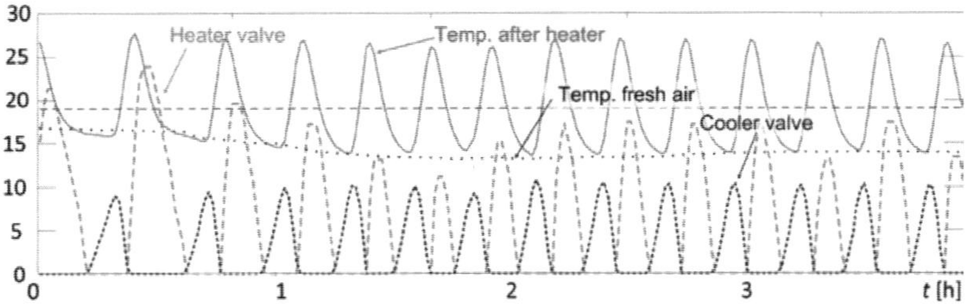

Fig. 4. Alternating heating and cooling as a result of defective control

This control scenario motivated to do an overall analysis of the control situation of the HVAC systems at site to compare the initial control situation with current operating modus, looking for potential to reduce the consumption of heating and cooling fluid. Based on the internal best practices of the HVAC systems known as floating set-point, reduced mode (e.g. night set up) and heat recovery (e.g. recirculation) the HVAC have been analyzed mainly on the focus to implement a controller on a floating set-point strategy for temperature control. Traditionally climate control is realized by PI(D) controllers. Heating and cooling are realized by means of heat exchangers. Control of heat exchangers is a well investigated topic. For example Bonivento et al. (2001) and Raul et al. (2013) compare classical PI(D) and modern model based predictive control of heat exchangers and show the advantage of predictive control. Bonivento et al. (2001) and Chalupa et al. (2010) apply GPC (Generalized Predictive Control). However PFC (Predictive Functional Control) is an easier realizable predictive control as no numerical optimization is necessary even when considering manipulated variable limits (Richalet and O'Donovan, 2009). As a lot of PFC application has been already used successfully in the industry (see e.g. Abdelghani-Idrissi et al., 2001; Arbaoui et al., 2007), PFC was selected for temperature control in this application.

4. MATHEMATICAL MODEL OF THE PLANT

A HVAC consists of the following parts: (1) air mixing chamber (2) cooler/heater and (3) room, whose temperature is to be controlled. Fig. 5 shows the scheme of a heat exchanger (heater or cooler). For simplicity both cases are treated together in a uniform way.

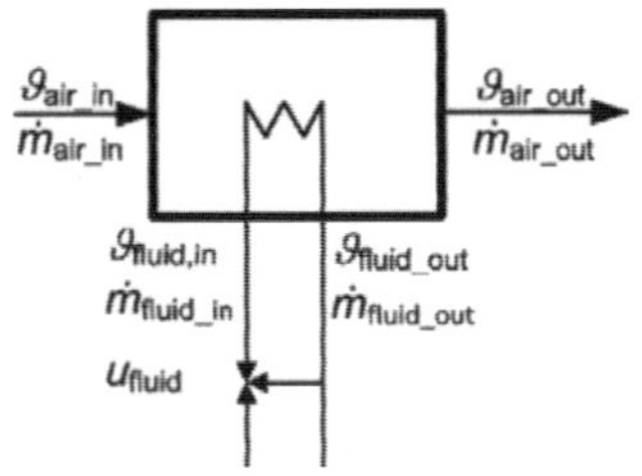

Fig. 5. Heater or also cooler

Both the heater and the cooler are heat exchangers. The following notations are used:

$\vartheta_{air\_in}(t)$: inlet air temperature , $\vartheta_{fluid\_in}(t)$: inlet temperature of heating or cooling fluid,

Journal of Physics: Conference Series **570** (2014) 032003 doi:10.1088/1742-6596/570/3/032003

$\vartheta_{\text{air\_out}}(t)$: outlet air temperature, $\vartheta_{\text{fluid\_out}}(t)$:outlet temperature of heating or cooling fluid.

The heat exchange process can be divided into two parts (Fig. 6):

- static part: symbolizing the stationary heat exchange process.
- dynamic part: symbolizing the time required for heat exchange process.

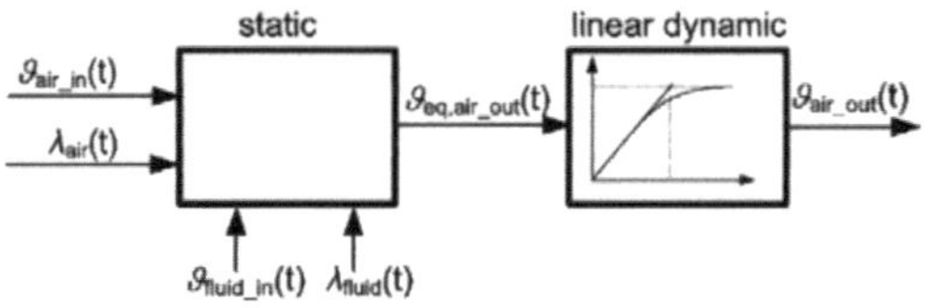

Fig. 6. Static and dynamic parts of the heat exchanger

The following static equations describe the process

$$\vartheta_{\text{eq,air\_out}}(t) = \lambda_{\text{fluid}}(t) \cdot \vartheta_{\text{fluid\_in}}(t) + \lambda_{\text{air}}(t) \cdot \vartheta_{\text{air\_in}}(t) \tag{2}$$

where

$\lambda_{\text{fluid}}(t)$: thermal convexity of (heating or cooling) fluid

$\lambda_{\text{air}}(t)$: thermal convexity of the air

The factors $\lambda_{\text{fluid}}(t)$ and $\lambda_{\text{air}}(t)$ represent the effect of controlling fluid flow and the air flow on the resulting (equivalent) temperature. They both depend approximately linearly on the product of the corresponding mass flows multiplied by the corresponding heating coefficients. The sum of the both thermal convexities is 1, consequently

$$\vartheta_{\text{eq,air\_out}}(t) = \lambda_{\text{fluid}}(t) \cdot \vartheta_{\text{fluid\_in}}(t) + [1 - \lambda_{\text{fluid}}(t)] \cdot \vartheta_{\text{air\_in}}(t) \tag{3}$$

The linear dynamic process can be approximated by the first-order differential equation with unity gain

$$T_{\text{HE}} \cdot \dot{\vartheta}_{\text{air\_out}}(t) + \vartheta_{\text{air\_out}}(t) = \vartheta_{\text{eq,air\_out}}(t) \tag{4}$$

with

T_{HE}: time constant of the heat exchanger

The advantage of introducing the convexity factors is that they influence linearly the output (and the equivalent) temperature. The fluid convexity can be defined between zero and unity. In the knowledge of the temperatures the fluid convexity can be calculated from (8)

$$\lambda_{\text{fluid}}(t) = \frac{\vartheta_{\text{eq,air\_out}}(t) - \vartheta_{\text{air\_in}}(t)}{\vartheta_{\text{fluid\_in}}(t) - \vartheta_{\text{air\_in}}(t)} \tag{5}$$

The fluid convexity is a function of the manipulated signal $u_{\text{fluid}}(t)$ of the valve.

In (2) the thermal convexity was used. The aim of introducing the thermal convexity (Abdelghani-Idrissi et al., 2001) is to eliminate the nonlinear function between the real manipulated signal and the physical variable (mass flow multiplied by thermal capacity coefficients) from a thermal apparatus like the heat exchanger. In this way linear control algorithm can be used for the nonlinear plant.

The air mixing chamber and the room can be modelled in a similar way by using the thermal convexity (Kreutz et al., 2013).

5. PFC TEMPERATURE CONTROL AND SIMULATION

The temperature controller is a cascade one, consisting of

- outer loop (room) temperature controller
- inner loop (heater/cooler) temperature controller.

The dynamics of the room, the cooler and the heater are all modelled by a first-order lag. The following time constants were identified based on experiments: mixer: 0.5 min, both heater and cooler: 3.33 min, room (working hall): 23.33 min. The model was validated with real-time weather data over the year. The nonlinear static characteristic is represented by the convexity equations.

The PFC control algorithm is described in (Richalet and O'Donovan, 2009) and (Kreutz et. al., 2013) in details. The prediction horizon for the PFC was n_p=1 sampling step (common only for first-order processes) and the sample time was Δt=100 ms.

The PFC algorithm is simple for first-order processes

$$y_p(k) = -a_p y_p(k-1) + K_p(1+a_p)u(k-1) \tag{6}$$

where u the input, y_p the process output and k the discrete time (relative to the sampling time) is. a_p and K_p are the process parameters. The process model is approximated by

$$y_m(k) = -a_m y_m(k-1) + K_m(1+a_m)u(k-1) \tag{7}$$

where the index "m" denotes the model contrary to the index "p" process. The manipulated signal is calculated for stepwise change y_r of the reference signal by

$$u(k) = k_0\left[y_r(k) - y_p(k)\right] + k_1 y_m(k); \qquad k_0 = \frac{1-\lambda_r^{n_p}}{K_m\left[1-(-a_m)^{n_p}\right]}; \qquad k_1 = \frac{1}{K_m} \tag{8}$$

PFC assumes an exponential decrease of the future control error with reduction factor λ_r. PFC ensures for apariodic processes a desired settling time of $t_{95\%} = -(3\cdot\Delta t)/\ln(\lambda_r)$.

The following constraints were considered:

- The valves are constrained between 0% and 100 %.
- The convexity factor is related to the temperatures and the flows. The flow of the heat exchanger fluid is related to the valve characteristic, which is in the most cases a non-linear function. The convexity factor is back-calculated when the constraints were achieved.

As heater and cooler should never work simultaneously, a split-range control was realized. Fig. 7 shows the control structure.

The control algorithm was written in Matlab and was tested by a simulated process also written in Matlab.

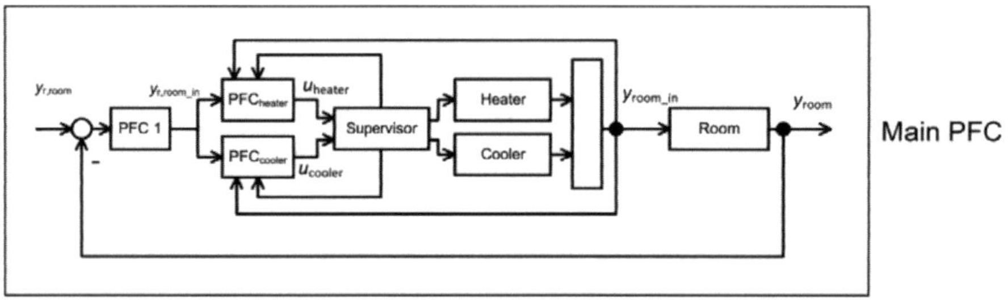

Fig. 7. Cascade split-range control structure

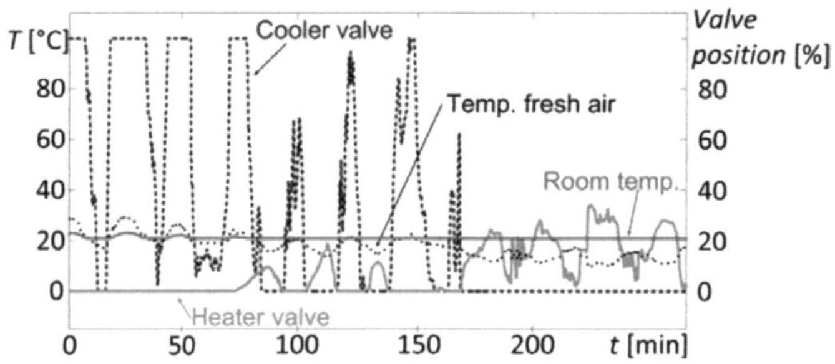

Fig. 8. Simulation 1: 100% fresh air, set-point 21°C

The simulated PFC control in Fig 8 represents the control of the initial parameters controlling to a constant set-point of 21°C with a flap position of 100% fresh air. At beginning of this control the cooler works at limit and the controlled variable cannot be forced to the set-point. This is due to the fact that the fresh air temperature is very high on a hot summer day in August and the cooler is at constructional limit.

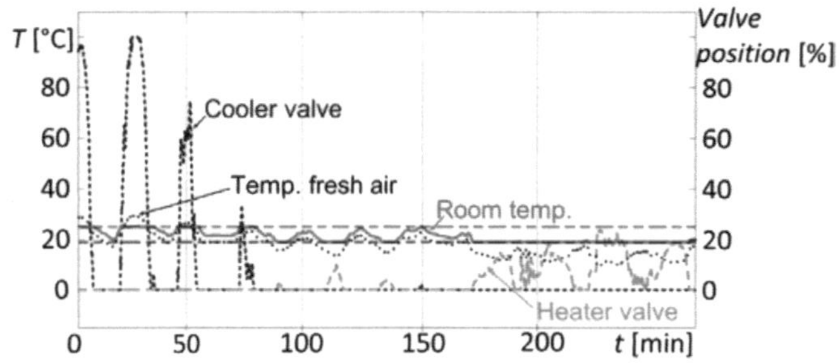

Fig. 9. Simulation 2: 100% fresh air, floating set-point from 19°C to 25°C

In Fig. 9 the floating temperature strategy is applied through the split into a heating set-point of 19°C and a cooling set-point of 25°C. The positive effect on the less activity of the valves results in the reduction of the deviation between the set-points and the fresh air temperature. The effect works stronger on the cooling performance while increasing the set-point from 21°C to 25 °C and still with a reduction of the valve activity, but less stronger on the heating performance decreasing from 21 °C to 19°C. The plot shows the floating of the temperature between the heating and the cooling set-points when no treatment by the heat exchangers is necessary. This cost-free modus depends mainly on the interactions between the fresh air temperature, the internal heat load and the due of fresh air.

6. IMPLEMENTATION OF THE CONTROL SOFTWARE

1. The program code was transferred to Simatic SCL (Structured Control Language) of Siemens AG. It was verified by a simulated process in a field PG (Programming Device). The PFC controller is a function block. Fig. 10 shows the main functions of the PFC controller block.

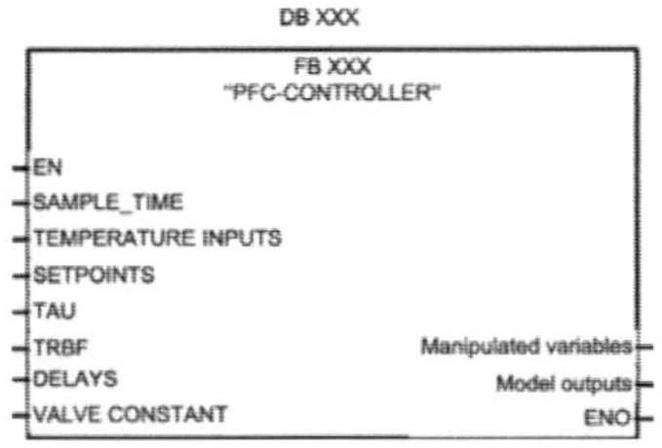

Fig. 10. PFC function block in Simatic S7

2. The control was on-line simulated in a test CPU and the data were recorded by a PLC analyzer and were transferred to an Excel data file. The structure of the control simulation is shown in Fig. 11.

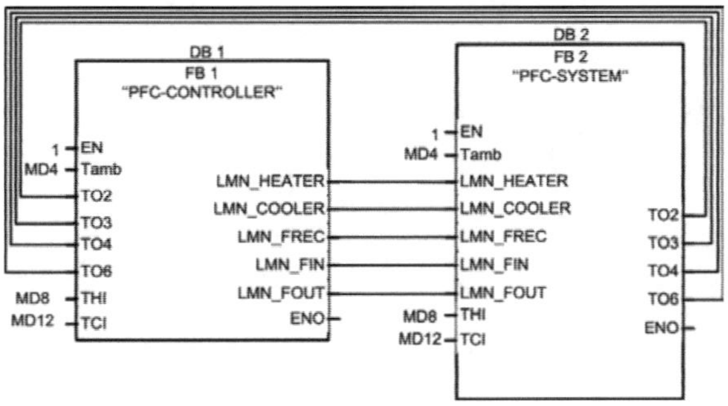

Fig.11. Simulated process in the PLC

The input variables were defined for test purposes via flags:

- MD4: weather data are passed through a flag,
- MD8: heater medium data are passed through a flag,
- MD12: cooler medium data are passed through a flag.

Fig. 12 shows the simulated control behavior to a stepwise set-point change.

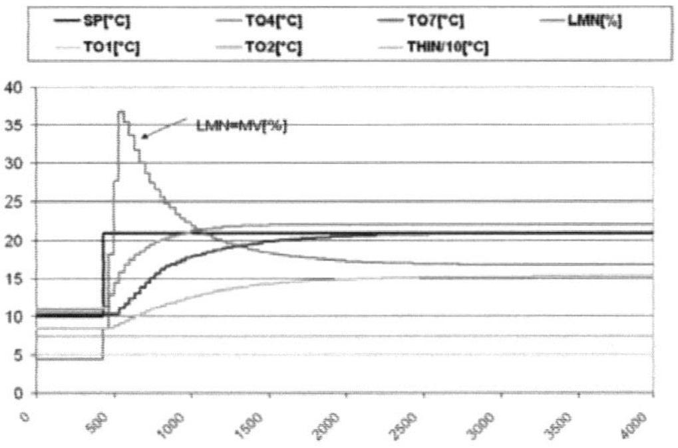

Fig.12. Step response PFC S7

7. INTEGRATING OF PFC TO PLC

The PFC functional block was integrated into the S7 project and linked to the necessary interfaces to switch between the PI(D) and the PFC control and to visualize the main parameters on the display. The process model quality was tested by calculating the model output compared with the measured temperature, see Fig. 13. (The main PFC controller contains the cascade and split-range loops.)

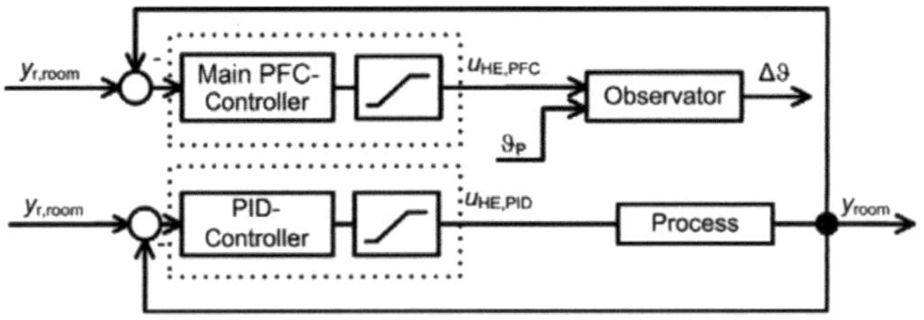

Fig. 13. Model check by temperature calculation

1. In the next step the calculated PFC controller output was compared with the manipulated signal of the PI(D) controller. If the PFC controller works well, then the two manipulated signals are near to each other (Fig. 14).

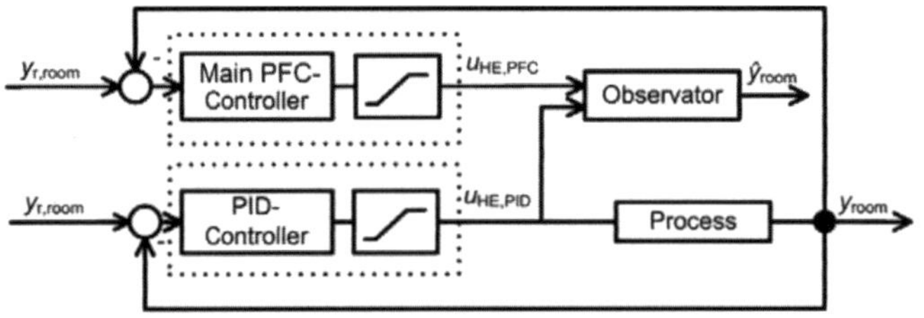

Fig.14. Check of the calculated manipulated signal

2. If both tests were absolved satisfactory then the control type can be switched from PI(D) to PFC. Because of the similar values of the both manipulated signals the switch will be bumpless (Fig. 15).

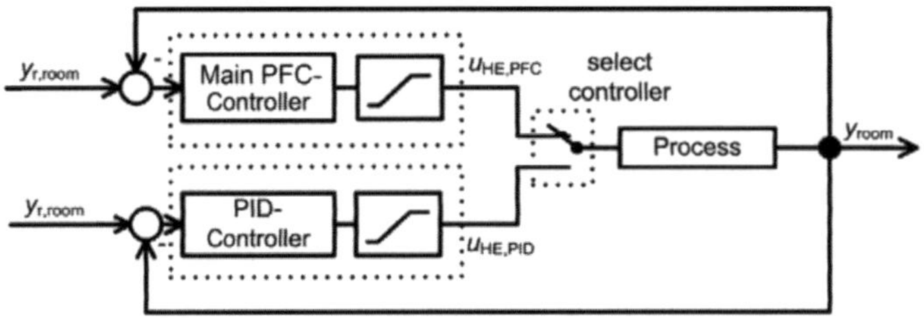

Fig. 15. Integration of PFC and PI(D) in the PLC

3. Fig. 16 shows the PFC controller parameters on the display. The adjustable parameters are:

- TAU: time constant,
- TRBF: closed loop settling time,
- AC/BC: transfer factor,
- R: delay,
- VE-Ventil-Stel: valve position.

Fig. 17 shows the selection of the controller type. The switch decided between the manipulated variables of the PFC and the PI(D), which generates the set-point for the control block of the heat exchanger (heater and cooler).

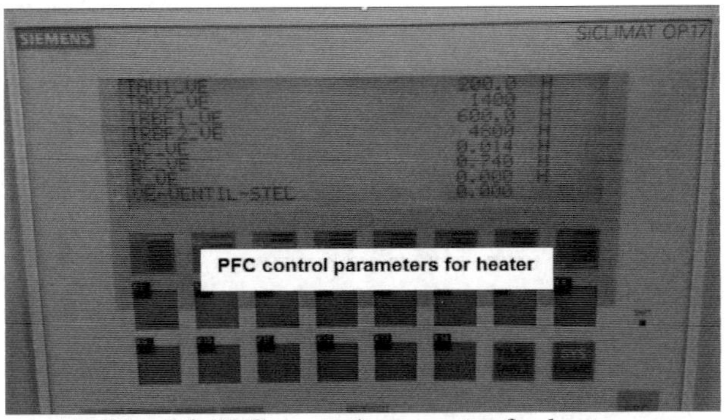

Fig. 16. PFC control parameter for heater

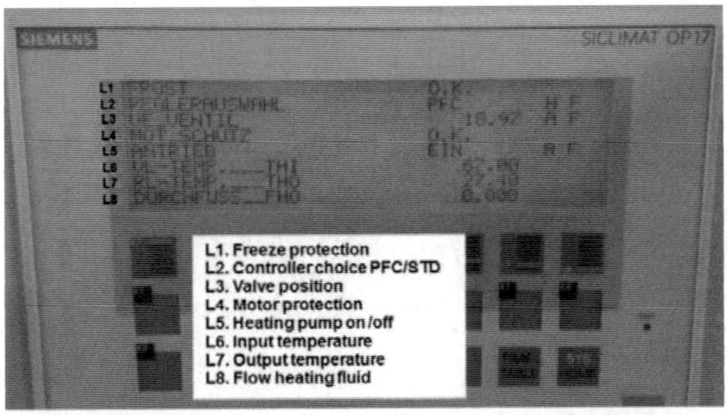

Fig. 17. Selection of controller structure for heater

8. REAL TIME CONTROL AND PERFORMANCE

After all steps of implementation and check were satisfactory, the control of the HVAC system was switched to PFC control. Fig. 18 and 19 shows the real-time control of PFC with an integrated floating set-point strategy. Fig. 17 shows the control at warm days in August and Fig. 18 at cold days in November.

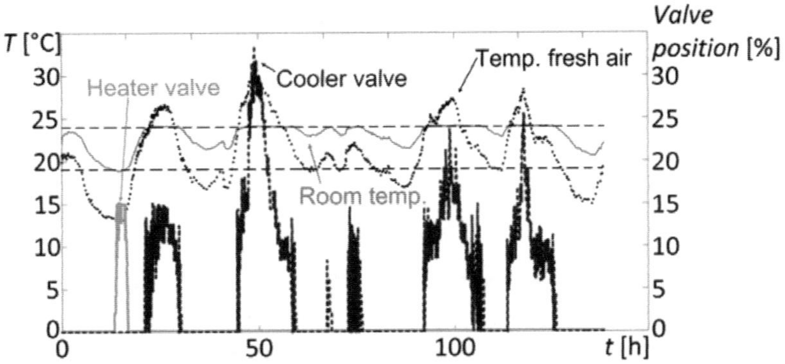

Fig. 18. Real-time control: PFC with floating set-point strategy in August

Journal of Physics: Conference Series **570** (2014) 032003 doi:10.1088/1742-6596/570/3/032003

For the conditioned zone it was possible to define the floating set-point range between a heating set-point of 19 °C and a cooling set-point of 25 °C coming also to the maximum tolerance limits. As shown in the control plots the control is very accurately on the set-point limits without any violation.

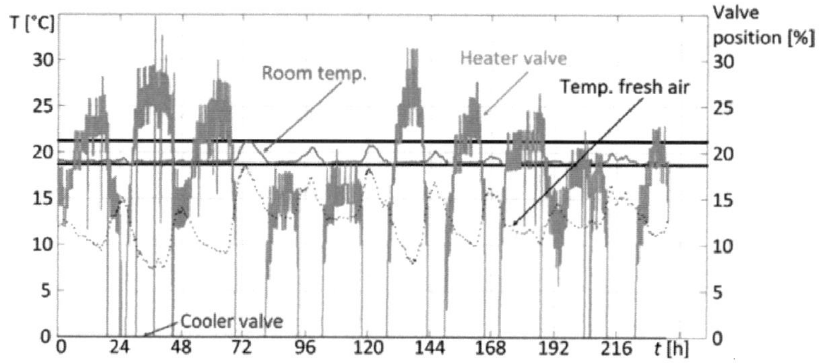

Fig. 19. Real-time control: PFC with floating set-point strategy in November

Therefore the full range could be applied. The performance was evaluated by the fact that the sporadic real-time measurements match very well with the simulated control in section 5. Therefore the performance can be evaluated by the calculation of the cumulative actions of the heating and cooling with the equations (9) and (10)

$$\textit{Cumulative actions heating valve} = \sum_{k=1}^{N} u_{\text{HE\_heating}} \tag{9}$$

$$\textit{Cumulative actions cooling valve} = \sum_{k=1}^{N} u_{\text{HE\_cooling}} \tag{10}$$

where $u_{\text{HE\_heating}}$ and $u_{\text{HE\_cooling}}$ are the heating and cooling valve movements and N is the number of simulations.

Fig. 20 shows the relative savings when controlling on a floating set-point between 19 and 25 °C compared to a control on one set-point of 21°C. For these calculations real weather data over a full year have been fed into the simulation program. The total benefit by reducing heating and cooling energy is here simulated with around 30%. Validated with real-time measurements calculating the energy consumption by means of the sensors placed on the heat exchangers (Section 2) a number of 25 % for the specific HVAC system can be presented. This energy reduction is mainly achieved by the change of the set-point limits, to be sure that PFC control not to violate the maximum limits.

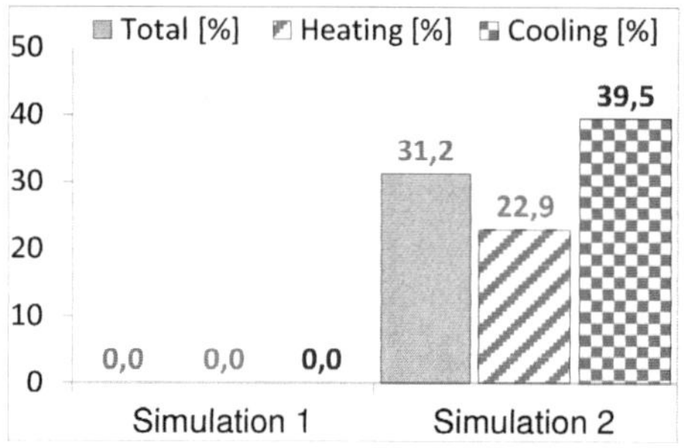

Fig. 20. Performance results with real weather data of a full year

The cumulative actions of the heat exchanger valve, giving the mass flow of the heat exchanger fluid, are directly proportional to the absolute energy consumption

$$W[\text{kWh}] \sim \dot{m}_{\text{fluid}} \sim \sum_{k=1}^{N} u_{\text{HE}} \tag{11}$$

9. CONCLUSIONS

The implementation of a controller in an industrial PLC is critical because it can damage the process to be controlled. The different steps should be defined and validated with great caution.

The first step is to model the process, to simulate it and to validate the accuracy. The next step is to translate it in SCL and comply with the syntax and arrays of the controller. This SCL code has to be validated, through a separate model building block. Then the new control software can be transferred to the CPU. The next step is to validate the model: to check that the process temperature fits the model temperature. As both controllers, PI(D) and PFC, propose control signals an elementary comparison has to be performed between the two - possible different - values. If there is no major off-set, PFC can takes over the control form PI(D). Due to this cautious procedure a bumpless switch between the controllers can be realized.

Product engineers and operators do not like if a process is disturbed artificially. Therefore the success of the implementation of a new control algorithm depends on the skill of the commissioning. Future savings can be prevented by a not careful operation, e.g. by a bumped switch to a new controller or by violation of the technological limits. This paper reports on a successful implementation of a predictive controller, a subject which is very important but scarcely treated in scientific works.

REFERENCES

[1]Abdelghani-Idrissi M.A., Arbaoui M.A., Estel L., Richalet J. (2001). Predictive functional control of a counter current heat exchanger using convexity property. Chemical Engineering and Processing, 40(5), 449-457.

[2]Arbaoui, M.A., Vernières-Hassimi, L., Seguin, D., Abdelghani-Idrissi, M.A. (2007). Counter-current tubular heat exchanger: Modeling and adaptive predictive functional control. Applied Thermal Engineering 27, 2332 -2338.

[3]Bonivento, C., Castaldi, P., Mirotta, D. (2001). Predictive control vs. PID control of an industrial heat exchanger, 11th *Mediterranean Conference* on *Control* and Automation, Dubrovnik, Croatia.

[4]Chalupa, P., Bobál, V., Kubalčík, M., Novák, J. (2010). Adaptive predictive control of through-flow heat exchanger. 18th Mediterranean Conference on Control and Automation, Marrakech, Morocco.

[5]Kreutz, M., Richalet J., Mocha, K., Haber, R. (2013). Predictive Functional Control (PFC) of a Heating Ventilation Air Conditioning (HVAC) Process. Nordic Process Control Workshop. Oulu, Finland.

[6]Raul, P.R., Srinivasan, H., Kulkarni, S., Shokrian, M., Shrivastava, G., Rhinehart, R.R. (2013). Comparison of model-based and conventional controllers on a pilot-scale heat exchanger ISA. Transactions, 52, 3, 391-405

[7]Richalet J. and O'Donovan, D. (2009). Elementary predictive functional control, Springer Verlag, Berlin.

Journal of Physics: Conference Series **570** (2014) 042001 doi:10.1088/1742-6596/570/4/042001

Application of advanced methods for the prognosis of production energy consumption

R Stetter[1], P Witczak[1], B Staiger[2], C Spindler[1], J Hertel[1]

[1] Hochschule Ravensburg-Weingarten, 88241 Weingarten, Germany
[2] Kirchner Konstruktionen GmbH, 88250 Weingarten, Germany

E-mail: stetter@hs-weingarten.de

Abstract. This paper, based on a current research project, describes the application of advanced methods that are frequently used in fault-tolerance control and addresses the issue of the prognosis of energy efficiency. Today, the energy a product requires during its operation is the subject of many activities in research and development. However, the energy necessary for the production of goods is very often not analysed in comparable depth. In the field of electronics, studies come to the conclusion that about 80% of the total energy used by a product is from its production [1]. The energy consumption in production is determined very early in the product development process by designers and engineers, for example through selection of raw materials, explicit and implicit requirements concerning the manufacturing and assembly processes, or through decisions concerning the product architecture. Today, developers and engineers have at their disposal manifold design and simulation tools which can help to predict the energy consumption during operation relatively accurately. In contrast, tools with the objective to predict the energy consumption in production and disposal are not available. This paper aims to present an explorative study of the use of methods such as Fuzzy Logic to predict the production energy consumption early in the product development process.

1. Introduction, background and overview

The energy necessary for the production of goods, e. g. the energy for raw material generation, for casting or for milling, can very often not be analyzed in depth in early stages of product development processes.,-This is because today's common tools used for today's product development, such as geometry generating CAD tools (computer aided design tools), do not have the ability to support designers and engineers in this endeavor [2]. The energy consumption in production and disposal is determined very early in the product development process by designers and engineers, for example, through the selection of raw materials, explicit and implicit requirements concerning the manufacturing and assembly processes, or by decisions concerning the product architecture. Today, designers and developers must more or less "blindly" decide, because it is impossible in today's industrial reality to predict the energy consumption in the production. It is hypothesized that intelligent tools and procedures, which allow a prognosis of the energy consumption in production, can shift the knowledge concerning production energy consumption into earlier phases and can increase the potential for energy savings (compare [3]). It is, for instance, only possible to change the architecture of a product i.e. the logical arrangement, early in the conception phase. Similarly, if a raw part is made by casting or cutting, the decision has to be made very early in the process and has immense consequences concerning the energy necessary for production. Figure 2 shows a representation, which is based on similar representations in the area of early determination of product properties (compare [3]), in which it is clearly recognizable that in the early phases of the product development, the energy consumption can be influenced significantly more through ~~by~~ product changes and which also results in considerably smaller change costs.

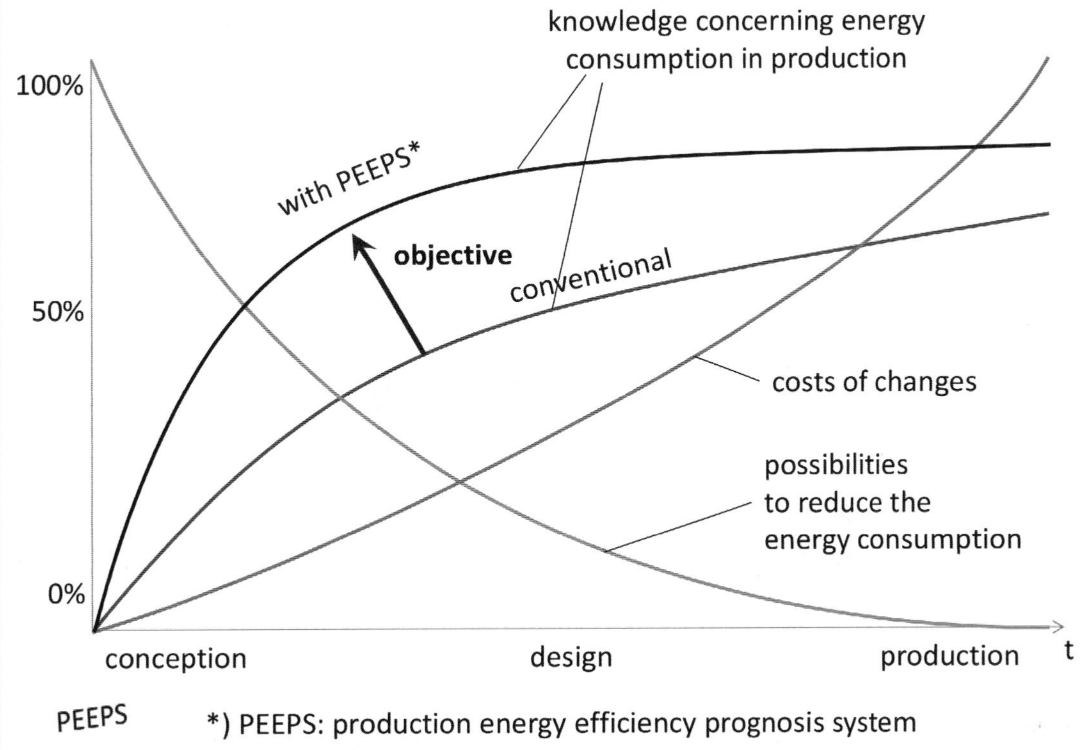

Figure 1. Early evaluation of production energy consumption.

The manufacturing and assembly processes of products consume a considerable amount of the overall energy consumed during the lifetime of a product, ranging from 20% for conventional products, such as cars (compare [4]), up to 80% for products with a large share of electronic components (compare [1]). In the production of goods, several processes lead to the consumption of energy. Figure 2 lists the different sources of energy consumption in the production of parts together with possibilities to achieve a prognosis of this energy.

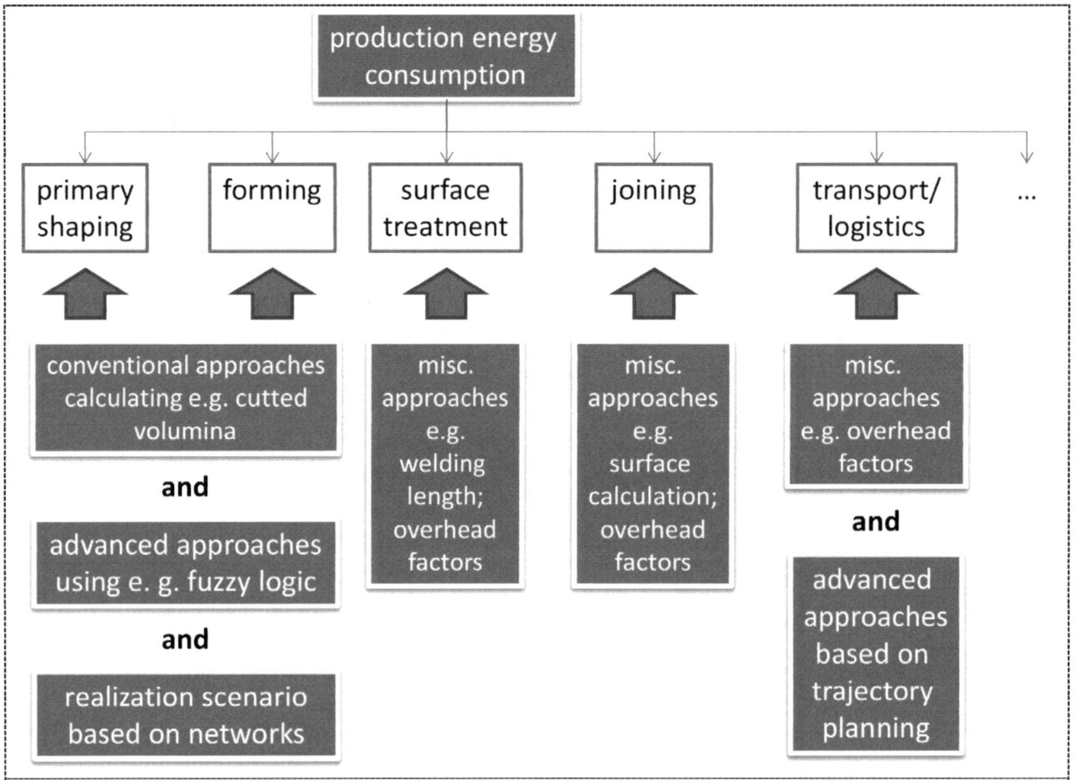

Figure 2. Early evaluation of production energy consumption.

For typical manufacturing operations, conventional approaches can be applied, which are, for instance, based on the volume to be removed by a cutting operation. One example of how this approach is carried out is described in Section 2. However, the application of such methods can be complicated by the complexity of today's products (compare [2]).Some alternatives are advanced approaches, for instance, based on fuzzy logic (see Section 3). Such approaches are explored in this publication together with realization scenarios based on work done through networks of engineers and companies. The processes of surface treatment (e.g. painting), joining (e.g. welding) and logistics (e.g. transportation with vehicles of any kind) can be taken into consideration roughly by means of overhead factors estimating typical energies used in such processes. A promising possibility for a more refined prognosis is the application of trajectory planning methods in order to improve estimation capabilities; such approaches are also the emphasis of this publication (see Section 4). The vision is that all approaches can be combined in a future system, here referred to as PEEPS (Production Energy Estimation and Prognosis System).

2. Analytical approach

Calculations of the energy consumption in production can conventionally be based on certain volumes and/or weights of the components, or of certain sections of the components. For example, the milling volume for milling operations can be used to determine the energy necessary for this milling operation. Research in production technology can provide the necessary tables and equations to determine this energy, if only the milling volume and the milling operation are defined accurately enough . The geometry is developed today almost exclusively in three-dimensional CAD systems. In such systems, the volumes and weights of all components are available; therefore the possibility to couple future systems with CAD systems is very promising.

It is important to note that most of the current CAD systems (CREO, CATIA, NX,) build the products in a tree-like structure and add features to a certain "starting" geometry in order to achieve the final geometry. Usually, these features are representative of manufacturing operations, e.g. a hole might be the representative of a drilling operation. Therefore, these features are appropriate objects for analyzing the energy necessary for certain manufacturing steps. Many works of research have described energies for certain manufacturing operations, such as drilling, milling and grinding. Very often, the resulting equations have a general form similar to Equation 1, describing the ideal material removal energy (compare Deneka&Tönsdorf [6]):

$$E_c = k_{c1.1} \cdot \left(\frac{h}{h_0} \right)^{-m_c} \cdot \Delta V \qquad (1)$$

with $K_{c1.1}$ as specific heat capacity, m_c as XXX, H as density and ΔV as volume difference. The energy for manufacturing is usually higher because of several sources of losses and additional processes which can be necessary or superfluous. The influence of these losses can be covered using equation 2:

$$E_G = E_R + E_F + E_M + E_I + E_A \qquad (2)$$

with E_R as initial object creation energy, E_F as ideal final object creation energy, E_M as energy lost by the machine, E_I as energy for the infrastructure and E_A as assembly energy. The features which describe subtracted volume can be understood and described as an information object thus allowing object oriented programming. CREO 2.0 offers the possibility to program additional functionalities using the ProToolkit. Therefore it is possible to integrate the functionalities of PEEPS smoothly into CREO – the user will experience the tool as additional functionality of his/her well-known CAD system. This system is currently implemented and will be finished as a functional prototype at the end of 2014.

3. Application of Fuzzy Logic

The use of fuzzy logic creates a possibility to integrate human decision-making ability into technical systems, instead of using purely mathematical models (see e.g. [7], [8]). This section explains the application of fuzzy logic for estimating production energy consumption. For this explanation, the proposed procedure scheme is first explained (compare Figure 3).

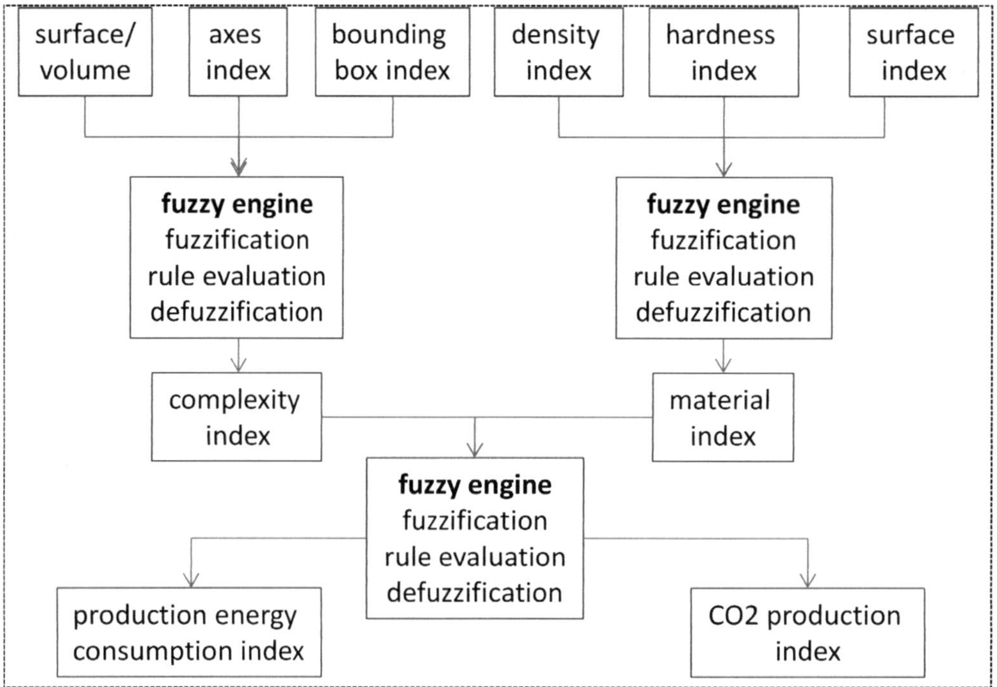

Figure 3. Main procedure scheme.

Secondly, the proposed input product characteristics are described. Thirdly, the relevant indexes are listed and elucidated. Finally, the fourth subsection is giving the results of a sample application.

3.1. Proposed procedure scheme

The inputs of a fuzzy logic system are the fuzzy variables. For the given system, six sensible fuzzy variables were found (compare Section 4.2). In the main part of the system, these fuzzy variables are first used in two fuzzy engines in order to determine the first indices (compare Section 4.3). The results, which, after defuzzification, are once again crisp values, are then given to a third fuzzy engine in order to find two important indices – the "production energy consumption index" and the "CO2 production index". Figure 2 depicts the proposed procedure scheme.

3.2. Definition of fuzzy variables

Defining the input variables is a very demanding task. In this case, an extensive survey in production engineering was carried out. The main idea behind the fuzzy variables goes back to research works by Achiche & Ahmed [9]. In this research, comparable indices were used in a fuzzy engine in order to investigate 3D shapes and the way they are perceived. One example for an indicator is the surface/volume ratio; this first indicator is mainly for the complexity of a product and is the ratio between surface and volume. Simple forms which can be produced with a small amount of energy, such as simple cylinders, are characterized by a small surface/volume ratio. This ratio is defined as shown in Equation 3:

$$\frac{\text{surface}}{\text{volume}}\text{ratio} = s / v = \frac{\text{component surface}\left[\text{mm}^2\right]}{\text{component volume}\left[\text{mm}^3\right]} \tag{3}$$

In a similar manner, five other indicators are defined in order to capture certain characteristics of the respective product.

3.3. Definition of output indices

Two main influences characterize the energy consumption in production: the complexity of the component, leading to many subsequent energy consuming production and transportation processes, and the material of the component – the density, hardness and other material properties, which can lead to a multiplication of energy expenditure for the production of the respective component. On the one hand, the complexity index summarizes the influences of other indices which indicate the complexity of a product's component. On the other hand, the influences of indices which indicate the material difficulty of a component are summarized in the material index.

Both influences need to be combined in order to assess the energy necessary for production – thus forming the production energy consumption index. In a similar manner, the CO2 production index can also be generated.

3.4. Exemplary application and results

For an exemplary application, four different parts from the CAD training and developments for a formula student race car were chosen. The data was taken from the CAD model and for some of the entities, sensible estimations were made. The main parameters of the four parts were given to the initial two fuzzy engines and the first two indices were determined. The results in Figure 3 show clear distinctions.

The high complexity index of the housing is definitely a consequence of the hollow structure. Such structures are relatively light and have a small volume but require a lot of energy in order to produce. On the other hand, the low material index of the housing is influenced by the rather low hardness of this part and because of no necessary surface treatment – in contrast, for instance, to the piston with elaborate surface treatment for the running socket.

In the second fuzzy engine, these results were combined leading to the production energy consumption index (Figure 4 – right side).

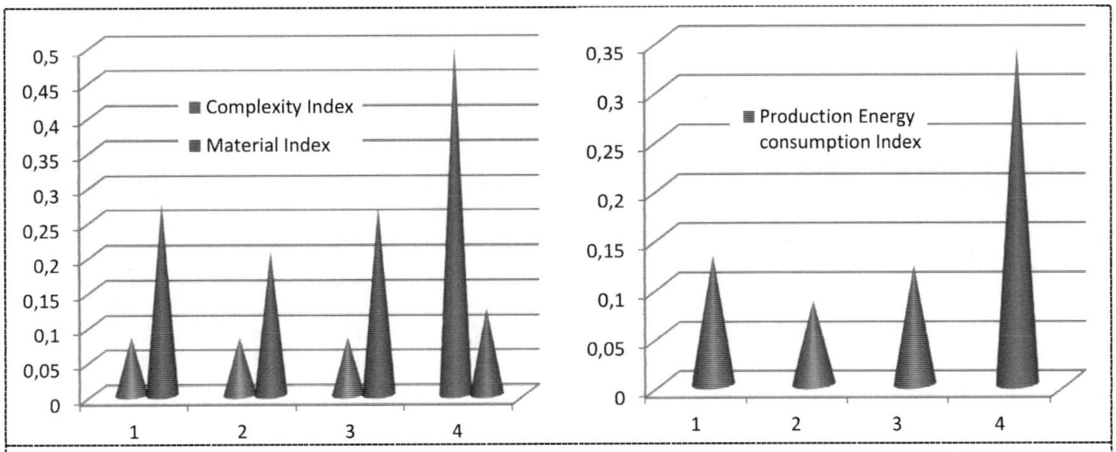

Figure 4. Results of the sample application.

The results show clear differences. For instance, part 4 is a hollow housing with a complex geometry, but will be produced from a soft material.

4. Realization Scenario

One approach for applying the proposed method in industry is the formulation of rules (compare [2]). The authors believe that such a procedure will be a good start to enhancing the prognosis possibilities for production energy consumption. However, in the long run, intelligent, learning systems may lead

Journal of Physics: Conference Series **570** (2014) 042001 doi:10.1088/1742-6596/570/4/042001

to superior performance. The process of realizing this (Figure 5) is generally comparable to a data mining approach.

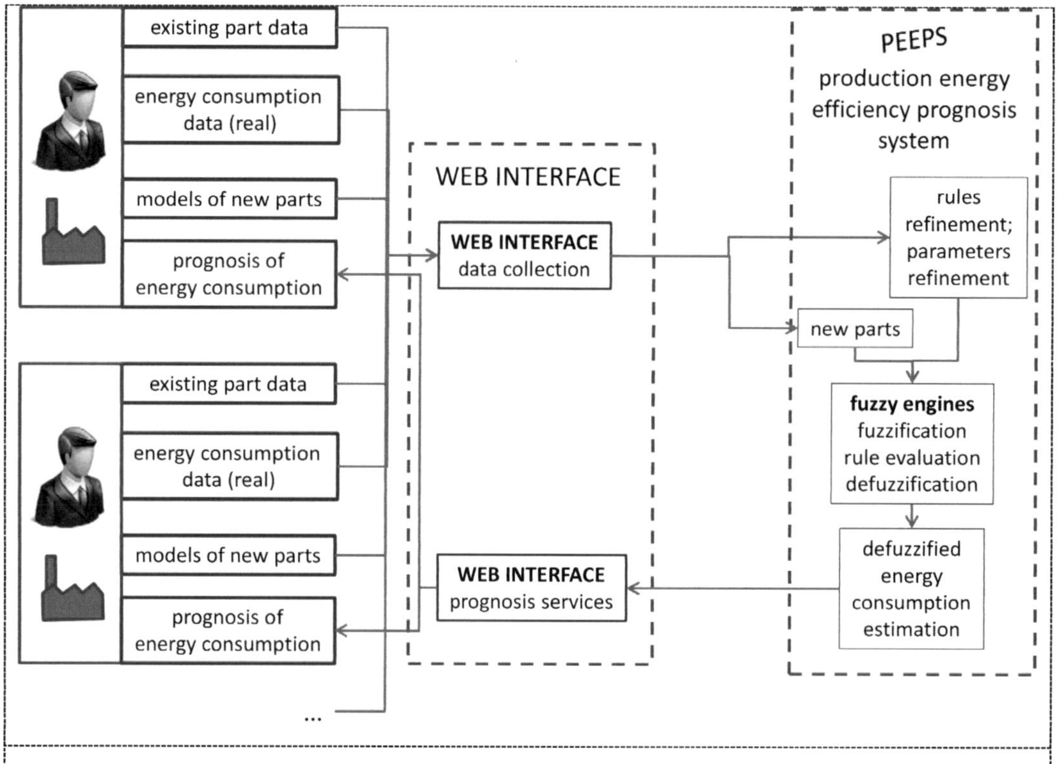

Figure 5. Web-based realization scenario.

Today, the information systems of leading companies could assign the energy consumption of the whole production segment to the single product components with relative accuracy. In the simplest case, where only one component is built in serial production, the overall energy consumption could just be divided by the number of components built. Today, companies have the technology to determine how much energy is needed to produce a certain component, as well as all spatial information concerning this component in CAD system. However, these companies cannot know WHY they need this amount of energy, and therefore cannot produce a prognosis for future components. The idea presented here is that many companies will enter this kind of information (component description and production energy consumption) into a web application. The component data could be entered in the fuzzy system as described in Section 3. The results can then be compared to the real results also provided by the companies. The difference between the results of the prognosis and the real consumption can then be used to modify rules and parameters of the fuzzy system. In this manner, a continuously improving prognosis system can be built. Companies which provide their information about existing products could also use the web application to get an accurate estimate of the energy consumption of future product components (compare Figure 4).

5. Framework for trajectory planning

This section is aimed at logistics and transportation and the prognosis of the energy consumption of these processes. It starts with a short description of the situation of an engineer who is planning the geometry of components and the architecture of a new product. The energy consumption during the production of the product is, amongst other things, dependent on the transportation processes [8]. In order to estimate the energy for these transportation processes, the first main piece of information is the distance to be covered between the different stations where manufacturing, assembling or surface treatment processes are performed. Consequently, the engineer would need to plan the whole manufacturing, assembling and surface treatment process. For products which are not completely new (which is the case for more than 90% of all products), these processes are captured in the enterprise resource planning (ERP) systems, such as SAP, and can be used for the sake of the estimation. The next step would be a mapping of the stations to physical locations in order to measure the distances between the stations [11]. Again, for products which are not completely new, these locations are available in the layout planning tools, which are usually part of the digital manufacturing and production systems, such as DELMIA. Today, these tools are connected in the IT infrastructure of many companies by certain connectors, such as DESC. A component of a production energy efficiency prognosis system could obtain this information from the interfaces of a digital manufacturing and production system, and could transfer them into a form useable by other components of this system (Figure 6).

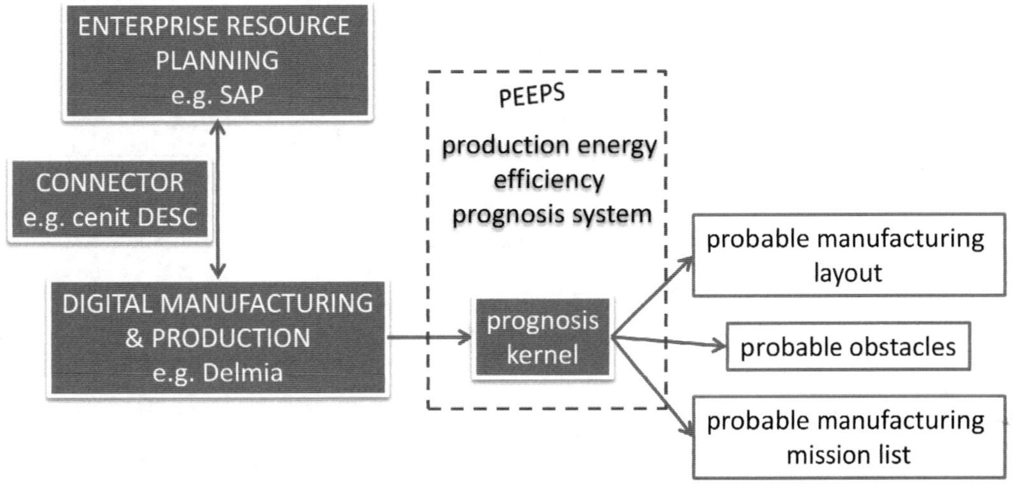

Figure 6. Prognosis kernel.

As stated above, one important entity of information is the list of manufacturing, assembling and surface treatment processes, which can be referred to as a mission list. Essentially, a mission list (ML) is a vector with as many elements as necessary operations:

$$ML = \left[PSP_i\right]_{i=0}^{i=n}\text{; for } i = 1,\ldots,n \tag{1}$$

with PSPi representing production steps with a respective specific ID, and n being the necessary number of such production steps. The next task is the assignment of production steps to production stations. This assignment can be done in a 2 x n assignment matrix (AM):

$$AM = \left[PSP_i \; PST_i\right]_{i=0}^{i=n}\text{; for } i = 1,\ldots,n \tag{4}$$

with PSP_i representing production steps with a respective specific ID, and PST_i being production stations with a respective, specific ID. As stated above, a further important entity of information is the arrangement of the production station, captured in a production layout. Here, only a probable layout can be hypothesized during prognosis. A probable production layout matrix (PPLM) is a n x m-matrix using an appropriate grid:

Journal of Physics: Conference Series **570** (2014) 042001 doi:10.1088/1742-6596/570/4/042001

$$\mathrm{PPLM} = \left[\mathrm{PPST}_{ij}\right]_{i,j=0}^{i,j=n,m} \text{; for } i = 1,\ldots,n \text{ and } j = 1,\ldots,m \tag{5}$$

with $PPST_{ij}$ being the diagonal elements, which describe probable production stations with their respective ID. For the trajectory generation, information about obstacles in the production environment is also necessary. This can also be generated from information in the digital manufacturing and production system and formulated in an obstacle matrix. An obstacle matrix (OM) is a n x m-matrix using an appropriate grid:

$$\mathrm{OM} = \left[\mathrm{OP}_{ij}\right]_{i,j=0}^{i,j=n,m} \text{; for } i = 1,\ldots,n \text{ and } j = 1,\ldots,m \tag{6}$$

with OP_{ij} being the diagonal elements which describe probable obstacles. Here, a value of "0" denotes "no obstacle" at this position of the grid and "1" denotes "obstacle present" at this position of the grid. Additionally, regions of danger, particularly security around obstacles, can be indicated in this matrix (compare). With the listed information, a second section of a production energy efficiency prognosis system could generate probable trajectories (compare Seybold et al. [10]) and a third section of this system could then estimate the energy consumption for instance based on the traveled distance, probable velocities and rolling friction, probable air drag, probable inclination and probable acceleration (Figure 7).

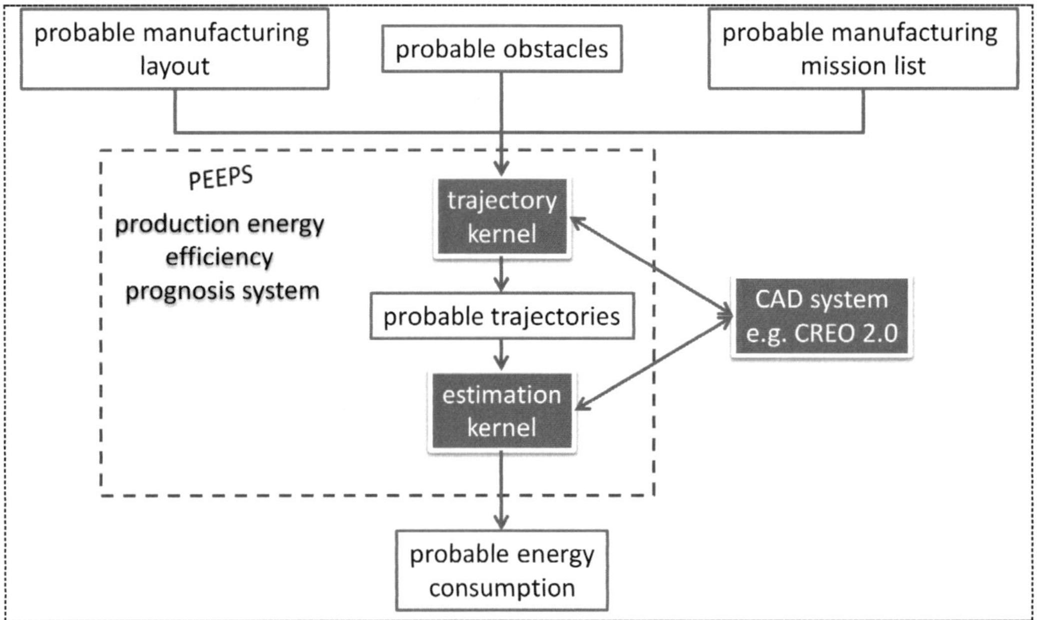

Figure 7. Trajectory kernel and estimation kernel.

The result of this step is an estimation of the probable energy consumption in production; in the final vision of the project, all intermediate steps are carried out automatically by the production energy efficiency prognosis system, allowing the engineer to test several product component and product architecture alternatives. On this basis, he may consciously choose the most sustainable alternative with the smallest energy consumption possible; this possibility is currently not available to engineers in industrial practice.

6. Conclusion

Product development engineers currently have nearly no support concerning the consumption of energy in the later production when they are determining product architecture and product geometry. This paper explored approaches which presents a basis to develop tools which can provide engineers with a prognosis of the energy consumption in production very early in the product development process. One main possibility is analytical approaches based on product geometry and choice of material. Fuzzy logic can be used to estimate a production energy consumption index, based on six indices which can relatively easy be collected in a CAD system which contains extended information, such as material and surface hardness – information which is important for the product drawing anyway. An energy consumption prognosis of the transportation processes in manufacturing and assembly can, amongst other things, be realized by means of trajectory planning based on probable manufacturing layouts and probable obstacle locations. Today, the information about these steps is present in the numerous systems which build up the digital factory. However, the connection with and processing of this information is missing. The presented research is a first step into the direction of enhanced prognosis possibilities. In the further course of the project, detailed mathematical formulations and verifications are planned to provide assistance to engineers who must make far-ranging decisions in the early phases of product design.

Acknowledgments

This project is supported by a grant from the European Regional Development Fund and the Ministry of Science, Research and the Arts of Baden-Württemberg, Germany (information under: www.rwb-efre.baden-wuerttemberg.de).

References

[1] Williams, E.D.: Revisiting energy used to manufacture a desktop computer: hybrid analysis combining process and economic input-output methods.. Conference Record 2004 IEEE International Symposium on Electronics and the Environment, pp. 80- 85.

[2] Stetter, R; Witczak, P., Prasad, P. K.; Staiger, B.: Fuzzy Logic for Production Energy Efficiency Estimation. Submitted to the International Conference on Diagnostics of Processes and Systems (DPS 2013).

[3] Bernard, R.; Stetter, R.: Early Determination of Product Properties. In: Hubka, V. (Hrsg.); et al.: Proceedings of ICED 97, Tampere. Zürich: Edition Heurista, 1997, S. 2/675 - 2/680.

[4] Schröder, C. "Green Carbody Technologie" will 50 Prozent Energie in der Produktion einsparen. In: ATZ-Online, 22.06.2009.

[5] Stetter, R.: Prognose des Energieverbrauchs in der Produktion. In: Forschungsreport für den Maschinenbau in Baden-Württemberg. Bingen: Public, 2013.

[6] Denkena, B.; Tönsdorf, H.K.: "Spanen" Berlin Heidelberg: Springer, 2011.

[7] Ross, T.: Fuzzy logic with Engineering applications. New York: McGraw-Hill, 1995.

[8] Kasabov, N.: Fuzzy sets and Fuzzy logic. In: Foundations of Neural Networks, Fuzzy Systems, and Knowledge Engineering. MIT Press, 1996.

[9] Achiche, S. and Ahmed, S.: Investigating the Relationships between Quantitative and Qualitative Properties of 3d Shapes using Fuzzy Logic Models. In: Proceedings of the International Conference on Research into Design, ICoRD'09, 2009.

[10] Seybold, L., Pieczynski, A., Paczynski, A, Krokowicz, J., Stetter R.: Development of a Trajectory Kernel for Autonomous Vehicles. In: Bokor, J. (Ed.): Proceedings of the 9th Workshop on Advanced Control and Diagnosis, ACD'2011, 17. and 18. November 2011, Budapest, Ungarn.

[11] Stetter, R.; Paczynski, A.; Witczak, P.; Staiger, B.: Advanced Trajectory Planning for Production Energy Estimation. In: Pomiary Automatyka Robotyka nr 2/2014, pp. 70-77, ISSN 1427-9126.

Analysis of Usefulness of a Fuzzy Transform for Industrial Data Compression

Anna Sztyber

Warsaw University of Technology, Faculty of Mechatronics, Institute of Automatic Control and Robotics, Boboli 8, 02-525 Warsaw, Poland

E-mail: a.sztyber@mchtr.pw.edu.pl

Abstract. This paper presents the first part of an ongoing work on detailed analysis of compression algorithms and development of an algorithm for implementation in a real industrial data processing system. Fuzzy transforms give promising results in an image compression. The main aim of this paper is to test the possibility of an application of the fuzzy transforms to the industrial data compression. Test are carried out on the data from DAMADICS benchmark. Comparison is provided with a piecewise linear compression, which is nowadays the standard in the industry. The last section contains discussion of the obtained results and plans for the future work.

1. Introduction

Nowadays, in industrial plants huge amount of data is measured, processed, and archived, so an issue of data compression arises. The question is not only how to compress collected sensor data, but also whether they should be compressed at all?

The cost of storage devices for data archiving is decreasing, but it is still nonzero. There are also problems of processing large data sets and network traffic.

On the other hand, advanced techniques of data mining and data processing are becoming more common, so there is an issue of the impact of the data compression on the results of calculations.

The archival process data can be used for:

- calculation of statistics, like means and standard deviations,
- balance calculations,
- feature extraction,
- control loop monitoring,
- building cause-effect process models,
- tuning diagnostic systems,
- identification of models (fuzzy, neural, parametric) for fault detection,
- synthesis of control algorithms, like model predictive controller.

Due to the author specialisation the above list contains mainly tasks connected with fault diagnosis and process monitoring and is probably not complete.

Analysis of the impact of the compression on data driven process analyses was presented in [1]. Authors point out that standard compression methods change means, standard deviations, non-linearity indices, and Harris index. Moreover the compression can change signal spectrum, introduce nonlinearities, and dump higher frequency oscillations. Even a noise reduction can be a flaw, because it changes the variation and Harris index. Authors of [1] insist, that calculations should be carried out using the original data. There are also analyses of the impact of the data compression on control loop monitoring [2], pattern matching [3], and PCA models [4].

All above considerations regards the standard industrial compression algorithms. Nowadays, the standard is a piecewise linear trending. Among the most widespread solutions there are:

- Boxcar [5]–archived are only data points, which differ more than the predefined threshold from previous recorded value;
- Backward Slope [5]–signal is approximated by a piecewise linear function, the line slope is determined by the two previously recorded values. Archived are only data points, which differ more than the predefined threshold from the calculated slope;
- Boxcar-Backward Slope–combination of the two methods, data point must exceed both thresholds to be archived;
- Swinging Door [6]–signal is approximated by a piecewise linear function, data points are limited by two line segments with specified maximal distance;

In all methods data points are archived with a time stamp. All the standard methods are very simple and easy to implement, but not very sophisticated. Nowadays, data compression is a large scope of research, mainly due to the multimedia applications. So the question arises, if some solutions can be adapted to improve the industrial data compression. In this paper a fuzzy transform is analysed.

The paper is structured as follows. Definitions of fuzzy transforms are recalled in Sect. 2. Results of calculations are described in Sect. 3, and results for piecewise linear compression in Sect. 3.1. Discussion of the results is presented in Sect. 4.

2. Fuzzy Transform

Fuzzy sets were defined to represent uncertain information [7]. The fuzzy logic is used in the industry for synthesis of fuzzy controllers, fuzzy modelling (Wang-Mendel or Takagi-Sugeno-Kanga models), and to deal with uncertain information in diagnostic systems.

The fuzzy transforms have applications in the image compression in order to avoid JPEG blocking artifacts [8–10]. Main idea of a fuzzy compression is to cover function domain by the set of fuzzy sets, called a fuzzy partition, and represent function by a smaller set of transform coefficients. There is one coefficient for each fuzzy set in the partition. The coefficients are the best in the mean square sense with respect to the selected partition.

Definitions of fuzzy partitions and fuzzy transforms will be cited after [8].

Definition 1 (Fuzzy partition [8]) *Let $x_1 < \ldots < x_n$ be fixed nodes within $[a, b]$, such that $x_1 = a$, $x_n = b$ and $n \geq 2$. We say that fuzzy sets $A_1, \ldots, A_n$, identified with their membership functions $A_1(x), \ldots, A_n(x)$, defined on $[a, b]$, form a **fuzzy partition** of $[a, b]$ if they fulfil the following conditions for $k = 1, \ldots, n$:*

(i) $A_k : [a, b] \to [0, 1], A_k(x_k) = 1,$

(ii) $A_k(x) = 0,$ *if* $x \notin (x_{k-1}, x_{k+1}),$

(iii) $A_k(x)$ *is continuous,*

(iv) For $k = 2, \ldots, n$ $A_k(x)$ *monotonically increases on* $[x_{k-1}, x_k]$*, and for* $k = 1, \ldots, n-1$ $A_k(x)$ *monotonically decreases on* $[x_k, x_{k+1}]$*,*

(v) $\forall x \in [a, b] \Sigma_{k=1}^{n} A_k(x) = 1.$

The partition is called uniform if points $x_1, \ldots, x_n$ are equidistant. Exemplary uniform fuzzy partitions are shown in Fig. 1 and Fig. 2.

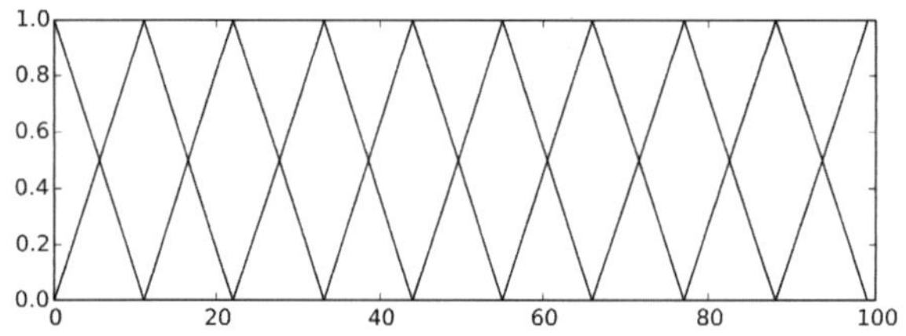

Figure 1. Uniform fuzzy partition of $[0, 100]$ with triangular shaped functions.

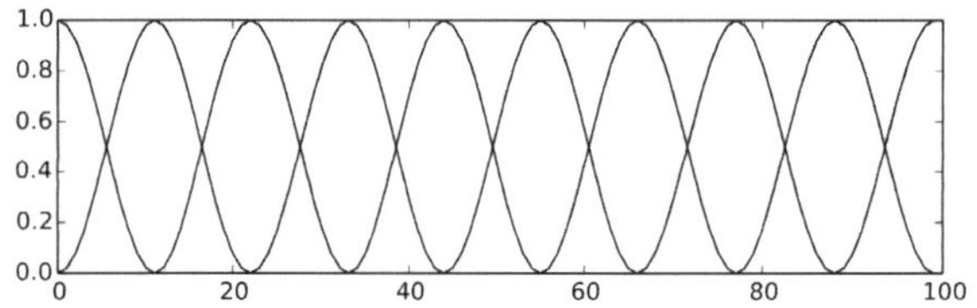

Figure 2. Uniform fuzzy partition of $[0, 100]$ with sinusoidal shaped functions.

The fuzzy transform, called also F-transform, is a linear transformation. The transform coefficients are weighted mean values of the transformed function with respect to the fuzzy partition. Discrete fuzzy transform and inverse discrete fuzzy transform are defined as follows.

Definition 2 (Discrete F-transform [8]) *Let a function f be given at nodes $p1, \ldots, p_m \in [a, b]$, $A_1, \ldots, A_n$ be basic functions which form a fuzzy partition of $[a, b]$. We say that the n-tuple of real numbers $[F_1, \ldots, F_n]$ is the **discrete F-transform** of f with respect to $A_1, \ldots, A_n$ if:*

$$F_k = \frac{\Sigma_{i=1}^{m} f(p_i) A_k(p_i)}{\Sigma_{i=1}^{m} A_k(p_i)}.$$

Definition 3 (Inverse discrete F-transform [8]) *Let function $f(x)$ be given at nodes $p1, \ldots, p_m \in [a, b]$, and $[F_1, \ldots, F_n]$ be the discrete F-transform of f w.r.t. $A_1, \ldots, A_n$. Then the function*

$$f_{F,n}(p_i) = \Sigma_{k=1}^{n} F_k A_k(p_i),$$

*defined at the same nodes, is the **inverse discrete F-transform**.*

For each data point p_i only two membership functions have nonzero values, so the fuzzy transform can be calculated in time $O(m)$, were m denotes the number of data points.

There is also a possibility of using some t-norm instead of weighted average [9, 10], but it introduces additional artifacts into the recovered signal.

3. Results

Numerical experiments with fuzzy compression were carried out on real industrial data from DAMADICS benchmark [11]. Measurements of juice pressure at valve inlet ($P51\_05$, in the following abbreviated to $P1$) were used. The range of values was 0-1000kPa. The data was collected each second, overall batch from one day of 86400 points was processed.

Firstly, tests were carried out with uniform fuzzy partitions. Results are shown in Fig. 3. Original signal is shown in the upper plot ($P1$). The signal was sampled with high speed, so large compression ratios can be considered. Signals $P1_{FTT1}$, $P1_{FTT2}$, $P1_{FTT3}$ were reconstructed respectively from 200, 2000 and 8000 fuzzy transform coefficients.

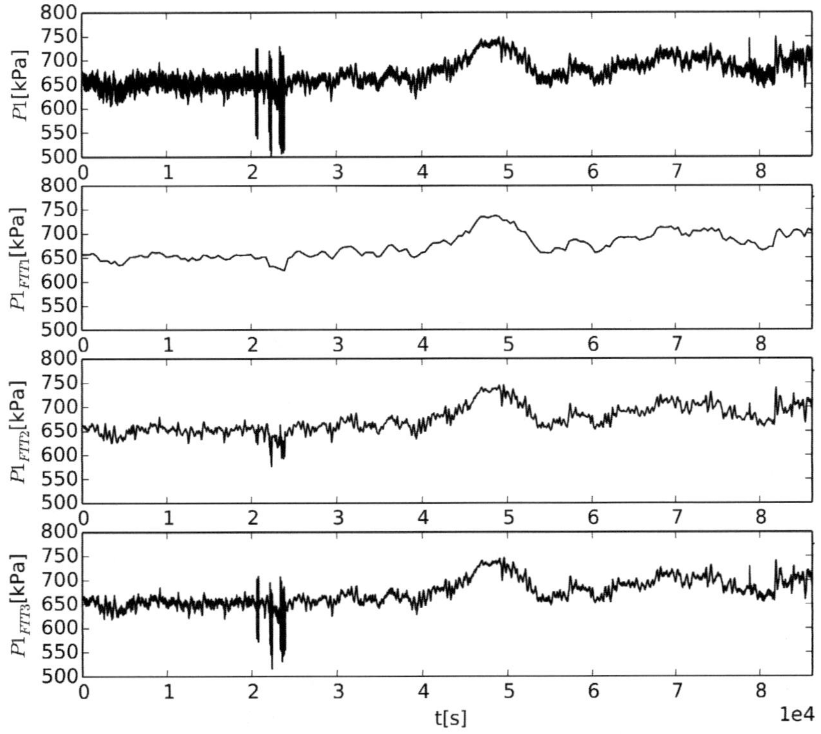

Figure 3. Fuzzy compression with uniform, triangular shaped fuzzy partition.

Results of compression are summarised in Tab. 1. The results contain mean squared error mse, maximal error $maxe$, and difference of a mean of original signal and mean of recovered signal $emean$. All errors are in kPa. Number of the fuzzy transform coefficients is denoted by n. Results of transformation with sinusoidal shaped functions are included in the table, denoted respectively by $P1_{FTS1}$, $P1_{FTS2}$, and $P1_{FTS3}$. Results for sinusoidal shaped membership functions are slightly better.

Table 1. Results of fuzzy compression with uniform fuzzy partition.

Signal	$P1_{FTT1}$	$P1_{FTT2}$	$P1_{FTT3}$	$P1_{FTS1}$	$P1_{FTS2}$	$P1_{FTS3}$
n	200	2000	8000	200	2000	8000
mse	10.13	6.61	4.24	9.98	6.52	4.14
$emax$	135.6	106.6	44.7	135.3	105.9	42.0
$emean$	0.01	0.0002	0.0002	0.01	0.0001	0.0002

Difference between a signal compressed with triangular and sinusoidal shaped membership functions can only be seen while zooming in (Fig. 4). Both methods introduce their characteristic artifacts. With the triangular shaped membership functions decompressed signal is piecewise linear.

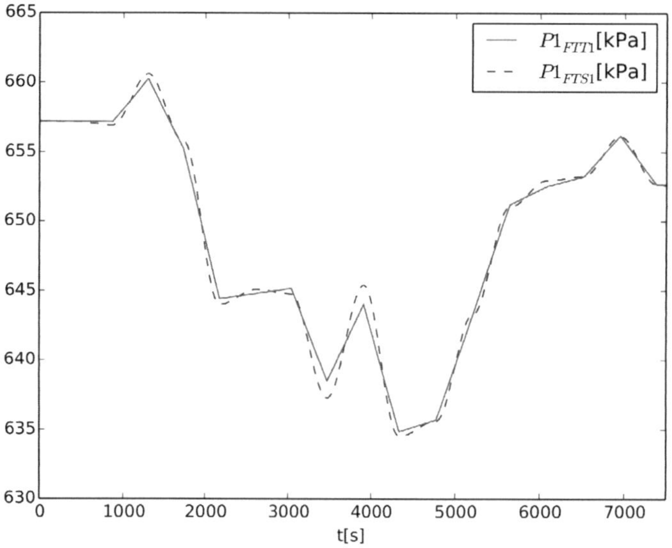

Figure 4. Signals compressed with triangular ($P1_{FTT1}$) and sinusoidal ($P1_{FTS1}$) shaped membership functions.

The fuzzy transform has smoothing properties, therefore some quickly varying signal components are eliminated. This causes large maximal error. On the other hand, distraction in signal mean is negligible. The question arises, if it is possible to better maintain quickly varying signal components. The natural answer is to use a non-uniform partition. This approach was used for the image compression in [9].

Next part of the tests was carried out with the non-uniform fuzzy partitions. The points forming partition $x_1, \ldots, x_n$ were placed with density approximately proportional to the original signal variation. Example of the non-uniform partition is shown in Fig. 5. There is a larger number of fuzzy functions for quickly varying part of the signal. For clarity only 25 sets were drawn.

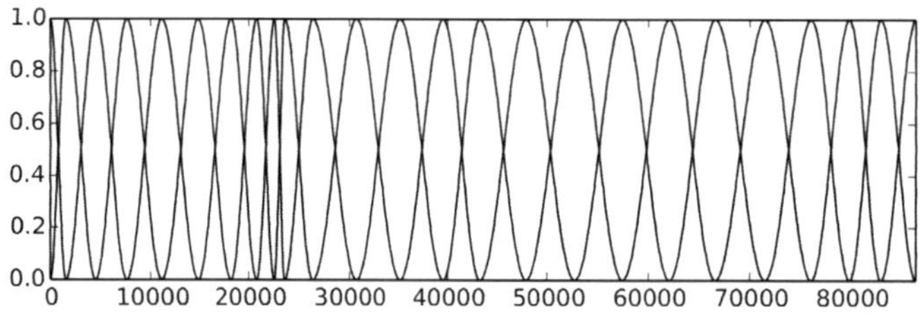

Figure 5. Example of non-uniform fuzzy partition with sinusoidal shaped functions.

Signals compressed with non-uniform partition are shown in Fig. 6. Results are listed in Tab. 2. Method for finding non-uniform partition proportional to signal variance can return slightly different size of the obtained partition then desired, so actual numbers are provided in the table. Signals $P1_{FTT1var}$, $P1_{FTT2var}$, and $P1_{FTT3var}$ were compressed with triangular membership functions, and $P1_{FTS1var}$, $P1_{FTS2var}$, and $P1_{FTS3var}$ with sinusoidal.

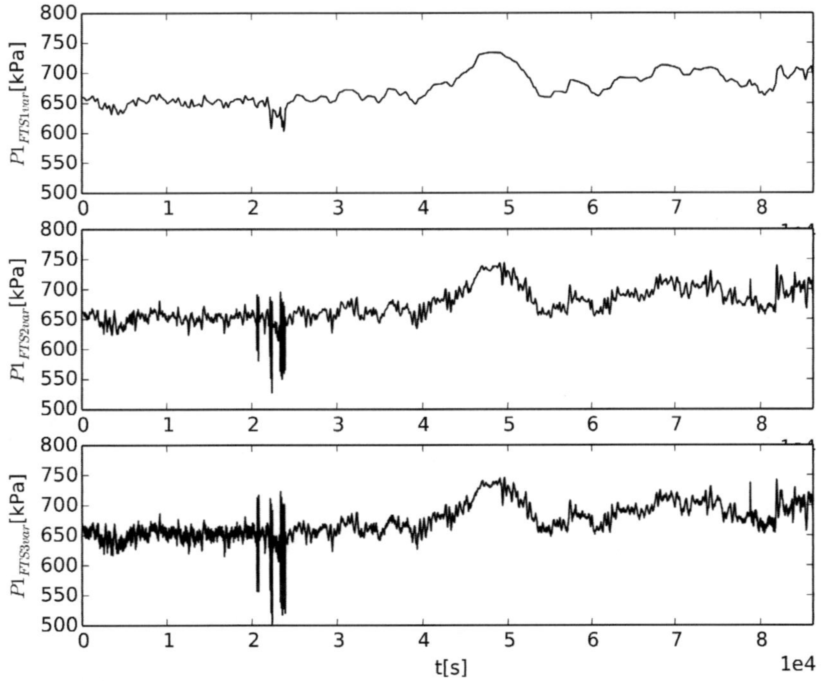

Figure 6. Fuzzy compression with non-uniform, sinusoidal shaped fuzzy partition.

Table 2. Results of fuzzy compression with non-uniform fuzzy partition.

Signal	$P1_{FTT1var}$	$P1_{FTT2var}$	$P1_{FTT3var}$	$P1_{FTS1var}$	$P1_{FTS2var}$	$P1_{FTS3var}$
n	201	2002	7995	201	2002	7995
mse	9.99	5.23	3.83	9.87	5.12	3.76
$emax$	116.4	64.0	26.1	116.9	60.8	25.2

Results for the non-uniform partition are clearly better then for the uniform partition, but there is one drawback. For uniform partition only the transform coefficients must be written into the archived file. In case of the non-uniform partition time stamp for each coefficient is needed.

3.1. Standard compression

The obtained results should be compared with the standard compression methods. For this purpose program was written to simulate a compression of a system OSISoft PI. The compression has two steps: the first is a Boxcar algorithm, and the second is described as a Swinging Door algorithm, however the provided description [12, 13] is not exactly consistent with the original algorithm [6]. Purpose of the program was not to precisely mimic PI compression, but to have some representation of the piecewise linear compression.

Algorithm needs two parameters: $ExDev$ - threshold for the Boxcar algorithm, and $ExComp$ - threshold for the Swinging Door algorithm. It is not straightforward to obtain desired compression ratio using this two parameters, so numbers of recorded points will be slightly different then numbers of transform coefficients.

Results are presented in Fig. 7. Signals $P1_{PI1}$, $P1_{PI2}$, and $P1_{PI3}$ were recovered respectively from 201, 2002 and 7995 data points. Numerical results are summarised in Tab. 3. The number of recorded points is denoted by n.

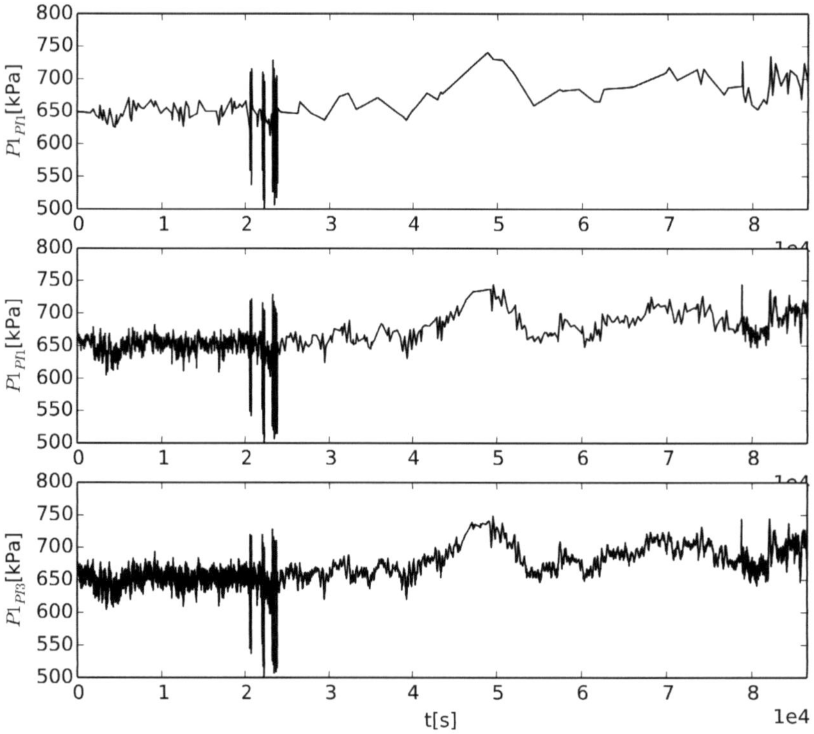

Figure 7. Compression by the piecewise linear approximation.

Table 3. Results of the compression by the piecewise linear approximation.

Signal	$P1_{PI1}$	$P1_{PI2}$	$P1_{PI3}$
n	201	2002	7995
mse	10.86	5.57	3.37
$emax$	44.09	21.2	14.9
$emean$	1.16	0.05	0.02

The piecewise linear compression gives smaller maximal error. Mean square errors are comparable. Fuzzy transforms gives slightly better results for higher compression ratios. Fuzzy compression gives good results in the recovery of the signal mean.

4. Discussion

Fuzzy transform is a possible method of the data compression. In the terms of the mean square error it gives results comparable with standard methods. There is a wide variety of other methods, that can be applied to this problem. In this section characteristic features of fuzzy transform will be summarised, so it can be compared with existing approaches.

The fuzzy compression has following advantages:

- very little error of mean of the decompressed signal, so it is suitable for balance calculations,
- good preservation of lower frequency signal components,
- simple signal transformation,
- during decompression each signal sample is calculated using only two transform coefficients, so there is a possibility to decode only fragment of the signal (advantage over Fourier, Discrete Cosine, and Wavelet Transform),
- compression ratio depends on the number of membership functions in the fuzzy partitions and is known in advance (advantage over piecewise linear trending).

The fuzzy transform has also smoothing and denoising properties. It is a good feature in most applications. Visual appearance of the signal is better. On the other hand, in some applications reduction of the variance can be a drawback [1].

In the fuzzy transform compression ratio is chosen by the user of the system. In the piecewise linear trending user defines thresholds for Boxcar and Swinging Door algorithms. Relation between this parameters and compression ratio depends on the processed signal and is hard to determine, therefore often parameters values are inappropriate, and signal is over-compressed or under-compressed [14, 15].

The fuzzy compression has following disadvantages:

- dumping of higher frequencies,
- large maximal error.

Maximal error in the fuzzy compression is not known in advance. This feature is unacceptable in the safety critical measurements. In the piecewise linear trending maximal error is determined by the thresholds.

5. Conclusions

In this paper analysis of the fuzzy transform as a tool for the industrial data compression was presented. The fuzzy transform has a number of promising features, but obtained results, in the author opinion, are not good enough.

Many authors [2–4, 16, 17] indicates that other transforms like Fourier Transform, Discrete Cosine Transform and particularly Wavelet Transform work well in the data compression. Therefore, the next step will be to test this solutions.

Another important issue is a problem of assessment of a compression algorithm. In this paper rather naive or straightforward measures of compression quality were considered. Mean square error is the most popular measure, but it does not take into account signal dynamics. Therefore, there is a need to determine a set of measures, that will cover all spectrum of desired compression algorithm features.

It could also be noted, that very wide variety of compression ratios was considered. In large industrial plants it is impossible to tune compression of each measurement separately, therefore there is a need for a tool for automatic selection of compression parameters.

Acknowledgments
The work was partially financed from the Innotech project of the National Centre for Research and Development, number: Hi3/227511 and by grant of Dean of Faculty of Mechatronics.

References
[1] Thornhill N F, Choudhury S and Shah S L 2004 *Journal of Process Control* **14** 389–398
[2] Sivalingam S and Hovd M 2011 *Control Automation (MED), 2011 19th Mediterranean Conference on* pp 594–599
[3] Singhal A and Seborg D E 2003 *American Control Conference*
[4] Imtiaz S A, Shoukat Choudhury M A A and Shah S L 2007 *Industrial & Engineering Chemistry Research* **46** 481–491
[5] Hale J C and Sellars H L 1981 *Chemical engineering progress.* **77** 3843
[6] Bristol E 1990 *ISA National Conference Proceedings* pp 749–754
[7] Zadeh L 1965 *Information and Control* **8** 338 – 353 ISSN 0019-9958 URL `http://www.sciencedirect.com/science/article/pii/S001999586590241X`
[8] Perfilieva I 2006 *Fuzzy Sets and Systems* **157** 993 – 1023
[9] Nobuhara H, Pedrycz W, Sessa S and Hirota K 2006 *Information Sciences* **176** 2526 – 2552
[10] Hirota K and Pedrycz W 2002 *Fuzzy Sets and Systems* **126** 325 – 335
[11] Bartyś M, Patton R, Syfert M, de las Heras S and Quevedo J 2006 *Control Engineering Practice* **14** 577 – 596 ISSN 0967-0661 a Benchmark Study of Fault Diagnosis for an Industrial Actuator A Benchmark Study of Fault Diagnosis for an Industrial Actuator URL `http://www.sciencedirect.com/science/article/pii/S0967066105001796`
[12] Benvenuto P 2012 Pi 101, module 10: Compression algorithm URL `http://community.osisoft.com/index.php?/topic/1061-pi-101-module-10-compression-algorithm/`
[13] OSIsoftLearning 2012 Osisoft: Exception and compression full details URL `https://www.youtube.com/watch?v=89hg2mme7S0`
[14] Alsmeyer F 2006 *16th European Symposium on Computer Aided Process Engineering and 9th International Symposium on Process Systems Engineering* (*Computer Aided Chemical Engineering* vol 21) ed Marquardt W and Pantelides C (Elsevier) pp 1533 – 1538
[15] Pettersson J and Gutman P O 2004 *Journal of Process Control* **14** 431 – 439 ISSN 0959-1524 URL `http://www.sciencedirect.com/science/article/pii/S0959152403000969`
[16] Watson M, Liakopoulos A, Brzakovic D and Georgakis C 1995 *American Control Conference, Proceedings of the 1995* vol 2 pp 1265–1269
[17] Watson M J, Liakopoulos A, Brzakovic D and Georgakis C 1998 *Industrial & Engineering Chemistry Research* **37** 267–274

Combustion distribution control using the extremum seeking algorithm

A Marjanovic[1], M Krstic[2], Z Djurovic[1], G Kvascev[1] and V. Papic[1]

[1]School of Electrical Engineering, University of Belgrade, Belgrade, Serbia
[2]Department of Mechanical and Aerospace Engineering, University of California, San Diego, La Jolla, USA

E-mail: amarjanovic@etf.bg.ac.rs

Abstract. Quality regulation of the combustion process inside the furnace is the basis of high demands for increasing robustness, safety and efficiency of thermal power plants. The paper considers the possibility of spatial temperature distribution control inside the boiler, based on the correction of distribution of coal over the mills. Such control system ensures the maintenance of the flame focus away from the walls of the boiler, and thus preserves the equipment and reduces the possibility of ash slugging. At the same time, uniform heat dissipation over mills enhances the energy efficiency of the boiler, while reducing the pollution of the system. A constrained multivariable extremum seeking algorithm is proposed as a tool for combustion process optimization with the main objective of centralizing the flame in the furnace. Simulations are conducted on a model corresponding to the 350MW boiler of the Nikola Tesla Power Plant, in Obrenovac, Serbia.

1. Introduction

Thermal power plants are in the group of large-scale systems with many complex processes taking place inside of them. There are a number of papers dedicated to the modeling, control and fault detection to these relevant processes [1], [2]. As such, the optimization of the combustion process is a very important task. Proper control algorithm could significantly increase the efficiency of the process, which is one of the main goals in this sort of systems. At the same time, the consequence of the poor control algorithm is a low safety and reliability of the system. Namely, prolonged exposure to high temperatures inside the furnace contributes to the short lifetime of the components and can result in different kind of failures. Over 30% of all the failures in the boilers are caused by high temperatures. Whether the change is instant or gradual over time, high temperature can lead to unwanted deformations and material cracks [3].

Inevitable by-products of the combustion process are the deposits of soot and slag on the walls of the boiler components. Soot and slag are the products of molted ash, which is located on the surfaces that absorb heat by emission. Large amount of soot and slag at the bottom section of the boiler contribute to higher temperatures and deposits at the top section of the boiler. These deposits decrease the heat transfer together with the coefficient of efficiency. Therefore, it is necessary to maintain the temperature inside the boiler below the melting temperature of the ash, so that it can be disposed using the ash transport system. Additional harmful products of combustion are oxides of nitrite, sulfur and carbon. Recent pollution standards are setting high objectives when it comes to these oxides. There are many ways of dealing with this problem. Some of them involve fuel additives, alternative fuels and low NOx burners and others involve optimal control algorithms. An important instrument for process simulation and therefore, an important tool for the analysis of influence of different fuels, geometry and firing, is the Computational Fluid Dynamics (CFD) [4], [5]. However, since CFD employs large amounts of data it is not convenient for online procedures.

That is one of the reasons for which we propose the use of non-model based extremum seeking control (ESC) approach. Although the basics of extremum seeking were set in the early 1920s, it was not until the end of 20th century that it secured an important role in both theory and practice [6], [7], [8], [9]. Widely known application of ESC is the combustion timing control for HCCI engines [10], antilock braking systems (ABS) design [11], increasing the efficiency in photovoltaic systems [12], etc. The aim of the ESC is to design a control algorithm which drives the parameters of the system towards their optimal value. For some systems the question of optimality is easily formulated, such as minimizing the net power output [13]. For others, objective function is a result of thorough analysis of system behavior and can be defined using the available measurements. In those cases, the model of the system is not needed in the optimization process, which is usually highlighted as an advantage of this approach. However, such control does not guarantee adequate transient performance. On the other hand, criterion function includes model and other dynamic related parameters of the system [14]. The main drawback of this approach is its sensitivity to model uncertainties.

The paper is organized as follows: Section 2 provides description of the thermal power plant boiler necessary for the combustion process understanding. Formulation of the combustion distribution optimization based on ESC is given in Section 3. Section 4 presents main results and comments on the performance of the proposed gradient extremum seeking based control algorithm implemented on the model of the boiler in thermal power plant Nikola Tesla, in Serbia.

2. Combustion process description

Thermal power plant Nikola Tesla A (TENT A) is located in Obrenovac, Serbia. It consists of six units with overall nominal power of 1650 MW. Lignite coal with low and variable calorific value (5000 kJ/kg to 9000 kJ/kg) is used as a regular fuel in all six blocks. Auxiliary oil fuel is used in the firing process and for supporting fire with low calorific value coal firing. The unwanted by-products of the firing process are large amounts of ash and slag, as well as the oxides of carbon and nitrite which contribute the increase in environmental pollution.

This paper focuses on the identification of the combustion process in the unit A6 with 350 MW of nominal power. The main control task is to maintain the reference block power, while controlling the fresh steam pressure in front of the turbine. One way to achieve this is to control the amount of coal and fresh air inserted into the boiler. Block A6 has six mills, as shown in Figure 1. Therefore, the required amount of coal is divided onto six parts and the operators have certain degree of freedom during this partitioning. The mills are positioned as to form a tangential configuration of the furnace. Figure 1. also shows the displacement of the six oil fuel burners used to support fire during ignition and some unexpected events.

Lignite coal is transported from the bunker, using doser and mill feeders. The amount of transported coal is proportional to the doser speed. After being partially dried in the recirculation channel, it is fed to the mill. Final coal drying is performed inside the mill, together with pulverization. The so-called air mixture is further sent to the mill separator. There, under the influence of inertia, insufficiently pulverized particles are extracted and returned back to the mill for re-pulverizing. From the mill separator, aero-mixture is transported throughout the channels to the burners. The combustion air is introduced using the fresh air fan. Heated air is divided into three streams: primary, secondary and tertiary. Primary air is fed to the recirculation channel and it is used for regulation of the aero-mixture exit temperature. Secondary air is transported to the burners and blown into the furnace, separately from aero-mixture. One part of the tertiary air is used for the additional combustion bar and the other is used for cooling during burner ignition.

During the reconstruction phase, a distributed control system was introduced to the block A6, which enabled additional possibilities in terms of control process quality. Data monitoring and storing is performed using SCADA system as well as the novel sensor system consisting of 42 pyrometers and several thermographic cameras. The pyrometers are distributed over five different levels (heights). Depending on the physical conditions of the furnace and the experience of the engineers working with it, the number of pyrometers varies with height. The levels at 17 m and 51 m are equipped with 4 pyrometer units per level, while there are 12 pyrometers per level at 25 m, 38 m and 43 m of height (Figure 2). The pyrometers are two-color processing units, meaning that the measurements are obtained at two close wavelength ranges (0,96μm and 1,05μm). This provides two significant information about the one-color and two-color temperature. The one-color temperature is the effective temperature and the two-color temperature is approximately the maximum temperature on the optical path. The nature of the two-color temperature estimation eliminates the influence of unburned coal particles, smoke, steam, etc. On the other hand, these particles affect the one-color temperature. Therefore, the difference between these two temperature estimations is a measure of combustion process efficiency.

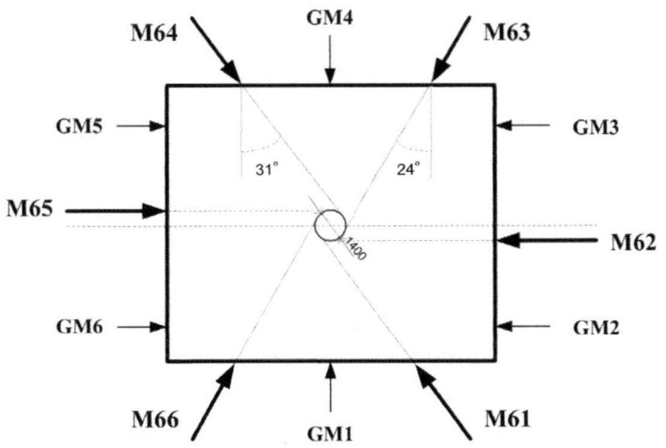

Figure 1. Displacement of the mills (M61 to M66) and oil fuel burners (GM1 to GM6) in block A6 of thermal power plant Nikola Tesla, Obrenovac, Serbia.

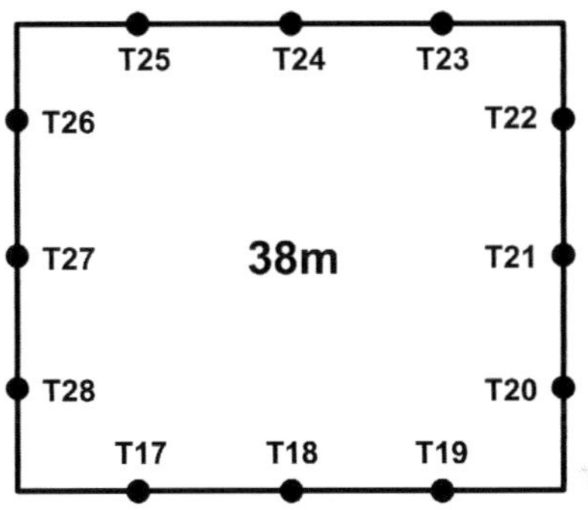

Figure 2. Configuration of the pyrometer sensor system at 38 m of height: three per each side of the boiler.

Previous detailed analysis of the system showed certain correlation between the mills loads and the temperature distribution inside the furnace. Furthermore, important conclusions have been drawn with respect to the relation of coal distribution over mills and the position of the flame using the visualization of the spatial temperature distribution. These results are an important step in obtaining optimal control law for the system. The combustion process is one very important and critical process. Therefore, in order to obtain and test the desired control algorithm, we first conducted a system identification procedure [15]. The system itself is exposed to different kinds of disturbances (variable calorific fuel value, non-consistent coal granulation and others). Used parameter estimation method is adaptive and provides the time-varying parameter estimation. At the same time it is robust to different kinds of disturbances present in the system. The constructed model is simplified as much as possible, but at the same time it describes all the important phenomena in the behavior of the system. As described in [15], this simplification implies the model with $M = 6$ inputs d_m (mill loads) and $N = 12$ output temperatures T_n (temperatures at 38m of height):

$$T_i(k) = -\sum_{n=1}^{N} a_{in} T_n(k-1) + \sum_{m=1}^{M} b_{im} d_m(k-1) + \zeta(k), \tag{1}$$

Journal of Physics: Conference Series **570** (2014) 052001 doi:10.1088/1742-6596/570/5/052001

where a_{in} and b_{im} are the model parameters obtained from the system identification procedure, $i = \overline{1, N}$ denotes the index of the output temperature and $\zeta(k)$ is the process noise. Equation (1) can be shown as a linear regression form

$$T_i(k) = W^T(k)X_i(k) + \zeta(k), \tag{2}$$

where $W^T(k) = [-T_1(k-1) \ldots - T_n(k-1)\ d_1(k-1) \ldots\ d_2(k-1)]$ is the regression vector of the model and $X_i(k) = [a_{i1} \ldots a_{in}\ b_{i1} \ldots b_{im}]$ is the parameter vector. Note that in general case the parameters are not constant and that the time index has been omitted due to notation simplicity.
Additionally, cross-correlation analysis shows antagonistic activity of the opposite mills. Figure 3. illustrates the influence of two mills M61 and M64 onto the difference of temperatures on the right side of the boiler [16]. Namely, this difference increases with increasing the load of mill M61, meaning that the flame slowly shifts towards the rear side of the furnace. On the other hand, the increase of mill M64 load shifts the flame towards the front side of the furnace. This is a very significant result because it can contribute to further control algorithm structure simplification.

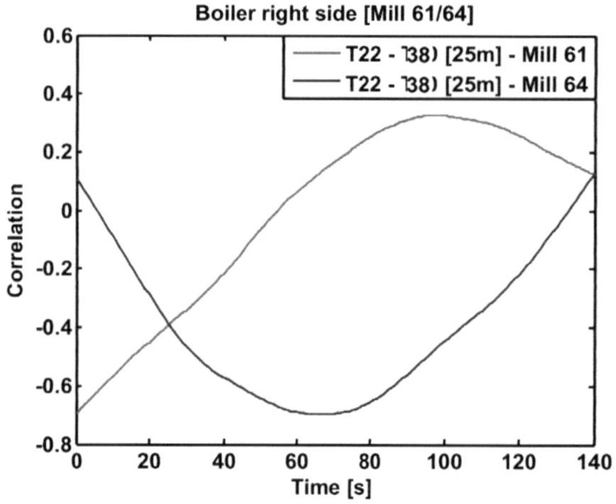

Figure 3. Influence of mills M61 and M64 onto the temperature difference on the right side of the furnace

3. Controller design

3.1. Optimization problem formulation

Previous section provides detailed description of the combustion process modelling. As explained, the process behaviour is essentially dictated by the pair of opposite mills rather than each mill separately. Therefore, there are three instead of six system inputs from control point of view. Furthermore, the overall coal requirement is the result of an external control algorithm and is not discussed in this paper. Let us assume that each mill pair should contribute to the combustion process equally, meaning that the total amount of coal is divided evenly onto these three pairs. Optimization problem is then reduced to finding the coal distribution within each pair. In other words, the control task is to determine the coal division ratio between each two mills, which would help optimize the combustion process. Let us denote by $n_1, n_2,$ and n_3 the ratio between d_1 and d_4, d_2 and d_5, d_3 and d_6, respectively (corresponding to the M61 to M66 in Figure 1).

Next important step is formulation of an adequate optimization criterion. As suggested, the idea is to centralize the flame inside the furnace. One solution to this problem is to ensure the approximately same temperatures on the opposite sides of the furnace. Keeping in mind the notation in Figure 2, we propose the following criteria

$$J_n = \sum_{i=17}^{22} (T_i - T_{i+k})^2, \quad (3)$$

where parameter $k = 6$ provides the summation of diametrically opposite temperature differences. Minimization of this criteria leads to an elliptical flame configuration. In the perfect scenario, all three ratios equal to one, meaning the uniform coal distribution onto six mills, would result in a minimum value of optimization function (around zero). In practice this is rarely the case, since there are many parameters influencing the combustion: coal burning process, current state of the mills, etc. For practical reasons, the ratio within each pair of mills is limited to $n_i \in [0.8,1.2]$, $i = 1,2,3$. This imposes a correction of the criteria (3) in order to include the constrains:

$$J = J_n + J_c, \; J_c = \sum_{i=1}^{3} C \cdot \left[max\left(\frac{n_i - n_{max}}{n_{max}}, 0, \frac{n_{min} - n_i}{n_{min}}\right)\right]^2, \quad (4)$$

where $n_{min} = 0.8$, $n_{max} = 1.2$ and parameter $C = 5 \cdot 10^5$ is selected as to insure the significance of the constraints in the overall criteria. The correction factor is equal to zero for distribution ratio within bounds and increases with distance from this range. Finally, the optimization problem reduces to minimization of the criteria (4) with respect to n_1, n_2 and n_3.

3.2. Extremum seeking control

As a tool for optimization, we propose the use of the extremum seeking approach which is widely used in optimization theory in many available forms. Paper employs the use of perturbation signals in optimization process. The basic structure for the multivariable nonlinear maps in presence of dynamics is shown in Figure 4. It is important to emphasize that the perturbation frequencies should ensure a sufficient time scale separation between the ES loop and the plant. Additionally, the perturbation signal $S(t)$ and demodulation signal $M(t)$ are chosen in such a way to ensure convergence of the algorithm. For an input vector consisting of three variables $n = [n_1 \quad n_2 \quad n_3]^T$

$$S(t) = [a_1 \, sin(\omega_1 t) \quad a_2 \, sin(\omega_2 t) \quad a_3 \, sin(\omega_3 t)], \quad (5)$$

$$M(t) = \left[\frac{2}{a_1} sin(\omega_1 t) \quad \frac{2}{a_2} sin(\omega_2 t) \quad \frac{2}{a_3} sin(\omega_3 t)\right], \quad (6)$$

where perturbation frequencies should satisfy several conditions: 1. each two should be different $(\omega_i \neq \omega_j)$, 2. the ratio of each two should be rational $(\omega_i/\omega_j \in Q)$, 3. the sum of each two should not be equal to the third one $(\omega_i + \omega_j \neq \omega_k)$, $i, j, k = 1,2,3, i \neq j \neq k$.

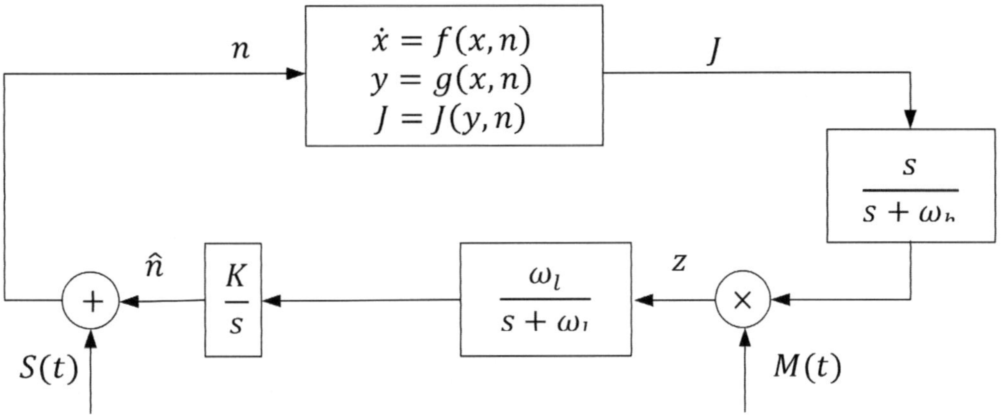

Figure 4. Gradient based ES algorithm for a dynamic multivariable nonlinear map.

Journal of Physics: Conference Series **570** (2014) 052001 doi:10.1088/1742-6596/570/5/052001

Perturbation amplitude a and diagonal matrix K are small parameters responsible for the stability and the convergence speed of the algorithm. There are several papers dealing with the analysis of stability of ES control for nonlinear dynamic systems [8], [17], which was not the focus of this paper. The two filters decrease the effect of the perturbation signals. Their role is to eliminate the DC component of the optimized signal J (high-pass filter) and to isolate a DC component of the product signal z necessary for gradient estimation (low-pass filter). Thus, the filters should have slower time scale than the periodic perturbations and therefore the plant itself. If all the conditions are met, the algorithm converges minimizing the criteria (4) obtained by the optimal value of input vector n^*.

4. Experimental results

In order to verify the proposed control approach, we employ the obtained model of the system, as described in Section 2. We introduce the reference speed to the algorithm, which represents the overall needed doser speed for all six mills. As indicated before, the speed of the doser is proportional to the coal amount requirement. Normally, this information is provided by the external control loop in charge of the desired fresh steam pressure. Dictated overall doser speed is given in Figure 5.

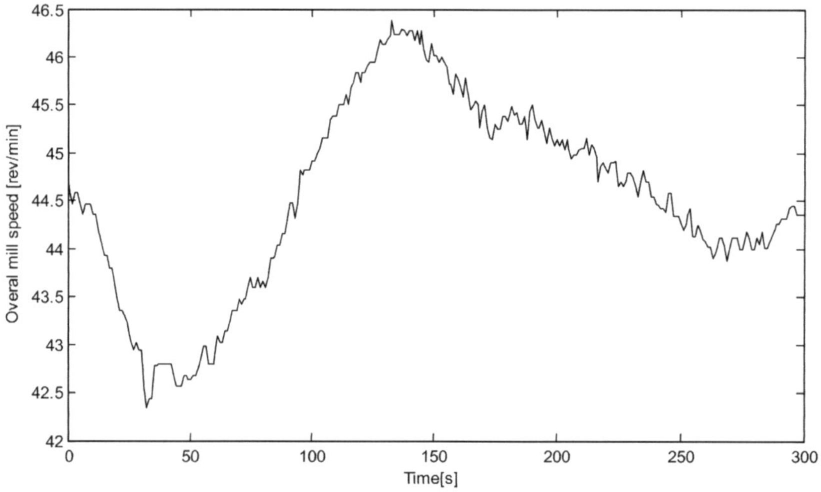

Figure 5.Overall doser speed proportional to the overall coal load requirements.

The goal of the control algorithm is to regulate the flame inside the furnace, by minimizing the preset criteria (4). The reference speed is divided onto three pairs of dosers and the objective of the ES structure is to estimate the optimal ratios within these pairs. The adopted perturbation frequencies are 0.8 rad/s, 0.6 rad/s and 0.5 rad/s. Parameters $a = 0.002$ and a diagonal matrix K are carefully chosen. These parameters were determined empirically. Imposing the constraints on the control signal would normally limit the selection of the appropriate parameter and there for the speed of the algorithm in order to stay within the desired bounds. Introduction of these constraints throughout the penalty function gives us more freedom in that sense. Also, the filter cut-off frequencies $\omega_l = [0.8, 0.5, 0.3]\ \mathrm{rad/s}$ and $\omega_h = 0.6$ are chosen as to satisfy the requirements from Section 3. Figure 7 illustrates the result of the optimizing procedure. Note that in the beginning all three ratios are in steady state close to one, meaning that they all contribute roughly the same in the control process. After the disturbance, introduced in the model parameters, criterion function experiences a peak (Figure 6), which is timely eliminated. At the same time, control signals experience a translation to a new steady state. In the first pair load is no longer equally distributed over mills. Namely, mill M61 is in charge of providing around 55% and mill M64 around 45% of the coal required from that pair.

Mills M63 and M65 encounter similar scenario, where as the ratio of mills M62 and M66 is still close to one. Penalty constraints function provides the control signals within the desired bounds, although the algorithm itself is reasonably fast (around 70 sec needed to achieve new steady state). The influence of these ratios on the coal distribution over mills, meaning the doser speeds are depicted in Figure 8. ES algorithm manages to adapt and slide the control parameters towards new values which result with the minimum objective function. Obviously, the ES approach is successful in these simulations and represents a good basis for further research in terms of additional combustion distribution optimization.

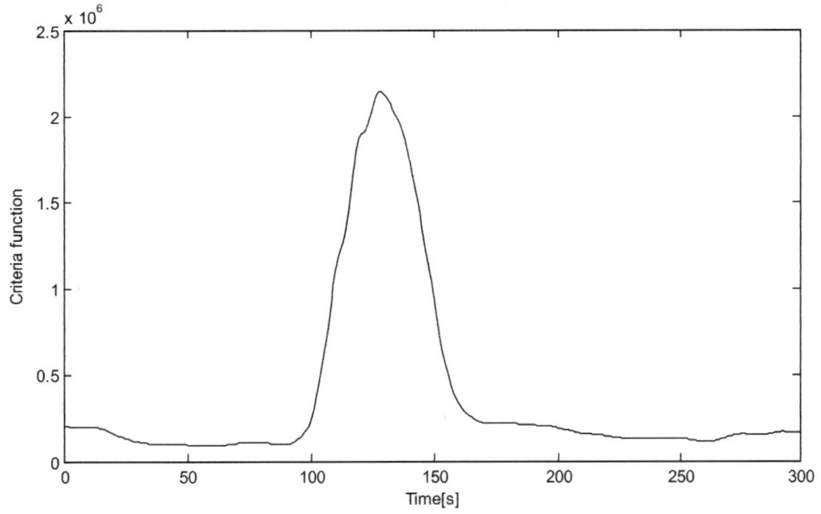

Figure 6. Algorithm criteria function.

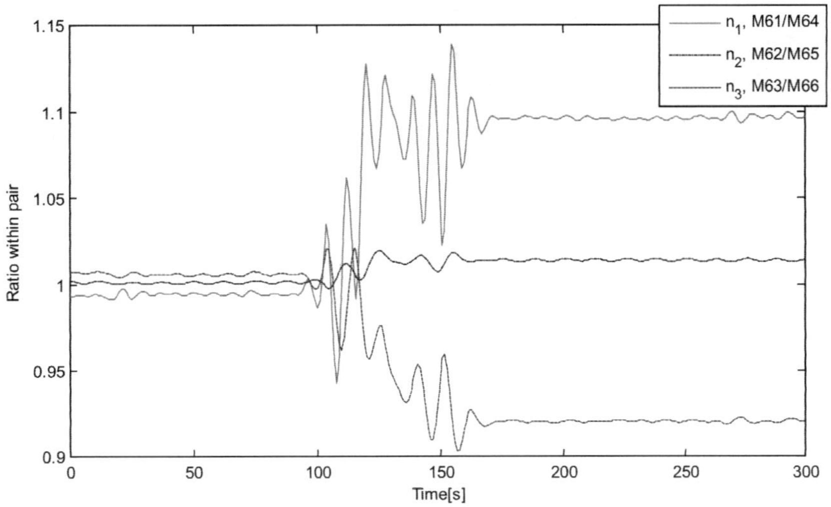

Figure 7. Control signals: ratios within each mill pair.

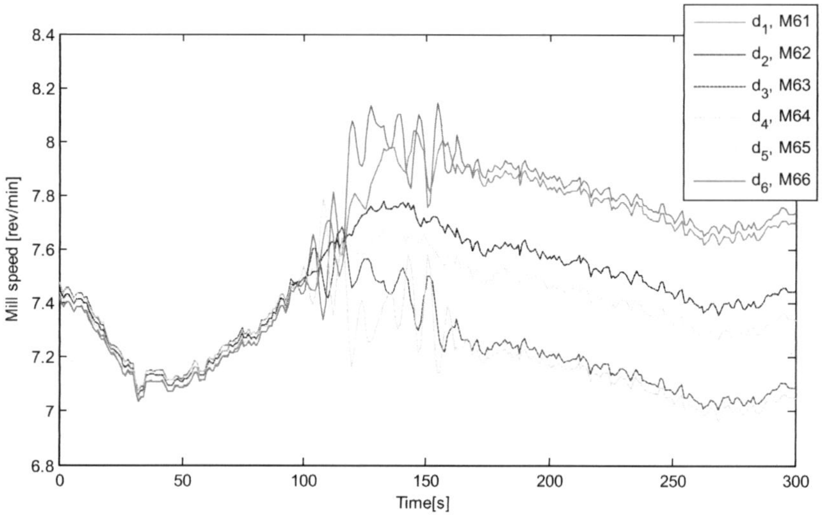

Figure 8. Model input signals: doser speeds.

5. Conclusion

The paper considers an application of extremum seeking in optimization of combustion distribution in thermal power plant furnaces. First we describe how the temperature spatial distribution depends on the firing and the coal partitioning over the mill circles. Next we provide a possible solution for flame geometry regulation in the form of an optimal control problem with an objective function which includes the set of constraints regarding the control signals. Finally, we introduce the proposed control algorithm to the model of the combustion process. There are many disturbances taking place inside the furnace and as a result of such disturbances the flame was shifted towards the front of the furnace. The results show good algorithm performance in terms of the temperature distribution control which provides ellipsoid flame configuration. Additionally, introduction of constraints into our objective function enables aggressive extremum seeking parameters and hence better convergence properties. Future work involves further algorithm tuning in terms of increasing the flame symmetry and more freedom when it comes to coal dissipation over mill pairs.

Acknowledgements

The authors gratefully acknowledge the financial support from the research "RObust Decentralised Estimation fOr large-scale systems (RODEO), MAE-PGR00152.

References

[1] Yin C, Caaillat S, Harion J. Baudoin B and Perez E 2002 Investigation of the flow, combustion, heat-transfer and emissions from a 609 MW utility tangentially fired pulverized-coal boiler *Fuel* **81** 997–1006

[2] Flynn D 2003 *Thermal Power Plant Simulation and Control* (London, UK The: Institution of Electrical Engineering)

[3] French D N 1993 *Metallurgical failures in fossil fired boilers 2nd edition* (New York: John Wiley & Sons, Inc)

[4] Saripalli R, Wang T and Day D 2005 Simulation of combustion and thermal flow in an industrial boiler *Proc. 27th Industrial Energy Technology Conf.* (New Orleans, LA)

[5] Korytnyi E, Saveliev R, Perelman M and Chudnovsky B 2009 Computational fluid dynamic simulation of coal fired utility boilers: An emerging tool *Fuel* **88** 9–18
[6] Krstic M 2000 Performance improvement and limitations in extremum seeking control *Syst. Control Lett.* **3** 313–26
[7] Nesic D 2009 Extremum seeking control: Convergence analysis *Eur. J. Control* **15** 331-47
[8] Krstic M and Wang H 2000 Stability of extremum seeking feedback for general nonlinear systems *Automatica* **36** 595–601
[9] Wang H, Yeung S and Krstic M 2000 Experimental application of extremum seeking on an axial-flow compressor *IEEE Trans. Control Syst. Technol.* **8** 300–9
[10] Killingsworth N, Aceves S, Flowers D and Krstic M 2007 Extremum seeking tuning of an experimental HCCI engine combustion timing controller *American Control Conf. ACC '07* (New York, NY, USA) pp. 3665-70
[11] Zhang C and Ordonez R 2007 Numerical optimization-based extremum seeking control with application to ABS design *IEEE Trans. Autom. Control* **52** 454–67
[12] Nwesaty W, Bratcu A and Hably A 2013 Extremum seeking control techniques applied to photovoltaic systems with multimodal power curves *Int. Conf. on Renewable Energy Research and Appplication ICRERA 2013* (Madrid, Spain)
[13] Chang Y and Moura S 2009 Air flow control in fuel cell systems:An extremum seeking approach *American Control Conf. ACC '09* (St. Louis, MO, USA)
[14] Guay M and Zhang T 2003 Adaptive extremum seeking control of nonlinear dynamic systems with parametric uncertainties *Automatica* **39** 1283–93
[15] Marjanovic A, Vujnovic S, Papic V and Todorov P 2014 Robust parameter estimation of the thermal power plant combustion process *1st Int. Conf. on Electrical, Electronic and Computing Engineering IcETRAN '14* (Vrnjacka Banja, Serbia)
[16] Marjanovic A, Jakovljevic M, Kvascev G and Đurovic Z 2013 Combustion process control based on flame visualization *Power turbines Europe '13* (Lisbon, Portugal)
[17] Tan Y, Nesic D and Mareels I 2006 On non-local stability properties of extremum seeking control *Automatica* **42** 889-903

A process for providing positive primary control power by wind turbines

V Marschner[1], J Michael[2] and J Liersch[2]

[1]Technische Universität Berlin, Straße des 17. Juni 135, 10623 Berlin, Germany
[2]Key Wind Energy GmbH, Cicerostraße 37, 10709 Berlin, Germany

E-mail: v.marschner@keywind.de, j.michael@keywind.de, j.liersch@keywind.de

Abstract. Due to the increasing share of wind energy in electricity generation, wind turbines have to fulfil additional requirements in the context of grid integration. The paper examines to which extent wind turbines can provide positive control power following the related grid code. The additional power has to be obtained from the rotating flywheel mass of the wind turbine's rotor. A simple physical model is developed that allows to draw conclusions about appropriate concepts by means of a dynamic simulation of the variables rotational speed, torque, power output and rotor power. The paper discusses scenarios to provide control power. The supply of control power at partial load is examined in detail using simulations. Under partial load conditions control power can be fed into the grid for a short time. Thereby the rotational speed drops so that aerodynamic efficiency decreases and feed-in power is below the initial value after the control process. In this way an unfavourable situation for the grid control is produced, therefore the paper proposes a modified partial load condition with a higher rotational speed. By providing primary control power the rotor is delayed to the optimum rotational speed so that more rotational energy can be fed in and fed-in power can be increased persistently. However, as the rotor does not operate at optimum speed, a small amount of the energy yield is lost. Finally, the paper shows that a wind farm can combine these two concepts: A part of the wind turbines work under modified partial load conditions can compensate the decrease of power of the wind turbines working under partial load conditions. Therefore the requested control power is provided and afterwards the original value of power is maintained.

1. Introduction

Wind energy covers a significant part of the demand for electricity in many countries, especially in Denmark (34 percent of the electricity consumption in the year 2013, cf. [1]), Portugal (25 percent), Spain (21 percent) and Germany (8 percent). The increasing proportion of the wind energy requires to involve wind turbines in the grid control and grid balancing. The expansion of renewable energies leads to a displacement of classical controllable thermal power plants. In addition, many renewable energy sources require a very powerful grid control due to fluctuating input.

By the reason of the limited storability of electrical energy (cf. Erdmann, Zweifel [2], p. 294) is the central task of grid control in maintaining the balance of electricity generation and demand. Imbalances are equalized with control power. This paper examines the extent to which wind turbines can provide positive control power.

The grid load is changing constantly subjected by connecting, disconnecting and controlling of electrical consumers. In power plants, which generate electrical energy by driving a generator causes a change in the grid load a different counter torque on the generator. Due to the change in torque arises a new rotational speed. A stationary frequency deviation results because the speed is proportional to the grid frequency. So, an oversupply of power generation leads in an increase of the grid frequency. An active power deficit causes a frequency reduction. Therefore the drive power is controlled, since even small changes in the grid load lead to unacceptable changes in the grid frequency (cf. Heuck, Dettmann [3], p. 62 - 71).

For the exact balancing of the grid frequency is used a further control loop. These one eliminates the steady-state error of the rotational speed control. In this case the grid frequency is used as a control variable instead of the rotational speed. The grid frequency is a result of the generator speed and is a secondary value. Consequently this control loop is called a secondary control. In this way the speed control is called primary control. The concept of internal – speed controller – and external controller – frequency controller – is a classical cascade control (cf. figure 1). These models represent the simplest case and are valid for stand alone grids with one power plant. In interconnected power grids set the secondary control the power plant output.

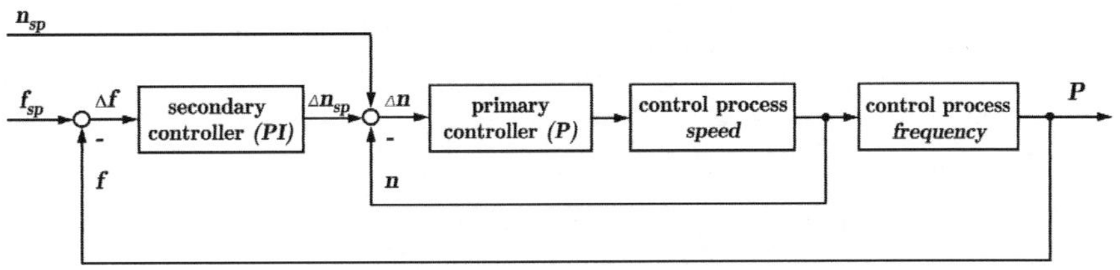

Figure 1. Control of the grid frequency

The frequency control in the power supply is performed by the controlled feeding of primary control power (cf. [4]). Conventional power plants achieve the power increase by increasing the fuel supply. This approach is not possible for wind turbines because the plants utilize the entire potential of the wind. The concepts developed in this work rely substantially on the inertia of the rotating components of the wind turbine.

The flywheel mass of the turbogenerator also plays an important role in classical power plants. In distinction to modern wind turbines the speed of the turbogenerator in power plants is proportional to the grid frequency. The grid-supporting effect of this inertia is assigned to spinning reserve (cf. [5]).

The profitability is not considered in this paper. The provision of primary control reserve of renewable energies would get paid the same way as the provision of primary control reserve of conventional power plants.

2. Operation concepts of wind turbines for providing primary control power

2.1. Providing primary control power at partial load operation

At partial load operation, wind turbines mostly work at this operation point, the entire energy supplied by the wind is converted into electrical energy, considering the efficiency of the wind turbine. Variable-speed turbines equalize the drive train loads by storing wind energy from wind fluctuations as rotational energy. To increase the feed-in capacity, it is possible to take advantage of the kinetic energy of the rotating parts in the opposite process. By imposing additional power from the generator, the output torque of the generator shaft is increased. The rotor experiences a higher counter torque and the rotational speed decreases. The stresses of the modules remain within the allowed range because the rated power is not exceeded. The following cases can be distinguished:

(i) The output power is increased by a certain value, while the rotational speed must not be less than the specified value. Afterwards the rotor accelerates according to the prevailing wind conditions.

(ii) The output power is increased and provided without considering the rotational speed. A superior device informs the control system of the wind turbine about the end of the process. The following situations can occur:

   (a) The required control power can be delivered completely. At the end of the process the rotor has a residual speed and accelerates to operating speed.

   (b) The required control power can not be provided completely, since the total kinetic energy of the rotor was converted. The rotor stops.

2.2. Providing primary control power at modified partial load operation

In partial load operation a reduction in rotor speed results in a lower power output, assuming previously the wind turbine was at the maximum power point. After an acceleration phase the rotor reaches the initial rotational speed and the initial power. Thus, the power output collapses after the control operation. Therefore we suggest a modified operation process for the partial load range, that prevents power losses at the end of the procedure. To obtain this desirable property, the partial load speed will be increased. This has two advantages and one disadvantage:

+ The rotor has a higher rotational energy due to the higher rotational speed. There is a bigger amount of energy stored, which can be used for the delivery of control power.
+ Because the rotor is operating at a higher rotational speed than the optimal speed, the power coefficient decreases. While the delivery of control power, the rotor speed returns to the optimum speed. The power coefficient increases – even the higher power output can be maintained.
– The previous fact implies directly the disadvantage: In this mode, the rotor does not work optimal, and loses some energy yield. On the other hand, the system obtains the desired primary control capability.

The modified operation leads to the scenario:

(iii) In the modified partial load operation primary control power is delivered as long as the optimum speed is reached.

3. Modelling

The different concepts are evaluated using simulations based on a physical model of the wind turbine. The following assumptions are made for modelling:

- Bearings, gearbox, generator and inverter have no losses.
- During the control process there is a constant wind speed.
- Generator and inverter can handle all speeds.
- The rotor blades are the result of an ideal design by Schmitz (Gasch, Twele [6], p. 202).

Under these assumptions, the model leads to the ideal case – it calculates the maximum control power that can be delivered.

3.1. Setting up the model equations
A wind turbine is the basis for the power conversion considerations. The focus is on the modelling of the rotor dynamics. A dynamic energy balance is expedient.

$$\frac{dE}{dt} = P_{in} - P_{out} \tag{1}$$

The storage term $\frac{dE}{dt}$ describes the system's energy change over time. Especially the energy of the rotating components of the wind turbine will be investigated. The balancing is done over the change of rotational energy $\frac{dE_{ROT}}{dt}$. The input power P_{in} is the power extracted from the wind – the rotor power. The output power P_{out} is the electric power produced by the generator – the generator power.

$$\frac{dE_{ROT}}{dt} = P_{rotor} - P_{gen} \tag{2}$$

The wind turbine's rotor is capable of converting a part of the winds energy. This energy conversion depends on the rotor design and the rotor angular velocity ω. The generator power can also be expressed as a rotation against a torque M with the angular velocity ω.

$$\frac{dE_{ROT}}{dt} = P_{rotor}(\omega) - M_{gen}\,\omega \tag{3}$$

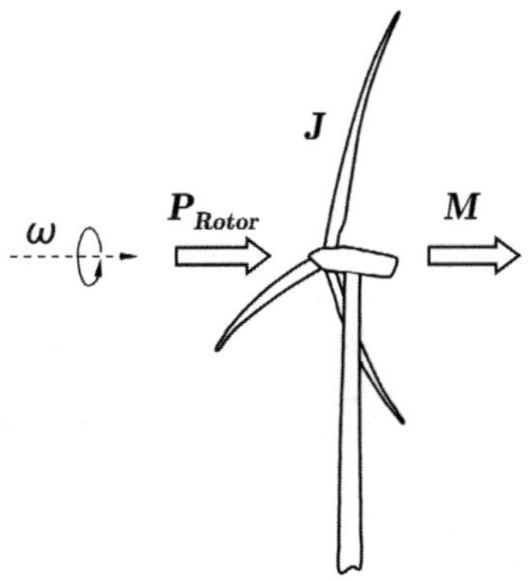

Figure 2. Model for rotor dynamics

The rotational energy of the rotor of the wind turbine is determined by its moment of inertia J and its angular velocity ω.

$$E_{ROT} = \frac{1}{2} J \omega^2 \tag{4}$$

These reflections lead to a differential-algebraic system of equations (DAE). It consists of a differential equation – equation 5 – and an algebraic constraint – equation 6.

$$\dot{E}_{ROT} = P_{rotor}(\omega) - M_{gen}\,\omega \tag{5}$$

$$E_{ROT} = \frac{1}{2} J \omega^2 \tag{6}$$

3.2. Solution of the model equations

The model equations are solved with the explicit Euler method. The discretisation turns the differential quotient of the rotational energy into a difference quotient:

$$\dot{E}_{ROT} \approx \frac{E_{ROT}(k+1) - E_{ROT}(k)}{\triangle t} \tag{7}$$

The numerical solution of the system can be obtained by repeatedly performing the following two equations.

$$\boxed{\begin{aligned} &E_{ROT}(k+1) = E_{ROT}(k) + (P_{rotor}(\omega, k) - M_{gen}(k)\,\omega(k))\,\triangle t && (8)\\ &\omega(k+1) = \sqrt{\frac{2E_{ROT}(k+1)}{J}} && (9)\end{aligned}}$$

For step $k = 0$ the rotational energy is calculated with the angular velocity at the initial step $\omega(k = 0)$:

$$E_{ROT}(0) = \frac{1}{2} J \omega(0)^2 \tag{10}$$

All following steps $k \geq 1$ are calculated with the equations 8 and 9 . In this way the results for the discrete-time functions of the rotational energy $E_{ROT}(k)$ and the angular velocity $\omega(k)$ can be calculated:

$$E_{ROT}(1) = E_{ROT}(0) + (P_{rotor}(\omega, 0) - M_{gen}(0)\,\omega(0))\,\triangle t \tag{11}$$

$$\omega(1) = \sqrt{\frac{2E_{ROT}(1)}{J}} \tag{12}$$

$$E_{ROT}(2) = E_{ROT}(1) + (P_{rotor}(\omega, 1) - M_{gen}(1)\,\omega(1))\,\triangle t \tag{13}$$

$$\omega(2) = \sqrt{\frac{2E_{ROT}(2)}{J}} \tag{14}$$

$$\ldots$$

The simulation variables – the rotational energy and the angular velocity – have been calculated for the discrete steps $k = 0, 1, 2, \ldots$. The shift to time-related variables is done via the step size $\triangle t$:

$$t = k\,\triangle t \tag{15}$$

3.3. Calculation of the rotor power

The rotor power $P_{rotor}(\omega)$ includes the aerodynamic aspect in the equation of the rotational energy. It represents the power obtained by the rotor depending on the respective rotor speed:

$$P_{rotor}(\omega) = c_P(\omega)\, P_{wind} \tag{16}$$

The power coefficient c_P is usually expressed over the tip speed ratio λ for a wind turbine. The exact graph of this function is determined by the profile type of the rotor blades. The tip speed ratio is obtained directly from the angular velocity, assuming constant wind speed.

Variable-speed wind turbines are always working with the maximum power coefficient in the partial load range. Upon delivery of primary control power, the speed drops off and the rotor operates at a lower power coefficient $c_{P,\,control}$. By pitching the rotor blades, it is possible to return to a higher power coefficient $c_{P,\,control,\,pitch}$.

$$c_{P,\,control} < c_{P,\,control,\,pitch} < c_{P,\,max} \tag{17}$$

Besides rotational speed, the rotor power also depends on the pitch angle. In a characteristic diagram several c_P-λ-curves for different pitch angles are listed (cf. figure 3).

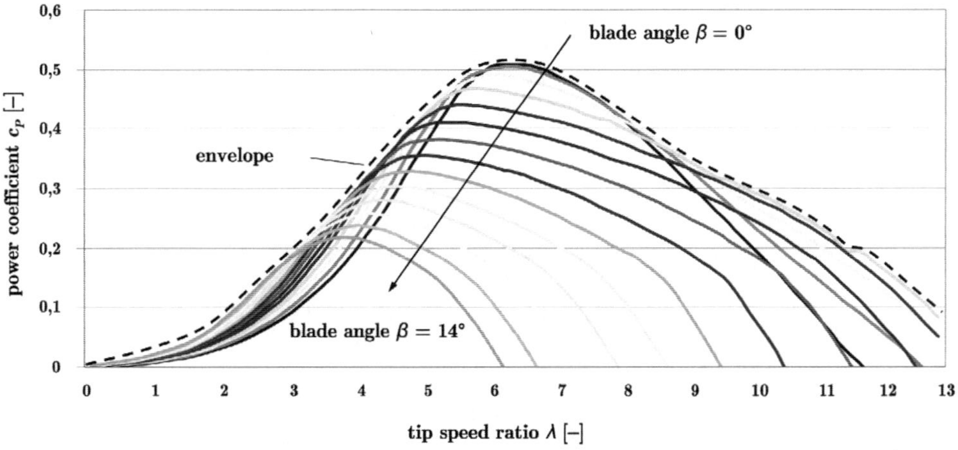

Figure 3. c_P-λ-characteristic diagram for different pitch angles β.

To maintain the characteristic diagram in figure 3 a rotor was designed according to Schmitz. Then c_P-λ-curves for different pitch angles were determined using the blade element method. The execution of the two methods, design and rotor blade element method, was performed according to Gasch, Twele ([6], p. 205 - 221).

The blade element method returns the desired characteristic diagram in the form of a table (cf. table 1). In each simulation step, the tip speed ratio is determined for the present rotational speed by the calculation algorithm, which searches in the corresponding row of the characteristic diagram for the highest c_P. The column returns the corresponding pitch angle β.

	$\beta = 0°$	$\beta = 1°$	$\dots$	
λ_1	$c_P(\lambda_1,\, 0°)$	$c_P(\lambda_1,\, 1°)$	$c_P(\lambda_1,\, \dots)$	$\dots$
λ_2	$c_P(\lambda_2,\, 0°)$	$c_P(\lambda_2,\, 1°)$	$c_P(\lambda_2,\, \dots)$	$\dots$
$\vdots$	$c_P(\dots,\, 0°)$	$c_P(\dots,\, 1°)$	$c_P(\dots,\, \dots)$	$\dots$
	$\vdots$	$\vdots$	$\vdots$	$\ddots$

Table 1. Characteristic diagram of the power coefficient c_P.

4. Evaluation of selected scenarios for the primary control capability of wind turbines

Based on the introduced model the possible contribution of wind turbines in grid control will be examined. The model parameters are adapted to a wind turbine of 2 MW class. For the following simulations the wind speed, unless otherwise stated, is set to $8\,\mathrm{m\,s^{-1}}$.

4.1. Simulation of primary control capability at partial load

An increase in the power output of wind turbines is possible at partial load only by reclaiming energy of rotation, if the turbine were did not operated throttled previously. The induced drop in rotational speed shifts the operating range of the rotor to unfavourable power coefficients. To return to the optimal operating point, the rotational speed is returned to the initial value after the delivery of the additional energy. This is accomplished by the opposite process. The generator output power is reduced below the currently available rotor power ($P < P_{rotor}$). Under this condition, the rotor accelerates and thereby increases its rotational energy. It is not possible to skip with the phase with reduced power after the delivery of the additional power P_A and continue to operate the turbine at the initial power P_0. The output power could only be provided at the expense of rotational energy. Finally, the system would stop.

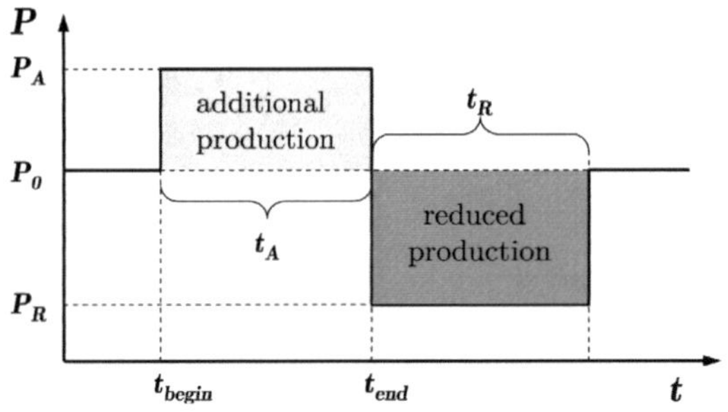

Figure 4. Additional power and reduced power at partial load

In figure 4 the profile of the output power at partial load is shown. The bold solid line shows a possible profile. At the time t_{begin} the infeed power increases from P_0 to P_A. The energy that is delivered additionally in the power grid is just the yellow coloured area under the power curve. At the end of increased output at time t_{end} follows a phase with reduced infeed power P_R. Compared to normal operation, the infeed of energy that results from the difference between violet and yellow area is eliminated.

In figure 5 the amount of additional power P_A is varied. The duration of the additional power is 10 seconds. The subsequent reduction in performance is at 10 % of the initial power P_0. The duration of the power reduction is variable. Once the initial speed n_0 is reached, it is switched back to their initial power P_0.

The rotational energy (figure 5 a) is converted directly into additional grid infeed power (figure 5 d) which is the requirement (cf. figure 4). The rotational speed (figure 5 b) drops significantly with increasing additional energy. Also the rotor power wanes (figure 5 c).

With the help of the torque profile (figure 5 e) the drive train load can be detected. The loads are not investigated in detail. Some scope to the rated loads does exist in the partial load operation. Power changes of 50 percent of the rated power are possible in 1 to 2 seconds in practice (cf. Heier [9], p. 314).

The pitch angle (figure 5 f) is gained from the characteristic map. The higher the additional power, the more pitch usage is required.

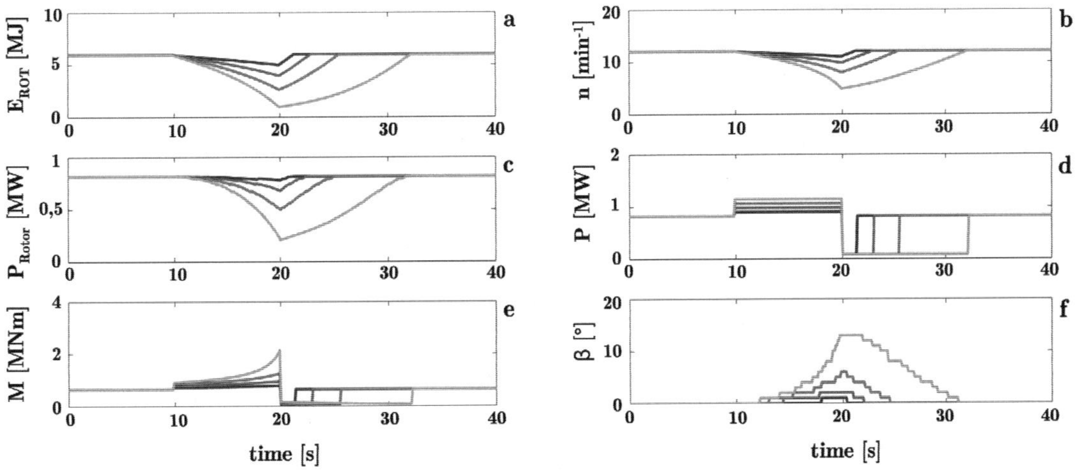

Figure 5. Simulation for different additional powers: +10 % (blue), +20 % (green), +30 % (red) and +40 % (cyan)

The quantity of electricity, which is lost during the rotor acceleration phase, is always greater than the amount that is obtained while the rotor deceleration. Once the rotor leaves the optimum speed range due to the deceleration, the aerodynamic efficiency decreases. Even with the assumed constant wind speed, the energy conversion of the rotor is reduced by a smaller power coefficient c_P. Thus, in the acceleration phase, there is initially a lower rotor power that gradually increases with increasing speed. Upon reaching the optimum rotor speed, the rotor power corresponds with the power at the initial time.

This simple approach is also proposed in the previous paper (cf. [7], p. 433-434). If the allowed grid frequency range is deceeded, the wind turbines will increase their infeed power at the expense of their rotational speed. There is no yield loss in normal operation. In the examined case a wind farm does provide frequency support during a short term failure. The drop of grid frequency is limited by the control reserve of the wind farm. The acceleration phase of the wind turbines, characterized by lower infeed power, takes place when the full generating capacity is available again. In this particular case, it is shown that even the simple approach can be valuable for the grid balancing.

Another approach proposes the throttling by a pitch angle (cf. [8]). If the grid balance requires positive control reserve, the pitch angle may reduce and the infeed power increases. Of course this approach leads to a continuously loss of energy yield.

4.2. Simulation of primary control capability at modified partial load

The modified operation mode at partial load consists, as explained, of an increase in the rotor speed $n > n_{opt}$. The progress of the infeed power of the wind turbine during delivering control power is shown in figure 6. While the rotor is delayed to the optimum angular velocity $\omega_{opt} = \lambda_{opt}\, v/r$ the generator releases the rotational energy stored in the speed difference $E_{ROT} = 1/2\, J\,(2\,\pi)^2(n^2 - n_{opt}^2)$. This amount of energy is called additional energy II. During deceleration to the optimum speed increases the aerodynamic efficiency of the rotor, the feed-in can be increased durably. The infeed amount obtained by this procedure is called additional energy I. Hence this energy is abandoned in favour of true primary control properties during normal operation.

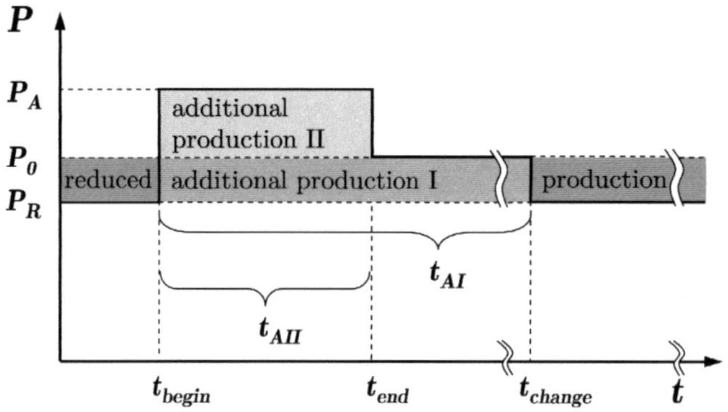

Figure 6. Additional power and reduced power at modified partial load

The operating points above the optimum speed are in the c_P-λ-characteristic diagram right side of the maximum power coefficient. The power coefficient decreases further with increasing speed enhancement and with increasing tip speed ratio. Due to the increased rotational speed more rotational energy is available. In consequence of the discrepancy to the maximum power coefficient, the feed-in can be raised permanently. Figure 7 shows the time characteristics of simulation variables for different speed increases. The advantageous properties increase with speed enhancement. However, the permanent yield loss increases further.

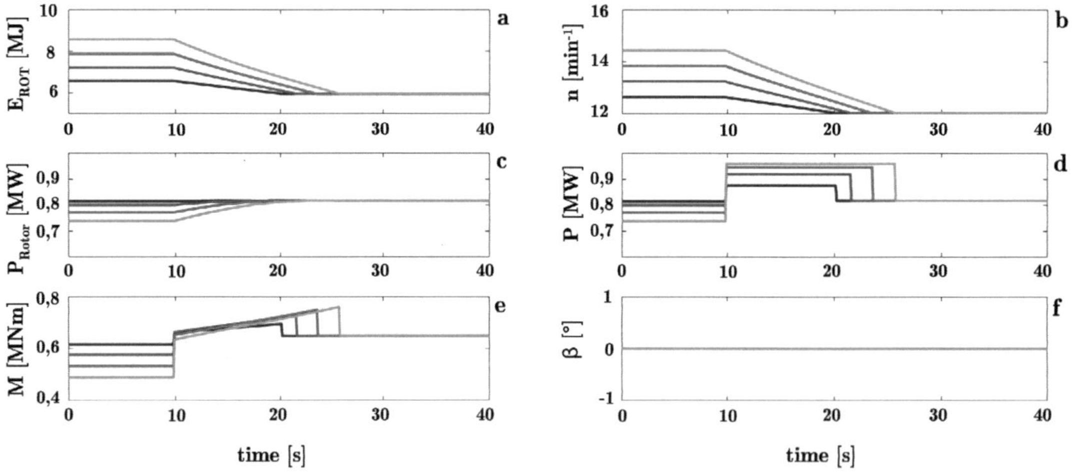

Figure 7. Speed enhancements of : +5 % (blue), +10 % (green), +15 % (red) and +20 % (cyan)

The stored rotational energy (figure 7 a) increases with the speed-raising. During the delivery of the control power, the speed is decelerated to the optimum speed (figure 7 b). The rotor power leaves the throttled level and reaches the optimal value (figure 7 c). After the delivery of the additional energy I and II the grid infeed power hits the unthrottled level (figure 7 d).

4.3. Combining both concepts in a wind farm

In a wind farm, the two mentioned concepts can be combined so that the disadvantages are weakened, and a maximum of the benefits remain. The idea is to operate only a part of the wind turbines of a wind farm at the modified partial load (mpl). The other part continues to work in the normal partial load (pl) without yield losses. When control power is requested all plants provide rotational energy as grid feed. The modified operated wind turbines then have higher power output, because the operating point is now in the optimum. For the remaining wind turbines it is necessary to reduce the power output below the initial level to prevent a further drop in speed. The configuration should be based on the premise that the increased power of the turbines in the modified operation compensates the reduced performance of the other turbines. As a result, there is a total power output (t), that can provide control power and then maintaining the initial level (figure 8).

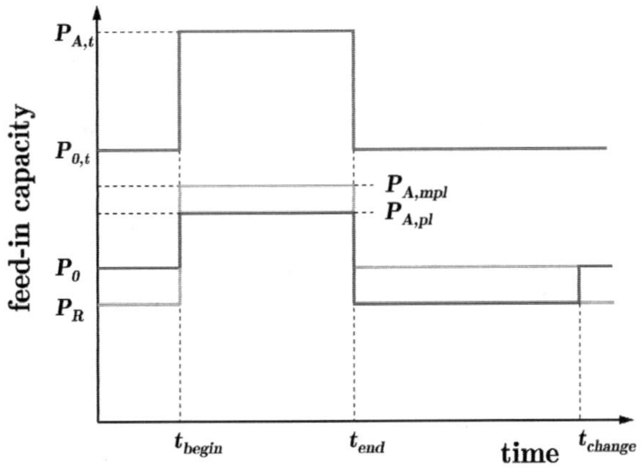

Figure 8. Infeed power of a wind farm at the request of control power

Table 2 shows an example design for two wind turbines: One turbine operates at partial load and another one at modified partial load. If 1 % energy yield loss is tolerated, the wind farm can be provided with a primary control capability. The primary control makes it possible to raise the infeed power for 20 seconds to 105.5 %. The example is for a wind speed of 7 $\mathrm{m\,s^{-1}}$. The speed of the modified operated turbine is increased by 9.5 percent. It may not be ignored that the values were determined with the idealized model. There are also losses and restrictions in practice, which have negative effect on the absolute values. Nevertheless the suggested approach achieves the goal – providing control power.

	normal mode		control process 20s		balancing 57s		normal mode	
	[kW]	[%]	[kW]	[%]	[kW]	[%]	[kW]	[%]
WT pl	547	100	564	103	537	98	547	100
WT mpl	537	98	591	108	547	100	537	98
total	**1084**	99	**1155**	105.5	**1084**	99	**1084**	99

Table 2. Control process in a wind farm, $v = 7\,\mathrm{m\,s^{-1}}$

In the example the balancing process is finished after 57 seconds – the turbine operating at partial load is accelerated back to the optimum speed. It should be considered that the table 2 is valid for a fixed wind speed. At the time t_{change} (figure 8) both groups change to intended normal operation. Afterwards control power can be delivered again. The control capability can be be used extensively because it is free of charge.

Journal of Physics: Conference Series **570** (2014) 052002 doi:10.1088/1742-6596/570/5/052002

5. Conclusion

Wind turbines can stabilize the grid frequency with short-term, but instantaneously available control power. If the frequency difference requires a long-term use of control power, wind turbines bridge the time left until the complete activation of the classical primary control power. In both cases, wind turbines make an important contribution to grid stability.

The support of other controllable power plants is necessary in the simple approach. The benefits for the grid control can be reached on two different ways: (1) The total infeed power can be equalized using the inertia of the wind turbines. In case of short term power system failures the wind turbines support the grid frequency by converting moderately rotational energy into additional infeed power. (2) If the classical control power is available, wind turbines can raise aggressively the infeed power to counteract the drop of grid frequency. The classical power plants increase the power output additionally in the context of primary control to compensate the reduced infeed power of wind turbines in the rotor acceleration phase.

The modified version is conditionally capable to participate in the grid control. The available control power depends on the load of the wind turbines. Thus, the wind turbines have a control reserve which is gaining in importance at a high share of wind energy in the energy mix.

Figure 9 (cf. Heier [9], p. 337) shows the control structure of a variable speed and grid connected wind turbine. At high wind speed the power limiting and controlling of the rotational speed is achieved by the pitch system. At partial load operation the speed is controlled by the inverter. During normal operation a balance between delivered power by the rotor and the taken power by the generator is always set. If control power has to be delivered, this equilibrium condition is temporarily suspended by a special controller. In figure 9 the new controller is illustrated by dashed blocks and connections. This controller is also connected to the pitch system. As shown, the pitch control improves the aerodynamic behaviour during decreasing speed. If the grid frequency or the gradient of the grid frequency leaves the allowed range, the operation management enables the control power.

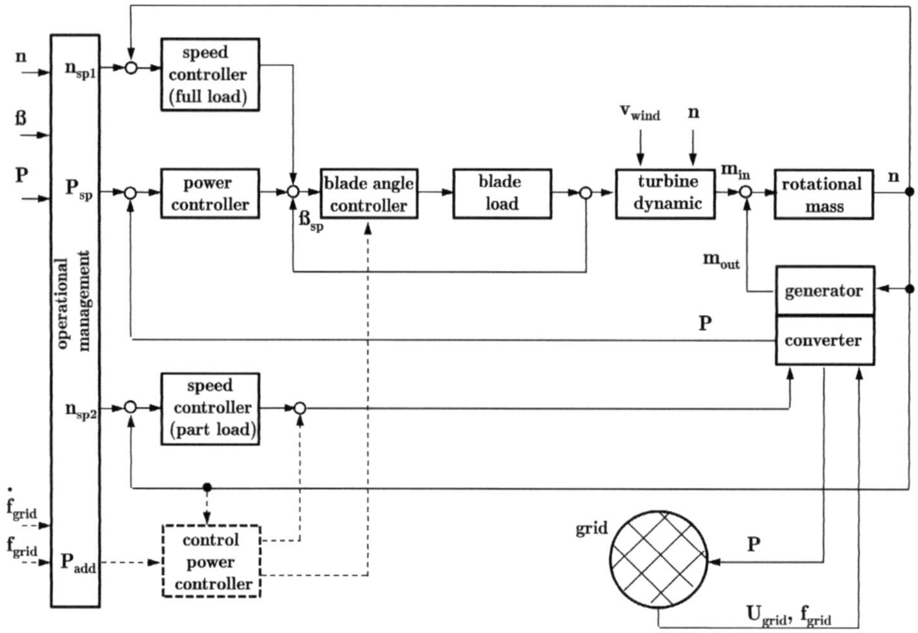

Figure 9. Advanced control structure for wind turbines

This paper shows how wind turbines can effectively participate in primary control of the electrical power grid. The quality of the control power from wind turbines is particularly high. In contrast to thermal power plants, the control power can be activated instantaneously (cf. Kurth, Kallina [10]). Thus grid frequency dips can be eliminated at their appearance. The implementation can be done with a powerful control algorithm and would be a further important step for the integration of renewable energy into the power system.

6. Future tasks

To obtain the control capability of wind turbines at partial load operation, only minor or no constructional changes are necessary. A great task consists in the determination of legal framework conditions for the participation of wind energy in the primary control (cf. [11]). Furthermore the consideration of the occurring dynamic loads is required as part of the type approval.

The primary control capability can be acquired with a powerful control software. The model predictive control (MPC) is considered as a particularly suitable method. The mechanical and aerodynamic model of the wind turbine rotor is the basis for the MPC to predict the future behaviour. The power conversion of the rotor is a highly non-linear process. Furthermore, during the increase in power output constraints need to be considered. For example, the available control power can be maximized. For both requirements, the handling of a non-linear processes and consideration of constraints, the MPC is suitable.

Especially the response to the varying wind speed requires a high-performance control algorithm. So far, disturbance variables were not considered. A decrease in wind speed during the control process implies another challenge for future work. For reasons of grid stability wind turbine downtime has to be minimized.

In addition to the control task of an individual wind turbine, the coordination of all plants of a wind farm will be another objective. Especially in large wind farms many possible variations of providing control power exist, due to the large number of wind turbines. For example, for true primary control capability the wind park can be divided into several subsets (cf. figure 8 and table 2). One subset supplies control power, while the other subsets are in the balancing phase. In general, the objective of the largest possible available control power at the lowest possible yield loss faces an optimization problem. To balance and control the amount of control power kept available at a certain prevailing wind speed should be the focus of upcoming research work.

References

[1] Roney J M, Earth Policy Institute 2014 *Denmark, Portugal and Spain Leading the World Wind Power* `http://www.earth-policy.org/data_highlights/2014/highlights46`
[2] Erdmann G and Zweifel P 2008 *Energieökonomik*, Springer-Verlag, Berlin Heidelberg
[3] Heuck K, Dettmann K D and Schulz D 2010 *Elektrische Energieversorgung*, 8. edition, Vieweg+Teubner, Wiesbaden
[4] Amprion GmbH *Technische Anforderungen an Kraftwerke für den Anschluss in unterlagerten 110-kV-Netzen*, `http://www.amprion.net/systemdienstleistungen-regelenergie`
[5] Weissbach T 2009 *Netzdynamikverhalten und die Rolle des Netzselbstregeleffekts*, Universität Stuttgart
[6] Gasch R and Twele J 2013 *Windkraftanlagen*, 8th edition, Vieweg+Teubner, Wiesbaden
[7] Morren J, de Haan S W H, Kling W L and Ferreira J A 2006 *Wind Turbines Emulating Inertia and Supporting Primary Frequency Control*, IEEE Transactions on Power Systems, VOL. 21 NO. 1
[8] Prillwitz F, Holst A and Weber H 2004 *Unterstützung der Primärregelung durch Windenergieanlagen*, Universtität Rostock
[9] Heier S 2009 *Windkraftanlagen, Systemauslegung, Netzintegration und Regelung*, Vieweg+Teubner, Wiesbaden
[10] Kurth M and Kallina G *Möglichkeiten zur Verbesserung des Regelverhaltens von Dampfkraftwerksblöcken*, `http://www.steag-energyservices.com/fileadmin/user_upload/steag-energyservices.com/downloads/veroeffentlichungen/V_Regelverhalten_von_Dampfkesseln.pdf`
[11] Brauns S, Jansen M, Jost D, Siefert M, Speckmann M and Widdel M 2014 *Regelenergie durch Windkraftanlagen*, Fraunhofer IWES, Kassel

Study of the Time Response of a Simulated Hydroelectric System

S. Simani, S. Alvisi, M. Venturini

Dipartimento di Ingegneria, Università degli Studi di Ferrara. Via Saragat 1E – 44122 Ferrara (FE), Italy

E-mail: {silvio.simani,stefano.alvisi,mauro.venturini}@unife.it

Abstract. This paper addresses the design of an advanced control strategy for a typical hydroelectric dynamic process, performed in the Matlab and Simulink environments. The hydraulic system consists of a high water head and a long penstock with upstream and downstream surge tanks, and is equipped with a Francis turbine. The nonlinear characteristics of hydraulic turbine and the inelastic water hammer effects were considered to calculate and simulate the hydraulic transients. With reference to the control solution, the proposed methodology relies on an adaptive control designed by means of the on–line identification of the system model under monitoring. Extensive simulations and comparison with respect to a classic hydraulic turbine speed PID regulator show the effectiveness of the proposed modelling and control tools.

1. Introduction

Hydroelectric plants convert hydraulic energy into useful energy (mainly electric and mechanical energy). Hydropower is, in fact, the most widely adopted form of renewable energy in the world today, accounting for approximately 16% of global energy production [1], *i.e.* 3673.1 TWh of energy are consumed from hydropower in various countries [2]. With increasing demand for electricity, and concern about reducing fossil fuel consumption, hydropower is likely to continue to play a key role in global energy production. Indeed, changing conditions in the power market have led to an increase in the demand of peak energy generation, short response time and fast frequency changes. Hydroelectric power plants thus need to be operated accounting for different load conditions. More in general, in the operation of hydropower systems the occurrence of variations in the flow is frequently experienced, being true either in routine operation, either in accidental or exceptional unforeseen events. The turbine operations such as start–up, load acceptance, load rejection and shutdown may result in hydraulic transients which can cause large pressure and sub–pressure oscillations in turbine hydraulic systems and must be evaluated to avoid mechanical failures. Consequently, there is a need for accurate simulation of transient flow in hydroelectric power plants and even though the basic technology in a plant has not changed much, powerful computers and software now allow the construction of virtual models and simulators of hydropower systems. Within this framework, Matlab and Simulink represent an interactive tool for modelling, simulating, and analysing dynamic systems which have been successfully applied also for nonlinear dynamics investigations in hydroelectric processes [3]. Furthermore, power plants are usually equipped with particular control systems to ensure stable operation. The satisfactory operation of a power system requires a frequency control that keeps it within acceptable limits when the system undergoes significant load variation. The design of proper control systems for hydraulic turbines remains a challenging and important problem.

This paper considers the simulation and the control of a typical hydroelectric power plant, which has a high water head and a long penstock with upstream and downstream surge tanks, and is equipped with a Francis turbine [4]. The simulation model was developed in the Matlab and Simulink environments [5, 6]. The nonlinear characteristics of hydraulic turbine and the inelastic water hammer effect were considered to calculate and simulate hydraulic transients. The proposed control system is compared to a standard hydraulic turbine speed governor PID described and tuned in [3].

In the proposed control system, an electric servomotor is used as a governor [7, 8]. These solutions are preferable for the control of hydropower systems as they have a simple design, require less maintenance and are less expensive than conventional governors. Regarding the compensation strategy, several methodologies have been proposed in the literature [9]. For example, conventional control schemes are documented in [9]. Alternative control methods can be based *e.g.* on optimal control theory, which was exploited for adjusting the blade angles to achieve maximum operating efficiency at a given load [10]. Frequency control strategies were proposed *e.g.* in [11], whilst intelligent approaches relying *e.g.* on adaptive network based fuzzy inference system were addressed in [12].

With reference to the control strategy proposed in this work, the application of model identification mechanisms in connection with model–based adaptive control design has gained increasing attention [13]. The strategy suggested in this paper belongs to the field of adaptive control, which has undergone significant development in recent years. The aim of this approach is to solve the problem of controller design for instance where the characteristics of the process under investigation are not sufficiently known, or change over time. Since a mathematical model is a description of system behaviour, accurate modelling for a complex system is very difficult to achieve in practice. On–line or adaptive parametric model identification schemes represent an alternative for developing experimental models for complex systems. Because of these considerations, this paper suggests the implementation an adaptive controller based on an iterative identification scheme, used for the on–line estimation of the controlled process. While the process time–varying parameters have been estimated, the time–varying controller parameters are computed on–line, in order to maintain the required control performances. Thus, instead of exploiting complicated analytical nonlinear models, it is suggested to describe the plant under investigation by a parameter–varying linear model, whose variables are obtained by an on–line identification procedure. Note that the control strategy proposed here was exploited in a different framework, as described *e.g.* in [14, 15].

Extensive simulations under different working conditions are performed for the considered hydroelectric dynamic process. The achieved results indicate that the proposed methodology is effective to accurately describe the hydroelectric power plant nonlinear dynamics as well as to design a hydraulic turbine speed control system. The proposed simulation system may also be employed for preliminary designs or assessments of hydropower projects.

The paper has the following structure. Section 2 provides an overview of the hydroelectric dynamic process for modelling and control purposes. The proposed adaptive controller design and the tuning strategy are presented in Section 3.1, with reference to the identification of a discrete–time parameter–varying linear model. The on–line identification strategy exploited in this work for obtaining the input–output description of the considered hydraulic process is discussed in Section 3. The achieved results summarised in Section 4 show the performances of the control schemes, compared to a classic control strategy relying on a standard PID regulator. Section 5 concludes the paper by highlighting the main achievements of the work, open problems, and further investigations.

2. Hydraulic System Description

A general diagram of a hydroelectric power plant with two surge tanks is shown in Fig. 1 [3].

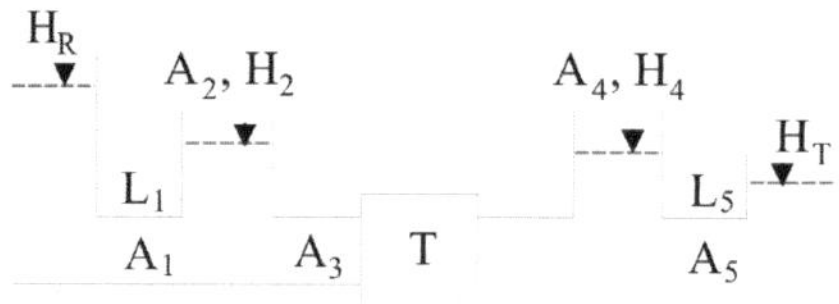

Figure 1. Layout of the simulated hydropower plant.

It consists of a reservoir with water level H_R, an upstream water tunnel with cross-section area A_1 and length L_1, an upstream surge tank with cross–section area A_2, and water level H_2. This is followed by a downstream surge tank with cross–section area A_4 and water level H_4, and a downstream tail water tunnel with cross–section area A_5 and length L_5. Moreover, the penstock between hydraulic turbine and two surge tanks has a cross–section area A_3 and length L_3. Finally, a tail water lake with water level H_T can be seen at the far right in Fig. 1. Reservoir water level H_R and tail water lake water level H_T are considered as constants.

Eqs. (1) and (2) express the non–dimensional flow rate and water pressure in terms of the corresponding relative deviations:

$$\frac{Q}{Q_r} = 1 + q \tag{1}$$

$$\frac{H}{H_r} = 1 + h \tag{2}$$

where q is the flow rate relative deviation, whilst h the water pressure relative deviation,

According to [3], with reference to a pressure water supply system, the Newton's second law for a fluid element inside a tube and the conservation mass law for a control volume, which accounts for water compressibility and tube elasticity, can be written. Under the assumption that the penstock is short or medium in length, water and pipeline can be considered incompressible and rigid, respectively. Therefore, only the inelastic water hammer effect can be taken into account. As shown in [3], the general expression can be simplified as:

$$\frac{h}{q} = -T_w\, s - H_f \tag{3}$$

Equation (3) represents the flow rate deviation and the water pressure deviation transfer functions for a simple penstock, where h is the water pressure relative deviation, q the flow rate relative deviation, H_f the hydraulic loss, s represents the Laplace operator, and T_w is the water inertia time given in Eq. (4):

$$T_w = \frac{L\, Q_r}{g\, A\, H_r} \tag{4}$$

The parameters that define the water inertia time T_w in Eq. (4) are the penstock length L, the rated flow rate Q_r, the gravity acceleration g, the cross–section area A, and the rated water pressure H_r. The typical hydroelectric power plant sketched in Fig. 1 can be divided into three sections: the upstream water tunnel, the penstock and the downstream tail water tunnel. The flow rate deviation and water pressure deviation transfer functions of the three sections are expressed in the following.

The upstream water tunnel connects the reservoir to the upstream surge tank. Since the inlet of upstream water tunnel is in reservoir and the water pressure deviation of the inlet is constant during hydraulic transients, the transfer function of the flow rate deviation and the water pressure deviation of the outlet of the upstream water tunnel can be expressed in the form:

$$\frac{h_1}{q_1} = -T_{w_1}\, s - H_{f_1} \tag{5}$$

The downstream tail water tunnel connects the downstream surge tank to the tail water lake. It is assumed that the outlet of the downstream tail water tunnel is in tailwater lake and the water pressure deviation of the outlet is constant. Therefore, the transfer function of flow rate deviation and the water pressure deviation of the inlet of downstream tail water tunnel has the form:

$$\frac{h_5}{q_5} = -T_{w_5}\, s - H_{f_5} \tag{6}$$

Usually, the water inertia in the draft tube is considered within the penstock. Thus, the transfer function of flow rate deviation (the subscript t refers to the turbine) and the water pressure deviation of the penstock can be written as:

$$h_t = h_2 - h_4 + h_3 \tag{7}$$

where:

$$\frac{h_3}{q_3} = -T_{w_3}\, s - H_{f_3} \tag{8}$$

The equations for the surge tanks are derived from the continuity of flow at the two junctions, where the hydraulic losses at orifices of surge tanks are neglected:

$$\begin{cases} \frac{A_2\, H_r}{Q_r}\frac{dh_2}{dt} = q_2 = q_1 - q_3 \\ \frac{A_4\, H_r}{Q_r}\frac{dh_4}{dt} = q_4 = q_3 - q_5 \end{cases} \tag{9}$$

The surge tank filling time can be expressed as:

$$T_s = \frac{A\, H_r}{Q_r} \tag{10}$$

The mathematical model and the performance curves of the Francis turbine considered in this paper were tuned in order to be compatible with the characteristics of the hydraulic system considered in [3]. The values of the main parameters of the hydraulic system and the Francis turbine at rated conditions are reported below:

- Reservoir water level H_r: 400 m;
- Water flow rate Q_r: 36.13 m^3/s;
- Turbine power P_r: 127.6 MW;
- Turbine rotational speed n_r: 500 rpm.

Under these assumptions, the rated value of the efficiency η_r is equal to 90%, while the turbine–rated torque is 2437 kNm. The approach used to derive the non–dimensional performance curves adopted in this paper is different from the one followed in [3] and outlined in the following.

A second order polynomial curve, reported in Eq. (11), is used to relate the non–dimensional water flow rate Q/Q_r to the non–dimensional rotational speed n/n_r. The non–dimensional parameter G, which varies in the range $0 \div 100\%$, represents the wicked gate opening.

$$\frac{Q}{Q_r} = G\left[a_1\left(\frac{n}{n_r}\right)^2 + b_1\left(\frac{n}{n_r}\right) + c_1\right] = f_1\,(n, G) \tag{11}$$

The curve at $G = 100\%$ (*i.e.* fully open wicked gate) is reported in Fig. 2, together with the curve at $\eta = 0\%$, so that the operating region allowed for the Francis turbine is defined. The water flow rate Q can be calculated by means of Eq. (11) for any operating point, as a function of the current rotational speed n and wicked gate opening G.

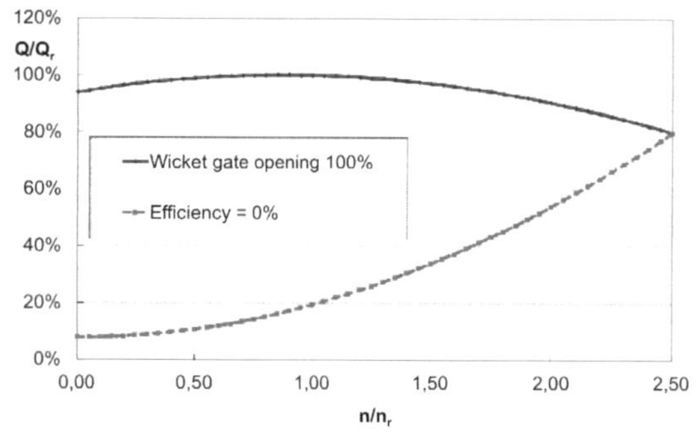

Figure 2. Non–dimensional water flow rate Q/Q_r vs. non–dimensional rotational speed n/n_r.

For the sake of simplicity, the turbine efficiency is assumed constant and equal to the rated value η_r (*i.e.* 90%). To account for the efficiency variation with the electric load, the turbine efficiency may be also expressed as a function of non–dimensional rotational speed (not considered in this paper). The non–dimensional turbine torque M is given by Eq. (12), as a function of the water flow rate Q, the water level H and the rotational speed n. According to the dependencies shown in Eq. (11), the turbine torque M is a function of the water flow rate Q, the rotational speed n and wicked gate opening G.

$$\frac{M}{M_r} = \frac{\frac{Q}{Q_r}\frac{H}{H_r}}{\frac{n}{n_r}} = f_2\left(Q, n, G\right) \tag{12}$$

Finally, as in [3], the relations of Eqs. (13) – (16) express all the non–dimensional parameters for the turbine in terms of the corresponding relative deviations. Note that, from the definition provided in Eq. (16), only negative values are allowed for y.

$$\frac{Q}{Q_r} = 1 + q_t \tag{13}$$

$$\frac{H}{H_r} = 1 + h_t \tag{14}$$

$$\frac{n}{n_r} = 1 + x \tag{15}$$

$$G = 1 + y \tag{16}$$

where q_t represents the turbine flow rate relative deviation, h_t the turbine water pressure relative deviation, x the turbine speed relative deviation, and y the wicket gate servomotor stroke relative deviation.

If the generator unit supplies an isolated load, then the dynamic process of the generator unit considering the load characteristic can be represented as:

$$\frac{x}{m_t - m_{g0}} = \frac{1}{T_a\, s + e_g} \tag{17}$$

where m_{g0} is the load torque, T_a the generator unit mechanical time, and e_g the load self–regulation factor.

In modern hydroelectric power plants, conventional PID control laws are applied to control the hydraulic turbine speed, where the control signal u is the sum of three elements of proportional, integral, and differential gain of hydraulic turbine speed deviation (error) x, which can be thus expressed as:

$$u = x\left(K_p + \frac{K_i}{s} + \frac{K_d\, s}{1 + T_n\, s}\right) \tag{18}$$

where K_p represents the proportional gain, K_i the integral gain, K_d the derivative gain, T_n the derivative filter time constant. Section 4 will analyse and compare the performance of this standard PID regulator designed in [3] with respect to the alternative control strategy proposed in this paper and described in Section 3.

Finally, with reference to the servomechanism of the process, by neglecting small time constants, the relationship between the control signal u and the wicket gate servomotor stroke y can be expressed by means of a first–order equation:

$$\frac{y}{u} = \frac{1}{T_y\, s + 1} \tag{19}$$

where T_y is the wicket gate servomotor response time.

3. Linear Parameter Varying Modelling for Control

This section describes the approach exploited for obtaining the mathematical description of the hydraulic system model, which is used for the design of the control strategy. In particular, the on–line identification scheme relying on the Least–Square Method (LSM) with adaptive directional forgetting discussed in [16] enhances the design procedure of the proposed adaptive controller, as shown in Section 3.1. This strategy represents an improvement with respect to classical LSM [17], and LMS with exponential forgetting [16].

The LSM with adaptive directional forgetting is thus used for the discrete–time on–line identification of processes that are described by the following transfer function $G(z)$:

$$G(z) = \frac{A(z^{-1})}{B(z^{-1})} = \frac{b_1\, z^{-1} + \ldots + b_m\, z^{-m}}{1 + a_1\, z^{-1} + \ldots + a_n\, z^{-n}}\, z^{-d} \tag{20}$$

where a_i, b_i, m, n, and d represent the unknown parameters and the structure of the model, defining the polynomials $A(z^{-1})$ and $B(z^{-1})$, whilst z is the discrete–time complex variable.

The estimated output of the process at the step k, $\hat{y}(k)$ is computed on the basis of the previous process input u and output y signals, according to (21):

$$\hat{y}(k) = -\hat{a}_1\, y(k-1) - \ldots - \hat{a}_n\, y(k-n) + \hat{b}_1\, u(k-1-d) + \ldots + \hat{b}_m\, u(k-m-d) \tag{21}$$

where $\hat{a}_i$ and $\hat{b}_i$ are the current estimations of process parameters at the step $(k-1)$. This equation can be written also in a vector form, as shown in (22) [17]:

$$\begin{aligned} \hat{y}(k) &= \Theta_{k-1}^T\, \Phi_k \\ \Theta_{k-1} &= \left[\hat{a}_1,\, \ldots,\, \hat{a}_n,\, \hat{b}_1,\, \ldots,\, \hat{b}_m\right]^T \\ \Phi_k &= \left[-y(k-1),\, \ldots,\, -y(k-n),\, u(k-1-d),\, \ldots,\, u(k-m-d)\right]^T \end{aligned} \tag{22}$$

The vector Θ_{k-1} contains the model parameter estimates computed at the step $(k-1)$, whilst the vector Φ_k contains the output and input values for the computation of the estimated output $\hat{y}(k)$. These vectors are used for describing the following identification method exploited for the on–line parameter estimation [16].

The LMS with adaptive directional forgetting scheme [16] is able to change the forgetting coefficient with respect to changes of the input and output signals. This method has been selected in this paper since it seems to improve the performances of LSM with exponential forgetting. Moreover, the main disadvantage of pure recursive LSM is the absence of any signal weighting. This feature is important when the actual process parameters change over time.

Thus, in the estimation scheme using *LSM with adaptive directional forgetting*, the process parameters are updated using the recursive expression:

$$\Theta_k = \Theta_{k-1} + \frac{C_{k-1}\,\Phi_k}{1+\xi_k}\left(y(k) - \Theta_{k-1}^T\,\Phi_k\right) \tag{23}$$

where:

$$\xi_k = \Phi_k^T\,C_{k-1}\,\Phi_k \tag{24}$$

In each step k, the matrix C is updated according to (25):

$$C_k = \begin{cases} C_{k-1} - \frac{C_{k-1}\,\Phi_k\,\Phi_k^T\,C_{k-1}}{\epsilon_k^{-1}+\xi_k}, & \text{if } \xi_k > 0 \\ C_{k-1}, & \text{if } \xi_k = 0 \end{cases} \tag{25}$$

with:

$$\epsilon_k = \varphi_k - \frac{1-\varphi_k}{\xi_{k-1}} \tag{26}$$

The forgetting coefficient φ_k is updated as follows:

$$\varphi_k = \frac{1}{1+(1+\rho)\left\{\ln(1+\xi_{k-1}) + \left[\frac{(\nu_{k-1}+1)\,\eta_{k-1}}{1+\xi_{k-1}+\eta_{k-1}} - 1\right]\frac{\xi_{k-1}}{1+\xi_{k-1}}\right\}} \tag{27}$$

where:

$$\nu_k = \varphi_k\,(\nu_{k-1}+1) \tag{28}$$

$$\eta_k = \frac{(y_k - \Theta_{k-1}^T\,\Phi_k)^T\,(y_k - \Theta_{k-1}^T\,\Phi_k)}{\lambda_k} \tag{29}$$

$$\lambda_k = \varphi_k\left[\lambda_{k-1} + \frac{(y_k - \Theta_{k-1}^T\,\Phi_k)^T\,(y_k - \Theta_{k-1}^T\,\Phi_k)}{1+\xi_{k-1}}\right] \tag{30}$$

are auxiliary variables.

The on–line identification method recalled here without using numeric filters was directly implemented in the Matlab® and Simulink® environments, as described in [13]. Note that the initial values of the parameters Θ_0, C_0, φ_0, λ_0, ρ, and ν_0 have to be properly selected.

Finally, once the time–varying parameters of the discrete–time linear model approximating the behaviour of the whole continuous–time nonlinear dynamic process described in Section 2 have been computed at each time step k, the adaptive controller is designed as described in Section 3.1.

Journal of Physics: Conference Series **570** (2014) 052003 doi:10.1088/1742-6596/570/5/052003

3.1. Adaptive Control Design

This section describes the adaptive control strategy used in connection with the on–line estimation scheme presented in Section 3. In more detail, with reference to the particular benchmark under diagnosis recalled in Section 2, a Ziegler–Nichols PI adaptive controller for processes of second order ($n = 2$) is exploited [13].

With reference to (20), with $n = m = 2$ and $d = 0$, the transfer function of the time–varying controlled system has the form:

$$G(z) = \frac{b_1\, z^{-1} + b_2\, z^{-2}}{1 + a_1\, z^{-1} + a_2\, z^{-2}} \tag{31}$$

whose parameters estimated using the on–line identification approach shown in Section 3 are:

$$\Theta_k = \left[\hat{a}_1,\, \hat{a}_2,\, \hat{b}_1,\, \hat{b}_2\right]^T \tag{32}$$

Therefore, the control law corresponding to a discrete–time Ziegler–Nichols PI adaptive controller has the form:

$$u(k) = K_p \left[e(k) - e(k-1) + \frac{T_s}{T_I}\,\frac{e(k) - e(k-1)}{2}\right] + u(k-1) \tag{33}$$

where $e(k)$ is the tracking error, *i.e.* $e(k) = r(k) - y(k)$, with $r(k)$ the set–point or reference signal, T_s the sampling time. The controller variables K_p and T_I are computed from the time–varying model parameters Θ_k. This type of control law can be transformed into its feedback representation:

$$u(k) = q_0\, e(k) + q_1\, e(k-1) + u(k-1) \tag{34}$$

The controller parameters q_0 and q_1 (or K_p and T_I) are computed using following relations [18]:

$$\begin{aligned} q_0 &= K_p\left(1 + \tfrac{T_s}{2\,T_I}\right), & q_1 &= -K_p\left(1 - \tfrac{T_s}{2\,T_I}\right) \\ K_p &= 0.6\, K_{P_u}, & T_I &= 0.5\, T_u \end{aligned} \tag{35}$$

where the variables K_{P_u} and T_u are the critical gain and the critical period of oscillations, respectively. Also these variables are functions of the time–varying model parameters Θ_k

In particular, when considering a second order model described by its estimated parameters $\hat{a}_2$, $\hat{a}_1$, $\hat{b}_2$, and $\hat{b}_1$, the variables K_{P_u} and T_u required by the Ziegler–Nichols method can be computed at each time step k from the following relations [13]:

$$\begin{cases} K_{P_u} = \frac{\hat{a}_1 - \hat{a}_2 - 1}{\hat{b}_2 - \hat{b}_1} \\ T_u = \frac{2\pi T_s}{\arccos\gamma}, \quad \text{with } \gamma = \frac{\hat{a}_2\,\hat{b}_1 - \hat{a}_1\,\hat{b}_2}{2\,\hat{b}_2} \end{cases} \tag{36}$$

In this way, the adaptive discrete–time linear controllers (33) or (34) are designed on the basis of the time–varying linear model (31) estimated via the on–line identification scheme from the data of the continuous–time nonlinear dynamic process described in Section 2.

4. Simulation Results

This section shows the achieved results regarding the design and the application of the adaptive controller to the sampled data acquired from the hydraulic model simulator. Note that the adaptive discrete–time controller is connected to the continuous–time controlled system via analog to digital and digital to analog converters. The simulated system of the hydroelectric power plant described in Section 2 has been developed in the Matlab and Simulink environments.

As shown in Fig. 3, the complete model consists of three subsystems. First, the speed governors and servomechanism, in which turbine speed dead zone, valve saturation, and limitation are implemented. This subsystem provides the implementation of both the classic PID governor and the adaptive solution proposed in Section 3. As shown in Fig. 3 (a), through the parallel structure in Simulink, it is also possible to compare the performances of the two regulators. Note that the adaptive controller is connected to the controlled system via analog to digital and digital to analog converters, which are not shown in Fig. 3. Second, the hydrodynamics system (b) contains the tunnels, the penstock, and surge tanks. Third, the turbine generator and the network (c) represent the generator unit operating in isolation.

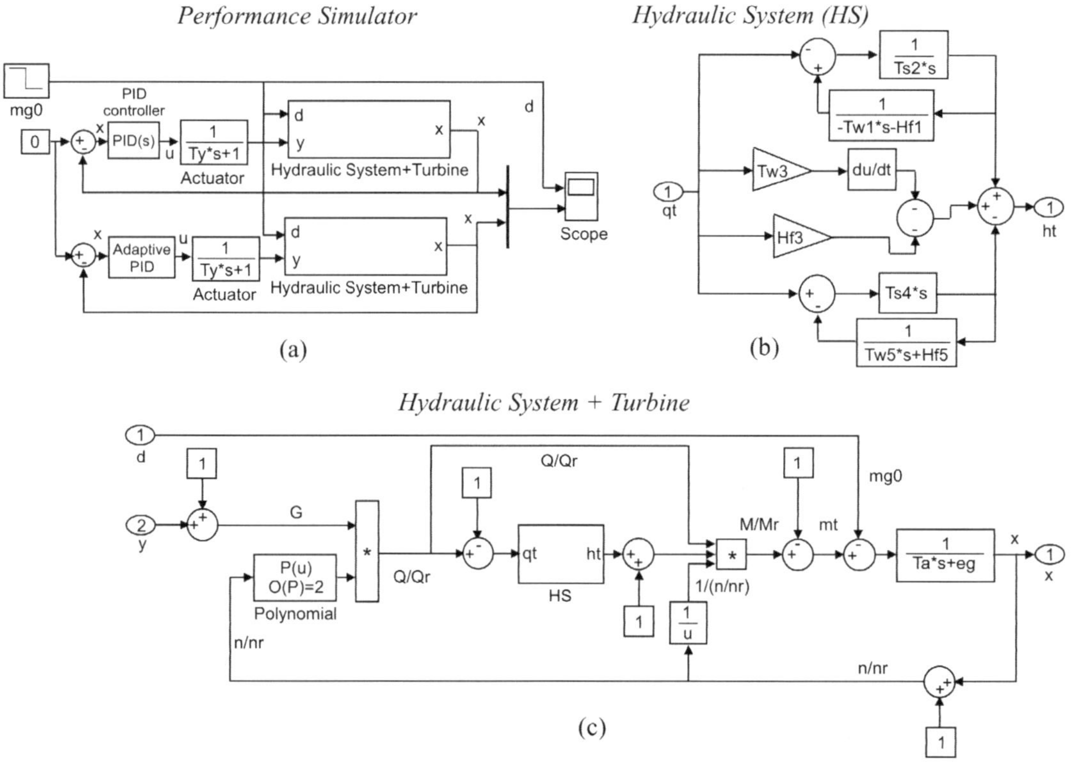

Figure 3. Simulink simulated systems: (a) speed controllers, servomechanism and hydrodynamic systems; (b) hydraulic system; (c) turbine, generator unit and network.

The simulation model with two surge tanks and a Francis turbine described in Fig. 3 should reflect the realistic behaviour of a hydroelectric power plant in the presence of large hydraulic transients after full load rejection m_{g0}. Usually, the most severe hydraulic transients will happen after full load rejection. Therefore, all of these simulations are performed on full load rejection operating conditions. The parameters of the model are summarised in Table 1.

The turbine speed governor plays a very important role in hydraulic transients caused by load changes. The classic PID controller proposed in [3] required an optimal tuning of its gains, and in this way only the dynamic performance of the generator unit can be improved. Moreover, in order to get the best dynamic performance, it is necessary to set different optimal PID gains on different operating conditions for turbine speed governor.

Table 1. Simulated model parameters [3].

$H_{f_1} = 0.0481$	$H_{f_3} = 0.0131$	$H_{f_5} = 0.0047$	$T_c = 20$	$T_{s_2} = 476.05s$	$T_{s_4} = 5000s$
$T_{w_1} = 3.22s$	$T_{w_3} = 0.83s$	$T_{w_5} = 0.1s$	$a = -0.08$	$b = 0.14$	$c = 0.94$
$T_a = 5.9s$	$T_y = 0.5s$	$e_g = 0$	$K_d = 1.00$	$K_i = 0.20$	$K_p = 1.00$

On the other hand, a different method is proposed in Section 3 for the on–line modelling technique oriented to the design of the adaptive controller relying on the time–varying identified parametric model. Therefore, the output y of the continuous–time nonlinear hydraulic model described in Section 2 is approximated by a time–varying discrete–time prototype of the type of (31) with one input. Note that the model on–line estimation scheme presented in Section 3 is tested using different data sets. Therefore, the time–varying model parameters are estimated in order to minimise the so–called model–reality mismatch. In this way, the on–line estimated time–varying linear prototype should be able to provide the optimal fitting of all the working conditions of the hydraulic power plant, without the need of changing the controller parameters as required by the classic PID governor. Using this identified prototype, the model–based approach for determining the adaptive controller shown in Section 3.1 is exploited and applied to the hydraulic simulator described in Section 2. Thus, according to Section 3.1, the parameters of the adaptive controller have been computed. In particular, the identified time–varying prototype consist of a second order model; thus, the adaptive regulator parameters in (33) or (34) are computed analytically at each time step k.

It is worth highlighting the strategy applied for achieving the required adaptive characteristics. With reference to (33) or (34), the adaptive controller parameters are tuned via the Ziegler–Nichols rules, applied to the time–varying linear model, and by considering different working conditions. Therefore, the optimal controller performances with respect to set–point variations are thus enhanced. In this way, if both the model on–line parametric identification and the regulator tuning procedure are properly performed, the parameter adaptation mechanisms will lead to optimal adaptive properties. Moreover, by means of the proposed adaptation mechanisms, the suggested design scheme is able to maintain good control performances, even when conditions, different from those considered in [3], are simulated.

In the following, the suggested adaptive controller, together with the classic PID governor, have been implemented and compared in the Matlab and Simulink environments. The initial conditions for the on–line estimation algorithm (23)–(30) are listed in Table 2. Regarding the standard PID governor parameters in Eq. (18), they were selected as $K_p = 1.0$, $K_i = 0.2$, $K_d = 1.0$, and $T_n = 0.01$ [3].

Table 2. Adaptive controller initialisation parameters.

$\Theta_0 = [0.1, 0.2, 0.3, 0.4]^T$	$C_0 = 10^9\, I_4$	$\varphi_0 = 1$	$\lambda_0 = 0.001$	$\rho = 0.99$	$\nu_0 = 10^{-6}$

The controller capabilities have been assessed in simulation by considering different load torque m_{g0} values. Fig. 4 shows the dynamics of the m_{g0} signal for the cases of $+100\%$ and -100%, which represent the turbine start–up and shutdown conditions, respectively. In particular, the start–up phase is assumed to last 300 s (due to the large size of the considered Francis turbine), while the shutdown maneuver takes just 30 s, to simulate an unplanned emergency shutdown. On the other hand, smaller m_{g0} variations analysed in the following have been modelled as step functions.

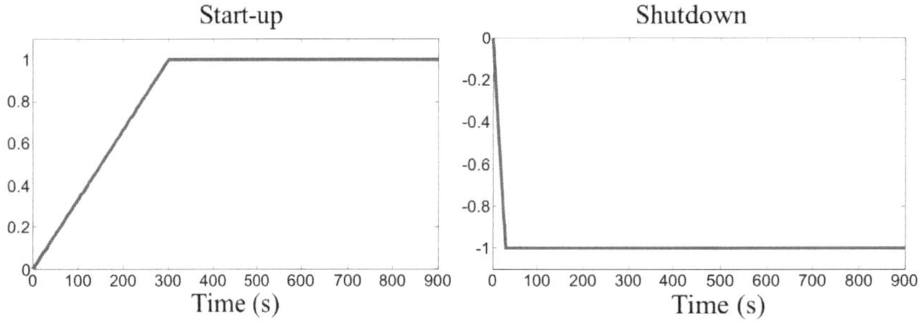

Figure 4. m_{g0} ramp functions for the turbine start–up and shutdown.

As an example, the results reported in Fig. 5 show that, even though both regulators can keep the relative deviation of the rotational speed null (*i.e.* the rotational speed constant) in steady–state conditions, the performances of the adaptive regulator of Eq. (34) are clearly better than those achievable by means of the standard PID governor of Eq. (18).

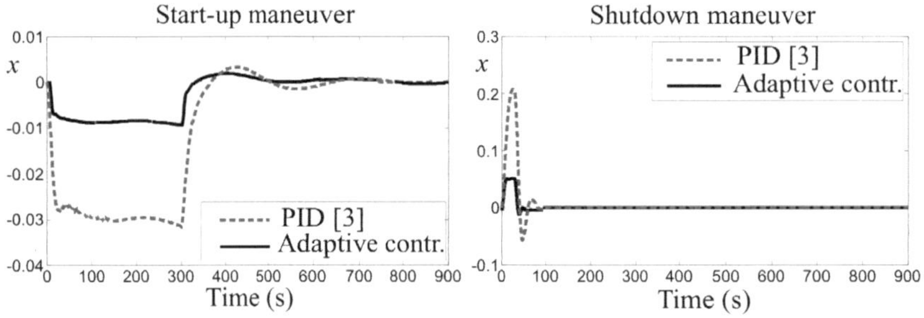

Figure 5. Turbine speed relative deviations x when the load torque m_{g0} changes in start–up and shutdown conditions.

In order to provide a comparison of the obtained performances, Table 3 summarises the achieved results in terms of per–cent Normalised Sum of Squared tracking Error ($NSSE$) values defined as:

$$NSSE\% = 100\sqrt{\frac{\sum_{k=1}^{N} x^2(k)}{N}} \tag{37}$$

that are computed with reference to the signal x for both the controllers for N samples. Moreover, the per–cent undershoot $\sigma\%$, overshoot $S\%$ and the settling time T_s have been also estimated, for different values of the load torque m_{g0}.

According to these simulation results, good properties of the suggested adaptive controller are highlighted, and they are better than the standard PID governor. Some further comments

Table 3. Controllers' $NSSE\%$, T_s, $\sigma\%$, and $S\%$.

Controller Type	m_{g0}	$NSSE\%$	T_s	$\sigma\%$	$S\%$
Classic PID	+1%	0.04%	33.16s	0.38%	0.005%
	−1%	0.04%	31.39s	0.005%	0.39%
	+10%	0.39%	58.13s	3.75%	0.05%
	−10%	0.41%	27.57s	0.04%	3.95%
	+100%	1.86%	725.59s	3.15%	0.32%
	−100%	3.34%	76.05s	5.66%	20.73%
Adaptive Controller	+1%	0.037%	28.39s	0.33%	0.004%
	−1%	0.038%	27.13s	0.004%	0.36%
	+10%	0.38%	29.69s	2.37%	0.02%
	−10%	0.39%	18.64s	0.02%	1.78%
	+100%	0.62%	701.36s	0.96%	0.12%
	−100%	1.38%	54.08s	1.19%	5.76%

can be drawn in general regarding the capability of the adaptive controller. The $NSSE$ values are usually comparable or slightly lower for reduced variations of the load torque, while they are considerably lower in case of significant transient maneuvers (*i.e.* start–up and shutdown). On the other hand, though always lower, the settling time T_s is not significantly decreased and remains comparable to that obtainable by means of the PID [3]. This is probably due to the inherent dynamics of the simulated hydraulic systems. Similarly, the per–cent $\sigma\%$ and $S\%$ are decreased in all cases by using the adaptive controller, and this effect is highlighted when considering the most severe transients (*i.e.* start–up and shutdown).

5. Conclusion

This paper addressed the design of an advanced control strategy for a hydroelectric power plant modelled in the Matlab and Simulink environments. The hydraulic system consisted of a high water head and a penstock with upstream and downstream surge tanks and a Francis turbine. The nonlinear characteristics of hydraulic model were considered to simulate the hydraulic transients and to evaluate the behaviour of the proposed hydraulic turbine regulating system. The suggested control methodology was based on an adaptive control designed by means of the on–line identification of the system model under monitoring. Extensive simulations showed the main features of the proposed modelling and control tools, as well as highlighted its benefit mainly in correspondence of the most severe transients. The considered simulation tools could be also useful for preliminary designs or assessments of the most important features of hydropower installations and their control strategies.

References

[1] I. Working Group III Mitigation of Climate Change, "Special Report on Renewable Energy Sources and Climate Change Mitigation," tech. rep., Intergovernmental Panel on Climate Change, 2011.

[2] R. R. Singh, T. R. Chelliah, and P. Agarwal, "Power electronics in hydro electric energy systems – A review," *Renewable and Sustainable Energy Reviews*, vol. 32, pp. 944–959, April 2014.

[3] H. Fang, L. Chen, N. Dlakavu, and Z. Shen, "Basic modeling and simulation tool for analysis of hydraulic transients in hydroelectric power plants," *IEEE Trans. Energy Convers.*, vol. 23, pp. 424–434, Sept. 2008.

[4] M. Popescu, D. Arsenie, and P. Vlase, *Applied Hydraulic Transients: For Hydropower Plants and Pumping Stations*. Lisse, The Netherlands: CRC Press, Jan. 2003.

[5] C. D. Vournas and G. Papaionnou, "Modeling and stability of a hydro plant with two surge tanks," *IEEE Trans. Energy Convers.*, vol. 10, pp. 368–375, Jun.

[6] D. N. Konidaris and J. A. Tegopoulos, "Investigation of oscillatory problems of hydraulic generating units equipped with francis turbines," *IEEE Trans. Energy Convers.*, vol. 12, pp. 419–425, Dec. 1997.

[7] *Power System Control & Stability*, vol. 111 of *IEEE Series Power Eng.* River Street, Hoboken, NJ: John Wiley and Sons, Inc., 2003.

[8] M. Hanmandlu and H. Goyal, "Proposing a new advanced control technique for micro hydro power plants," *Electrical Power and Energy Systems*, vol. 30, 2008. DOI: doi:10.1016/j.ijepes.2007.07.010.

[9] J. A. Laghari, H. Mokhlis, A. H. A. Bakar, and H. Mohammad, "A comprehensive overview of new designs in the hydraulic, electrical equipments and controllers of mini hydro power plants making it cost effective technology," *Renewable and Sustainable Energy Reviews*, vol. 20, pp. 279–293, 2013.

[10] P. Schniter and L. Wozniak, "Efficiency based optimal control of kaplan hydro generators," *IEEE Trans. Energy Conver.*, vol. 10, pp. 348–353, June 1995.

[11] O. P. Malik, G. S. Hope, G. C. Hancock, L. Zhaohui, Y. Luqing, and W. E. I. Shouping, "Frequency measurement for use with a microprocessor based water turbine governor," *IEEE Trans. Energy Conver.*, vol. 6, pp. 361–366, Sept. 1991.

[12] M. Djukanovic, M. Novicevic, D. J. Dobrijevic, B. Babic, J. Dejan, and Y. H. P. Sobajic, "Neural–net based coordinated stabilizing control for the exciter and governor loops of low head hydropower plants," *IEEE Trans. Energy Conver.*, vol. 10, pp. 760–767, Dec. 1995.

[13] V. Bobál, J. Böhm, J. Fessl, and J. Machácek, *Digital Self–Tuning Controllers: Algorithms, Implementation and Applications*. Advanced Textbooks in Control and Signal Processing, Springer, 1st ed., 2005.

[14] S. Simani, "Application of a Data–Driven Fuzzy Control Design to a Wind Turbine Benchmark Model," *Advances in Fuzzy Systems*, vol. 2012, pp. 1–12, November 2nd 2012. Invited paper for the special issue: Fuzzy Logic Applications in Control Theory and Systems Biology (FLACE) . ISSN: 1687–7101, e–ISSN: 1687-711X. DOI: 10.1155/2012/504368.

[15] S. Simani and P. Castaldi, "Data–Driven and Adaptive Control Applications to a Wind Turbine Benchmark Model," *Control Engineering Practice*, vol. 21, pp. 1678–1693, December 2013. Special Issue Invited Paper. ISSN: 0967–0661. PII: S0967–0661(13)00155–X. DOI: http://dx.doi.org/10.1016/j.conengprac.2013.08.009.

[16] R. Kulhavý, "Restricted exponential forgetting in real–time identification," *Automatica*, vol. 23, no. 9, pp. 589–600, 1987.

[17] L. Ljung, *System Identification: Theory for the User.* Englewood Cliffs, N.J.: Prentice Hall, second ed., 1999.

[18] K. J. Astrom and T. Hagglund, *PID Controllers.* Instrument Society of America, 1995.

Journal of Physics: Conference Series **570** (2014) 052004 doi:10.1088/1742-6596/570/5/052004

General model and control of an n rotor helicopter

A G Sidea[1,3], R Yding Brogaard[1,2,3], N A Andersen[1] and O Ravn[1]
[1] Department of Electrical Engineering, Technical University of Denmark, Elektrovej 326, 2800 Kongens Lyngby, Denmark
[2] FORCE Technology, Park Allé 345, 2605 Brøndby, Denmark

E-mail: agsidea@elektro.dtu.dk, ryb@force.dk

Abstract. The purpose of this study was to create a dynamic, nonlinear mathematical model of a multirotor that would be valid for different numbers of rotors. Furthermore, a set of Single Input Single Output (SISO) controllers were implemented for attitude control. Both model and controllers were tested experimentally on a quadcopter. Using the combined model and controllers, simple system simulation and control is possible, by replacing the physical values for the individual systems.

1. Introduction

Multi rotor helicopters are a type of unmanned aerial vehicles used for various purposes, ranging from entertainment flights to professional photography and surveillance. Although different designs are available, the most common is the one with an even number of rotors greater or equal to four, uniformly distributed on a circle in the horizontal plane. This is sometimes called "star configuration".

Multirotors are inherently unstable systems, requiring precise control in order to fly. To implement the required controllers, an accurate system model is needed. Due to system nonlinearity and aerodynamic relations, the system model is complex and different for each configuration.

Various studies related to multirotor modelling and position control were conducted. Starting from the simplest type of multirotors, quadcopters [1], to hexacopters [2] and octocopters [3], multirotors are widely analysed.

Nevertheless, most of the related work implemented models for a fixed number of rotors. In [4], various multirotor configurations are presented, but the modelling still focuses on an octocopter. [5] introduces a modular system that can have varying configurations and number of rotors. However, the modules are autonomous, constituting a cluster of individual rotors working together rather than a single multirotor.

This paper largely presents the results of the master thesis written by Rune Yding Brogaard [6]. The novelty of this work is that it presents a dynamic model that describes the general behaviour of the so-called star configuration. The used equations are derived as a function of number of rotors, denoted with n, which can be replaced with any even number greater or equal to four. This means the model is valid for any multirotor that has the given configuration.

Similarly, equations for flight control were derived to adapt to a variable number of rotors. Attitude control implies controlling each of the roll, pitch and yaw rotations. Having rotation angles as system inputs, the SISO controllers were designed to control the multirotor roll and pitch angles along with

[3] To whom any correspondence should be addressed.

roll, pitch and yaw velocity. With the help of the model, their optimal gains can be found and the multirotor can be given the desired attitude.

2. System equations

When considering multirotor equations, two sets of coordinate axes can be used. The first one, called helicopter-fixed frame (HF), moves together with the helicopter when it rotates, while the second one, called Earth-fixed frame (EF), remains fixed with respect to Earth. In hover, the helicopter fixed frame is the same as the earth frame. These axes can be seen in figure 1, together with the directions of rotation, roll ϕ, pitch θ and yaw ψ.

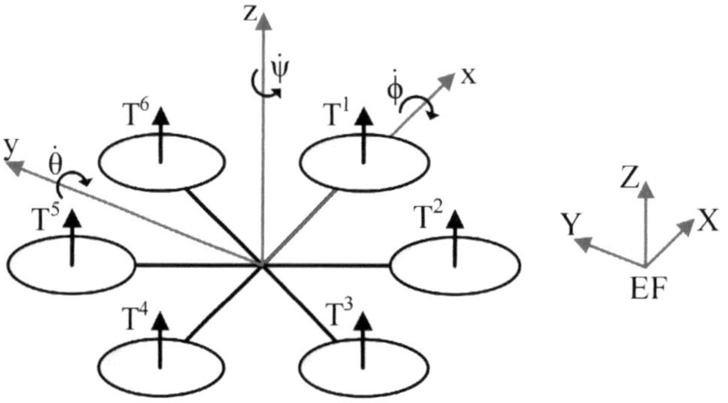

Figure 1. Hexacopter rotation angles in helicopter-fixed frame and Earth-fixed frame.

Before implementing a model for a multi rotor helicopter, formulae for each rotor have to be derived. To show which parameters are different for each rotor, the index i was used in formulae, where i ranges from 1 to n. The index was used as a superscript in order to avoid conflicts with common notations.

According to [7], the formulae for vertical thrust T^i and torque Q^i generated by each rotor have the form shown in (1) and (2). Also, the power required to produce T^i is given in (3).

$$T^i = C_T^i \rho A (\Omega^i R_r)^2 \tag{1}$$

$$Q^i = C_Q^i \rho A (\Omega^i R_r)^2 R_r \tag{2}$$

$$P^i = C_P^i \rho A (\Omega^i R_r)^3 \tag{3}$$

In the previous equations, ρ is the air density, A is the rotor disc area, Ω^i the rotor angular velocity and R_r the rotor radius. The formulae for thrust constant C_T^i , torque constant C_Q^i and power coefficient C_P^i from (4) , (5) and (6) are given in [7] as well.

$$C_T^i = \frac{a\sigma}{4}\left(\theta_t - \frac{\dot{z}^i + v_i^i}{\Omega^i R_r}\right) \tag{4}$$

$$C_Q^i = \frac{C_d}{8} + \frac{C_T^i}{\sigma}\frac{\dot{z}^i + v_i^i}{\Omega^i R_r} \tag{5}$$

$$C_P^i = k C_T^{i\,\frac{3}{2}} + \frac{\sigma C_d}{8} \tag{6}$$

Here, a is the lift slope of the blade, θ_t is propeller angle at the tip, $\dot{z}^i$ is the rotor velocity in z direction of HF (see figure 1), and C_d is the drag coefficient. Although the drag coefficient depends on factors such as Reynolds number or Mach number, it does not change in the used working range [7], hence it does not have an index. k is the induced power correction factor, which takes tip losses, wake swirl and non-uniform inflow into account. Solidity ratio σ is defined by the propeller blade area divided by rotor disc area.

Induced velocity, v_i^i, represents the velocity of air passing through each propeller area in hover. It can be found by using (7).

$$v_i^i = \left(\frac{T_h^i}{2\rho A}\right)^{\frac{1}{2}} \tag{7}$$

Hover thrust can be replaced with the formula in (8), where m is total helicopter mass and g is the gravitational acceleration.

$$T_h^i = \frac{mg}{n} \tag{8}$$

Equation (1) yields the formula for angular velocity:

$$\Omega^i = \frac{\sqrt{C_T \rho A T^i}}{C_T \rho A R_r} \tag{9}$$

By dividing (2) and (3), a relation between power and torque can be derived:

$$\frac{Q^i \Omega^i}{C_Q^i} = \frac{P^i}{C_P^i} \tag{10}$$

Since by definition $P = Q\Omega$, it results that $C_Q^i = C_P^i$. Taking this into consideration, from (9) and (3) the individual power for each rotor is:

$$P^i = \frac{C_Q^i T^i \sqrt{C_T^i \rho A T^i}}{{C_T^i}^2 \rho A} \tag{11}$$

After replacing thrust from (11) with the formula from (8), the power required for hovering is obtained:

$$P^i = \left(\frac{m^3 g^3 {C_Q^i}^2}{{C_T^i}^3 \rho A n^3}\right)^{\frac{1}{2}} \tag{12}$$

By dividing (2) by (1), the ratio between torque and thrust is obtained. It was shown in [6] that there is a linear relationship between the two, thus another coefficient can be introduced as follows:

$$C_{QT} = \frac{Q^i}{T^i} = \frac{C_Q^i R_r}{C_T^i} \tag{13}$$

3. Multirotor model

As mentioned previously, n has to be an even number and the rotors evenly distributed on the xy plane. In this case, the angle between each rod is $2\pi/n$, with the rod length from centre to rotor shaft $l \geq R_r/sin(\pi/n)$.

Due to system complexity, some assumptions need to be made to simplify the modelling process:

- The rotors are counted clockwise, with the first one being located on the *x* axis. They alternate between clockwise and counter-clockwise rotation.
- The multirotor centre of mass coincides with the origin of HF, which is situated at the intersection of rods.
- The structure is rigid, including the rotors (no blade flapping or coning).
- Propellers have an ideal blade twist.
- Thrust and torque are considered proportional, their ratio denoted with C_{QT}^{i}.

Figure 2 displays a block diagram of the system, where the control signals for rotation angles $(U_{\phi}, U_{\theta}, U_{\psi})$ and vertical thrust (U_T) are system inputs. To obtain the desired position, the correct input has to be given to each rotor. The rotors influence the roll, pitch, yaw rotations differently, depending on their geometrical positioning. Considering that they are distributed uniformly on a circle, the angles between them and the axes depend on the number of rotors. Thus, a general formula can be derived that describes each rotor's influence on the three rotations. This can be seen in (14), where *i* denotes the rotor number. The distributor converts the given input signals to the required rotor inputs, which are contained in vector *u*.

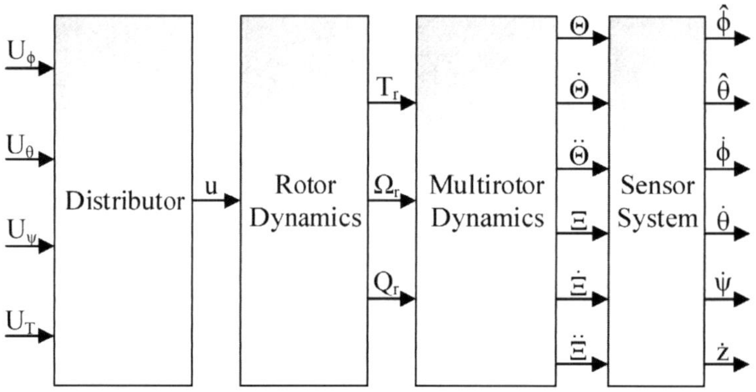

Figure 2. Block diagram of multirotor system.

$$u^i = U_T + U_\theta \cos\frac{2\pi(i-1)}{n} + U_\phi \sin\frac{2\pi(i-1)}{n} + (-1)^i U_\psi \tag{14}$$

For modelling the rotor, the first order transfer function in (15) is considered. Although a simplified version of the real function, it was proven to be accurate enough when comparing with experimental data. Time constant τ depends on motor response time and propeller inertia.

$$M_{tf} = \frac{1}{\tau s + 1} \tag{15}$$

The rotor dynamics block contains the elements in figure 3. C_{pwm} represents the relation between input signal and angular velocity. This differs from system to system, having to be determined through measurements. The C_{pwm} block outputs a vector containing the angular velocities of each rotor, which is used to calculate Ω_r and the vector T_r. The latter contains all thrusts corresponding to each angular velocity. f_T represents the relation from (1). From each thrust value, the corresponding torque is found by multiplying with C_{QT}.

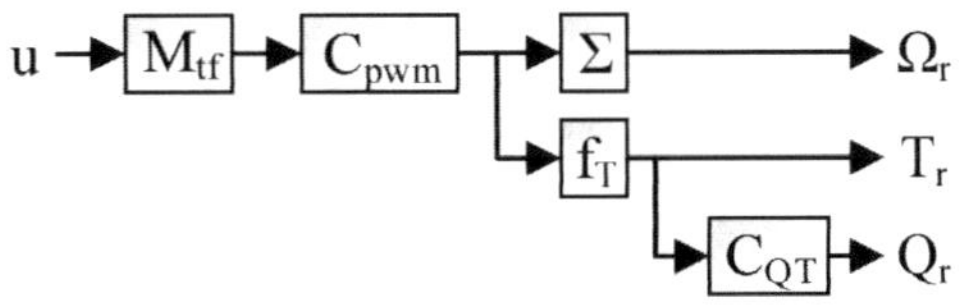

Figure 3. Rotor dynamics.

Ω_r has a single value, calculated as seen in (16). Since the clockwise rotating rotors have odd indexes, they will be subtracted from the sum. This results in a negative or positive value, depending on whether the clockwise or counter-clockwise rotating rotors spin faster.

$$\Omega_r = \sum_{i=1}^{n} (-1)^i \Omega^i \tag{16}$$

Multirotor dynamics can be divided in two types of movement: rotational and linear. To separate them, two vectors are used. $\Theta = [\phi\ \theta\ \psi]^T$ represents helicopter attitude, while $\Xi = [X\ Y\ Z]^T$ represents position in EF. Vectors $\dot{\Theta} = [\dot{\phi}\ \dot{\theta}\ \dot{\psi}]^T$, $\ddot{\Theta} = [\ddot{\phi}\ \ddot{\theta}\ \ddot{\psi}]^T$ contain rotational velocities and accelerations respectively. Corresponding vectors for linear movement are constructed using the same principle.

Taking into account the thrust forces, inertial and gyroscopic effects acting on the helicopter, the components for $\ddot{\Theta}$ can be written as:

$$\ddot{\phi} = \frac{\dot{\theta}\dot{\psi}(I_y - I_z) + J_r\dot{\theta}\Omega_r + l\sum_{i=1}^{n} T^i \sin\frac{2\pi(i-1)}{n}}{I_x} \tag{17}$$

$$\ddot{\theta} = \frac{\dot{\phi}\dot{\psi}(I_z - I_x) - J_r\dot{\phi}\Omega_r + l\sum_{i=1}^{n} T^i \cos\frac{2\pi(i-1)}{n}}{I_y} \tag{18}$$

$$\ddot{\psi} = \frac{\dot{\theta}\dot{\phi}(I_x - I_y) + J_r\dot{\Omega}_r + C_{QT}l\sum_{i=1}^{n}(-1)^i T^i}{I_z} \tag{19}$$

These equations were taken from [1] and generalized for *n* rotors. I_x, I_y, I_z are components of multirotor inertia on each of the axes. J_r is rotor inertia and l is multirotor rod length.
Similarly, position in EF can be found from:

$$\ddot{X} = \sum_{i=1}^{n} T^i \frac{\sin\psi\sin\phi + \cos\psi\sin\theta\cos\phi}{m} \tag{20}$$

$$\ddot{Y} = \sum_{i=1}^{n} T^i \frac{-\cos\psi\sin\phi + \sin\psi\sin\theta\cos\phi}{m} \tag{21}$$

$$\ddot{Z} = \sum_{i=1}^{n} T_i \frac{\cos\psi\cos\phi}{m} - g \tag{22}$$

The sensor system block takes attitude and position vectors, outputting parameters measured by sensors on board the multirotor. Accelerometer and gyroscope measurements are combined to give an estimation of roll and pitch angles ($\hat{\phi}$, $\hat{\theta}$). The gyro corrects errors resulting from sudden acceleration disturbances, while the accelerometer compensates for drift errors in the gyro [6].

4. Controller model

Because the system inputs are distributed as input signals to each rotor, each rotation around the coordinate axes can be controlled independently. Hence, a set of SISO controllers was implemented for roll, pitch and yaw velocities, as well as for roll and pitch angles. A yaw angle controller was not implemented due to the lack of a heading sensor on the physical system used for testing.

Considering the multirotor does not perform aggressive manoeuvres, it has only small deviations from hovering state. Thus, the system was linearized around this point. In Laplace domain, the linearized angular velocities are:

$$\dot{\phi}(s) = \frac{1}{s}\frac{l\sum_{i=1}^{n} T^i \sin\frac{2\pi(i-1)}{n}}{I_x} \tag{23}$$

$$\dot{\theta}(s) = \frac{1}{s}\frac{l\sum_{i=1}^{n} T^i \cos\frac{2\pi(i-1)}{n}}{I_y} \tag{24}$$

$$\dot{\psi}(s) = \frac{1}{s}\frac{J_r\Omega_r s - \Omega_r(0) + C_{QT} l \sum_{i=1}^{n}(-1)^i T^i}{I_z} \tag{25}$$

Both roll and pitch rotations are similar, therefore the controllers implemented for them have the same structure. The only difference between the two is inertia, which depends on the distances between the rotor and the *x* and *y* axes.

Proportional Derivative (PD) controllers were preferred for rotational velocity control due to their higher bandwidth. The steady state error is eliminated by the Proportional Integral Derivative (PID) controllers implemented for the angles.

For the two rotations, the closed loop transfer function of the PD velocity controller is:

$$G^{vel}(s) = \frac{K_P^{vel} LM(s)}{1 + K_P^{vel} LM(s) K_D^{vel}(s) H^{vel}(s)} \tag{26}$$

K_P^{vel} is proportional gain, while K_D^{vel} is the derivative one. *LM* is the linearized model of the system presented in figure 2 without the sensor block. H^{vel} represents the low pass filters of the gyros.

For angle control, G^{vel} was integrated to obtain rotation angle. Having that, the transfer function for angle control could be calculated as:

$$G^{ang}(s) = \frac{K_I^{ang}(s) K_P^{ang} G^{vel}(s)\frac{1}{s} H^{ang}(s)}{1 + K_D^{ang}(s) K_I^{ang}(s) K_P^{ang} G^{vel}(s)\frac{1}{s} H^{ang}(s)} \tag{27}$$

As in the previous case, K_P^{ang} is the proportional gain, K_D^{ang} the derivative one, whereas K_I^{ang} is integral gain. H^{ang} represents the angle estimation system.

Because no controller was implemented for the yaw angle, the steady state error had to be eliminated by using a PID controller for yaw velocity. Its closed loop transfer function is:

$$G^{vel}(s) = \frac{K_I^{vel}(s) K_P^{vel} LM(s) H^{vel}(s)}{1 + K_I^{vel}(s) K_P^{vel} LM(s) H^{vel}(s) K_D^{vel}(s)} \tag{28}$$

5. Model and controller verification

In order to test the designed model and controllers, measurements were performed on a physical system, namely a quadcopter from MikroKopter [8] mounted on a test rig. The specifications for the quadcopter can be found in Table 1. Due to the lack of documentation for some propeller parameters, the thrust and torque constants had to be determined experimentally. By varying rotor speed and measuring thrust and torque each time, relations expressing thrust and torque as a function of angular velocity were found. From these, the constant values were found to be: $C_T = 0.0158\,rad^{-2}$ and $C_Q = 0.023\,rad^{-2}$.

Table 1. Mikrokopter specifications.

Parameter	Value
Propeller radius, R_r	0.126 m
Propeller mass	0.008 kg
Motor mass	0.069 kg
Rod length, l	0.245 m
Rod mass	0.035 kg
Battery mass	0.238 kg
Flight Control Board mass	0.169 kg

The two setups used (figure 4) assured safety and consistency while performing tests. The poles used were high enough to ignore ground effect [9].

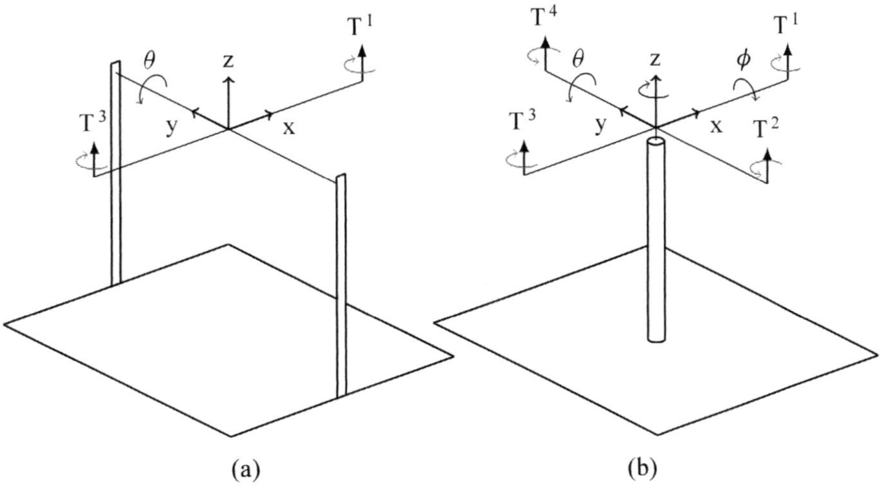

Figure 4. Experimental setup for testing: (a) roll and pitch rotations, (b) yaw rotation and free rotation along any axis.

Test rig (a) allowed rotation around the *x* or *y* axis by fixing the respective multirotor rods to two vertical poles. Test rig (b) allowed rotation around all axes due to the ball joint mounted at the end of the pole, but it was possible to lock it so that the quadcopter rotated only around *z* axis. Having all the equations derived previously, the quadcopter specifications were introduced in the model. The controller gains were chosen so that the phase margin would be higher than 60 deg. Their discrete transfer functions with sample time 1/487 s can be found in table 2.

In the beginning, each controller was tested separately, by allowing rotation only on one axis. System response was tested by sequentially applying a 0.4 rad/s step input to each angular velocity and 0.4 rad to each angle. After implementing the controllers on the quadcopter, the response to the same step input was measured.

Table 2. Discrete transfer functions of controllers for the tested quadcopter.

Controller	P	I	D
Roll velocity	0.447	–	$\frac{10-9.72z^{-1}}{1-0.72z^{-1}}$
Pitch velocity	0.447	–	$\frac{10-9.72z^{-1}}{1-0.72z^{-1}}$
Yaw velocity	4.467	$\frac{1-0.846z^{-1}}{1-z^{-1}}$	$\frac{10-9.02z^{-1}}{1-0.02z^{-1}}$
Roll angle	2.114	$\frac{1-0.9979z^{-1}}{1-z^{-1}}$	$\frac{10-9.777z^{-1}}{1-0.777z^{-1}}$
Pitch angle	2.114	$\frac{1-0.9979z^{-1}}{1-z^{-1}}$	$\frac{10-9.777z^{-1}}{1-0.777z^{-1}}$

Figure 5 presents the comparison between simulated and measured results for the angle controllers. As it is seen in the figure, simulated response was plotted with a black line, while measured values are represented with a red dotted line. All controllers implemented on the quadcopter are discrete, having a sampling time of 1/487 s.

For testing the PID angle controllers, a step input of 0.4 rad was given. The response has some overshoot due to the integral gain. While the simulation had a 0.196 s rise time and 9.2% overshoot, the real system had lower overshoot of 7.3% at the cost of longer rise time (0.243 s roll and 0.26 s pitch).

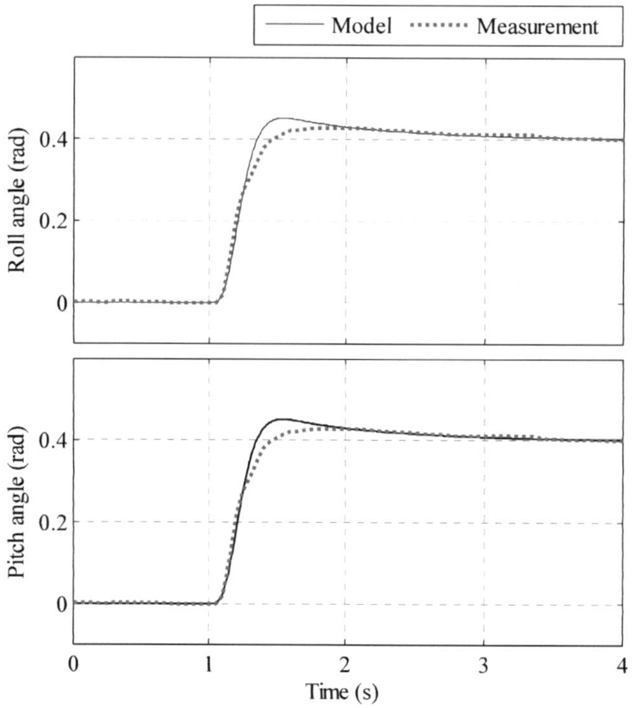

Figure 5. Step response of roll and pitch angle controllers.

The roll and pitch angle responses are very similar, which is no surprise, considering they have the same inertias due to an identical rotor displacement on the x and y axes.

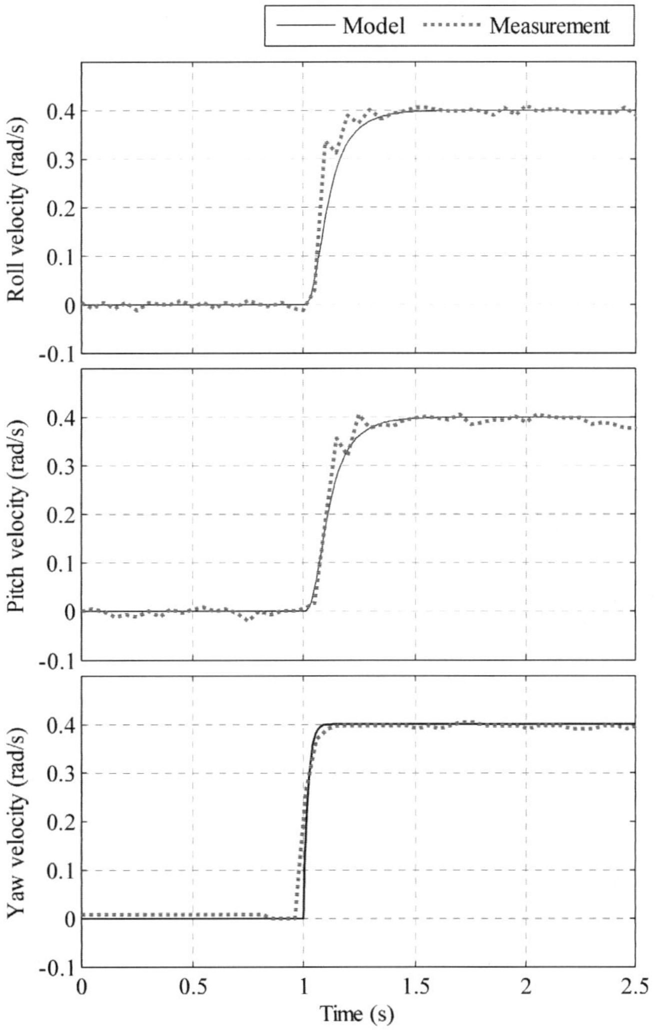

Figure 6. Step response of roll, pitch, yaw velocity controllers.

Figure 6 shows the step responses of the velocity controllers. Measured values follow the simulated ones with only slight deviations. The 0.21 s rise time of the real system for roll and pitch velocities is shorter than the simulated one of 0.26 s.

Yaw velocity has a faster response than the other two, having a simulated rise time of 0.042 s. Again, the measured values are very close to the simulation but the rise time is longer, being 0.056 s.

Although the inaccuracies could have been caused by some uncertainties in the model, the friction forces introduced by the test rigs also had a significant influence.

The previous results show individual responses of all controllers. A flying multirotor however is not restricted on any axis. For this reason, they had to be tested together, to see the influences they have on each other.

Journal of Physics: Conference Series **570** (2014) 052004 doi:10.1088/1742-6596/570/5/052004

Test rig (b) was used for this purpose. This time, the quadcopter could rotate freely on all axes, apart from the 0.5 rad angle limitation caused by the ball joint design. Another thing to be mentioned is that the centre of rotation of the ball joint was at 8 cm distance from the quadcopter's actual one.

A series of step inputs was given to the quadcopter at different times to test controller interactions. The responses of the roll and pitch angle controllers, as well as the one of yaw velocity controller can be seen in figure 7. Simulated results are plotted with continuous lines, while measured ones are represented with dotted lines.

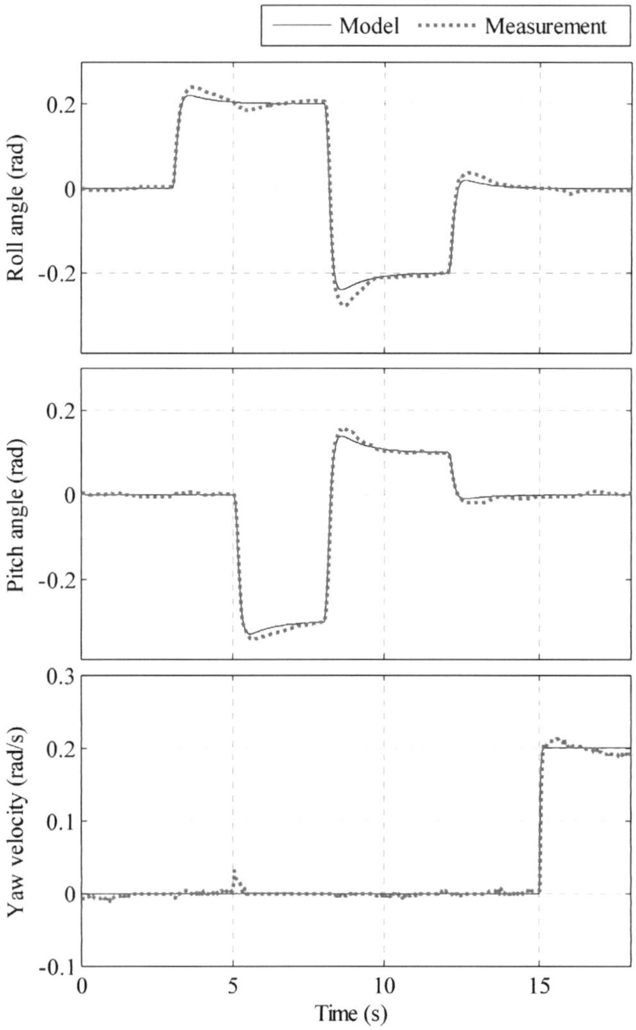

Figure 7. System response to multiple step inputs.

Before the measurement, all angles were set to 0. Step inputs in HF were given as follows:

- At 3 s, a 0.2 rad step was given to the roll angle. The real system had a higher overshoot than the simulated one. No notable influence towards the other rotations could be seen.
- At 5 s, pitch angle received a step of -0.3 rad. A slight decrease of the measured pitch and increase of the measured yaw velocity took place, but they quickly returned to their intended values.

- At 8 s, two steps were given: -0.4 rad for roll and 0.4 for pitch. These set the roll value to 0.1 and pitch value to -0.2 rad. Again, higher overshoot for the real system could be seen, but no changes in the yaw velocity.
- At 12 s, roll received 0.2 rad and pitch received -0.1 rad, returning both angles to 0.
- At 15 s, yaw velocity was given a step of 0.2 rad/s. The roll and pitch angles were not influenced by this change.

The controllers behaved nicely even for inputs very different from the ones required for stationary state. Their influence on each other was minimal, even when two steps were given at the same time. The higher overshoots of the real system were assumed to be caused mainly by the previously mentioned displacement of the centre of rotation.

6. Conclusions

The outcome of this work is a system model independent of the number of rotors that can be used for the configurations where the rotors are uniformly distributed on a circle. Each required parameter can be found by replacing the number of rotors, inertias, etc. in the formulae.

General equations for individual rotors were found and used in the model. Taking into account multirotor geometry and dynamics, the system was described, its equations written as a function of the number of rotors, n.

The designed attitude controllers are also independent of the number of rotors. After specifying the number of rotors and the other required values in the model, the optimal gains can be found for the controllers.

Due to the distributor, SISO controllers could be implemented for each rotation direction. PID controllers were used for roll and pitch angles. In the case of angular velocities, PD controllers were used for roll and pitch because of their higher bandwidth. Yaw velocity required a PID controller to eliminate the steady state error.

An experimental setup was used to see if the designed model and controllers apply to a quadcopter. The fact that the measured values closely follow the simulated ones proves that the model properly recreates the quadcopter physical system.

System simulations showed that the controllers offer a fast and stable response. Not only they perform well individually, but they do this even when combined, having negligible interference on each other.

The model has to be tested on other configurations, but it is expected to give similar results as the four rotor case, provided the correct values for inertias and other system parameters are used.

7. References

[1] Bouabdallah S and Siegwart R 2007 "Full control of a quadrotor" *Int. Conf. on Intelligent Robots and Systems* (San Diego, USA) pp 153-8

[2] Baránek R and Šolc F 2012 "Modelling and control of a hexa-copter" *13th Int. Carpathian Control Conf.* (High Tatras, Slovakia) pp 19-23

[3] Samano A, Castro R, Lozano R and Salazar S 2012 "Modeling and Stabilization of a Multi-Rotor Helicopter" *J. Intelligent and Robotic Systems* **69** 161-9

[4] Rinaldi F, Gargioli A and Quagliotti F 2013 "Proportional Integral Derivative and Linear Quadratic Regulation of a multirotor attitude: mathematical modelling, simulations and experimental results" *Int. Conf. on Unmanned Aircraft Systems* (Atlanta, USA) pp 433-42

[5] Oung R, Ramezani A and D'Andrea R 2009 "Feasibility of a Distributed Flight Array" *Joint 48th IEEE Conf. on Decision and Control and 28th Chinese Control Conf.* (Shanghai, China) pp 3038-44

[6] Brogaard R Y 2012 *Control of Multi Rotor Helicopter* MSc Thesis (Kongens Lyngby,

Denmark: Department of Electrical Engineering, Technical University of Denmark)

[7] Leishman J G 2006 *Principles of Helicopter Aerodynamics* (New York, USA: Cambridge University Press) 2nd ed.

[8] MikroKopter home page, http://www.mikrokopter.de/en/home

[9] Hayden J S 1976 "Effect of the Ground on Helicopter Hovering Power Required" *American Helicopter Society 32nd Annual National V/STOL Forum Proc.*

Modelling of Rotor-gas bearings for Feedback Controller Design

Lukas R. S. Theisen, Henrik Niemann

Department of Electrical Engineering, Technical Unversity of Denmark, Elektrovej, Bld. 326, 2800 Kgs. Lyngby, Denmark

E-mail: `lrst@elektro.dtu.dk`

Abstract. Controllable rotor-gas bearings are popular offering adaptability, high speed operation, low friction and clean operation. Rotor-gas bearings are however highly sensitive to disturbances due to the low friction of the injected gas. These undesirable damping properties call for controllers, which can be designed from suitable models describing the relation from actuator input to measured shaft position. Current state of the art models of controllable gas bearings however do not provide such relation, which calls for alternative strategies. The present contribution discusses the challenges for feedback controller design using the state of the art method, and an alternative data driven modelling approach is pursued based on Grey-Box system identification. The method allows development of models of the rotor-gas bearing suitable for controller design, which can be identified from data over the range of operation and are shown to accurately describe the dynamical behaviour of the rotor-gas bearing. Design of a controller using the identified models is treated and experiments verify the improvement of the damping properties of the rotor-gas bearing.

1. Introduction

Controllable gas bearings are popular for offering high speed operation at low friction using clean and abundant air as a lubricant. Design of such rotor-gas bearing systems has been the topic of a previous PhD project [3] from the Mechanical Engineering department at the Technical University of Denmark (DTU). A result of the work is a piezo actuated rotor-gas bearing test rig. A controller is however required to improve the poor damping properties, and a collaborative PhD has begun between the department of Mechanical Engineering and the Electrical Engineering control engineering group to further explore the design of such controllers. This paper provides an overview of the work so far of the collaborative PhD.

State of the art models of rotor-gas bearings [5, 4] rely on solving the Reynolds equation to model the pressure distribution of the fluid film. Morosi [4] included the effect of the piezo actuated valves into the Reynolds equation, and used it to develop a model of a short rigid rotor-gas bearing actuated by piezo valves based on this modified Reynolds equation (MRE). This model was used to manually tune the parameters of a decentralised controller. Recent work in print has however shown that alternative models of the flow in the valves are required to make the model reflect reality. The MRE requires iteratively solving causing an unknown analytical relation from input to the valves to displacement of the rotating shaft, which leaves the model unsuitable for design of model based controllers. This PhD project has therefore investigated development of such suitable models using system identification, e.g. in [8], which

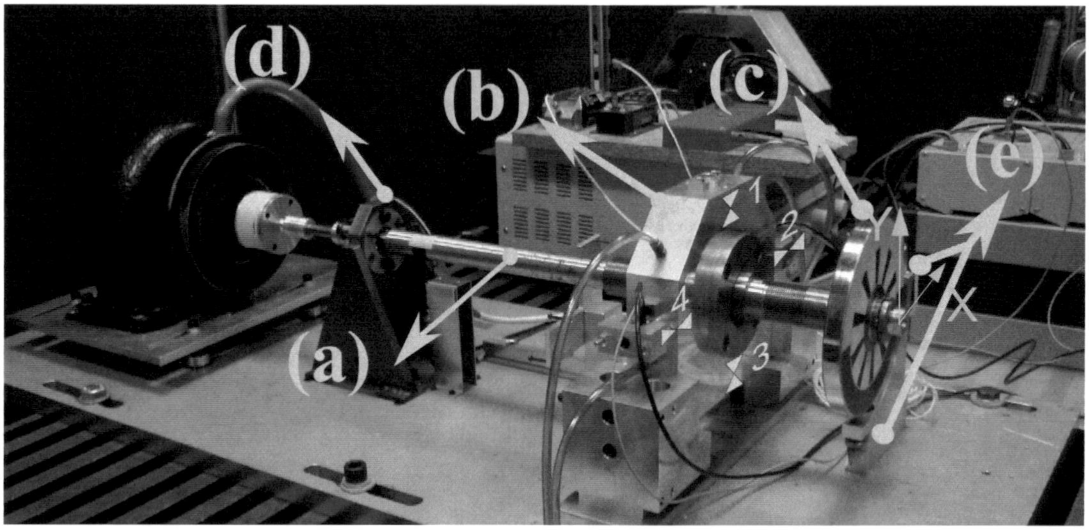

Figure 1. The experimental bearing setup. A flexible shaft (a) is supported by both a ball bearing (d) and the controllable gas bearing (b) with four piezo actuated valves. A disc (c) is mounted in one end to pre-load the journal and displacement sensors (e) measure the lateral movement of the disc.

showed that linear models identified from experimental data could describe the gas bearing behaviour. Models were identified over a range of rotational velocities, where the parameters were found using Grey-Box system identification. The present contribution makes use of the same bearing model and extends the results by also modelling the actuator dynamics.

The present contribution provides an overview of the progress in developing models and their usage for feedback controller design of the rotor-gas bearing from experimental data over the desired operational range determined by rotational velocity and injection pressure. The paper is structured as follows: A brief overview of the experimental test rig is given in Sec. 2, followed by an overview of the current state of the art model and its challenges. Section 4 shows formulation of the Grey-box model and its parameters are identified from experiments. The identified model is used in Sec. 5 to design a controller capable of increasing the damping properties of the rotor-gas bearing and experiments in Sec. 6 verify the capabilities of the designed controller.

2. Experimental Setup of Rotor-Gas Bearing

The experimental setup at hand is shown in Fig. 1: a turbine driven flexible shaft (a) is supported by both a ball bearing (d) and the controllable gas bearing (b), in which pressurised air is injected through four piezo actuated valves numbered as shown. The manometric injection pressure P_I of the pressurised air is measured by a mechanical gauge before splitting up to the four actuators. The absolute pressure in the valves P_{abs} is assumed to be the sum of the measured pressure P_I and the atmospheric pressure P_{atm}. A disc (c) is mounted in one end to pre-load the journal. The horizontal and vertical shaft deflections (e_x, e_y) are measured at the disc location using eddy current sensors (e) in the coordinate frame specified in the figure. The angular position of the shaft ϕ is measured by an optical encoder. The pressurised air generates a thin layer of fluid film in the $25\mu m$ thin gap between the shaft and the bearing housing. For a range of conditions, the fluid film generates restoring forces and thereby keeps the shaft levitating in a stable equilibrium. A more thorough description of the setup is available in [6].

The piezo-electric valves are subject to hysteresis and creep effect [1]. To counteract the hysteresis, decentralised PD-controllers are deployed, effectively reducing the position uncertainty by counteracting the hysteresis. The controlled valves are commanded reference positions $r_{p,i} \in [0;10]V$, and the corresponding valve positions $y_{p,i} \in [0;10]V$ are measured ranging from open valve ($0V$) to closed valve ($10V$). The valves are controlled pairwise as if there was just a single horizontal and a vertical valve commanded position reference $r(t) \triangleq [r_x(t), r_y(t)]^{\mathrm{T}}$, which is mapped as a reference to the individual valves according to the law from $r(t) \to r_p(t)$, similarly the lumped valve position vector $u(t)$ is defined as a function of the individual valve positions:

$$r_p(r_x(t), r_y(t)) = \begin{bmatrix} r_{p,1}(t) \\ r_{p,2}(t) \\ r_{p,3}(t) \\ r_{p,4}(t) \end{bmatrix} = \begin{bmatrix} r_0 + r_y(t) \\ r_0 + r_x(t) \\ r_0 - r_y(t) \\ r_0 - r_x(t) \end{bmatrix}, \qquad u(t) \triangleq \begin{bmatrix} u_x(t) \\ u_y(t) \end{bmatrix} = \begin{bmatrix} y_{p,2}(t) - y_{p,4}(t) \\ y_{p,1}(t) - y_{p,3}(t) \end{bmatrix} \quad (1)$$

This makes the valves cooperate and reduces the system from an over-actuated to a fully actuated. In addition the constant offset $r_0 = 5V$ ensures the largest dynamical range.

3. Rotor-gas Bearing Modelling Using Finite Element Models and the Modified Reynolds Equation

Current state of the art models of the rotor-gas bearing [5, 4] consist of two sub-models: A Finite Element model of the flexible shaft and a model of the thin layer of fluid film. This section provides an overview of these two models and their challenges.

3.1. The Modified Reynolds Equation

The Modified Reynolds equation [4] is used for modelling the behaviour of the thin layer of fluid film in the rotor-gas bearing. By a set of assumptions, a partial differential equation can be made, modelling the pressure p as a function of the fluid film thickness h, which varies along with the shaft position in the bearing and time t can then be set up:

$$\frac{\partial}{\partial \mathsf{y}}\left(\mathsf{p}\mathsf{h}^3\frac{\partial \mathsf{p}}{\partial \mathsf{y}}\right) + \frac{\partial}{\partial \mathsf{z}}\left(\mathsf{p}\mathsf{h}^3\frac{\partial \mathsf{p}}{\partial \mathsf{z}}\right) = 6\mu U\frac{\partial(\mathsf{ph})}{\partial \mathsf{y}} + 12\mu\frac{\partial(\mathsf{ph})}{\partial t} + 12\mathsf{p}V_I \quad (2)$$

where the fluid film coordinate frame $(\mathsf{x}, \mathsf{y}, \mathsf{z})$ chosen is: x the radial coordinate directed towards the centre of the shaft, y the circumferential coordinate and z being the axial coordinate. μ is the viscosity of the gas, U is the linear velocity of the rotating shaft at the bearing housing, V_I is the velocity profile of the injected gas assumed to be parabola shaped with a linear pressure drop along the length of the valves; Work in print however shows this does not model the flow well. The flow is assumed laminar, which is reasonable given the small thickness of the fluid film. The MRE has no known analytical solution, but discretisation in a fine grid allows an iterative solution to be found from a good initial solution guess. This provides the pressure profile, which upon integration provides the horizontal and vertical forces from the fluid film acting on the flexible shaft $F_{be} = [F_{X,be},\ F_{Y,be}]^{\mathrm{T}}$.

3.2. Finite Element Modelling of Flexible Shaft

The flexible shaft can then be modelled using a finite element (FE) method, where the shaft is divided into n_e sections, which can bend and rotate relative to each other around the $n_n = n_e + 1$ nodes connecting the sections giving four degrees of freedom (DOFs) per node. Given the geometry of the shaft and its material properties, the stiffness of each section can be approximated, the mass of each section can be calculated and the forces from each section acting on the other sections can then be expressed as a $n_n \cdot 4$ coupled differential equations with the linear and angular displacements from a horizontal and a vertical axis $q_F = [q_1,\ q_2, \dots, q_{4n_n}]^{\mathrm{T}}$

with corresponding time derivatives $\dot{q}_F$. The model is formulated using the mass matrix M_F, the stiffness matrix K_F, the gyroscopic matrix G_F and the damping matrix D_F:

$$M_F\ddot{q}_F(t) + (D_F - \Omega G_F)\dot{q}_F(t) + K_F q_F(t) = f_F(t), \quad q_F = [q_1, q_2, \dots, q_{4n_n}]^{\mathrm{T}} \tag{3}$$

where in the above $f_F(t)$ is the external forces acting on each node of the FE model from the fluid film, external disturbances etc.. Using the linearised stiffness and damping forces from the MRE bearing model, the FE model predicts the eigenfrequencies within $\pm 5\%$ in the stationary case over a wide range of injection pressures and rotational velocities.

The requirement of this iteratively solved pressure profile for every configuration however leaves the model unable to describe the relation from actuator input voltage to shaft displacement on a form suitable for controller design. The time dependent MRE still remains to be coupled to the FE shaft model and validated experimentally, before the relation from input voltage to shaft displacement can be approximated and put on an analytical form. This shows the need for alternative approaches to develop models for feedback controller design, which do not depend on the solution of the MRE.

4. A System Identification Approach - Data Driven Modelling

The modelling of rotor gas bearings still represents an open challenge. This section shows a heuristics based model approach, where basic knowledge from rotor-dynamics provides the basis for formulation of a model. The model parameters are then identified from experimental data, where the piezo-valves perturb the rotor-gas bearing. The method allows development of accurate models able to describe the rotor-gas bearing dynamics.

The measurement of the valve positions allows dividing the modelling in two: an actuator sub-model to model the valve dynamics from commanded valve reference position to valve position, and a bearing sub-model to model the relation from valve position to shaft deflection. The parameters of the identified models allow formulation of a global model.

4.1. Grey-Box Model of Gas Bearing

In a rotational range around the first two eigenfrequencies, the rotor gas bearing system consisting of the flexible shaft and the gas-bearing can be modelled as a 2 DOF coupled mass-spring damper system. Let $p = [e_x, e_y]^{\mathrm{T}}$ be the position vector consisting of horizontal and vertical shaft displacements, and denote time derivatives $\frac{d}{dt}(\cdot) = \dot{(\cdot)}$. The model then reads

$$\mathcal{M}\ddot{p}(t) + (\mathcal{D} - \Omega\mathcal{G})\dot{p}(t) + \mathcal{K}\,p(t) = \mathcal{B}_p u(t), \quad p = [e_x, e_y]^{\mathrm{T}}, \tag{4}$$

in which Ω is the rotation speed, $\mathcal{M}$ is the diagonal mass matrix, $\mathcal{D}$ is the damping matrix, $\mathcal{G}$ is the antisymmetric gyroscopic matrix, and $\mathcal{K}$ is the stiffness matrix, all with dimension 2×2. The right hand side should include external forces $f(t)$ acting on the rotor-gas bearing, which include: forces from mass unbalance, forces from the piezo valves and forces from external impacts. Section 4.4 shows how the mass unbalance response is filtered out. In [8] it was shown reasonable to assume the actuator force proportional to the valve position with gain $\mathcal{B}_p$, of dimension 2×2 and the model therefore reduces to Eq. 4. The model is reformulated to state space form to ease the Grey-Box modelling. A suitable choice of states is the deflection and velocity of the shaft $x \triangleq [e_x, e_y, \dot{e}_x, \dot{e}_y]^{\mathrm{T}}$. The measurement noise and errors from simplified model are modelled as additive noise $d(t)$ entering both output and states with a disturbance input gain $\mathcal{B}_d$ and the chosen state space formulation of Eq. (4) then reads:

$$\begin{aligned} \dot{x}(t) &= \mathcal{A}x(t) + \mathcal{B}u(t) + \mathcal{B}_d d(t), \quad x(0) = x_0 \\ p(t) &= \mathcal{C}x(t) + d(t) \end{aligned} \tag{5}$$

where the system-, input gain-, and output matrix are

$$\mathcal{A} = \begin{bmatrix} 0 & I \\ -K & -D \end{bmatrix}, \quad \mathcal{B} = \begin{bmatrix} 0 \\ B \end{bmatrix}, \quad \mathcal{B}_d = \begin{bmatrix} 0 \\ B_d \end{bmatrix}, \quad \mathcal{C} = \begin{bmatrix} I & 0 \end{bmatrix}, \tag{6}$$

where $D \triangleq \mathcal{M}^{-1}(\mathcal{D} - \Omega\mathcal{G})$, $K \triangleq \mathcal{M}^{-1}\mathcal{K}$, and $B \triangleq \mathcal{M}^{-1}\mathcal{B}_p$ are the matrices to be identified along with disturbance gain B_d, and initial value x_0, each with four parameters giving 20 unknowns in total. The parameters of (6) are identified by recasting the problem to a model parametrized in $\hat{\theta}_b \triangleq \{\hat{D}, \hat{K}, \hat{B}, \hat{x}_0, \hat{B}_d\}$ as $\mathfrak{M}_b(\theta)_b$. Each matrix $\hat{D}, \hat{K}, \hat{B}, \hat{B}_d$ has four elements denoted by small letters and subscripts xx, xy, yx, yy. The model then reads

$$\mathfrak{M}_b(\theta_b) : \begin{cases} \dot{x}(t) = \mathcal{A}(\theta_b)x(t) + \mathcal{B}(\theta_b)u(t) + \mathcal{B}_d(\theta_b)d(t), & x(0) = x_0(\theta_b) \\ p(t) = \mathcal{C}x(t) + d(t) \end{cases} \tag{7}$$

4.2. Grey-Box Model of Lumped Actuators

This section formulates a similar model of the lumped PD-controlled piezo valves. The closed loop horizontal and vertical lumped valve can each be modelled as a second order low-pass filter. The valve dynamics can be seen as transfer functions with two poles $p_{1,j}$, and $p_{2,j}$, where j refers to the horizontal valve x or vertical valve y and gain $\kappa_{a,j}$. Considering only the commanded reference position as input, the dynamics then read:

$$\begin{bmatrix} u_x(s) \\ u_y(s) \end{bmatrix} = \begin{bmatrix} H_{a,x}(s) & 0 \\ 0 & H_{a,y}(s) \end{bmatrix} \begin{bmatrix} r_x(s) \\ r_y(s) \end{bmatrix}, \qquad H_{a,j}(s) = \frac{\kappa_{a,j}}{\left(\frac{1}{p_{1,j}}s + 1\right)\left(\frac{1}{p_{2,j}}s + 1\right)} \tag{8}$$

in which $H_{a,j}(s)$ is the second order filter of the specified form. The model can be formulated in the same structure as Eq. 7, exploiting that cross coupling terms $k_{xy}, k_{yx}, d_{yx}, d_{xy}, b_{yx}, b_{xy}, b_{p,yx}, b_{p,xy}$ are zero. Reformulating this to a Grey-box model and estimating the initial valve states x_a and a similar disturbance gain d_a, the model then reads

$$\mathfrak{M}_a(\theta_a) : \begin{cases} \dot{x}_a(t) = \mathcal{A}(\theta_a)x_a(t) + \mathcal{B}(\theta_a)r(t) + \mathcal{B}_{d,a}(\theta_a)d_a(t), & x_a(0) = x_{a0}(\theta_a) \\ u(t) = \mathcal{C}_a x_a(t) + d_a(t) \end{cases} \tag{9}$$

Estimating the gains, poles, initial state and disturbance gain, the actuator model thus only has 12 parameters being:

$$\theta_a \triangleq [\underbrace{k_{a,xx},\ k_{a,yy},\ b_{a,xx},\ b_{a,yy},\ d_{a,xx},\ d_{a,yy}}_{\text{valve coefficients}},\ \underbrace{b_{da,xx},\ b_{da,yy}}_{\text{disturbance gains}},\ \underbrace{x_{a1,x0},\ x_{a2,x0},\ x_{a3,x0},\ x_{a4,x0}}_{\text{initial state}}]^{\mathrm{T}}, \tag{10}$$

and a similar actuator model $\mathfrak{M}_a(\theta_a)$ has been formulated.

4.3. Description of Experiments

The model should represent the rotor-gas bearing over the range of operational conditions, which are defined by two main characteristics: rotational speed Ω and injection pressure P_I, which can vary within $\Omega \in [0; 6]krpm$ and $P_I \in [3; 7]bar$ respectively. This identification over the whole operational range will be available in [7], where the coefficients are estimated from data sets collected from a grid of injection pressures and rotational velocities representing the range of interest. Here a selected example is chosen $\bar{\Omega} = 0rpm$ and $\bar{P}_I = 4.0bar$. During data collection all variables were sampled at f_s=5kHz. A pseudo random binary sequence commanded as reference for the lumped valves $r(t)$ ensured excitation of the system and hence identifiability of the parameters. The input stepped from -1 to 1 at random sampling instants. The lumped valve references $r(t)$ and measured lumped valve positions $u(t)$ were logged as input and output for the actuator sub-model and the lumped valve positions $u(t)$ and the shaft displacement $p(t)$ as input and outputs for the bearing model.

Journal of Physics: Conference Series **570** (2014) 052005 doi:10.1088/1742-6596/570/5/052005

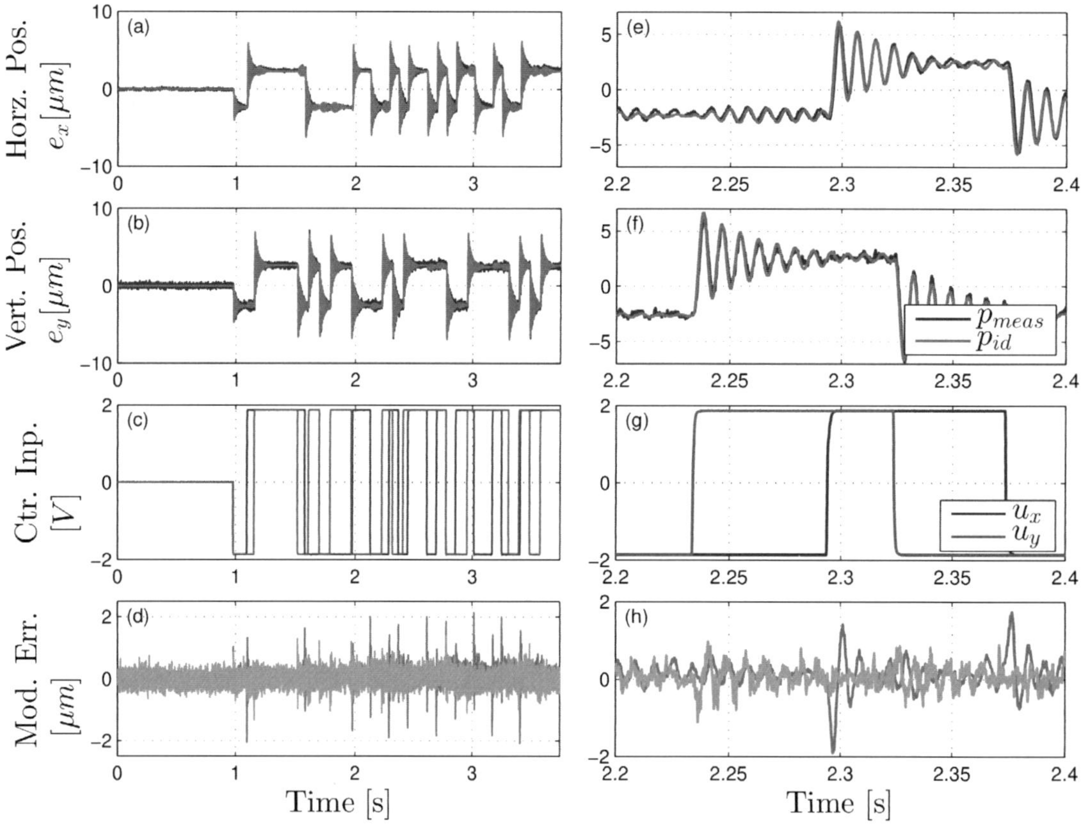

Figure 2. Example of an identification of data set collected at $P_I = 4bar, \Omega = 0rpm$. The valve positions $[u_x, u_y]$ exciting the bearing shown in c) excite the bearing dynamics causing deflection of the shaft. a) and b) show this measured deflection $p_{meas} = [e_x, e_y]$ and the predicted using the identified model $p_{id} = [\hat{e}_x, \hat{e}_y]$ subject to valve excitation. d) shows the residual between identified and measured response $\varepsilon = p_{meas} - p_{id}$. Subplots e, f, g and h show corresponding time zooms.

4.4. Prefiltering

Before identification, the data sets are prefiltered using a run-out filter $\mathcal{F}_r$, which filters out the response from surface unsmoothness and mass unbalance response. This is calculated from a data set collected at each operational condition, where the shaft is not excited, which allows generation of the response as function of encoder angle $\mathcal{F}_r(\phi)$. The offset of both inputs and outputs estimated as mean of the first 2000 samples are subtracted from the data sets. A median filter of size 3 is used to smooth out noise from the shaft position measurements.

4.5. Identification

The parameters of both the actuator sub-model and the bearing sub-model are identified using the prediction error method [2], and initial guesses of the parameters are obtained from previous identified models. The model update iterations were stopped when the relative improvement norm was less than 10^{-4} indicating convergence.

Both the actuator model $\mathfrak{M}_a$ and the bearing model $\mathfrak{M}_b$ are then identified from the respective data sets. The actuator sub-models are found to be fairly constant over the investigated range of interest, and a nominal actuator model G_{act} is chosen. The bearing model parameters however vary with both injection pressure and rotational velocity as expected from [3]. The model residuals are expected to be white noise, which is not the case as shown in Fig. 2 (d) and (h). The norm of the residual however is small indicating low importance of the residual dynamics.

Cascading of the bearing model and the actuator model provides the total rotor-gas bearing model: $G_p(s) = G_{bear}(s)G_{act}(s)$.

5. H_∞ Control

A suitable control strategy such as the mixed sensitivity-approach can improve the poor damping characteristics of the rotor-gas bearing. The identified model is used for design of such a controller.

The H_∞ controller $K_\infty(s)$ has been designed using the stacked requirements $||N||_\infty = \max_\omega \bar{\sigma}(N(j\omega)) < 1$;, where $N = [W_pS, W_uK_\infty S]^{\mathrm{T}}$. The controller is designed for a model identified at $P_I = 7bar, \Omega = 4000rpm$. The chosen weights W_p and W_u shown in Fig. 3 ensure an increase in damping without counteracting low frequency disturbances such as changes in equilibrium position due to changing operational condition. The controller obtained using the specified weights is reduced from 24 states to a fourth order controller $K_\infty(s)$ using Gramian-based input/output balancing. The reduction factor in sensitivity towards disturbances is determined from the output sensitivity calculated as:

$$S_o(s) \triangleq (\mathbb{I}_2 + G_p(s)K_\infty(s))^{-1} \tag{11}$$

The reassembled bearing model [7] is used to assess performance of the controller over the range of operation. This model is developed from identified models from data collected over a range of injection pressures ($P_I \in [3;7]bar$) and rotational speeds ($\Omega \in [0;6]krpm$) and describes the rotor-gas bearing behaviour over the wide range of operational conditions. Figure 4 shows the output sensitivity of the closed loop system for different randomly chosen operational points models within the operational range. The sensitivity is reduced in the desired frequency range from $[70 : 190]Hz$ by a factor three to nine, though at the cost of an increased sensitivity of a factor 1.2 at low frequencies, and a peak sensitivity around $280Hz$ of a factor 1.8, which is affordable.

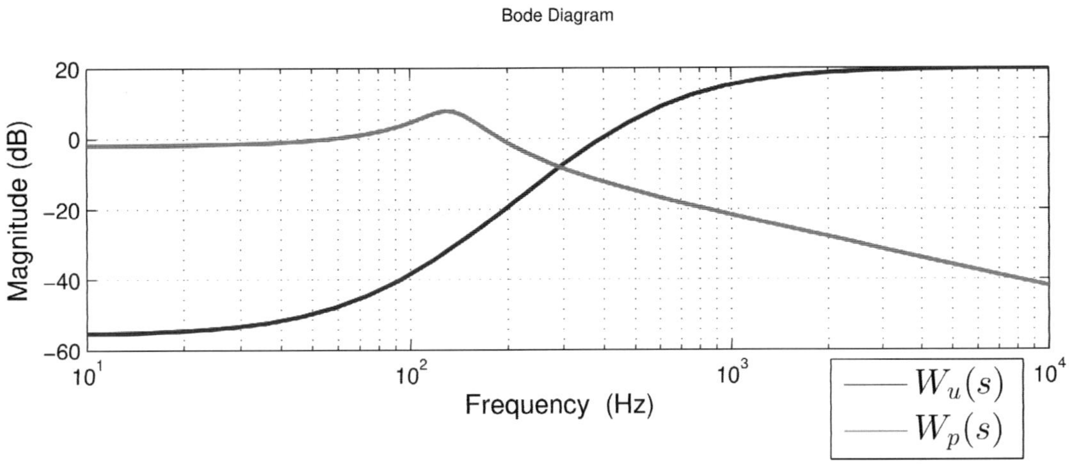

Figure 3. Weights $W_p(s)$ and $W_u(s)$ for mixed sensitivity controller design

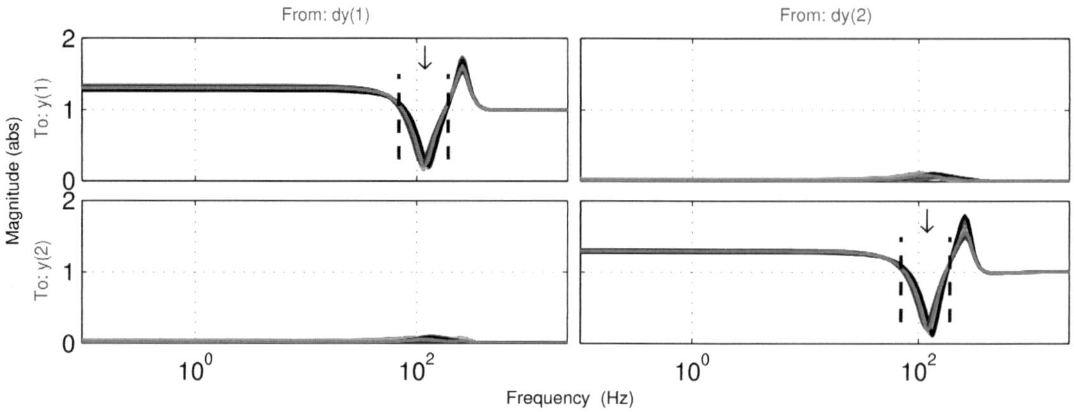

Figure 4. Output sensitivity $S_o(s)$ for $N_r = 30$ realizations of the rotor-gas bearing within the operational. The black arrow and lines mark the frequency interval where the sensitivity is reduced.

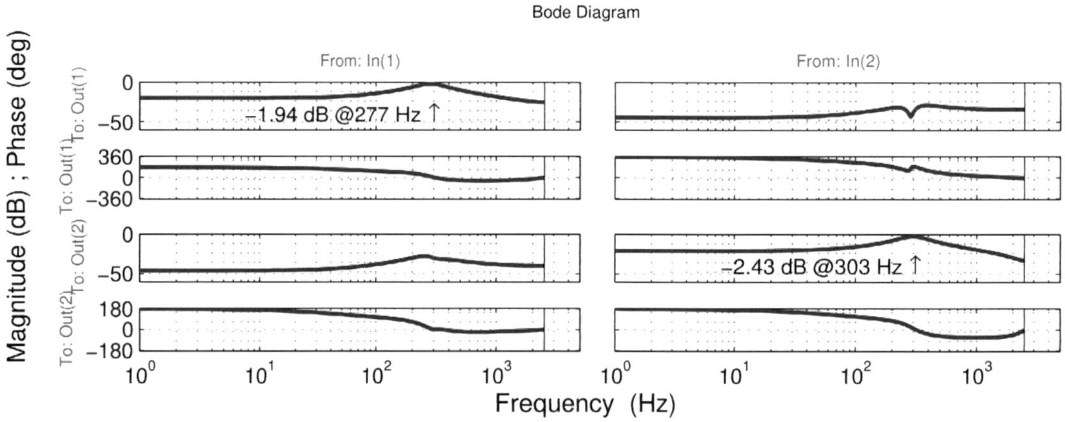

Figure 5. Frequency response of controller $K(z)$

Discretisation of the controller to $K_\infty(z)$ using a Tustin-approximation allows implementation on the rotor-gas bearing test-rig, and the frequency response in Fig. 5 shows the desirable capabilities of the controller: at low frequencies and DC the controller is not active, only in an interval around the critical frequencies is the controller active.

6. Experimental Results

The controller is implemented on the rotor-gas bearing system, and impacts are applied to the rotor-gas bearing both with the controller on and off. Figure 6 shows a horizontal impact response at $\bar{P}_I = 4bar$, $\bar{\Omega} = 0rpm$: using the robust controller, the horizontal damping is found increased from 0.0567 to 0.173 - a factor three. A vertical impact in Fig. 7 shows similar increase in damping. Horizontal and vertical impacts at higher pressure $\bar{P}_I = 7bar$ show a damping increase by a factor six from 0.0282 to 0.1668, which is within the expected range of damping increase predicted by the sensitivity function. Equivalent results can be obtained for non zero rotational velocities.

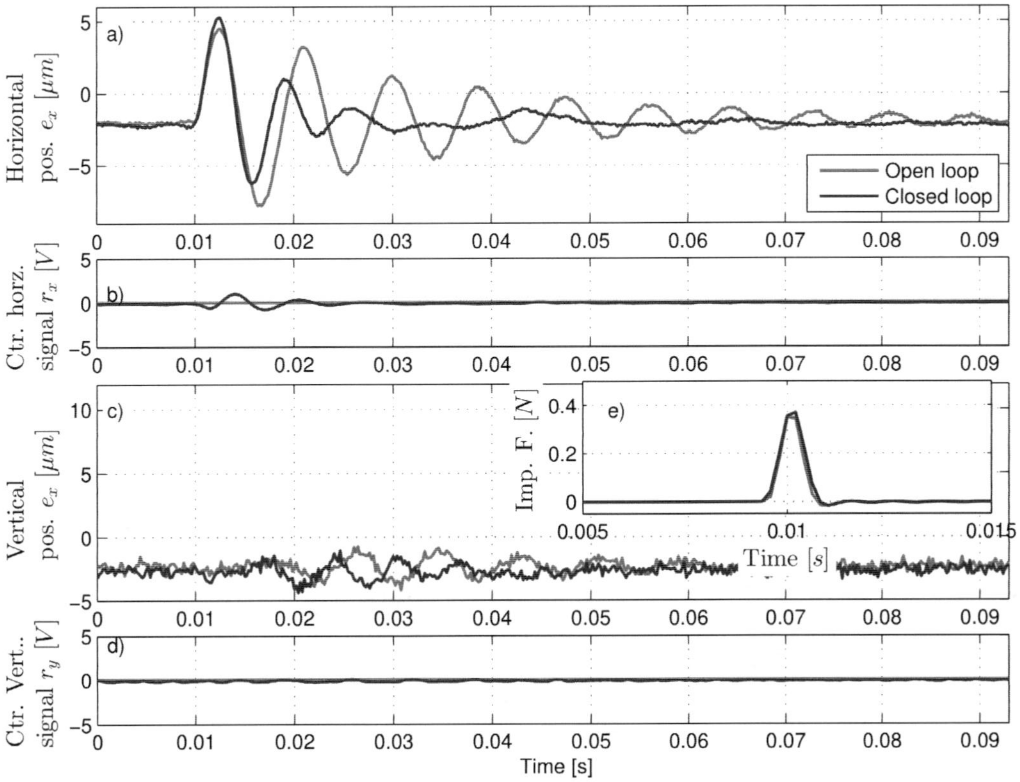

Figure 6. Horizontal impact response without and with the designed controller at$\bar{P}_I = 4bar$, $\bar{\Omega} = 0rpm$. Impact occurs close to time $t = 0.01s$. a and c show measured deflections, b and d show commanded valve positions and e shows the measured impact force.

7. Conclusion

Two Grey-box models were developed modelling the actuators and bearing dynamics for a rotor-gas bearing, and successfully used to identify model parameters describing the relation from commanded to measured valve position, and from valve position to shaft displacement of the rotor-gas bearing. A mixed sensitivity controller was designed to stabilise the rotor-gas, and its increase in damping was validated experimentally for chosen conditions.

Acknowledgments

The Danish Ministry of Science, Innovation and Higher Education is gratefully acknowledged for the support to the FTP research project 12-127502.

References

[1] K. K. Leang and S. Devasia. Iterative feedforward compensation of hysteresis in piezo positioners. *42nd IEEE Conference on Decision and Control, vols 1-6, Proceedings*, pages 2626–2631, 2003.

[2] L. Ljung. *System identification - Theory for the User.* Prentice-Hall, 1999.

[3] S. Morosi. *From Hybrid to Actively-Controlled Gas Lubricated Bearings Theory and Experiment.* PhD thesis, Technical University of Denmark, april 2013.

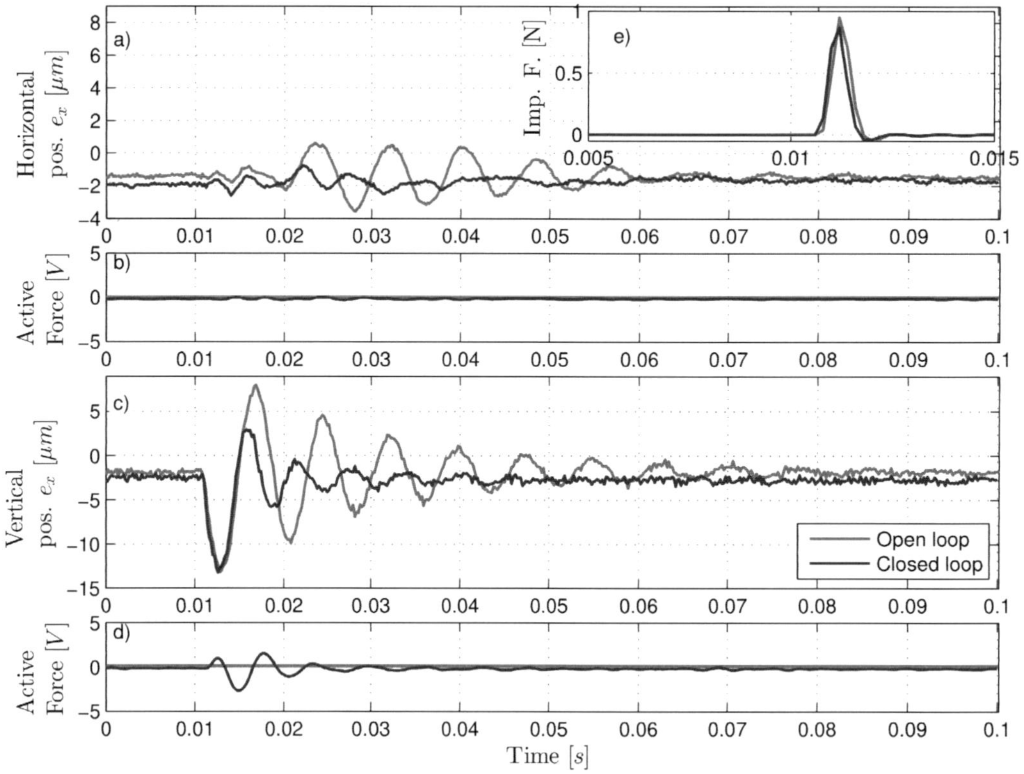

Figure 7. Vertical impact response without and with the designed controller at$\bar{P}_I = 4bar$, $\bar{\Omega} = 0rpm$. Impact occurs close to time $t = 0.01s$. a and c show measured deflections, b and d show commanded valve positions and e shows the measured impact force.

[4] S. Morosi and I. F. Santos. Active lubrication applied to radial gas journal bearings. Part 1: Modeling. *Tribology International*, 44(12):1949–1958, November 2011.

[5] S. Morosi and I. F. Santos. On the modelling of hybrid aerostatic-gas journal bearings. *Proceedings of the Institution of Mechanical Engineers, Part J: Journal of Engineering Tribology*, 225(7):641–653, June 2011.

[6] S. Morosi and I. F. Santos. Experimental investigations of active air bearings. In *ASME Turbo Expo 2012: Turbine Technical Conference and Exposition*. American Society of Mechanical Engineers, 2012.

[7] L. R. S. Theisen, H. Niemann, I. F. Santos, M. Blanke, and R. Galeazzi. Modelling and identification for control of rotor gas bearings. *Submitted to Mechanical Systems and Signal Processing*, 2014.

[8] L. R. S. Theisen, F. G. Pierart Vásquez, H. H. Niemann, I. F. Santos, and M. Blanke. Experimental grey box model identification of an active gas bearing. *Vibration Engineering and Technology of Machinery*, pages 963–976, 2014.

Multi-robot Task Allocation for Search and Rescue Missions

Ahmed Hussein[1], Mohamed Adel[2], Mohamed Bakr[2], Omar M. Shehata[2] and Alaa Khamis[3]

[1] Intelligent Systems Lab (LSI) Research Group, Universidad Carlos III de Madrid (UC3M), Av. de la Universidad 30, 28911, Madrid, Spain
[2] Robotics and Autonomous Systems (RAS) Research Group, German University in Cairo (GUC), 5th settlement - New Cairo, 11432, Cairo, Egypt
[3] Engineering Science Dept., Suez University, Egypt and Vestec. Inc., 125 Northfield Drive West, Waterloo, N2L 6N8, Canada

E-mail: ahussein@ing.uc3m.es; mohamed.adel03@gmail.com; mohamedbakrak47@yahoo.com; eng.omar.mahmood@gmail.com and akhamis@pami.uwaterloo.ca

Abstract. Many researchers from academia and industry are attracted to investigate how to design and develop robust versatile multi-robot systems by solving a number of challenging and complex problems such as task allocation, group formation, self-organization and much more. In this study, the problem of multi-robot task allocation (MRTA) is tackled. MRTA is the problem of optimally allocating a set of tasks to a group of robots to optimize the overall system performance while being subjected to a set of constraints. A generic market-based approach is proposed in this paper to solve this problem. The efficacy of the proposed approach is quantitatively evaluated through simulation and real experimentation using heterogeneous Khepera-III mobile robots. The results from both simulation and experimentation indicate the high performance of the proposed algorithms and their applicability in search and rescue missions.

1. Introduction

Extensive research nowadays is focusing on multi-robot systems (MRS). These systems offer many advantages as they have high potential to solve a multitude of problems such as industrial warehouses, surveillance and in search and rescue missions. MRS provide high reliability and performance of complex tasks even though they are simple in design [1]. It should also be noted that MRS possess the potential of producing poor results and/or non-deterministic performance if the design did not take into consideration the interaction and interference between the robots [2].

One of the main problems in MRS is task allocation which is the process of assigning tasks to the robots. It has to be done in order to avoid clashes between multiple robots and to increase the overall performance of the system. Therefore the allocation of the tasks to the proper robots strongly affects the performance of the system [3]. This problem is known as Multi-Robot Task Allocation (MRTA) problem.

In this paper, generic approaches are proposed to solve large-scale and heavily constrained MRTA problem. These approaches generate a solution in which the given robots are efficiently allocated to the given tasks to maximize the overall performance and minimize the total cost of the system. It takes into consideration the real world constraints of the system, the requirements of the tasks and the capabilities of the robots. Moreover, path-planning algorithms are implemented to enable the robots finding collision-free path from a start position to a given goal position, avoiding a collection of obstacles in any given environment.

The proposed algorithms are simulated for a search and rescue scenario with three heterogeneous Khepera-III mobile and five survivors distributed in different locations within the simulation field. The robots have different capabilities, one is equipped with a gripper, second is equipped with a camera and the third equipped with a laser range finder. The survivors have different requirements according to the needed capabilities of the robots. Thereafter, the algorithms are tested with the real Khepera-III mobile robots and experimented in an indoor arena and the survivors are represented by labeled objects. The results from both simulation and experimentation closely related which validate the performance of the proposed algorithms.

The remainder of this paper is organized as follows: Section 2 introduces the MRTA problem formulation followed by describing the proposed algorithms in section 3. Section 4 demonstrates the search and rescue scenario, simulation and real experimental work, followed by discussing the results in section 5. Finally, conclusion and future work are summarized in Section 6.

2. Multi-Robot Task Allocation Problem

MRTA problem addresses the question of assigning the task-to-robot in order to achieve the overall system goals [4][5]. The MRTA problem is a dynamic decision problem that varies in time with phenomena including environmental changes, the problem should be solved iteratively over time [6]. Thus, the problem of task allocation becomes more complex to tackle.

2.1. Formulation

Since the main objective is to solve the task allocation problem of MRS in real world applications. Therefore, through the different phases of the development of the proposed approaches, the central target was to introduce a generic approach that is capable of solving various MRTA problems of different features and challenges. This goal had to be taken into consideration during the formulation of the problem and therefore the use of the multiple traveling salesmen problem (mTSP) formulation after being extended and adapted to the MRTA problem requirements [2], see Figure 1.

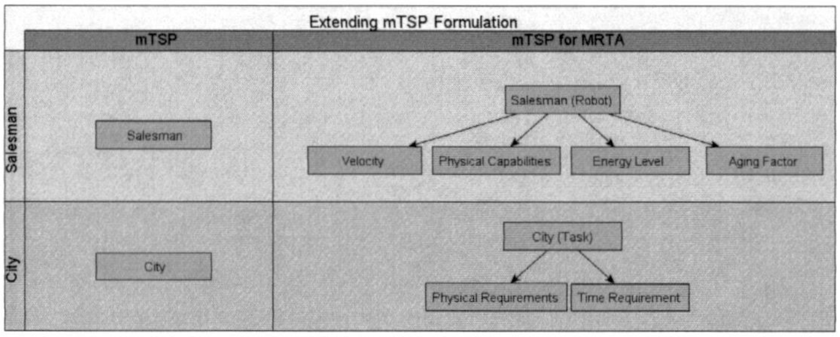

Figure 1. Extended mTSP formulation for MRTA Problem

3. Proposed Approach

MRTA approaches can be categorized as market-based approaches and optimization-based approaches. In the former, the problem is handled in the context of auctioning and in the later, the optimal is solved as an optimal assignment problem.A comparative study between these two approaches is presented in

[7]. In this paper, market-based approach is adopted due to its capability to handle complex heavily constrained MRTA problems.

3.1. Market-based approach with relinquishing and optimization

The market-based approach gained a considerable attention within the robotics research community because of several desirable features, such as the efficiency in satisfying the objective function, robustness and scalability [5]. Market-based multi-agent coordination approaches share a set of fundamentals. Auction theory provides detailed definitions for many of these fundamentals [8],[9]. The market-based approach is based on auctions, which consists of bidders and goods available for the bidding process. In the MRTA problem, the robots are considered as the bidders and the tasks are considered as the goods. The auction takes place and a centralized agent makes the final evaluation and the winner determination technique. The proposed approach is an combination of the Single-good Auctions and Contract Net Protocol (CNP) [2].

The proposed approach starts as follow:

- First step is the announcement stage, the centralized agent set a list of all the available tasks and define all their requirements in terms of skills and execution time.
- Second step is bidding stage, the robots start bidding on the available tasks taking into consideration the matching capabilities to the requirements of the tasks.
- Third step is the selection stage, the centralized agent evaluates all the bids and start the winning determination strategy.
- Fourth step is the assigning stage, the centralized agent announces the winner by assigning a robot for each task.

As described in [2], the MRTA problem is formulated as an instance of the mTSP. The the same objective function of the mTSP cannot be straightforwardly used as the objective function for the MRTA problem. Hence some variations are introduced to the objective function of the mTSP in order to be effectively used for the MRTA problem. For k sub-tours and t tasks in each sub-tour, the total traveling time is calculated as follows:

$$A = \frac{\sum_{i=1}^{t-1} \text{distance between}(\, subtour_{j_i}, subtour_{j_{i+1}}\,)}{\text{Velocity of Robot}_j}$$

$$B = \frac{\sum_{i=1}^{t} \text{task execution time}\ (\, subtour_{j_i}\,)}{\text{Aging factor of Robot}_j}$$

$$\text{f}(\mathbf{x}) = \underset{j \in \{1,2,\ldots k\}}{\arg\max}\ (A + B) \tag{1}$$

It is not necessary to have a task for each robot, but it is a must to have a robot for each task. To reach near optimal solutions, relinquishing process along with simple optimization technique were implemented as enhancements to the market-based approach.

The relinquishing process is an iterative process, where at the end of each iteration, each robot waives a random task from its assigned tasks to be available again in the tasks list. This ensure some diversification in the solution, because after relinquishing a random task the robot solution (sub-tour) is changed in which in the next iteration it may give a better bidding value to a different task therefore diversification is achieved. Additionally, in the end of the algorithm, and having all robots with assigned tasks from the market-based approach, an optimization technique is applied to the sub-tour considering the robot and the tasks in the sub-tour as a single traveling salesman problem. According to previous studies in optimization techniques [10], simulated annealing is selected to be the optimization technique for the sub-tours

Algorithm 1 is the proposed algorithm used to solve the MRTA problem in this work using the market-based approach including the relinquishing process and the optimization enhancements [2].

Algorithm 1: Market-based approach with relinquishing and optimization

```
Input: Tasks list tasks, Robots list robots, Distances between tasks distances, Number of
       Iterations n
Output: Best allocation bestAlloc
Best cost bestCost
Available tasks availTasks
Bidding list biddingList
Minimum bid minBid
Current allocation curAlloc
Current cost curCost
User define percentage r_o
availTasks ← tasks
for i ← 1 to n do
    while isEmpty(availTasks) do
        biddingList ← getBids(robots, availTasks, distances)
        minBid ← getMin(biddingList)
        curAlloc ← subTour(robots, availTasks, minBid)
    end
    availTasks ← relinquish(curAlloc, availTasks) // r ← getRandomNumber(1)
    if curCost < bestCost then
        bestAlloc ← curAlloc
    else
        if r_o > r then
            bestAlloc ← curAlloc
        end
    end
end
optiAlloc ← optimize(bestAlloc) //
```

3.2. Path Planning Algorithm

The market-based approach generate the allocation of the tasks to the robots. Then for the robots to navigate from their current position to the task position, path planning algorithm is required. There are different approaches to determine the path which the mobile robots should take to navigate to the desired location given the complete map of the obstacles and the starting position. In this work, Dijkstra and A* algorithms are implemented to generate the optimal path.

3.2.1. Dijkstra Algorithm Dijkstras algorithm is a graph search algorithm that solves the single-source shortest path problem for a fully connected graph with non-negative edge path costs, producing a shortest path tree. This algorithm discretizes the terrain map to blocks, in which each block is a node. Each node has 4 edges and 4 corners. The edges and corners are used to define relations between the nodes and the cost to move form one node to the other. The cost can be represented by the function $l(x,u)$, where its costs $l(x,u)$ to move from node x by applying the action u. Thus, the summation of the edges and corner costs along the path from the initial node to the goal node results the total cost of the plan. The mathematical definition can be calculated as follows:

$$l(x,u) = \begin{cases} 0, & \text{if } x \text{ is equal to current node or there is no common edge or corner} \\ 1, & \text{if } x \text{ has a common node or edge} \\ \infty, & \text{if } x \text{ contains an obstacle} \end{cases} \tag{2}$$

3.2.2. A Algorithm* A* is a special case of best-first informed search. Its main advantage over Dijkstra algorithm is that it uses the knowledge known about the problem in form of a heuristic function to prune the search.The mathematical definition can be calculated as follows:

$$f(x) = g(x) + h(x) \tag{3}$$

$g(x)$ denotes the cost-to-go from the starting state to the successor state. On the other hand, $h(x)$ denotes the cost-to-go from this specific successor state to the goal state. The total cost $f(x)$ is computed for each successor node, and the node with the least $f(x)$ is chosen as the next successor. Thus, by choosing the most appropriate successor node, the optimal path to the goal can always be guaranteed. The costs $g(x)$ and $h(x)$ are the distances between the nodes, which can be simply determined by calculating the straight distances between the nodes.

3.2.3. Trajectory Planning After planing the path it is essential to smooth the sharp edges of the path and generate the velocity states along the path from the smoothed path as follows:

- Spline are ordinary cubic parametric polynomial functions. Since there are four unknown coefficients, four conditions must be defined in order to define the shape of the spline in order to avoid large curves which will lead to collision with obstacles. The splines used are the *Cubic Bezier Splines* [11] which have the following equation.

$$S(t) = (-P0 + 3P1 - 3P2 + P3)\tau^3 + (3P0 - 6P1 + 3P2)\tau^2 + (-3P0 + 3P1)\tau + P0 \tag{4}$$

- Velocity generation is the process of linking the trajectories created by the splines to the time parameter in order to calculate the velocities and the acceleration of the robot. A velocity limiter is defined which is based on the actual maximum value for velocity, where V_{transl} is the transnational velocity, and V_{rot} is the rotational velocity.

$$V_{max} = V_{transl} + V_{rot} \tag{5}$$

From the kinematics of the differential drive model, the transnational velocity can be calculated according to the intrinsic parameters of the robot, where L and R are the left and right wheels' velocities respectively.

$$V_{transl} = \frac{V_{max}}{1 + \frac{L}{R}} \tag{6}$$

4. Experimental Work

In order to validate the proposed algorithms, a set of MRTA problem scenarios are tested. In each scenario, the algorithm is applied and the results are reported for further evaluation through the evaluation metrics. All scenarios have three main inputs, namely, the robots, the tasks and the distance matrix representing the tasks' location. The expected output for solving the MRTA problem is the best allocation that maximizes the performance of the system. On another level, path-planning algorithms are tested to generate the near optimal path from the start to the desired goal locations, given the map with the locations of all robots and tasks.

4.1. Evaluation Scenario

Since one of the main sources of complexity of MRS applications is the size of the robot team as well as the number of tasks to be executed, it was essential to test qualitatively and quantitatively the proposed algorithms for their capability of handling complex MRTA problems. Figure 2 presents the adopted search and rescue scenario for experiments in this study.

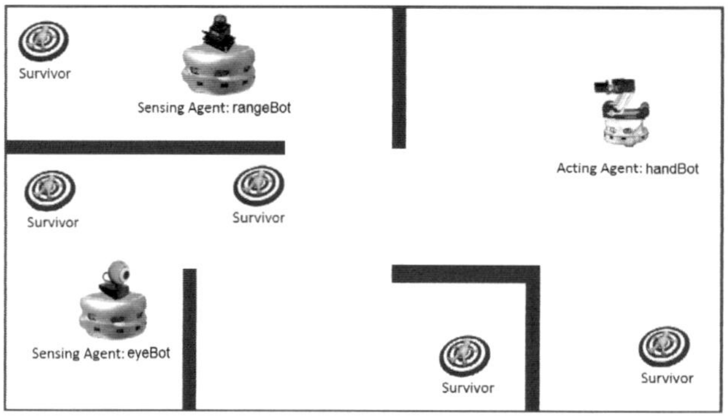

Figure 2. MRS Search and Rescue Scenario

In this proposed search and rescue scenario, five survivors (from some human-made or natural-made disaster for example) are located in the search area, these survivors have different requirements. Three robots are used in the search area to accomplish the search and rescue mission. Each robot has unique sensing and acting capabilities. In this study, the robots used are modeled by the Khepera-III mobile robots along with their peripherals. The first robot is equipped with a laser rangefinder sensor and denoted the name (rangeBot), the second robot is equipped with a gripper and denoted the name (handBot) and the third and last robot is equipped with a camera and denoted the name (eyeBot). The presented scenario is modeled in order to be able to find the best allocation of each task (survivor in this case) to the available robots. The model representation of this problem is illustrated in Figure (3).

Name	Requirements	Time Required [min]
Survivor 1	G	0.17
Survivor 2	G	0.17
Survivor 3	C	0.03
Survivor 4	LR	0.08
Survivor 5	C	0.03

Name	Skills	Velocity [m/min]	Battery Life [min]	Aging Factor
handBot	G	10.8	12	0.87
eyeBot	C	11.1	12	0.92
rangeBot	LR	8.6	12	0.81

G	Gripper
C	Camera
LR	Laser Range Finder

Figure 3. Search and Rescue Scenario Database Model

4.2. Evaluation Metrics

Evaluating the performance of mobile robots and assessing their behavior in different applications is still an open research field [12]. Although substantial progress has been made in defining the framework and standards for the evaluation process. These frameworks are usually defined according the priorities of the mission at hand. Comparisons are made between contrasting values such as: accuracy at goal versus the time taken to reach it and the length of the trajectory versus the risk of collision.

- One of the most important metrics used is the error between the reference trajectory and the actual trajectory. This shows how accurately the robots can track a given path. The trajectory is generated by the pathfinder algorithm discussed in 3.2. Instead of evaluating each state in the pose of the robot $[x, y, \theta]^T$ independently, the absolute euclidean distance defined in equation (7)was chosen as an error metric.

As a secondary part in evaluating the performance, the accuracy of the target localization process carried out by the rangeBot is considered, also the absolute Euclidean distance is used.

$$error = \sum_{i=1}^{n} \left(\sqrt{(x_{i,ref} - x_i)^2 + (y_{i,ref} - y)^2} \right) \tag{7}$$

Where (n) is the total number of points generated by the pathfinder. While $(x_{i,ref}, y_{i,ref})$ are the reference x and y positions at a given sample (i), and (x_i, y_i) are the actual positions of the robot calculated through odometery.

- The final performance index used is the risk factor with respect to the obstacle. Given a function $\beta(i)$ which calculates the Euclidean distance to the nearest obstacle at sample (i). The risk factor of the complete trajectory is given by the following equation (8):

$$RiskFactor = \sum_{i=1}^{n} \frac{1}{\beta(i)} \tag{8}$$

4.3. Experimental Setup

The proposed task allocation approach is implemented using Java. Java is used in the implementation as well as in applying the algorithm over the proposed scenario and in calculating the results of the evaluation metrics. Moreover, the experiments are simulated using a free software called Virtual Robotics Experimental Platform (V-rep) [13]. This simulation environment (Figure (4-a)) includes all the modules of the Khepera-III mobile robots that are used practically to test the approach. Setting up an experiment requires understanding of both the practical and the simulation models and identifying the communication method between agents in the MRS. Three robots with different capabilities have been configured, rangeBot is a mobile robot that mimics Khepera-III mobile robot with laser-range finder module, handBot mimics a model with gripper module and eyeBot mimics a model with camera module.

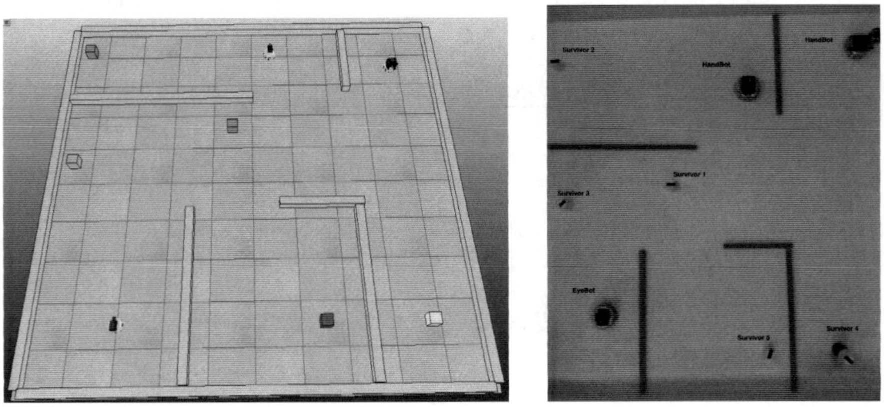

(a) v-rep Simulation - Initial Positions

(b) Khepera-III Mobile Robots Initial Positions

Figure 4. Virtual and actual arena.

The real experiments are conducted in a 2x2 m^2 indoor arena. The real robots are equipped with one of the peripheral modules, namely, laser rangefinder, gripper and camera. The laser rangefinder has a scan area of 240° with a maximum range of 4m. The sensor has an angular resolution of 0.36^o [14]. The measurements are used to determine the location of the target using *SPIKE* algorithm. The gripper can carry an object with maximum size of 50mm and maximum weight of 50gm. The camera is capable of capturing images with a resolution up to 640x480 pixels and acquisition speed up to 15 frames per second. The obstacles and survivors are placed in predefined locations similar to the simulation, as shown in Figure (4-b). The workstation used in the experiments runs a Windows operating system with a 1.7GHz processor and 6GB RAM.

5. Results and Discussion

In this study, the five survivors previously mentioned are modeled as five tasks where each task has both physical and time requirements. The robots are also modeled with their physical skills, operating velocities, their current energy level as well as their efficiency. Next step is to handover the developed problem model to the proposed algorithm to perform the required computations. The output of the algorithm is summarized in Figure (5), which represents the optimal allocation of the tasks to the robots based upon their requirements, skills as well as their initial conditions.

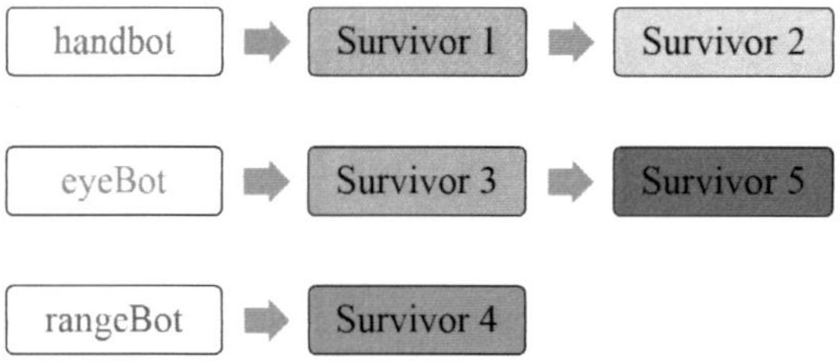

Figure 5. Search and Rescue Allocation

As can be seen in Figure (5), survivors 1 and 2 that required a gripper are assigned to the handBot robot, while survivors 3 and 5 that required a camera are assigned to the eyeBot robot, and finally survivor 4 that required a laser rangefinder is assigned to the rangeBot robot. The time chart in Figure (6) reflects the time taken by each robot for both traveling time and execution to finish its assigned tasks. Moreover, it shows the total time taken for the whole mission to be accomplished, which will be the maximum time of the three robots since the robots will be working in parallel.

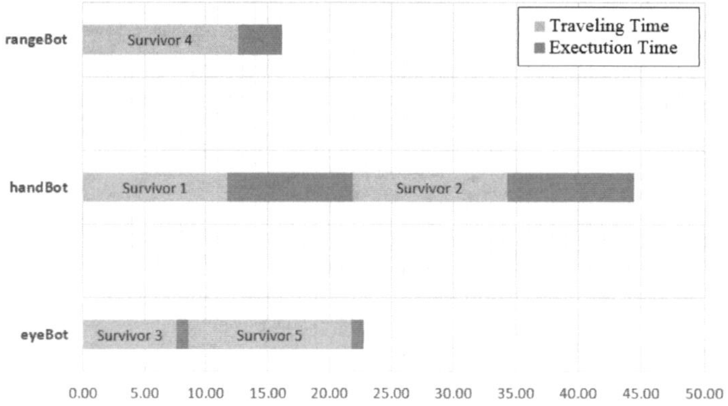

Figure 6. Search and Rescue Allocation

This scenario was implemented on the V-rep simulation software [13], as can be seen in Figure (4-a). Both market-based algorithm and path planning algorithm are implemented for each of the three robots to cooperate in solving this MRTA problem. The total simulation time was 1 minute and 23 seconds, in which the robots completed their assigned tasks to rescue their assigned survivors (represented by cube shapes in the simulation).

On another level, real life experiments are conducted using the three Khepera-III mobile robots (using the same initial conditions as the V-rep simulations), as can be seen in Figure (4-b). Robots are equipped with the designated peripheral modules; the laser rangefinder is installed on the rangeBot, camera is installed on the eyeBot and the gripper is installed on the handBot. Moreover, the survivors are modeled with white plastic cylinders with black line on top. The output from the market-based algorithm in terms of the best allocation of the tasks to the robots was integrated with the robots' code, along with the path planning algorithm. Thus, the robots can navigate through the map and locate the survivors, execute the assigned task and continue till all assigned tasks are completed.

The retrieved data from each robot after conducting the experiment include the actual trajectory evaluated from the on-board sensors using odometry, as well as the time consumed to navigate to the task location and the time consumed to finish it.

As an example for the obtained results, Figure (7) shows the rangeBot actual trajectory plotted against the reference trajectory in a two dimensional grid representing the actual map. The results of the *SPIKE* target localization algorithm showing the localized survivor.

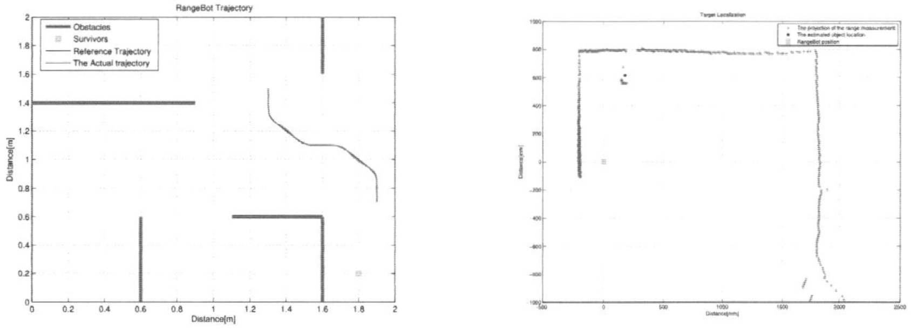

(a) Trajectory of the rangeBot. (b) Survivor Localization using the rangeBot

Figure 7. RangeBot results

The performance of the robots is evaluated according to the metrics defined in 4.2. The first index is the error between the final position and the desired goal. In which all three robots managed to be autonomously localized within the vicinity of the desired goal with maximum absolute error of $0.0065m$. Figure 5 shows both absolute mean and maximum errors along the whole trajectory from the starting position to the goal position. The risk factor value which is the minimum absolute distance between the robot and the obstacles is reflected in Figure (5). The error bars values reflect the variances in the measured values for repeating the experiment five times.

The results prove the ability of the proposed allocation and planning algorithm in managing a set of heterogeneous mobile robots in search and rescue scenario.

6. Conclusion and Future Work

This paper presented market-based approach to solve multi-robot task allocation (MRTA) problem. The developed algorithm is responsible for optimally assign a set of mobile robot to a set of tasks. Moreover, a path-planning and landmark extraction algorithms are adopted to enable the robots to navigate through any given map and to be able to localize the target in any of the cases. A search and rescue scenario has been used to qualitatively and quantitatively evaluate the performance of the allocation algorithm and its ability to handle capabilities matching constraints in heterogeneous environments. Both simulation and

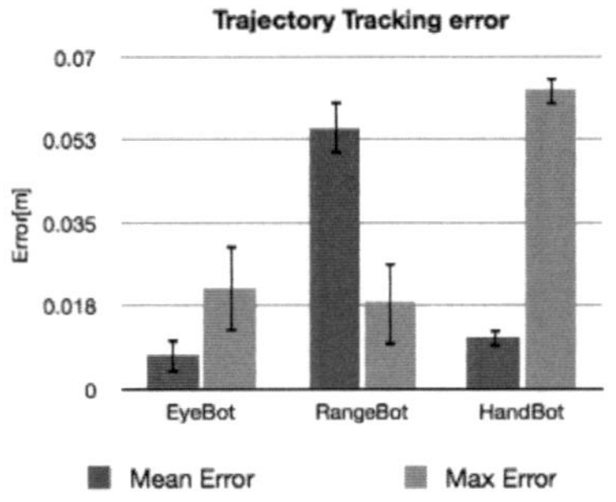

(a) Absolute error in distance between reference and actual trajectories

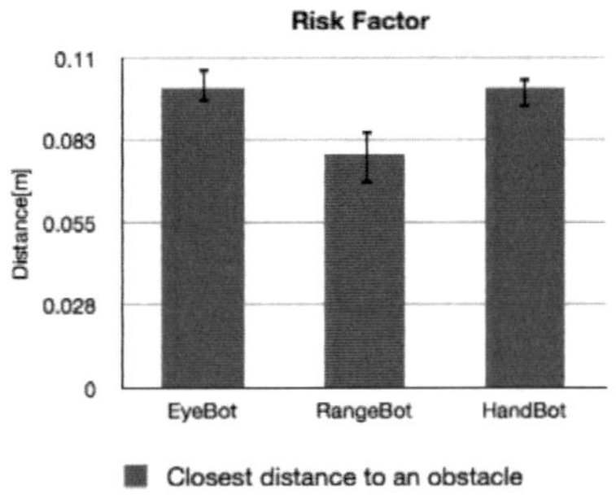

(b) Risk factor evaluation.

Figure 8. Evaluation metrics

experimental results reflect the efficacy of the proposed approach. Further investigation can be conducted for other aspects such as handling robot failure, alleviating communication burden, handling task-related constraints. Also, more examples can be investigated such as tight tasks that cannot be decomposed into single robot tasks or tasks with precedence constraints.

References

[1] L. Parker, "Multiple mobile robot systems," *Springer Handbook of Robotics*, pp. 921–941, 2008.
[2] A. Hussein and A. Khamis, "Market-based approach to multi-robot task allocation," *International Conference on Individual and Collective Behaviors in Robotics (ICBR)*, 2013.
[3] Y. Han, D. Li, J. Chen, X. Yang, Y. Hu, and G. Zhang, "A multi-robots task allocation algorithm based on relevance and ability with group collaboration," *International Journal of Intelligent Engineering and Systems*, vol. 3, no. 2, 2010.
[4] K. Lerman, C. Jones, A. Galstyan, and M. Mataric, "Analysis of dynamic task allocation in multi-robot systems," *The International Journal of Robotics Research*, pp. 225–241, 2006.
[5] F. Tang and L. Parker, "A complete methodology for generating multi-robot task solutions using asymtre-d and market-based task allocation," *IEEE International Conference on Robotics and Automation*, 2007.
[6] B. Gerkey and M. Mataric, "A framework for studying multi-robot task allocation," 2003.
[7] M. Badreldin, A. Hussein, and A. Khamis, "A comparative study between optimization and market-based approaches to multi-robot task allocation," *Advances in Artificial Intelligence*, vol. 2013, 2013.
[8] P. Klemperer, *(2004). Auctions: Theory and Practice.* Princeton University Press, 2004.
[9] V. Krishna, *Auction theory.* Elsevier, 2002.
[10] M. Badreldin, "A metaheuristic approach to multi-robot task allocation," Master Thesis, German University in Cairo, 2013.
[11] G. J. Yang and B. W. Choi, "Smooth trajectory planning along bezier curve for mobile robots with velocity constraints," *International Journal of Control and Automation*, vol. 6, no. 2, 2013.
[12] D. Calisi and D. Nardi, "Performance evaluation of pure-motion tasks for mobile robots with respect to world models performance evaluation of pure-motion tasks for mobile robots with respect to world models."
[13] "Virtual robotics experimental platform (v-rep), http://www.coppeliarobotics.com/." *Coppelia Robotics V-rep: Create. Compose. Simulate. Any Robot*, Web: 18 Aug. 2014.
[14] F. Lambercy and J. Tharin, "Khepera-iii laser range finder user manual," *K-Team S.A.*, vol. 1st, 2011.

Journal of Physics: Conference Series **570** (2014) 062001 doi:10.1088/1742-6596/570/6/062001

Current transformer model with hysteresis for improving the protection response in electrical transmission systems

Robert Matussek[1], Cezary Dzienis[1], Jörg Blumschein[1] and Horst Schulte[2]

[1] Siemens AG, IC SG Energy Automation, Berlin, Germany

[2] HTW Berlin, Dept. of Engineering 1, Control Engineering, Berlin, Germany

E-mail: robert.matussek@siemens.com

Abstract. In this paper, a generic enhanced protection current transformer (CT) model with saturation effects and transient behavior is presented. The model is used for the purpose of analysis and design of power system protection algorithms. Three major classes of protection CT have been modeled which all take into account the nonlinear inductance with remanence effects. The transient short-circuit currents in power systems are simulated under CT saturation condition. The response of a common power system protection algorithm with respect to robustness to nominal parameter variations and sensitivity against maloperation is demonstrated by simulation studies.

1. Introduction

This paper discusses current transformers (CTs) as an example of time-domain simulation of a nonlinear inductor. An overview of the basic principles of a conventional CT is given in section 2. To gain a deeper insight into the nonlinear and time-variant behavior a generalized steady-state analysis is carried out as well. The state-of-the-art methods and models of a nonlinear inductive element including hysteresis will be discussed in section 3. Different hysteresis models are referenced focussing on the practicability for a transient CT simulation. Especially the Preisach model of hysteresis is presented and several features are shown. Section 4 covers the CT differential equation in combination with a hysteresis model. Further requirements to the hysteresis model are pointed out. In section 5 the CT model is used to produce test cases for protection device test. The robustness of the particular power system protection algorithms is evaluated under the influence of CT saturation.

2. Principles of conventional CTs

Current transformers (CTs) are widely used in power plants, substations and in high-voltage as well as in medium-voltage industrial applications. Their purpose is to measure large load currents and also short-circuit currents with well-defined accuracy. Primary currents in the range of a few kiloamperes are transformed into measurable quantities of a few amperes. According to the winding ratio a primary current i_1 is transformed into a secondary current i_2 considering the current error:

$$i_\mu = w_1 \cdot i_1 - w_2 \cdot i_2 \tag{1}$$

Figure 1 shows the principle of a transformer. The primary and the secondary circuit are coupled via a ferromagnetic core. In order to achieve a realistic model of the dynamic behavior it is not sufficient to use a static characteristic between the magnetic flux density B and the magnetic field strength H. The utilization of a hysteresis model can yield additional effects like remanence and dynamical phase lag.

It is known that the CT accuracy is influenced by its physical dimensions:

- cross-section of the iron core A_{Fe}
- mean iron core length l_{Fe}
- width of the gap l_{Gap}

The maximum average permeability $\overline{\mu}$ can be accomplished with a closed iron core. Unlike split cores, a closed profile causes a high remanent flux. Commonly used core materials are silicon-based alloys, e.g. $FeSi$ 5 %. CTs with a closed iron core, such as X, P, PX, PS, TPS, TPX can be summarized in a common class referred to as TPX. All TPX CTs operate with very high accuracy within the linear range below the saturation region. The current error ε_i as well as the phase displacement δ_i [9, 10] are small (see figure 4). The dynamic operation at a higher level can drive the transformer into a remanence point. The remanent flux density B_R can take values close to 100 % of the saturation flux density B_S as can be seen in figure 2. According to equation 2 the dynamical magnetizing inductance L_M takes values from close to 0 in the saturation region up to a few $Henry$ in the linear region. The inductance L_M can be modeled either from the primary or from the secondary side. For the following considerations the inductance is placed on the secondary side. Commonly there is only one primary winding, so that w_1 equals 1 and the current transormation only depends on w_2. For reasons of completeness w_1 is respected anyway in the following equations. The quantity of L_M considerably depends on the CT class and the magnetic field strength H (operating point) and the corresponding gradient $\mu = {}^{dB}/_{dH}$.

$$L_M(i_\mu) = \frac{u_2}{{}^{di_\mu}/_{dt}} = \frac{w_2^2 \cdot A_{Fe} \cdot \mu(i_\mu)}{l_{Fe}} \tag{2}$$

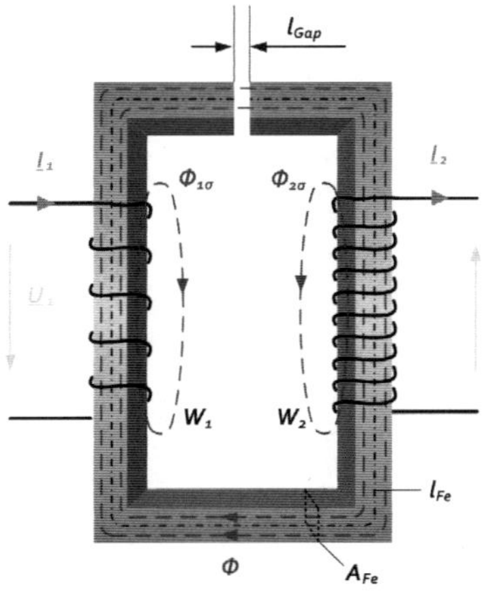

Figure 1. Magnetic coupled circuits by means of an iron core

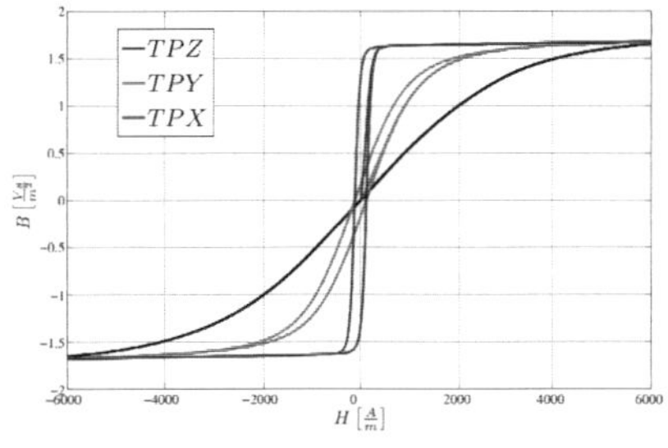

Figure 2. Hysteresis loops of the three CT classes TPX, TPY and TPZ

Inserting a gap into the iron core mainly causes two effects:

(i) The magnetizing characteristic will be linearized. The characteristic is linear even for great values of the magnetic field strength H.

(ii) The hysteresis loop will be tightened and thus the remanent flux density B_R will be decreased.

For this reason the TPY class of a CT is called *linearized core*. The maximum remanent flux is limited to 10 %. The TPZ class has a large gap and is therefore called *linear core*. It shows a negligible remanent flux. The gradient $\Delta B/\Delta H$ which is depending on the operating point in the B–H–plane is represented by dynamical magnetizing inductance in the equivalent circuit in figure 3. The power system can be modeled as an ideal current source due to the fact that there is no feedback from the CT to the power system. The primary losses in terms of copper loss R_{1v} and leakage field loss $L_{1\sigma}$ therefore become invalid. In case of static or digital relays the resistive part of the burden is dominant and the inductive part can be neglected. The internal resistance R_{CT} is commonly given by the manufacturer. It influences the operational overcurrent factor K_{scco} in case of a connected burden $\underline{Z}_b \approx R_b$ if it differs from the nominal burden $\underline{Z}_{bn}$. Figure 4 depicts current error and phase displacement. This phasor diagram is only valid for stationary, sinusoidal quantities. Also, it is assumed that the error current $\underline{I}_0$ shows these properties. In that case, the error current could be split into an active component and reactive component. However, this assumption cannot hold due to the fact that the magnetizing inductance is nonlinear. In the following, a hysteresis model will be used to simulate the magnetizing inductance in the whole operating range. By exploiting the nonlinear magnetizing current i_μ in combination with the dynamic operation of the hysteresis loop, the subdivision into active and reactive loss components is no longer useful. Furthermore, the dynamic behavior of a CT can be described by the secondary time constant:

$$T_s = \frac{\sum_{sec.} L}{\sum_{sec.} R} \approx \frac{L_M}{R_{CT} + R_b} \tag{3}$$

For small time constants the flux in the core can be reduced in short time. The primary current shows an exponentially decaying DC component. This component is damped by the CT, whereby the filter effect is enhanced by a small time constant (particularly TPZ class).

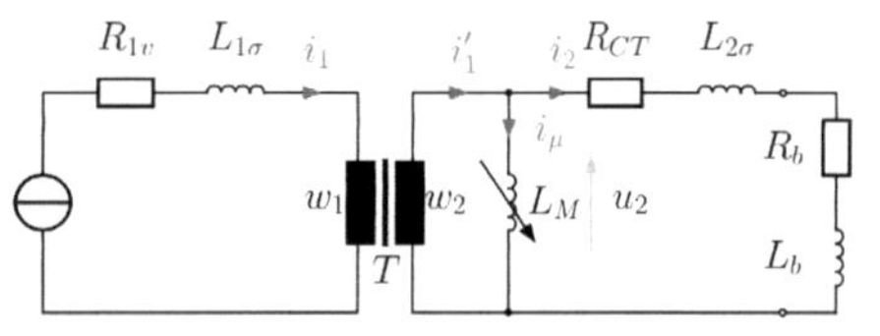

Figure 3. Equivalent circuit of a conventional CT

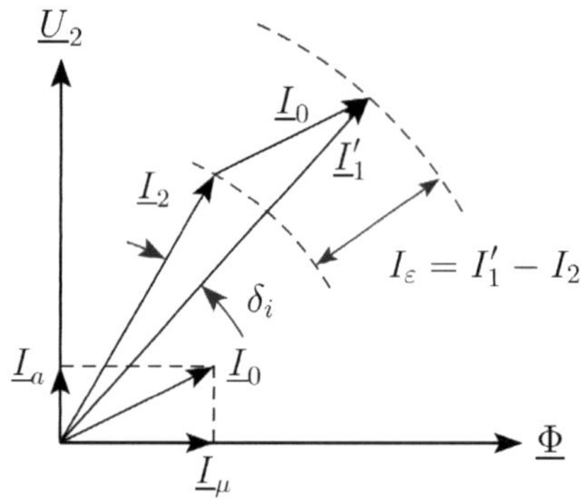

Figure 4. Stationary phasor diagram of a CT

3. Methods of modeling magnetic hysteresis

Different approaches to modeling magnetic hysteresis have been published and the most common are the Jiles-Atherton model [11] and the Chan-Model [2].

3.1. Jiles-Atherton model

The Jiles-Atherton model based on a first-order differential equation is closely related to Curie-Weiss theory of magnetism. It was first introduced in 1983 and it splits the magnetization M into reversible and irreversible portions. it has been proved to work well for stationary input quantities, but shows disadvantages under asymmetrical excitation as for offset minor loops. Even though approaches are available to improve the characteristic [3], still the model is less flexible concerning parametrization.

3.2. Chan model

The newer Chan model is used in *LTSpice IV*® software, a popular circuit simulator implementing the SPICE kernel. The model computes the Induction B using a simple algebraic equation with only three shape parameters. Asymmetrical input quantities lead to particular instabilities and to non-physical behavior. This disadvantage is fixed in *LTSpice IV* but the method is patented by the *Linear Technology Corporation* [5].

Journal of Physics: Conference Series **570** (2014) 062001 doi:10.1088/1742-6596/570/6/062001

3.3. Preisach Model

The scalar Preisach model [14] is based on the integration over an area S, which consists of a positive region S^+ and negative region S^-.

The Preisach density function $p(\alpha, \beta)$ must satisfy the condition

$$p(\alpha, \beta) = p(-\alpha, -\beta) \quad \forall \quad (\alpha, \beta) \in S \tag{4}$$

as well as

$$p(\alpha, \beta) \geq 0. \tag{5}$$

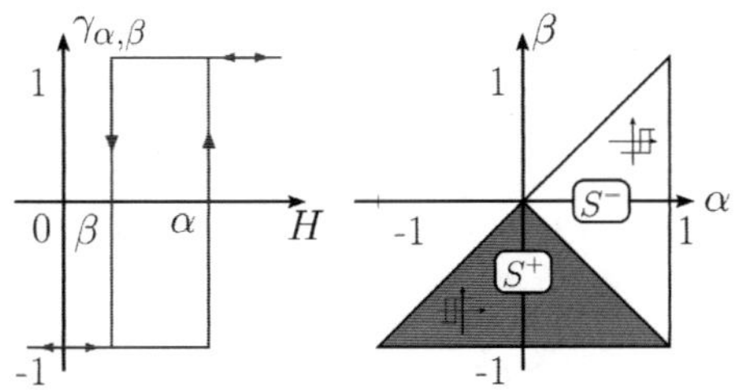

Figure 5. Fundamental hysteresis operator (left) and Preisach plane (right)

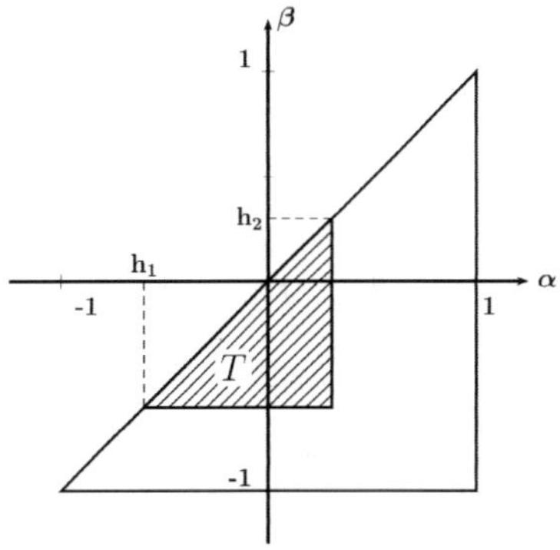

Figure 6. Definition of the Everett function

The input quantity of $H(t)$ leads to a specific proportion between the positive and the negative subarea. The resulting magnetization at each time instance is calculated by the *Preisach formalism*:

$$\begin{aligned} M &= \int_S p(\alpha, \beta)\gamma_{\alpha,\beta}(H(t))\, d\,\alpha\, d\,\beta \\ &= \int_{S^+} p(\alpha, \beta)\, d\,\alpha\, d\,\beta - \int_{S^-} p(\alpha, \beta)\, d\,\alpha\, d\,\beta \end{aligned} \tag{6}$$

At each time instance a magnetization M is determined. This scalar value depends not only on the current model input H itself, but also depends on the *magnetic history*:

$$m(h) = m \begin{bmatrix} h_1 & & h_3 & \cdots & & h_n \\ & h_2 & & \cdots & h_{n-1} & \end{bmatrix} \tag{7}$$

The alternating sequence of the magnetization's local extrema is stored in a descending order. Past extrema can be deleted by current values of the magnetic field strength. The memory vector $m(h)$ works like a LIFO-stack* [15].

* Last-In-First-Out

3.3.1. Everett function The Everett function defines, similar to the Preisach density function, the shape of the resulting hysteresis loop and the virgin curve. Mathematically the Everett function can be obtained by solving the defining equation

$$E(h_1, h_2)' = \underbrace{\int_{\alpha=h_1}^{\alpha=h_2} \int_{\beta=h_1}^{\beta=\alpha}}_{=T(\alpha,\beta)} p(\alpha, \beta)\, d\,\beta\, d\,\alpha \tag{8}$$

where

$$E(h_1, h_2) = \begin{cases} E(h_1, h_2)' & \text{if} \quad h_1 < h_2 \\ -E(h_2, h_1)' & \text{if} \quad h_1 \geq h_2 \end{cases} \tag{9}$$

In many cases it is not possible to find an antiderivate due to the fact that the Preisach density function itself has a complex structure [13]. The calculation of the magnetization M with the Everett function is carried out stepwise. The input values for the Everett function are the current field strength h and the last extremum stored in the memory vector $m(h)$:

$$M = M_n + \underbrace{E(h_n, h)}_{\Delta M} \tag{10}$$

Along the virgin-curve, a reversal point does not yet exist. Therefore the calculation method

$$M = \frac{1}{2} \cdot E(h, -h) \tag{11}$$

is applied. The selection of a suitable Everett function (see also [7]) allows the simulation of the three aforementioned CT classes. Both, the remanent flux B_R and the coercitive field strength H_c as well as the characteristic of the virgin curve are defined by the Everett function.

4. CT Modeling

The induced voltage on the secondary clamps of the CT

$$\begin{aligned} u_2 = & -w_2 \cdot \frac{d\,\Phi}{d\,t} & = -\frac{d\,\Psi}{d\,t} \\ = & -\frac{d\,\Psi}{d\,i_\mu} \cdot \frac{d\,i_\mu}{d\,t} & = -L_M(i_\mu) \cdot \frac{d\,i_\mu}{d\,t} \end{aligned} \tag{12}$$

and the magnetizing inductance $L_M(i_\mu)$ are related to the generalized law of induction. Further, Ampère's law in equation 13 can be applied:

$$i_\mu = H_{Fe} \cdot l_{Fe} + H_{Gap} \cdot l_{Gap} \tag{13}$$

If the core is closed and no gap exists, the second term in equation 13 will be zero. After summing up all resistors and inductors

$$R_2 = R_{CT} + R_b \tag{14}$$

$$, L_2 = L_{2\sigma} + L_b\ , \tag{15}$$

using the mesh equation for the secondary side

$$\sum_u = u_2 + R_2 \cdot i_2 + L_2 \cdot \dot{i}_2 = 0 \tag{16}$$

and, using the node rule

$$\sum_i = i_1 - i_\mu - i_2' = 0 \tag{17}$$

the differential equation of the CT finally takes the form

$$\dot{i}_\mu = \frac{\frac{w_1}{w_2} \cdot \left(i_1 \cdot R_2 + \dot{i}_1 \cdot L_2\right) - i_\mu \cdot R_2}{L_2 + \frac{w_2^2 \cdot A_{Fe} \cdot \mu}{l_{Fe}}} \tag{18}$$

The nonlinear relationship between the magnetic field strength $H \sim i_\mu$ and the magnetic flux density B will be determined from the hysteresis model at any time instance. A convenient hysteresis model takes the magnetic field strength H as an input and the flux density B as an output parameter. Internally, the model calculates the magnetization M which is the material-dependent part of the magnetic field strength. It can be written

$$M = \frac{B}{\mu_0} - H = \chi H \tag{19}$$

as a relationship between the magnetization M and the magnetic flux density B. As can be seen in figure 7 that there is a feedback loop from the permeability μ to the magnetizing inductance L_M. Previously, the permeability μ is calculated as the gradient $\Delta B/\Delta H$. The model is numerically solved using *Matlab/Simulink*®. Consequently, the dynamical system's behavior significantly depends on the nonlinearity and the hysteresis model. For this purpose the Preisach model has been chosen that reflect these physical characteristics. The inclusion of further effects like frequency-dependent eddy currents or excess losses [13] in the CT model is possible in general. Due to the fact that mainly low frequencies in the range of 50 Hz – 60 Hz are investigated these extensions can be neglected.

4.1. Computational effort

The Preisach model in its original form is computationally intensive. However, it can be made more efficient using a transform based on Everett's idea [6]. The need to solve the double integral (equation 6) at each time instance is transformed into the solution of a two-dimensional function. Thereby, the computational effort is enormously decreased.

4.2. Vanishing initial susceptibility

The susceptibility $\chi = dM/dH$ at the origin as well as in reversal points of the magnetic field strength H moves towards zero in the classical Preisach model. This behavior cannot be approved by physical measurements [8]. For this reason, the classical Preisach model is extended by the Product Preisach model [12].

Journal of Physics: Conference Series **570** (2014) 062001 doi:10.1088/1742-6596/570/6/062001

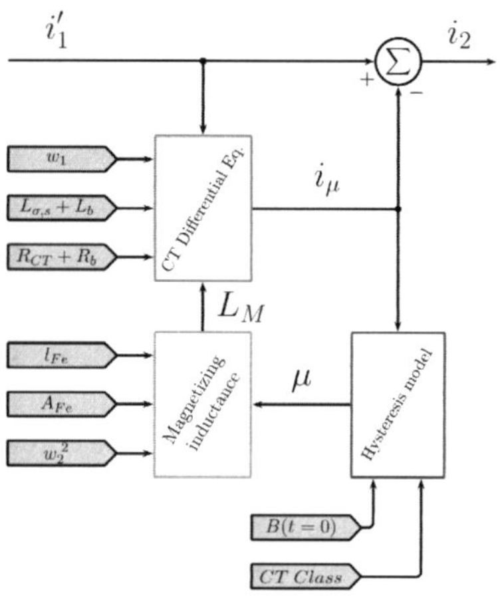

Figure 7. Generalized simulation scheme

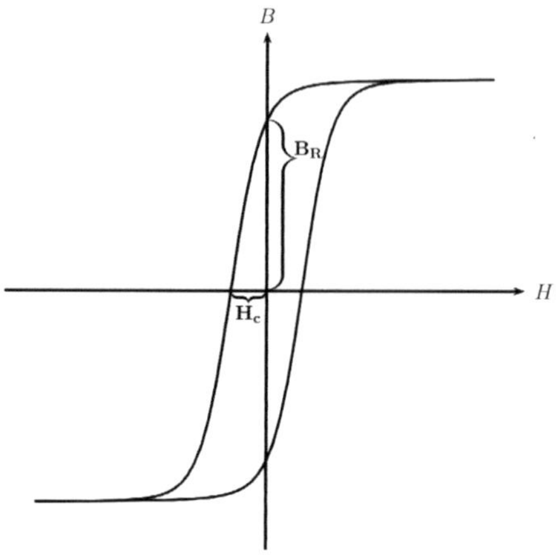

Figure 8. Major hysteresis loop parameters

4.3. Total saturation

Within the Preisach model, there is an maximum output limit where $S = S^+$ or $S = S^-$ according to figure 5. Beyond this certain point, the transducer is completely saturated and no further increase of the flux density B will take place. Depending on the magnitude of the current flowing through the magnetizing inductance this certain point might be passed. The procedure to cause a slightly increase in flux density beyond the saturation point is a multiplication of the calculated magnetization with the weighted Langevin function [1]. This only affects the region of great field strengths and smoothes the transition around the point of saturation.

5. Modeling of power system faults and the impact of CT saturation

A power system protection device operates as a real-time signal-processing-unit. It is located in substations and power plants. A protection device measures currents and voltages in the power system and a binary *trip signal* is given to the circuit breaker if a power system fault is detected in the protected equipment (line, transformer, busbar, etc.). Figure 9 shows the scheme of a protection device and its connections to the instrument transformers, the circuit breaker and an optional communication channel.

In the following, the CT model as described above will be used to process records of power system faults. The digital recording of currents (and voltages) is processed by the CT model and afterwards these signals are applied to two protection devices *Siemens 7SD522*, which are capable of

- Distance protection
- High-Speed-Distance protection (HSD)
- Differential protection

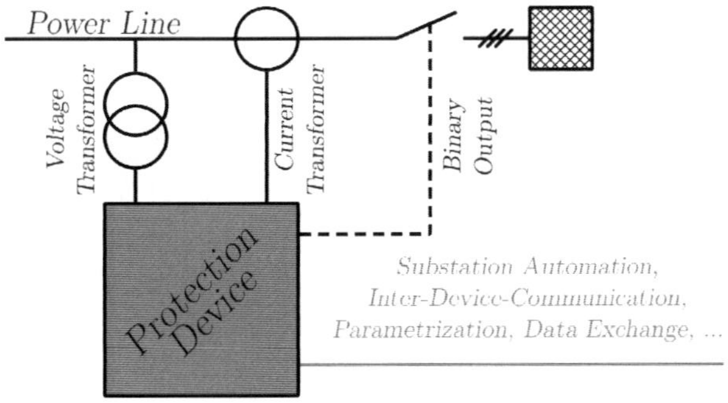

Figure 9. Simplified wiring scheme of a protection device

The mentioned protection functions require a current signal and must be robust toward CT saturation effects.

5.1. Internal faults
An internal fault is located on the protected line (or electrical equipment in general). The protection devices must under any circumstances determine an internal fault even though the current signal is distorted by CT saturation effects. A high voltage power system including the protected line has been simulated (Figure 10). It can deliver 3-phase currents and voltages at the stations A and B, which are at the same physical location as the protection devices. The fault position can be varied from 10 % to 90 % of the line length as seen from station A. Furthermore, the fault inception angle referred to the voltage can be varied. The angle controls the extent of the exponentially decaying DC component as a part of the fault current. Especially, angles close to the voltage's zero crossing point lead to a large DC component in the fault current. The simulated

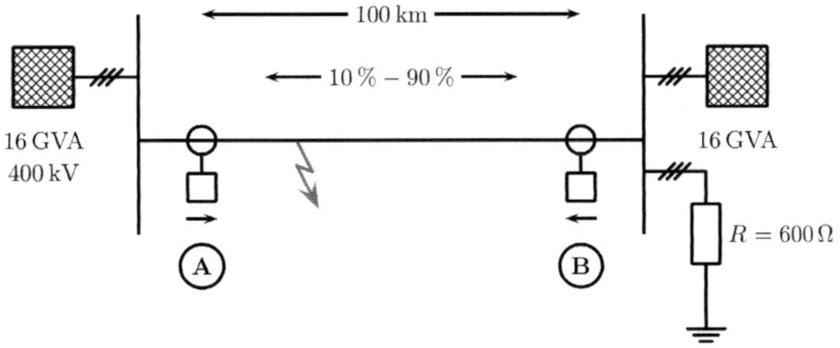

Figure 10. Test System 1 (Internal faults)

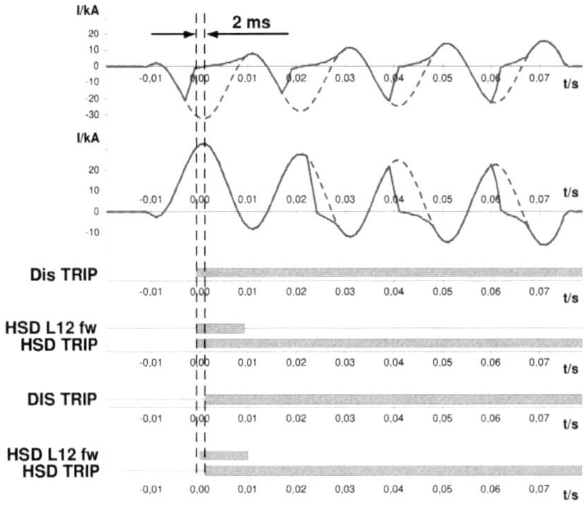

Figure 11. Fault record of internal fault with and without CT saturation

faults are identified as forward faults by both devices A and B. Depending on the fault distance from each station at least one device has to trip instantaneously. Figure 11 depicts the reaction on a power system fault with and without CT saturation. The L_1–L_2 fault is located at 20 % of the line in forward direction. The fault inception takes place at an angle of 150°. The CT is simulated with a ratio of $^{500\,A}/_{1\,A}$ as a TPX class CT. The initial remanence in both phases is set to −80 %. The two upper binary traces show the reaction of distance protection and HSD [4] without the influence of CT saturation, the current is ideally transformed (dashed curves). The two lower binary traces show the behavior under CT saturation condition (solid curves). It can be seen that the trip command is delayed only by 2 ms in this case.

5.2. External faults

An external fault is located outside of the protected line or the electrical equipment in general as can be seen in figure 12. The protection device must not trip under any circumstances even though the current signal is distorted by saturation effects and lead to an apparently increasing differential current. The single phase-to-earth fault shown in figure 13 is measured by both differential protection devices A and B. Station A again is equipped with an ideal CT, whereas in station B a class TPZ CT is used. Because of the huge transformer ratio of $^{5000\,A}/_{1\,A}$ the CT does not saturate. It can be seen that within the linear operating region of the CT of class TPZ the DC component is highly damped. As a result the difference current between the two stations (second analog trace) marks a great measurement error. The devices' reactions are depicted in the lower part. The calculated restraint current (green trace) overbalances the increasing differential current (red trace). In addition the fast charge comparison stage $I_{Diff>>}$ detects the measuring imprecision and an undesired overfunction can be avoided.

Journal of Physics: Conference Series **570** (2014) 062001 doi:10.1088/1742-6596/570/6/062001

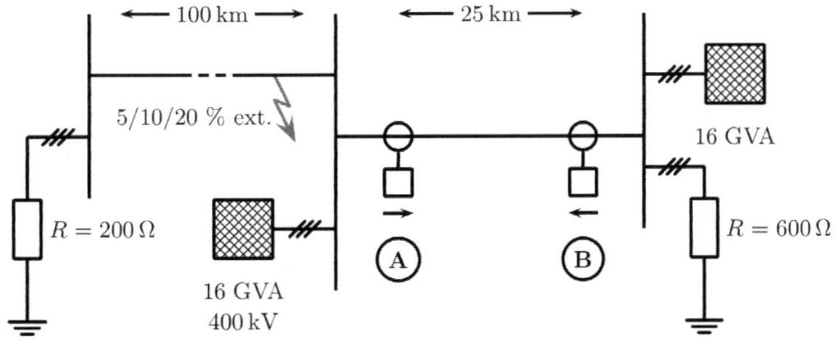

Figure 12. Test System 2 (External faults)

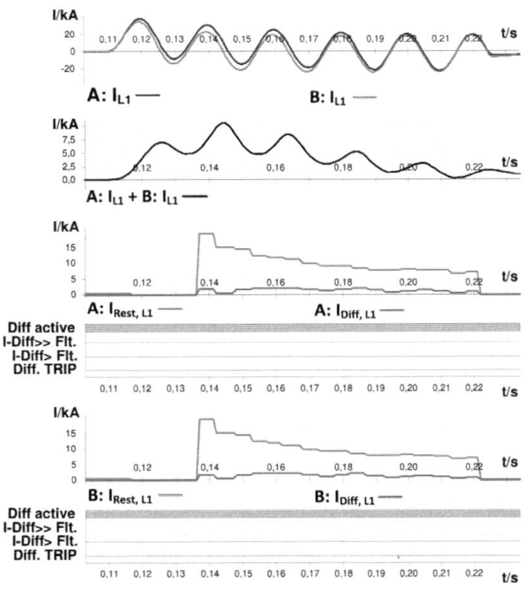

Figure 13. Fault record of an external fault at two stations

6. Conclusions

The described model is capable of simulating saturation effects in conventional CTs. Already existing real-time power system modeling units commonly integrate a simple $B-H$ characteristic neglecting the hysteresis effect. Due to the low computational effort, the presented model could be implemented in a variety of power system models and test environments. The CT model can process arbitrary primary currents and therefore has the ability to create test scenarios under the influence of CT saturation which are very complex. Especially, scenarios containing several duty cycles can be investigated based on the assumption of a magnetic history or an initial remanent flux. Concepts for the detection of current waveforms affected by CT saturation have to be robust against further effects like transformer inrush, frequency deviation and others. Also the nonlinear inductor model can be ported to further fields of investigation like electrical generators, motors and transformers.

References

[1] A. Benabou et al., eds. *Minor loops modelling with a modified Jiles–Atherton model and comparison with the Preisach model.* 18th International Symposium on Soft Magnetic Materials. Vol. 320. 20. Journal of Magnetism and Magnetic Materials, 2008, pp. 1034–1038.

[2] J.H. Chan et al. "Nonlinear transformer model for circuit simulation". In: *Computer-Aided Design of Integrated Circuits and Systems, IEEE Transactions on* 10.4 (1991), pp. 476–482.

[3] K. Chwastek. "Modelling offset minor hysteresis loops with the modified Jiles–Atherton description". In: *Journal of Physics D: Applied Physics* 42.16 (2009), p. 165002.

[4] C. Dzienis et al. "An experimental analysis of high-speed-distance protection". In: *Modern Electric Power Systems (MEPS).* Ed. by Proceedings of the International Symposium. 09/2010, pp. 1–8.

[5] M.T. Engelhardt. *Asymmetric minor hysteresis loop model and circuit simulator including the same.* US Patent 7,502,723. 2009.

[6] D. H. Everett. "A general approach to hysteresis. Part 4. An alternative formulation of the domain model". In: *Trans. Faraday Soc.* 51 (0 1955), pp. 1551–1557.

[7] P. Hejda and T. Zelinka. "Generalized Preisach Model Of Hysteresis - Theory And Experiment". In: *Czech. J. Phys.* (1990).

[8] O. Henze et al. "Hysteresis model and examples of ferromagnetic materials". In: *Magnetics, IEEE Transactions on* 39.3 (05/2003), pp. 1151 –1154.

[9] *Instrument transformers - Part 1: Current transformers.* Standard. IEC 60044-1. Switzerland, 02/2003.

[10] *Instrument transformers - Part 6: Requirements for protective current transformers for transient performance.* Standard. IEC 60044-6. Switzerland, 03/1992.

[11] D. C. Jiles and D. L. Atherton. "Ferromagnetic hysteresis". In: *Magnetics, IEEE Transactions on* 19.5 (09/1983), pp. 2183 –2185.

[12] G. Kádár et al. "Bilinear product Preisach modeling of magnetic hysteresis curves". In: *Magnetics, IEEE Transactions on* 25.5 (09/1989), pp. 3931 –3933.

[13] I.D. Mayergoyz. *Mathematical Models of Hysteresis and Their Applications: Second Edition.* Electromagnetism Series. Academic Press, 2003.

[14] F. Preisach. "Über die magnetische Nachwirkung". German. In: *Zeitschrift für Physik* 94 (5-6 1935), pp. 277–302.

[15] E.D. Torre. *Magnetic hysteresis.* IEEE Press, 1999.

Degradation Modelling for Health Monitoring Systems

R Stetter[1], M Witczak[2]

[1] Hochschule Ravensburg-Weingarten, 88241 Weingarten, Germany
[2] University of Zielona Góra, 65-417 Zielona Góra, Poland

E-mail: stetter@hs-weingarten.de

Abstract. Condition-monitoring plays an increasingly important role for technical processes in order to improve reliability, availability, maintenance and lifetime of equipment. With increasing demands for efficiency and product quality, plus progress in the integration of automatic control systems in high-cost mechatronic and critical safety processes, the field of health monitoring is gaining interest. A similar research field is concerned with an estimation of the remaining useful life. A central question in these fields is the modelling of degradation; degradation is a process of a gradual and irreversible accumulation of damage which will finally result in a failure of the system. This paper is based on a current research project and explores various degradation modelling techniques. These results are explained on the basis of an industrial product – a system for the generation of health status information for pump systems. The result of this fuzzy-logic based system is a single number indicating the current health of a pump system.

1. Introduction, background and overview

Several research activities referring to "health monitoring", "remaining useful life estimation" and "degradation modelling" focus on the behavior of a system during its lifetime. The results of these research activities are required for condition-based maintenance, which carries the potential to increase the safety, quality, and reliability, and to reduce the operating costs of a process. One aspect shows the enormous importance of these research activities: systems are very often a part of a superordinate system; if a system ceases operation because of degradation, this might lead to a shutdown of the whole superordinate system causing high costs for the operating company. One strategy to avoid such situations are regularly scheduled changes or upgrades of important systems. A lack of information of the condition of the system often leads to waste – fully functional units are replaced and destroyed. The methodologies for the given field can be divided into model-based and data-driven approaches [1].

In the model-based approach, a mathematical model of the system can be used to calculate an estimation of the outputs based on the current inputs. By comparing the model output and the real measurements a residual can be created from which information about degradation can be derived. With a good knowledge of underlying physics or the degradation structure, a very high level of accuracy of the model can be attained. However, it is often difficult to build good models for large scale systems, especially if the knowledge of underlying physics is limited. Alternatively, using data-driven methods, the information about degradation can be extracted from current and historical data through measurements. Currently, a huge amount of measurement data is collected and thus available from modern process control systems enabling such approaches. One key element of a system's

Journal of Physics: Conference Series **570** (2014) 062002 doi:10.1088/1742-6596/570/6/062002

"health monitoring" or "remaining useful life estimation" is degradation modelling, which is the main content of this paper.

2. Analytical approach

Most failures of engineering systems result from a gradual and irreversible accumulation of damage that occurs during a system's life cycle. This process is known as degradation [2]. In many applications, it can be very difficult to assess and observe physical degradation, especially when real-time observations are required [3]. Degradation modelling attempts to characterize the evolution of degradation signals. There are a significant number of research works that have focused on degradation models; a summary can be found in Zhou et al. [3]. In contrast to many publications which focus on the stochastic nature of damage, this paper focuses on the explanation of certain degradation phenomena, creating additional possibilities to model degradation. These models are usually easy to understand for product development engineers and thus ease implementation of such in an industrial context. In order to explain the degradation mechanism, a typical complex product – a car engine – is used (Figure 1).

Figure 1. Car engine – example for explaining degradation phenomena (Source: Chevrolet: D. Kimble)

For a detailed discussion of a complicated system, such as a car engine, a technique called "black-box" is frequently used in product development. In this technique only the inputs and outputs to a system are depicted. Figure 2 shows one possible black-box diagram for the engine, which is used to structure to following discussion.

2.1. Performance oriented degradation detection

One of the main indicators of degradation of a product is the performance of the respective system, which means the quantitative fulfillment of the central functions of the system. For an engine the performance is primarily shown by the speed and the torque of the output shaft (which can be multiplied to the power delivered by the engine); secondary indicators could be the ability of the engine to increase the speed within a time frame under a certain load and with certain connected moments of inertia.

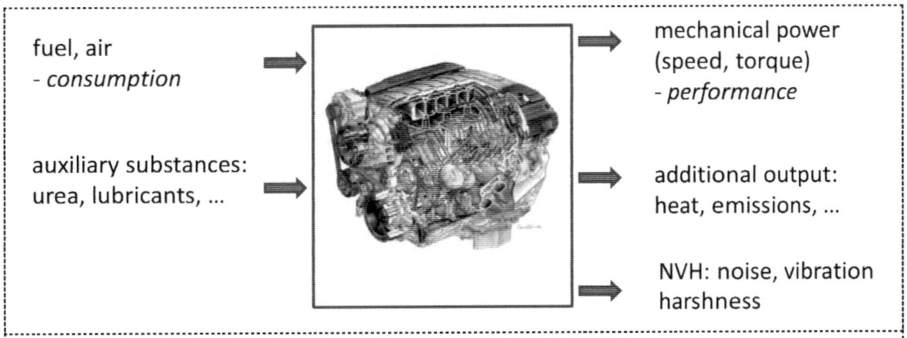

Figure 2. Black-Box diagram car engine.

In general it is sensible to define a dimensionless degradation indicator in relation to the initial performance:

$$DI_{Po} = \frac{\text{current performance}}{\text{initial performance}} \quad (1)$$

Here, DI_{Po} represents the degradation indicator which is performance oriented. For the engine this might be:

$$DI_{Po} = \frac{\text{current output power}\,[\text{kW}]}{\text{initial output power}\,[\text{kW}]} \quad (2)$$

It is important to note that many systems do not exhibit linear performance degradation. Figure 3 shows typical examples of performance degradation.

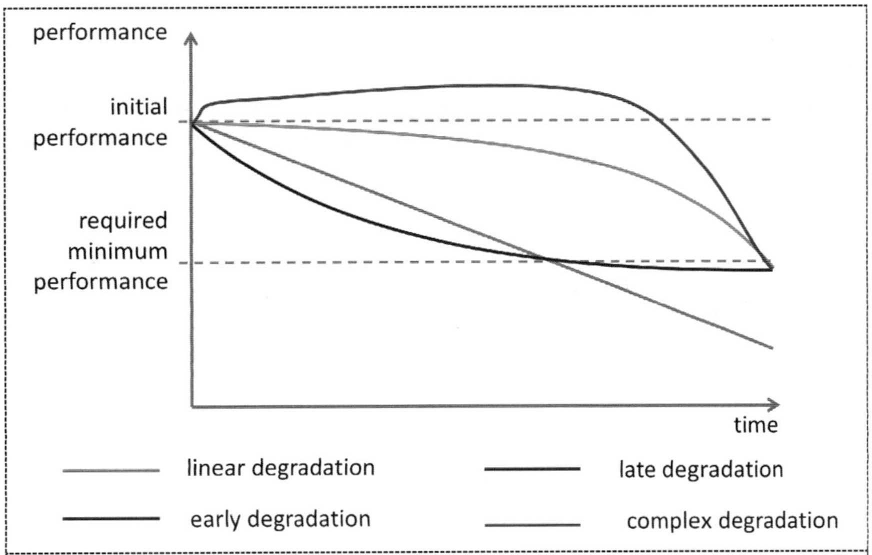

Figure 3. Typical performance degradation

The required minimum performance is usually within the range of 60% to 90% of the initial performance. Linear performance degradation would ease mathematical modelling, but is usually not realistic because damage is usually accumulated stochastically. Additionally, in real systems, the initial operation hours may lead to better friction behavior and reduced impedances (burn-in – see curve complex degradation). In later stages, certain layers of substance (for instance, friction-reducing layers) are partially eroded, leading to higher friction and lower performance. Through combinations of these effects, other performance degradation may occur (compare Figure 3). The complex degradation, in particular, will require intelligent treatment. Simple threshold mechanisms may allow a detection of the required minimum performance but will not allow foresight of the moment in time when the system will reach this threshold. A possible solution is polynomial degradation models based on empirical results together with specified margins. Figure 4 shows one example: a fourth degree polynomial degradation model.

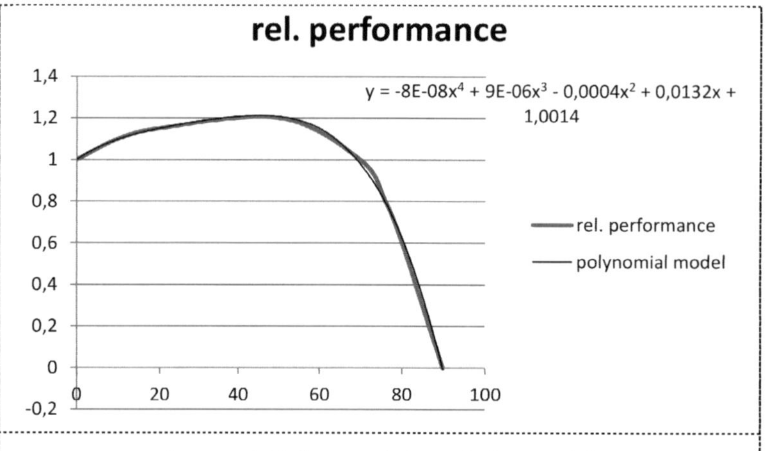

Figure 4. Polynomial degradation model

It is important to note that the performance very often cannot be directly measured. For engines today, the speed is usually measured by appropriate sensors, but torque sensors are still too costly to be implemented. The motor control systems usually contain some knowledge of the torque, but this "knowledge" results from mathematical models and other input signals, such as the amount of fuel injected. Because of this, virtual sensor techniques are often applied to allow performance oriented degradation detection.

2.2. Additional output oriented degradation detection

As visible in Figure 2, systems very often produce several outputs other than the desired outputs which are actually needed for the function of the product. For a car engine, these outputs are heat and emissions. Prime examples for degradation in an engine are worn piston rings. Worn piston rings result in lubricants in the combustion chamber. The burning of these lubricants leads to additional substances in the exhaust gases, which are sometimes even visible and can be easily measured. Similarly, a damaged cylinder head gasket may allow cooling fluid into the combustions chamber, also leading to additional unwanted substances in the exhaust gases. A dimensionless degradation indicator can be defined as:

$$DI_{AOo} = \frac{\text{initial additional output}}{\text{curent additional output}} \quad (3)$$

Here, DI_{AOo} represents the degradation indicator, which is additional output oriented.

2.3. Consumption oriented degradation detection

One important characteristic of degraded systems is decreased efficiency. For the engine example certain effects such as a reduced compression ration leads to inferior engine combustion and higher losses. Such effects are visible in a higher consumption; for the engine in a higher fuel consumption. A dimensionless degradation indicator can be defined as:

$$DI_{Co} = \frac{\text{initial consumption}}{\text{current consumption}} \quad (4)$$

Here, DI_{Co} represents the degradation indicator, which is consumption oriented. It is strongly connected to efficiency.

Therefore, another dimensionless degradation indicator can be defined as:

$$DI_{Eo} = \frac{\text{current efficiency}}{\text{initial efficiency}} \quad (5)$$

Here, DI_{Eo} represents the degradation indicator, which is efficiency oriented.

2.4. Auxiliary substance oriented degradation detection

Typical auxiliary substances found in car engines are lubricants. The condition of lubricant is usually a very good degradation indicator. A high amount of water in the lubricant is usually an indicator of issues with the cylinder head gasket. Particles in the lubricant may indicate wear of certain parts. Also, the color of a lubricant changes over time. This kind of degradation detection usually relies on a variety factors including the viscosity, light transmissibility and particle content of the lubricant. A dimensionless degradation indicator can be defined as:

$$DI_{ASo} = \frac{\left(\sum_{i}^{n} \frac{\text{current substance characteristic}_i}{\text{initial substance characteristic}_i} \right)}{n} \quad (6)$$

Here, DI_{ASo} represents the degradation indicator, which is auxiliary substances oriented. A substance characteristic can be a measurement of a property such as viscosity, and n is the number of such characteristics used to characterize the degradation behavior.

2.5. Noise/vibration/harshness oriented degradation detection

The concept of "noise, vibration, and harshness" (NVH), refers to research concerning the noise and vibration characteristics of systems, such as vehicles. A dimensionless degradation indicator can be defined as:

$$DI_{NVHo} = \frac{\text{initial noise / vibration / harshness indicator}}{\text{current noise / vibration / harshness indicator}} \quad (7)$$

Here, DI_{NVHo} represents the degradation indicator which, is noise/vibration/harshness oriented. The noise / vibration / harshness indicator can be a measurement of a vibration sensor.

It is important to note that it is sometimes it is possible to measure NVH even without specific sensors. Rad et al. [4] report that research concerning an online tool wear estimation algorithm has been developed for milling operations. The algorithm uses the AC current signal of the spindle for the detection process (due to its applicability and low acquisition cost in the industry) and applies a time-frequency analysis, gathering the effects of NVH on the main spindle.

2.6. Life-time oriented degradation detection

One of the most straightforward approaches is to observe the operation time of a system. Here, it is very often sensible to include the load of the system. For an engine, it might also make sense to calculate the delivered power multiplied by the operation time.

A dimensionless degradation indicator can be defined as:

$$DI_{LTo} = 1 - \frac{\int_0^{\text{current time}} \text{power}(t) \cdot dt}{\text{expected power} \cdot \text{operation time}} \quad (8)$$

Here, DI_{LTo} represents the degradation indicator which is life-time oriented.

2.7. Geometry oriented degradation detection

A typical example of a geometry degradation indicator would be worn pistons, which increase the size of the bore hole due to gradual removing of material during operation. A dimensionless degradation indicator can be defined as:

$$DI_{Go} = \frac{\text{current dimension}}{\text{initial dimension}} \quad (9)$$

Here, DI_{Go} represents the degradation indicator, which is geometry oriented. It is strongly connected to efficiency.

2.8. Selection of degradation indicators

In the previous sections, multiple methods of deriving degradation indicators were explained. It is a major challenge to choose the proper degradation indicators. Quite often, experienced engineers can make a sensible choice. An elaborate methodology is described by Lamoureux et al. [5]. Here, a complete methodology for the selection and validation of health indicators (similar to degradation indicators) in health monitoring systems design is introduced. This methodology, called integrated prognostics and health management, is based on the use of numerical modeling and uncertainties propagation to generate data in sufficient amount to validate health indicators. In order to simulate the distribution of health indicators, a well-known surrogate model called Kriging is utilized. Lamoureux et al. [5] have also defined some numerical key performance indicators as criterion for performing the validation of health indicators. A combination of the product oriented possibilities to find degradation indicators presented in this paper and the methodology of Lamoureux et al. [5] to validate them suggests an interesting possibility for the design of future health monitoring systems.

3. Health Monitoring System for Pumps

The industrial product example for this research is a three spindle screw pump [6]. A three spindle screw pump is a positive displacement pump that uses one or several screws to move fluids or solids along the screw(s) axis (Figure 5).

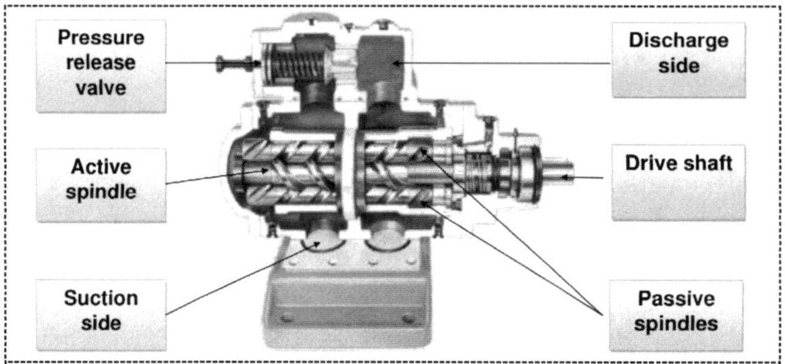

Figure 5. Three spindle screw pump.

Three-Spindle screw pumps are used for transport of viscous fluids with lubricating properties. They are suited for a variety of applications such as fuel-injection, oil burners, boosting, hydraulics, fuel, lubrication, circulating and feed and so on. The fuzzy system, which aims to generate a health status information (compare [7], [8]) is based on several inputs which may either come directly from sensors, from information from the frequency converter, which is supplying the motor driving the pump, from an online model of the pump system (compare [9] and **Fehler! Verweisquelle konnte nicht gefunden werden.**) or from a life counter. The inputs and their sources are summarized in Table 1.

Table 1. Inputs and their sources

Inputs	Source
Output pressure	Sensor reading (direct)
Pressure Pulsation	Sensor reading (indirect)
Torque	Frequency Converter
Speed	Frequency Converter
Fluid Temperature	Sensor reading (direct)
Life count	Life counter
Q_V – Leakage Loss	Online model
P_r – Friction Loss	Online model

This information is fuzzified, the rules are applied, and the defuzzification takes places. These steps result in crisp values for certain failure modes (Table 2).

Table 2. *Failure modes*

Failure Modes	Explanation/Example
Vibration	
Line blockage	e.g. clogged filter
Frequency converter failure	e.g. overspeed
Housing failure	e.g. cracks due to ageing
Pump overheat	
Worn spindles	abrasion on spindles
Worn bearings	increased friction

These crisp values are then concentrated to one single health status number by means of an unique algorithm, which is described in detail in Section 5 of this paper. An overview of the process is given in Figure 6.

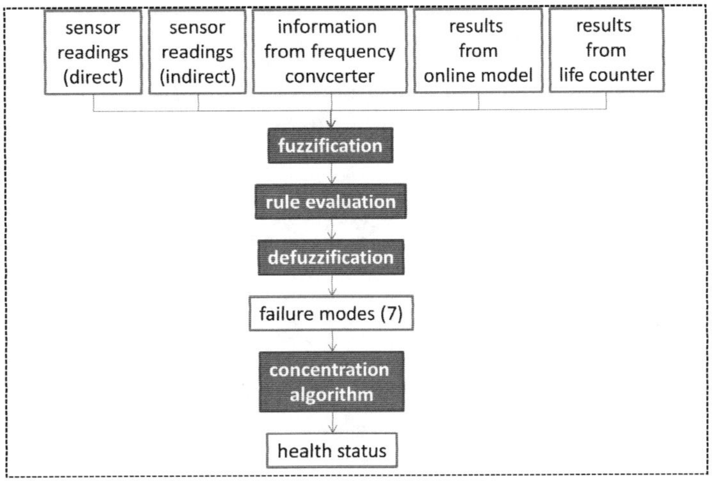

Figure 6. Process for determining the health status.

Pressure Pulsations (see Table 1) are the fluctuations in the basic pressure/head being developed by the pump. These pulsations can sometimes be very severe and cause damage to the piping or other components in a hydraulic system (Figure 7).

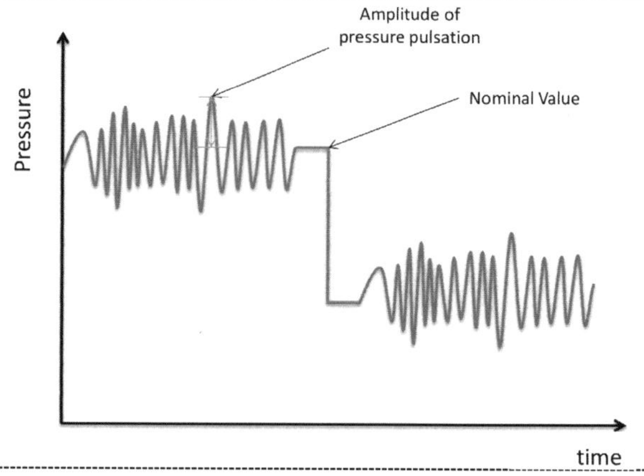

Figure 7. Pressure pulsation.

This indicator uses a special kind of noise/vibration/harshness oriented degradation detection (compare section 2.5). The amplitude of the pressure pulsation is a consequence of the vibrations within the pump.

P_r – the Frictional Power Loss Coefficient is the loss of energy or "head" that occurs in pipe flow due to the viscous effects generated by the surface of the pipe. The friction power of a pump, which is considered as power loss, can be described as in Equation 10:

$$P_{r2} = P_{r1} \cdot \sqrt{\frac{vis_2}{vis_1}} \cdot \left(\frac{n_2}{n_1}\right)^{1,5} \quad (10)$$

The Leakage Loss Coefficient Q_v can be described as in Equation 11:

$$Q_v = Q_{v10} \cdot \left(\frac{\Delta P}{10}\right)^x \sqrt{\frac{vis_1}{vis_2}} \quad (11)$$

Q_{v10} and vis_1 are coefficients or reference values proposed by leading pump manufacturers. Both indicators show the efficiency of the pump; so consumption oriented degradation detection is applied (compare section 2.3).

4. Degradation Indicators

One important input to the system which serves as an indicator for degradation, is the Pressure Pulsation (compare section 3). For this indicator the following graph gives a clear idea of the range of pressure pulsation and speed for the three states "New", "Trend" and "Worn" pump. If the pressure pulsation ranges from 2 % -10 % with speed ranges from 2900 rpm-1000 rpm, respectively, then the condition of the pump is new. If the pressure pulsation at 1000 rpm is raised nearly 15%, this indicates the "Trend" condition (indicating the beginning degradation of the pump and thus serving as a pre-warning) pump condition. For "Worn" pump, Pressure pulsation will be approximately 20% at the speed of 1000 rpm (Figure 8).

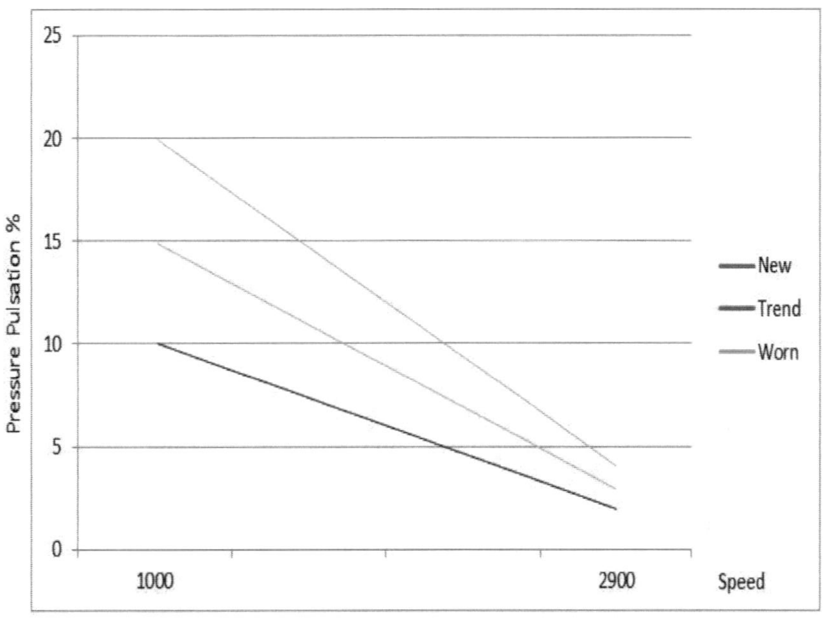

Figure 8. Degradation indicator pressure pulsation.

A second powerful indicator is the Frictional Power Loss Coefficient (compare Section 3). Increased frictional power loss is frequently caused by overloading and insufficient lubrication. New, conditioned pumps have a theoretical frictional power loss coefficient, P_{rth}. A recalculation of P_r after some time that shows friction being 75% higher than the originally calculated value of P_{rth} indicates a "Trend" state. If the friction is 150% higher than the initial value of P_{rth}, then the pump is said to be in the "Worn" state (Figure 9).

As a third indicator, the Leakage Loss Coefficient can be used (compare Section 4). Increased leakage losses may be a result of abrasive flow causing wear of spindles in the pump. New pumps have a theoretically calculated leakage loss coefficient, Q_{vth}. If the leakage loss is 30 % higher than the calculated theoretical value of Q_{vth}, then the condition of the pump is "Trend".

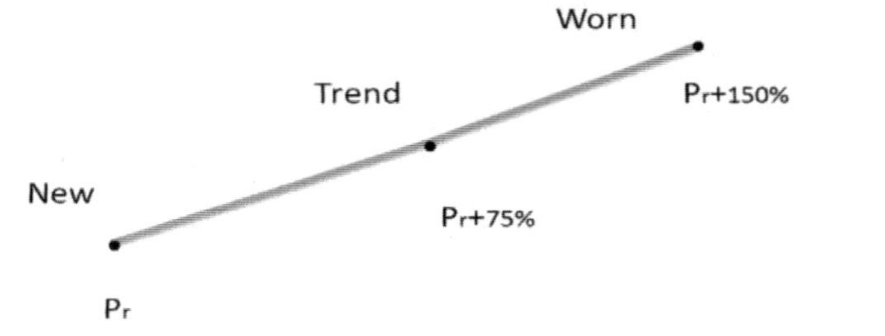

Figure 9. Degradation indicator friction power.

For "Worn" pumps, the leakage loss can be 100% higher than the calculated value of Q_{vth} (Figure 10).

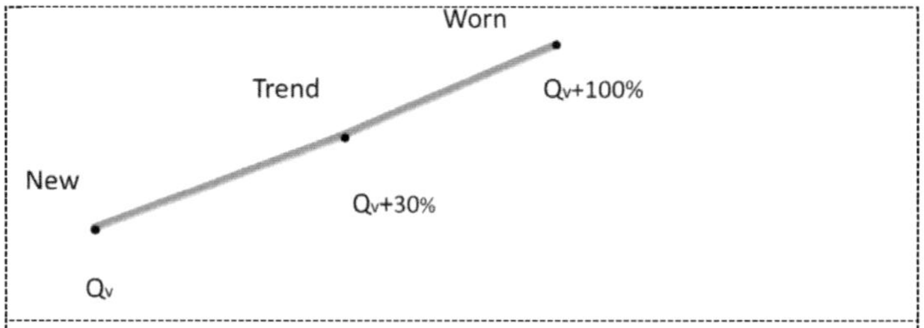

Figure 10. Degradation indicator leakage loss.

Figure 11 gives an overview of the degradation of a pump.

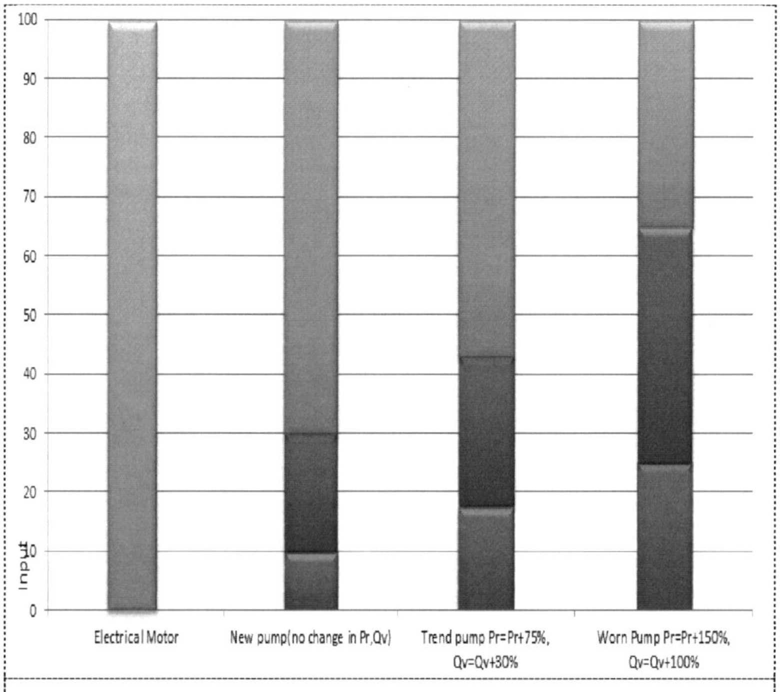

Figure 11. Overview of degradation indicators.

5. Calculation of the Health Status Number

A useful health status number can be calculated through an analogous process to FMEA [7], [8]. The severity of a failure mode plays an important role in the sensible health status, as not all failure modes have the same effects, and, therefore, the same severity. Here, a variable "importance" can be used. The occurrence can be understood, in this case, as the single failure mode output from the fuzzy model - the variable can, for instance, be named "fuzzy output". The third parameter "compensatability" is optional: analogous to "detection", some failure modes can be compensated by means of a control system. Equation 12 explains how the pump health status is calculated based on the fuzzy outputs obtained.

$$\text{Health status } (\text{HS}) = 1000 \cdot n - \sum_{i=1}^{n} \text{importance(I)}_n \times \text{fuzzy output(FO)}_n \times \text{compensatability(C)}_n \quad (12)$$

Here, n is the number of failure modes. The sources of this equation are described in detail in [7]. The health status calculation includes a threshold setting to amplify bigger failures and attenuate smaller failures [11]. This will avoid exaggeration of the effect of failure modes. Figure 12 shows a flow chart describing the algorithm for this threshold.

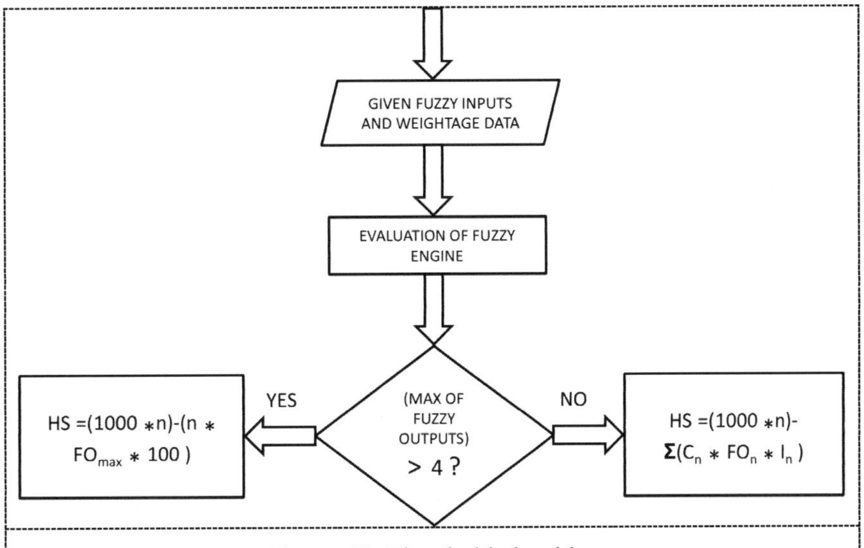

Figure 12. Threshold algorithm.

6. Validation

An intuitive set of test cases with 15 contrasting operating modes was formed to test behavior of the model. The results are shown in the Figure 13.

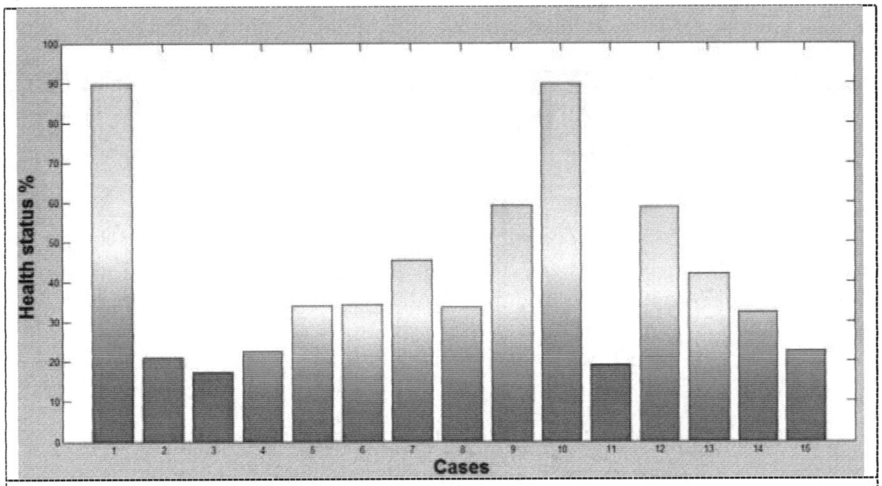

Figure 13. Health status results for certain test cases.

In Figure 14, the 3rd and the 11th case show a very low health status number due to the failure mode "frequency converter failure" and "highly worn bearings". Cases 1 and 10 have a good health status because of their normal operating conditions. Cases 9 and 12 show medium health. Case 8 is for highly worn spindles. High speed and torque failure mode are depicted in Case 15. It is evident from Figure 13 that fuzzy mapping gives a meaningful representation of the health status of the pump in the form of a unique number or as percentage.

7. Summary

The main goal of this project, which is the basis the research described in this paper, is the development of a prognosis of future control and diagnosis technologies for pumps and intelligence in pumps. Experts in the field agree that in future health status analyses, information about the components is desirable in order to reduce shut-down times, plan availability and service, and extend the lifetime of systems. In this paper, the main focus was on the modelling of degradation. Several degradation mechanisms were explained and initial indicators were defined. Three different degradation indicators were explained with an application example. Further work will investigate additional degradation indicators and expand the mathematical modelling possibilities.

References

[1] Louen, C.; Ding, S. X.; Kandler C.: A new Framework for Remaining Useful Life Estimation Using Support Vector Machine Classifier. In: Proceedings of SysTol 2013, Nice, 9th – 11th October 2013.

[2] Bogdanoff, J. L. and Kozin, F.: Probabilistic Models of Cummulative Damage. Wiley, New York (1985).

[3] Zhou, R.; Serban, N.; Gebraeel, N.: Degradation Modeling Applied to Residual Lifetime Prediction Using Functional Data Analysis. In: The Annals of Applied Statistics 2011, Vol. 5, No. 2B, 1586–1610.

[4] Rad, J. S.; Hosseini, E.; Zhang, Y.; Chen, C.: Online Tool Wear Monitoring and Estimation Using Power Signals and S-Transform. In: Proceedings of SysTol 2013, Nice, 9th – 11th October 2013.

[5] Lamoureux, B.; Mechbal, N.; Massé, J.-R.: Selection and Validation of Health Indicators in Prognostics and Health Management System Design. In: Proceedings of SysTol 2013, Nice, 2013.

[6] Stetter, R.: Implementation of advanced control and diagnosis. In: Aubrun, C.: Proceedings of the European Workshop on Control Engineering for Industry, CEIND'2013, 15. and 16. May 2013, Nancy, France.

[7] Kleinmann, S., Stetter, R., Mohan, S., Dabrowska, A., Leonardo, D., Koller-Hodac, A.: Advanced Diagnosis of Industrial Pump Systems. In: Proceedings of the 8th SAFEPROCESS 2012: IFAC International Symposium on Fault Detection, Supervision and Safety for Technical Processes, 29th to 31st August 2012, Mexico City, Mexico.

[8] Kleinmann, S., Kubendra Prasad, P. K., Stetter, R.: Optimization of a Pump Health Monitoring System using Fuzzy Logic. In: Proceedings of SysTol 2013, Nice, 9th – 11th October 2013.

[9] Kleinmann, S.; Stetter, R.; Abdul Jalal, M. F.: Praxisorientierte Modellierung von Pumpensystemen für Monitoring, Regelung und Diagnose. In at – Automatisierungstechnik 59 (2011) 5, pp. 309 – 322.

[10] Kleinmann, S., Dabrowska, Koller-Hodac, A., Leonardo, D., Stetter, R.: Model of a combined pump and drive system for advanced control and diagnosis. In: The Archive of Mechanical Engineering. Volume LVII, Number 4 / 2010, pp S. 405 - 414. 2010.

[11] Kleinmann, S., Leonardo, D., Mohan, S., Stetter, R., A., Koller-Hodac, A.: Modelling, Control and Monitoring of Industrial Pump Systems. In: Niemann, H. (Ed.): Proceedings of the 10th Workshop on Advanced Control and Diagnosis, ACD'2012, 8. and 9. November 2012, Copenhagen, Denmark. Williams, E.D.: Revisiting energy used to manufacture a desktop computer: hybrid analysis combining process and economic input-output methods.. Conference Record 2004 IEEE International Symposium on Electronics and the Environment, pp. 80- 85.

European Workshop on Advanced Control and Diagnosis IOP Publishing
Journal of Physics: Conference Series **570** (2014) 062003 doi:10.1088/1742-6596/570/6/062003

Properties of input-output Hammerstein-bilinear structure with application to an industrial air handling unit

T Larkowski[1], I Zajic[1], K J Burnham[1] and L Koszalka[2]

[1] Control Theory and Applications Centre, Coventry University, Coventry, UK

[2] Computer Systems and Networks, Wroclaw University of Technology, Wroclaw, Poland

E-mail: `aa6665@coventry.ac.uk`

Abstract. When developing mathematical models, especially for control, the practical interest lies in relatively simple extensions of linear structures that offer improved modelling capabilities. In this paper a discrete-time input-output Hammerstein-bilinear structure is introduced and its properties are discussed in detail. It consists of a cascade connection of a static nonlinearity followed by a dynamic bilinear system. By combining advantages of constituent subsystems the Hammerstein-bilinear structure allows for both an input dependant dynamic behaviour (particular property of bilinear systems) and an increased flexibility of the steady-state characteristic (particular property of Hammerstein models) to be obtained simultaneously. Modelling capabilities of such structure are evaluated on an air-handling unit that is a part of an industrial heating, ventilation and air-conditioning system.

1. Introduction

Most, if not all, physical systems are inherently nonlinear, hence they can be satisfactorily approximated by standard linear modelling techniques only over a rather narrow range of operation. A need for more flexible structures that allow models to be adequate over wider operating ranges stems, therefore, rather naturally and has become a prominent and an important topic of research in the control community, see for instance [1], [2] and [3]. When developing mathematical models, especially for control, the practical interest lies in relatively simple extensions of linear structures that offer improved modelling capabilities and whose properties are well understood. Among others, the frequently considered nonlinear models include: Hammerstein systems (HS) [4], Wiener systems (WS) [5], bilinear systems (BS) [6] and Lure models [7]. A HS comprises of a static nonlinearity followed by a dynamic linear subsystem, whilst a WS model is a dual of a HS with the order of these elements reversed. A Lure model also consists of the same two elements but with a static nonlinearity located in the feedback. A BS is a structure where the output is a function of a product between the output and the input (or state, in general). Each of these models can be used to mimic certain qualitative behaviour of the process of interest. A detailed comparison and discussion of their specific features can be found in [8]. Although relatively simple, by providing enhanced modelling capabilities and permitting design of efficient dedicated control schemes, these structures have been demonstrated to be of a great pragmatic utility in many industrial control problems, see [9], [10], [11] and [12]. In this paper, a discrete-time input-output Hammerstein-bilinear system (HBS) structure is considered that consists of a cascade connection of a static nonlinearity followed by a dynamic BS. Such a model structure combines advantages of its constituent subsystems allowing for both an input dependant dynamic behaviour (particular property of BS) and an increased flexibility of the steady-state characteristic (particular property of HS) to be obtained simultaneously. Although the more general class of nonlinear autoregressive with exogenous input (NARX) models, to which the HBS belongs, has received interest in the literature, see for instance [13], [7], [3], the HBS on its own has been given limited attention. This paper aims at improving this situation by focusing directly on the HBS structure and its properties. Also, an exemplary application to industrial problem is presented.

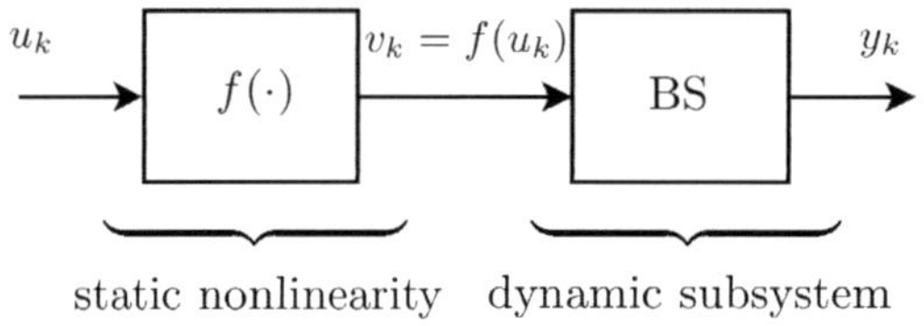

Figure 1. Diagrammatic representation of a Hammerstein-bilinear structure.

2. Definition of Hammerstein-bilinear structure

HBS structure is a member of a general family of NARX models, more specifically, it belongs to a sub-class of models retaining affinity with respect to the output signals, i.e. so-called output affine models [14]. It is constructed as a cascade connection of a static (memoryless) nonlinearity followed by a dynamic time-invariant BS, see Figure 1, and is given by

$$y_k = \sum_{j=1}^{n_a} a_j y_{k-j} + \sum_{i=1}^{n_b} b_i v_{k-i} + \sum_{j=1}^{n_a}\sum_{i=1}^{n_b} \eta_{ij} v_{k-i} y_{k-j}, \qquad v_k = f(u_k) \tag{1}$$

with the bilinearity defined as a product between system output y_k and the intermediate input variable v_k, and $f(\cdot)$ being a general scalar static nonlinear function. To facilitate subsequent identification it is advantageous to restrict $f(\cdot)$ to functions that are smooth, linear in the parameters and have continuous derivative. Such functions that are commonly used are polynomials, see [4], [15] and [12], where HS models are considered. (Because the focus is placed on properties of the HBS structure, additional signals modelling disturbances are not considered.) The input u_k, before entering the dynamic BS and yielding the overall system output y_k, undergoes a nonlinear transformation through $f(\cdot)$ to form the intermediate input signal v_k. Not all bilinear coefficients must necessarily be present in (1), hence a particular structure can be obtained by setting selected η_{ij} to zero. Driven by control pragmatism, considerations are restricted here to structures whose dynamic part is not purely bilinear, i.e. not all b parameters are simultaneously null. The HBS can be interpreted as a generalisation of both of its constituent subsystems, i.e. the HS and BS structures. In particular, a BS is obtained from (1) by setting $u_k = v_k$, i.e. by selecting $f(x) = x$, which gives

$$y_k = \sum_{j=1}^{n_a} a_j y_{k-j} + \sum_{i=1}^{n_b} b_i u_{k-i} + \sum_{j=1}^{n_a}\sum_{i=1}^{n_b} \eta_{ij} u_{k-i} y_{k-j}. \tag{2}$$

Similarly, a HS is obtained by setting $\eta_{ij} = 0 \ \forall \ i, j$ in (1), which leads to

$$y_k = \sum_{j=1}^{n_a} a_j y_{k-j} + \sum_{i=1}^{n_b} b_i v_{k-i}, \qquad v_k = f(u_k). \tag{3}$$

Also, a linear structure can be obtained by imposing both restrictions simultaneously, i.e. $\eta_{ij} = 0 \; \forall i, j$ and $u_k = v_k$.

The HS structure is of particular interest when modelling processes with nonlinear actuators such as flow controlling valves. Also, more generally, it is useful in situation where the dynamic behaviour can be approximated sufficiently well via a linear model, whilst nonlinearity is manifested in the overall steady-state characteristic, see [4] and [15]. A BS model structure is useful when dealing with applications where there is heat exchange and/or transfer of heat is involved, see [16] and [17]. As well, there are examples of BS arising naturally that can be found when modelling various chemical processes, where, quite commonly the exogenous inputs are flow rates.

Because the HBS structure comprises HS and BS as its special cases it is applicable in situations where both the steady-state characteristic and the dynamics are nonlinear. The HBS can, alternatively, be expressed as the following NARX difference equation

$$y_k = \sum_{j=1}^{n_a} a_j y_{k-j} + \sum_{i=1}^{n_b} b_i f(u_{k-i}) + \sum_{j=1}^{n_a} \sum_{i=1}^{n_b} \eta_{ij} f(u_{k-i}) y_{k-j}. \tag{4}$$

By postulating that the function $f(\cdot)$ is linear in the parameters, denoted by α, the difference equation (4), becomes bilinear in both i.e. input-output and its coefficients, i.e. a, b, η and α. This property significantly facilitates the parameter estimation problem. Analogously to the classification of input-output BS, HBS structures can be divided into three main categories: i) subdiagonal $\eta_{ij} = 0 \; \forall j > i$, ii) superdiagonal $\eta_{ij} = 0 \; \forall j < i$ and iii) diagonal $\eta_{ij} = 0 \; \forall j \neq i$. Note that from a physical point of view the class of superdiagonal HBS appears to be a somewhat peculiar because it suggests that future values of input are correlated with past values of system output. The HBS representation (4) can be re-expressed such that the resulting structure exhibits input (or, alternatively, output) dependency of the parameters. The corresponding system with input dependent parameters a_j^k is given by

$$y_k = \sum_{j=1}^{n_a} a_j^k y_{k-j} + \sum_{i=1}^{n_b} b_i f(u_{k-i}), \qquad a_j^k = a_j + \sum_{i=1}^{n_b} \eta_{ij} f(u_{k-i}). \tag{5}$$

In terms of a quasi-transfer function representation, the HBS can be expressed as

$$y_k = G(q^{-1}) f(u_k), \tag{6}$$

where the input dependent quasi-transfer function of the constituent BS is given by

$$G(q^{-1}) = \frac{y_k}{v_k} = \frac{\sum_{i=1}^{n_b} q^{-i} b_i}{1 - \sum_{j=1}^{n_a} q^{-j} a_j^k} = \frac{\sum_{i=1}^{n_b} q^{-i} b_i}{1 - \sum_{j=1}^{n_a} q^{-j} \left[a_j + \sum_{i=1}^{n_b} \eta_{ij} f\left(u_{k-i}\right)\right]}. \tag{7}$$

From a structural point of view, an advantage resulting from the interpretation of the constituent BS model as a linear time-varying system is that standard well understood notions from classical linear system theory such as system time-constants, damping/natural frequency and steady-state gain are to large extent retained. As a consequence, many existing methods developed for linear systems, including both control and identification, can be readily extended and transferred to the BS case.

3. Steady-state properties

Considering (6)-(7) the steady-state characteristic of the HBS can be expressed as a combination of the steady-states of the constituent static nonlinearity and the BS. Therefore, before discussing steady-state properties of the HBS structure, the steady-state behaviour of the component subsystems is analysed first in detail.

The steady-state characteristic of the overall HBS is given by

$$y_{ss} = g_{BS}(v_{ss})v_{ss} \tag{8}$$

where the subscript ss denotes the steady-state value. The steady-state gain g_{BS} of the component BS is given by

$$g_{BS}(v_{ss}) = \frac{y_{ss}}{v_{ss}} = \frac{\sum_{i=1}^{n_b} b_i}{1 - \sum_{j=1}^{n_a} \left[a_j + \sum_{i=1}^{n_b} \eta_{ij} f(u_{ss})\right]} = \frac{\bar{b}}{1 - \bar{a} - \bar{\eta} v_{ss}}, \tag{9}$$

where $\bar{a} = \sum_{j=1}^{n_a} a_j$, $\bar{b} = \sum_{i=1}^{n_b} b_i$ and $\bar{\eta} = \sum_{i=1}^{n_a} \sum_{j=1}^{n_b} \eta_{ij}$. It is observed that the steady-state characteristic of the HBS is, firstly, nonlinear, and, secondly, dependent on both the input and the form of $f(\cdot)$. Also, note that $\bar{a} + \bar{\eta} v_{ss} \neq 1$ for y_{ss} to remain bounded.

3.1. Steady-state behaviour of Hammerstein systems

In the case of HS, see (3), the steady-state gain is equal to the static nonlinearity $f(\cdot)$ scaled by a constant gain g_L of the linear dynamic subsystem, i.e.

$$y_{ss} = g_L f(u_{ss}), \qquad g_L = \frac{y_{ss}}{v_{ss}} = \frac{\sum_{i=1}^{n_b} b_i}{1 - \sum_{j=1}^{n_a} a_j} = \frac{\bar{b}}{1 - \bar{a}} \tag{10}$$

with the assumption that $\bar{a} \neq 1$. In contrast to the BS model, the HS structure can exhibit IM, if the static input nonlinearity is non-strictly monotonic. However, similarly to the BS structure, it is also inherently incapable of exhibiting OM. The steady-state asymptotes depend on a particular form of the static input nonlinearity $f(\cdot)$, hence cannot be characterised for a generic case.

3.2. Steady-state behaviour of bilinear subsystem

Depending on a value of $\bar{\eta}$ the steady-state of a BS, see (2), can exhibit either saturation or 'burst' characteristics. Whilst a constant steady-state gain is obtained if $\bar{\eta} = 0$, in which case the BS is so-called static-linear, see [14]. Note that a particular case for which the condition $\bar{\eta} = 0$ is satisfied occurs when $\eta_{ij} = 0 \; \forall i, j$, i.e. when the BS is reduced to a linear system. An illustrative plot showing representative steady-state characteristics of an arbitrary BS for a range of different values of $\bar{\eta}$ is given in Figure 2(a). By limiting consideration only to the quadrant where both $u_{ss} \geq 0$ and $y_{ss} \geq 0$, the three darker curves above the black dashed line correspond to the cases where $\bar{\eta} > 0$, i.e. a so-called positive bilinearity (burst - typical of exothermic chemical processes). Whilst the remaining three brighter curves below the black dashed line correspond to $\bar{\eta} < 0$, i.e. a so-called negative bilinearity (typical of many industrial systems with saturation nonlinearities) [11].

A crucial property of a model is the extent to which it can mimic the steady-state characteristics of a process. In this connection, the existence of a steady-state multiplicity is of interest. Following [2], the input (output) multiplicity is defined as the property where a single steady-state output (input) corresponds to multiple steady-state inputs (outputs). The input multiplicity (IM) property appears to be more frequently encountered phenomena than the output multiplicity (OM) and is related to the notion of an 'optimal' or 'sweet' point in the process steady-state curve [8]. However, the OM has also been encountered in many industrial problems, see [7] and the references therein. For the IM to exist the steady-state characteristic must not be strictly monotonic or, equivalently, must not be invertible.

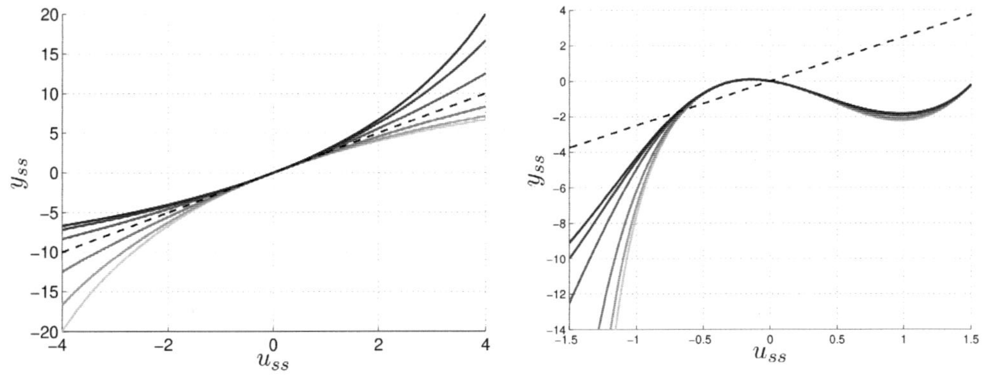

(a) An arbitrary BS. (b) An arbitrary HBS.

Figure 2. Illustrative plots showing representative steady-state characteristics for a range of different values of $\bar{\eta}$ parameter. The dashed black line corresponds to a linear system case.

Considering the constituent BS, it is observed that neither IM nor OM are possible. This fact follows from a property that the corresponding steady-state curve is strictly monotonic, which, for instance, can be useful when designing nonlinear compensators, see [18] and [11]. However, for completeness, it is worth noting that generally, in fact, the IM is feasible in a somewhat degenerate case when y_{ss} is constant. Recalling the previously postulated assumption that $\bar{a} + \bar{\eta}u_{ss} \neq 1$, this condition is satisfied if $\bar{b} = 0$ or $u_{ss} = 0$ or $\bar{a} = 1$. Similarly, a degenerate OM can also be exhibited in general when u_{ss} is constant. The steady-state input as a function of a steady-state output is given by

$$u_{ss} = \frac{(1 - \bar{a})y_{ss}}{\bar{b} + \bar{\eta}y_{ss}}, \tag{11}$$

where it is assumed that $\bar{b} + \bar{\eta}y_{ss} \neq 0$. Considering (11), the degenerate OM is present if $\bar{a} = 1$ or $y_{ss} = 0$ or $\bar{b} = 0$. Driven by utility for control, these degenerate forms of IM and OM are, however, deliberately excluded in the remainder of this paper, hence BS structures are treated as not being able to exhibit any form of multiplicity. Also, it can be verified that the horizontal asymptote of the steady-state characteristic, obtained for $u_{ss} \to \pm\infty$, is given by $-\bar{b}/\eta$, whilst the vertical asymptote, obtained for $y_{ss} \to \pm\infty$, is given by $(1 - \bar{a})/\eta$.

3.3. Steady state behaviour of Hammerstein-bilinear structure

Since composed of HS and BS models, the HBS inherits properties of the constituent structures allowing for an extended range of potential steady-state characteristics to be obtained, when compared to those feasible for BS. In particular, this includes an important capability of exhibiting IM. This property is illustrated in Figure 2(b) for an arbitrarily parametrised HBS with $f(\cdot)$ selected as a third order polynomial and a range of different values of $\bar{\eta}$. Similarly as in the case of HS, the asymptotes of the steady-state characteristic of the overall system cannot be characterised for a generic case. Furthermore, the HBS structure can behave as a static-linear system in two cases. Firstly, if the function $f(\cdot)$ is linear, hence the HBS is reduced effectively to a BS submodel. In this case the condition for static-linearity is satisfied when $\bar{\eta} = 0$. Secondly, if the static nonlinear function $f(\cdot)$ is selected as an inverse of the steady-state characteristic corresponding to the component BS, cf. (11), i.e. $v_{ss} = \frac{(1-\bar{a})u_{ss}}{\bar{b}+\bar{\eta}u_{ss}}$ assuming $\bar{b} + \bar{\eta}u_{ss} \neq 0$. Note that because the steady-state characteristic of BS is strictly monotonic its inverse exists.

4. Dynamic properties

Similarly as when discussing behaviour in steady-state, dynamic properties of the HBS structure depend on its constituent components and hence dynamic properties of HS and BS structures are analysed first.

4.1. Dynamics of Hammerstein subsystem

One of the important attributes characterising the class of HS structures is that the eigenvalues of the corresponding linearised HS model are constant thus independent of an operating point. Therefore, the qualitative behaviour of HS models do not change with the reference and is determined exclusively by the linear subsystem, see [2], [1]. A direct consequence is that HS structures are not desirable when modelling processes with operating point dependant dynamics, such as those with a time-varying time-constant. Also, it is interesting to note that if the static nonlinearity is swapped with the dynamic linear subsystem leading to a WS, the same observation is no longer valid, see [15]. Consequently, HS are useful in cases where the system dynamics is reasonably linear and nonlinearities can be attributed to the steady-state behaviour.

4.2. Dynamics of bilinear subsystem

Consider the quasi-transfer function representation of a BS, which is obtained from (7) by imposing $v_k = u_k$, i.e.

$$G(q^{-1}) = \frac{\sum_{i=1}^{n_b} q^{-i} b_i}{1 - \sum_{j=1}^{n_a} q^{-j} \left[a_j + \sum_{i=1}^{n_b} \eta_{ij} u_{k-i}\right]}. \tag{12}$$

It is observed from (12) that both static and dynamic behaviour is input dependant. Consequently, the (time-varying equivalent) poles of the BS structure are determined by the instantaneous roots of the denominator in (12). This can be illustrated by a simple, yet representative, example of a first order diagonal BS model expressed in a quasi-transfer function form as $G(q^{-1}) = \frac{q^{-1}b_1}{1-q^{-1}(a_1+\eta_{11}u_{k-1})}$ with the time-varying discrete pole located at $a_1 + \eta_{11}u_{k-1}$. Considering, for simplicity, the case of $a_1+\eta_{11}u_{k-1} \geq 0$ and utilising a direct relationship between the continuous-time and discrete-time plane, the corresponding time-constant of a continuous-time counterpart is given by $-T_s/\ln(a_1 + \eta_{11}u_{k-1})$, where T_s denotes the sampling time. It can be deduced that with the expression $a_1 + \eta_{11}u_{k-1}$, hence the input, increasing in magnitude, the corresponding time-constant increases, i.e. the system dynamics becomes faster, and vice-versa.

Figure 3(a) shows this phenomenon for the introduced exemplary first order diagonal BS. Considering top right panel it is observed that, first, because $\eta_{11} < 0$ the BS exhibits a saturation steady-state nonlinearity and, second, that the dynamics become faster with an increasing magnitude of the input. This is verified by the upper right and the lower left panels, where the changing location of a discrete pole over time is plotted. It is seen that the pole moves towards the origin with the increasing magnitude of the input. The lower right panel shows the corresponding location of the time-constant of the continuous-time counterpart. It is observed that the equivalent time-constant decreases, proportionally to an inverse of the logarithm, confirming that the system becomes progressively faster. Moreover, it can be verified that if $a_1 + \eta_{11}u_{k-1} < 0$ the analysed BS will exhibit dynamics with damped oscillations.

Since the dynamics is affected by the input, the stability of BS structure is input dependent. This property is clearly observed in the case of the introduced exemplary first order diagonal BS model, where it is required that the time-varying pole stays within the unit circle, i.e. $|a_1 + \eta_{11}u_{k-1}| < 1$. This condition implies that the magnitude of the input must be restrained to $\left(\frac{1-a_1}{\eta_{11}}, \frac{-1-a_1}{\eta_{11}}\right)$. Also note that $|a_1| < 1$ for stability of the autoregressive linear part. With reference to [19] and [20], the corresponding sufficient conditions for a general BS structure to

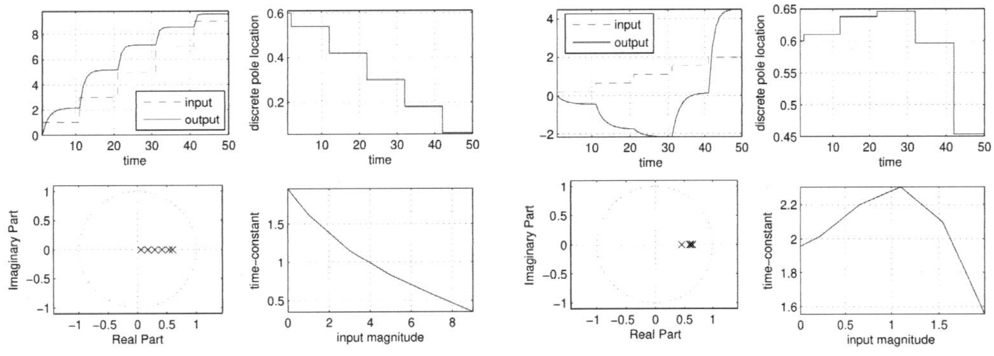

(a) First order diagonal BS. (b) Exemplary HBS.

Figure 3. Illustrative plots showing dependence of system dynamics on input.

remain bounded are given by

$$|p_i| < 1, \qquad M_u < \frac{\Pi_{i=1}^{n_a}(1-|p_i|)}{\sum_{i=1}^{n_a}\sum_{j=1}^{n_b}|\eta_{ij}|}, \tag{13}$$

where M_u denotes a finite magnitude of the input and p_i correspond to the n_a poles of a linear part of the BS structure, i.e. roots of the polynomial $1 - \sum_{j=1}^{n_a} q^{-j}a_j$.

4.3. Dynamics of Hammerstein-bilinear structure

Dynamic properties of the HBS are defined by the two component substructures. Note that this observation is in stark contrast to a HS structure, where the static nonlinearity has no impact on the overall qualitative system behaviour. In the case of the HBS structure such dependency is present because the dynamic part is input dependent, hence, is implicitly influenced by the nonlinear function that transforms it. Consequently and differently to the BS case, denominator coefficients of the HBS structure, cf. (7), do not directly depend on the instantaneous values of the input u_k but on the instantaneous values of the intermediate input v_k. This property increases the potential range of feasible dynamic behaviour that can be obtained, in particular, a range of possible dependency of system poles on the input is wider than that of the BS structure.

This observation is illustrated in Figure 3(b), where it is shown that the dependency of the discrete pole location on the input magnitude may well be of a non-monotonic nature - a phenomenon infeasible for BS models. The HBS structure comprises the arbitrarily parametrised exemplary first order diagonal BS defined previously followed by an arbitrary third order polynomial, i.e. $f(x) = \sum_{i=1}^{3} \alpha_i x^i$. It is observed that the dynamics of the system, becomes slower initially only to get faster once the input magnitude exceeds approximately 1.1. Consequently, the relationship between the time-constant of the corresponding continuous-time counterpart and the input magnitude is also distinctively dissimilar, which is observed in the lower right panel. Such a non-monotonic behaviour of the apparent time-constant was reported in [4] when modelling a distillation column with input and output being percentage change in a reflux flow and in the top composition, respectively.

The discussed reference dependent dynamics of the HBS structure, cf. equation (4), can be demonstrated explicitly by linearisation around a working point, denoted (u_k^*, y_k^*), and by examining subsequently the poles of the resulting linearised counterpart, i.e.

$$\Delta y_k = \sum_{j=1}^{n_a} \frac{\partial y_k}{\partial y_{k-j}}\bigg|_{(u_k^*, y_k^*)} \Delta y_{k-j} + \sum_{i=1}^{n_b} \frac{\partial y_k}{\partial u_{k-i}}\bigg|_{(u_k^*, y_k^*)} \Delta u_{k-i}, \tag{14}$$

where the deviation variables are denoted with Δ, e.g. $\Delta y_k = y_k - y_k^*$. The partial derivatives in (14) are given by

$$\frac{\partial y_k}{\partial y_{k-j}} = a_j + \sum_{i=1}^{n_b} f(u_{k-i})\eta_{ij}, \qquad \frac{\partial y_k}{\partial u_{k-i}} = \frac{\partial f(u_{k-i})}{\partial u_{k-i}} \left(b_i + \sum_{j=1}^{n_a} y_{k-j}\eta_{ij} \right). \tag{15}$$

The transfer function of the deviations from the working point is expressed as

$$\frac{\Delta y_k}{\Delta u_k} = \frac{\sum_{i=1}^{n_b} q^{-i} \left[\frac{\partial f(u_{k-i})}{\partial u_{k-i}}\Big|_{(u_k^*)} \left(b_i + \sum_{j=1}^{n_a} y_{k-j}^* \eta_{ij} \right) \right]}{1 - \sum_{j=1}^{n_a} q^{-j} \left[a_j + \sum_{i=1}^{n_b} f(u_{k-i}^*)\eta_{ij} \right]}. \tag{16}$$

Considering (16), it is seen that the local dynamics of the HBS is governed by the roots of the denominator, whose coefficients are functions of both the working point and $f(\cdot)$. Note that a similar observation is valid with respect to the numerator coefficients, too. This shows, therefore, that, first, the dynamics of the HBS structure is reference dependent and, second, that it also depends on the selection of $f(\cdot)$.

Conditions for stability of the HBS structure can be inferred from the analogy to the BS, see (13), where the crucial difference is that M_u is substituted with M_v, denoting a finite magnitude of the intermediate input signal v_k. Only when $f(\cdot)$ is strictly monotonic, by using the relation $u_k = f^{-1}(v_k)$, this condition can be expressed explicitly in terms of the magnitude M_u of the actual input that excites the HBS structure.

Table 1. Overview of properties ($\surd$ - presence, $\times$ - absence) of selected model structures.

Structure / Property	IDS	IM	OM	IDD
linear	$\times$	$\times$	$\times$	$\times$
Hammerstein	$\times$	$\surd$	$\times$	$\times$
Wiener	$\times$	$\surd$	$\times$	$\surd$
bilinear	$\surd$	$\times$	$\times$	$\surd$
Hammerstein-bilinear	$\surd$	$\surd$	$\times$	$\surd$
Lure	$\surd$	$\times$	$\surd$	$\surd$

5. Discussion of properties of Hammerstein-bilinear structure with relation to other nonlinear models

Crucial properties of all discussed structures are juxtaposed in Table 1 that is partially based on that given in the survey paper [14]. Linear model structures are also included for completeness. The following abbreviations are used: IDS - input dependent stability, IM - input multiplicity, OM - output multiplicity, IDD - input dependent dynamics. Generally, HS structure can be treated as a first step towards an increase of modelling flexibility of ordinary linear structures. It allows a nonlinear steady-state gain with IM to be obtained but the transient behaviour remains unchanged over the entire operating range. The next step could be to consider the WS structure which additionally offers the IDD. However, it is worth emphasising, citing [8], that 'this behaviour is near the limit of the Wiener model's qualitative capability' and achievable at a cost of complex static elements that are undesirable in practice. In contrast, BS are inherently flexible enough to exhibit strong IDD, which comes at a price of IDS, thus a need of restricting magnitude of the input, cf. (13). Also, contrary to both HS and WS, BS structures lack possibility of IM and can exhibit only a single family of steady-states curves, i.e. hyperbolic, as shown in Figure 2. It must be added that although structurally WS possess capability of IM, to facilitate identification and control, it is, in fact, usually assumed that the static nonlinearity is invertible. This means that the potential for IM is inevitably lost, see [21], [22], [2], [15] for such cases. Moreover, WS models seem to be less popular than for instance HS structures, which is likely due to increased difficulty in their identification [1]. HBS structures add to the repertoire of BS the potential of IDS and widen variety of feasible steady-states and dynamics, at the same time retaining capability of a strong IDD. In applications where the process exhibits OM other nonlinear structures, such as Lure models, are required. Note that Lure models, similarly as HBS, are capable of exhibiting IDS and IDD, but in contrast to HBS are unable of IM. In [14] nonlinear structures with static nonlinearities in the forward path, such as HS and WS, are classified as being 'mildly' nonlinear, whilst BS, due to IDS, are placed in the 'intermediately' nonlinear class. Lure models, because of OM, are categorised as 'strongly' nonlinear. Following this reasoning, the HBS structures, although remain in an 'intermediately' nonlinear class, can be considered as a one step further towards a strongly nonlinear class of models.

6. Industrial case study

In this section the HBS structure is used for modelling an air handling unit, which is a part of an industrial heating, ventilation and air-conditioning system used to maintain specific tight and stable environmental conditions during the production of blood glucose test strips. The main goal of modelling is subsequent control synthesis and tuning to optimise the overall energy consumption.

The system is represented schematically in Figure 4 and consists of a standard air-to-water heat exchanger, where an inflow air is cooled down by cold water. In nominal conditions the inflow air, cold water mass-flow and the inflow cold water temperature are all approximately constant. Therefore, for modelling purposes only two inputs, i.e. inflow air temperature and valve stem position, and a single output of interest, i.e. outflow air temperature, are considered. The main motivation for considering the HBS structure for this particular application follows from first principles modelling conducted previously in [23] and [24]. These investigations have uncovered that, firstly, it is advantageous to account explicitly for the nonlinear valve characteristic and approximate it by a HS-type nonlinearity. Secondly, that the dynamics of the air handling unit is operating point dependant and can be modelled by a BS. Thirdly, that the bilinear product term has physically meaningful significance and stems directly from energy

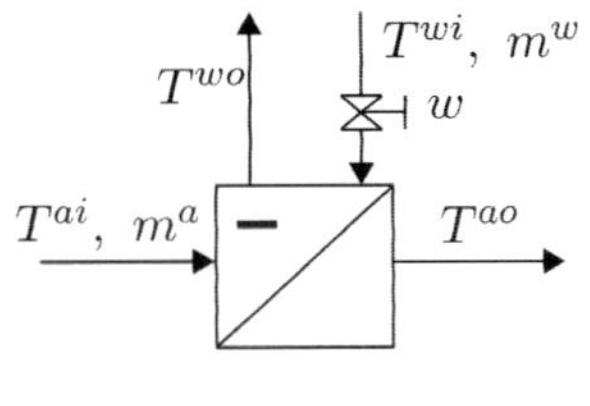

Figure 4. Schematic representation of the air-handling unit, where: w - valve stem position [%], T^{ai} - inflow (on coil) air temperature [K], T^{ao} - outflow (discharge) air temperature [K], T^{wi} - inflow cold water temperature [K], T^{wo} - outflow cold water temperature [K], m^w - cold water mass-flow rate [kg/s], m^a - inflow air mass-flow [kg/s].

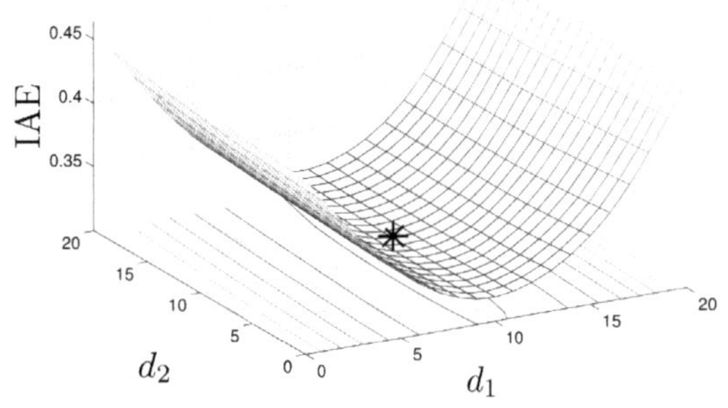

(a) The best solution at $(10, 10)$ is indicated.

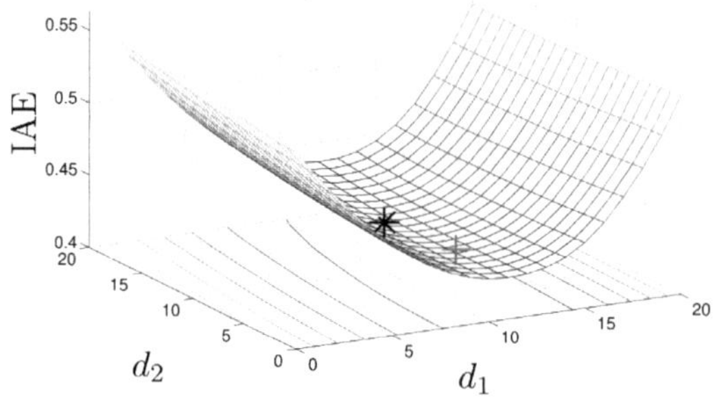

(b) The value corresponding to the best model estimated from the identification data set at $(10, 10)$ is indicated by a star, whilst the minimum of the surface $(11, 5)$ is marked by a cross.

Figure 5. (a) - identification and (b) - validation results in terms of the IAE criterion plotted in function of delays corresponding to the two inputs.

balance equations. Consequently, the following HBS model structure is proposed to model the air-handling unit

$$T_k^{ao} = a_1 T_{k-1}^{ao} + b_1 v_{k-1-d_1} + \eta_{11} T_{k-1}^{ao} v_{k-1-d_1} + b_1^* T_{k-1-d_2}^{ai}, \qquad v_k = \sum_{l=1}^{5} \alpha_i w_k^l, \tag{17}$$

where d_1 and d_2 denote unknown transport delays. Although properties of single-input single-output HBS structures were discussed in this paper exclusively, their extension to two-input single-output case defined by (17) should be straightforward, especially because only w_k is transformed through the static polynomial nonlinearity of 5th order.

Since both equations in (17) are separately linear in their parameters, this property is exploited by making use of a so-called bilinear parametrisation method, see [25]. This approach has been applied for the purpose of HS identification in [26] and identification of errors-in-variables problems in [27], where it has been shown to possess numerically appealing properties. Its consistency and convergence properties when applied to HS have been examined in [28].

Two sets of data sampled at 5s, referred to as the identification and validation data set, are used giving 7,800 and 9,000 data points, respectively. Valve position is scaled to 10 to ease plotting, also all data is normalised by subtracting bases that are assumed to be the corresponding first values of signals. The identification algorithm is stopped when a maximal number of iterations, selected as 100, is reached or when $\|\hat{\theta}_i - \hat{\theta}_{i-1}\|_2 < 10^{-6}$, where $\hat{\theta}_i$ is the estimate of the parameter vector at i-th iteration. To quantify the identification results two performance criteria are used, i.e. the normalised integral absolute error (IAE) and the coefficient of determination (R_T^2), defined, respectively, as follows $\text{IAE} = \frac{1}{N}\sum_{k=1}^{N} |y_k - \hat{y}_k|$ and $R_T^2 = 100\left(1 - \|y - \hat{y}\|_2^2 / \|y - \bar{y}\|_2^2\right)$, where $\hat{y}_k$ is the identified model output, y and $\hat{y}$ are vectors built from the actual output and the model output, respectively, and $\bar{y}$ denotes the mean value of y. The notation $\|\cdot\|_2$ stands for the square norm of a vector and N is number of data samples.

Because not only the parameters but also the transport delays are unknown, the identification has to be repeated for all combinations of d_1 and d_2 in a predefined range. Results obtained in terms of the IAE criterion with both delays restricted to [0, 20] samples are shown in Figure 5(a), whilst the corresponding validation results are presented in Figure 5(b). The smallest value of IAE=0.316, marked in Figure 5(a) by a star, is obtained at $(d_1, d_2) = (10, 10)$. For completeness,

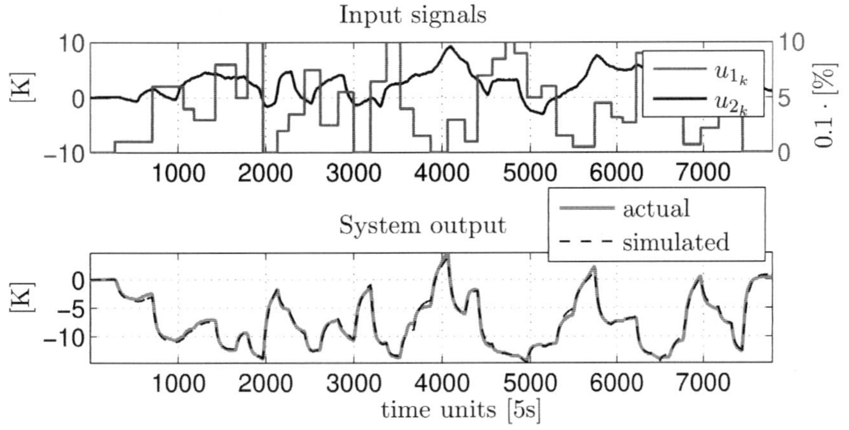

(a) Identification dataset, $R^2_T = 99.2\%$.

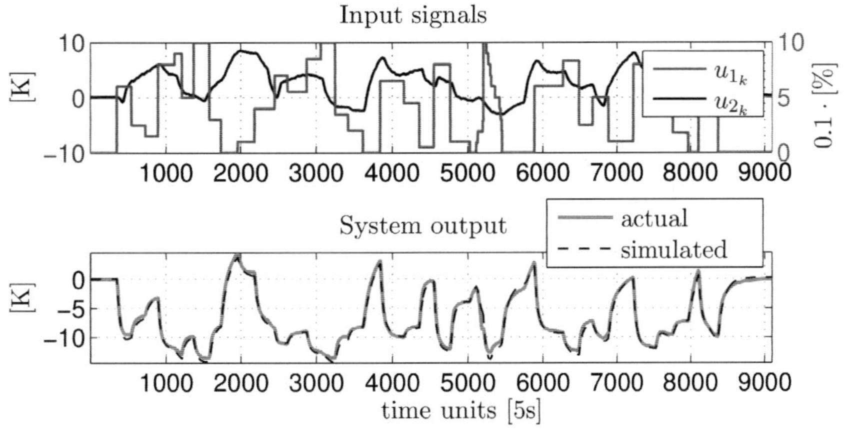

(b) Validation dataset, $R^2_T = 98.6\%$

Figure 6. Results showing performance of the identified HBS model.

the value corresponding to $(10, 10)$ is also marked in Figure 5(b), where the actual minimum at $(11, 5)$ is denoted by a cross. Considering the coordinates of the minima in both figures it is noted that whilst the first coordinate is similar, the second differs quite significantly. However, by analysing shape of surfaces, it is observed that in both cases they resemble three dimensional parabolas with a gradient visibly greater in the direction of d_1 than d_2. This means that the selection of d_1 is considerably more important than d_2. Consequently, the apparent discrepancy in the second coordinate between the two data sets is of minor significance. This observation is supported also by a relatively small difference of the validation IAE values corresponding to the points $(10, 10)$ and $(11, 6)$ which are 0.431 and 0.428, respectively. Therefore, it is $d_1 = d_2 = 10$ that is selected. Visual results in terms of the simulated output together with the input signals used are given in Figure 6(a) and 6(b) for the identification and validation data set, respectively. It can be observed that in both cases the model output matches the actual system output very well. The corresponding values of the R^2_T criteria are 99.2% and 98.6%, for the identification and validation data set, respectively, indicating that on average only approximately 1% of output variation remains unexplained by the identified model. The estimated parameters of the model

Journal of Physics: Conference Series **570** (2014) 062003 doi:10.1088/1742-6596/570/6/062003

(17) together with corresponding standard deviations are: $\hat{a}_1 = 9.846 \cdot 10^{-1} \pm 8.215 \cdot 10^{-5}$, $\hat{b}_1 = -4.977 \cdot 10^{-3} \pm 2.536 \cdot 10^{-5}$, $\hat{b}_1^* = -1.697 \cdot 10^{-4} \pm 1.715 \cdot 10^{-6}$, $\hat{\eta}_{11} = 8.813 \cdot 10^{-3} \pm 8.304 \cdot 10^{-5}$, $\hat{\alpha}_1 = 6.233 \cdot 10^{-1} \pm 2.629 \cdot 10^{-2}$, $\hat{\alpha}_2 = 9.918 \cdot 10^{-2} \pm 2.296 \cdot 10^{-3}$, $\hat{\alpha}_3 = -3.752 \cdot 10^{-3} \pm 6.810 \cdot 10^{-5}$, $\hat{\alpha}_4 = 4.886 \cdot 10^{-5} \pm 8.208 \cdot 10^{-7}$, $\hat{\alpha}_5 = -2.104 \cdot 10^{-7} \pm 3.449 \cdot 10^{-9}$. It is observed that the standard deviations are all at least of an order smaller than the values of the corresponding estimated parameters. These results indicate high confidence in the estimates and hence, in turn, also in the identified HBS model.

7. Summary

A class of Hammerstein-bilinear systems (HBS) has been introduced and its properties of main practical interest for modelling have been analysed and discussed in detail. An industrial case study has demonstrated application of HBS structures to an exemplary real-world problem.

Acknowledgments

Authors would like to thank Mr Dean Hill from Abbott Diabetes Care UK for providing data used in Section 6.

References

[1] Pearson R K 1995 *J. of Process Control* **5** 197–211
[2] Pearson R K and Pottmann M 2000 *J. of Process Control* **10** 301–315
[3] Aguirre L, Corrêa M and Cassini C 2002 *IEE Proc.- Part D: Control Theory and Applications* **149** 343–348
[4] Eskinat E, Johnson S H and Luyben W L 1991 *AIChE J.* **37** 255–268
[5] Ikonen E and Najim K 2001 *Proc. of the Institution of Mechanical Engineers, Part I: J.of Systems and Control Engineering* **215** 15–27
[6] Dunoyer A, Balmer L, Burnham K J and James D J G 1996 *Proc. of 13th IFAC World Congress* vol F (San Francisco, USA) pp 347–352
[7] Pottmann M and Pearson R K 1998 *American Institute of Chemical Engineers* **44** 131–140
[8] Pearson R K 1999 *Discrete-time dynamic models* (Oxford University Press, New York, USA)
[9] Meadows E and Rawlings J 1997 *Model predictive control* (Prentice Hall, Englewood Cliffs, NJ) chap 5, pp 233–310
[10] Ekman M 2005 *IEEE Trans. on Control* vol 13(1) pp 162–168
[11] Martineau S, Burnham K J, Haas O C L, Andrews G and Heeley A 2004 *Control Engineering Practice* **12** 457–464
[12] Rogers E, Freeman C T, Lewin P L and Owens D H 2009 *20th Int. Conf. on Systems Engineering* pp 1–12
[13] Leontaritis I J and Billings S A 1985 *Int. J. of Control* **41** 303–328
[14] Pearson R K 2003 *J. of Process Control* **13** 1–26
[15] Aguirre L A, Coelho M C S and Correa M V 2005 *IEE Proc. of Control Theory and Applications* **152** 349–356
[16] Mohler R R 1973 *Bilinear Control Processes: With Applications to Engineering, Ecology and Medicine* (*Mathematics In Science and Engineering* vol 106) (UK: Academic Press)
[17] Mohler R R and Kolodziej W J 1980 *IEEE Trans. on Systems, Man, and Cybernetics* vol 10 pp 683–688
[18] Wellstead P E and Zarrop M B 1991 *Self-Tuning Systems: Control and Signal Processing* (England: John Wiley & Sons)
[19] Lee J and Mathews V J 1994 *IEEE Trans. on Signal Processing* **42** 1871–1873 ISSN 1053-587X
[20] Johnson K K and Sandberg I W 1998 *IEEE Trans. on Signal Processing* **46** 2056–2058 ISSN 1053-587X
[21] Wigren T 1993 *Automatica* 1011–1025
[22] Kalafatis A, Arifin N, Wang L and Cluett W L 1995 *Chemical Engineering Science* 3693–3701
[23] Zajic I, Larkowski T, Hill D and Burnham K J 2011 *Proc. of 18th IFAC World Congress* (Milan, Italy) pp 4576–4581
[24] Zajic I, Larkowski T, Sumislawska M, Hill D and Burnham K J 2011 *Proc. of 21st Int. Conf. on Systems Engineering* (Las Vegas, USA) pp 59–61
[25] Ljung L 1999 *System Identification - Theory for the User* 2nd ed (New Jersey, USA: Prentice Hall PTR)
[26] Le F, Markovsky I, Freeman C T and Rogers E 2010 *Control Engineering in Practice* **18** 396–407
[27] Larkowski T, Linden J G and Burnham K J 2009 *Proc. 6th Int. Conf. on Informatics in Control, Automation and Robotics* (Milan, Italy) pp 142–147
[28] Bai E W and Li K 2010 *Automatica* **46** 1891–1896 ISSN 0005-1098

Practical issues in modelling the temperature for the control of smart buildings

Domenico Gorni, Antonio Visioli

Dipartimento di Ingegneria dell'Informazione, Brescia University, Via Branze 38 2513 Italy,
Dipartimento di Ingegneria Meccanica Industriale, Brescia University, Via Branze 38 2513 Italy

E-mail: d.gorni001@unibs.it, antonio.visioli@unibs.it

Abstract. In this paper we propose a technique for modelling the temperature of the rooms of a smart building for control purposes. The approach consists in determining an (black-box) ARMAX model based on the first-principle thermodynamics equations of the system. Then, the model parameters are estimated by initially employing an ad hoc procedure and then by using a self-calibration method when the performance deteriorates. The selection of the user defined parameters in the overall procedure is thoroughly discussed by means of simulation results in order to highlight the practical effectiveness of the methodology.

1. Introduction

According to The World Business Council for Sustainable Development (WBCSD) the building sector consumes more than 40% of the world energy use with the resulting carbon emission more than the transportation sector [1]. A new study on energy efficiency in buildings indicates that the global building sector needs to reduce of 60% its energy consumption before the 2050 to help fulfilment of the global climate change targets. Furthermore, the current energy prices are at an historically high level; thus, a reduction of energy waste in Heating, Ventilation and Air-Conditioning (HVAC) systems needs to be provided. Engineers and architects have influenced the energy use of the new buildings through the design of the envelope, the HVAC systems selection and the operation sequences specification. However, when the building has been finished, the consumption of energy is decided mainly by its control, maintenance and by its occupants. Accurate, simple models of the building thermodynamics provide bases for intelligent and optimal control [6, 18], which have potential to reduce energy consumption of HVAC systems [9, 13]. The most important fact, from a control engineering point of view, which decreases the efficiency of the HVAC systems, is that they are set to operate at designed thermal loads. However, these loads are time-varying and a lot of different parameters affect their variation. Furthermore, another important goal of the intelligent buildings is to maintain, or even improve, the users' comfort [4, 5]. Modelling the temperature of s building has been an active research field for decades, many software packages have been developed such as EnergyPlus, Modelica, TRNSYS, etc. [11, 17]. They use validated thermodynamic models to describe as accurately as possible the thermal processes in buildings. However, these models are too complex to be used for the purpose of designing model based controllers. Hence, simplified models have been proposed in order to provide a good approximation of the thermal behaviour of smart buildings. In particular, physical based models, for instance RC-networks, describe the

thermal system as a network of resistors and capacitors connected in series and in parallel form [10, 11, 12, 15]. In order to simplify the model, different kinds of grey-box, semi-physical models that are partially based on physical equation and partially on empirical knowledge, have been developed [14]. To further decrease the complexity of the modelling task, black-box models, such as ARMAX or neural-networks models [8, 7, 16], have been used to build a relations between inputs and outputs of the system by using measurements only.
In this paper we propose a methodology for the determination of a (simple) model of the temperature in the rooms of a smart building with the purpose of using it for the implementation of model-based control algorithm (for example, model predictive control strategies). First, the appropriate structure of an ARMAX model is determined by considering the first principle thermodynamics equations that describe the thermal behaviour of each single room in a house. Then, the procedure for the estimation of the parameters consists of a performing an ad hoc experiment in the first day and then by applying a self-calibration technique when the modelling error increases above a given threshold. Practical issues related to the implementation of the overall procedure, and in particular related to the choice of user-defined parameters, are discussed in order to provide useful guidelines for an effective use of the method.

2. Model description

The thermodynamic fundamental equation that describes the variation of the temperature of a single mass is:

$$Q = c_s m \frac{dT}{dt} \tag{1}$$

where Q [W] is the total amount of heat exchanged, c_s $\left[\frac{J}{kgK}\right]$ is the specific heat, m [kg] is the mass, T [K] is the temperature and t [s] is the time. By applying this equation to a room it is possible to describe the variation of its temperature by taking into account simplifying assumptions, such as: (i) the air inside the room is supposed to be at the same temperature; (ii) the room is considered a closed system with no mass exchange with the external environment.

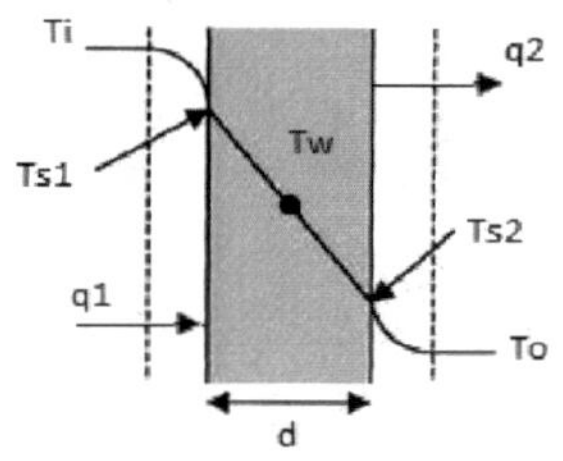

Figure 1. Temperature gradient and heat flow through a wall between two different rooms

The thermal flow can be transmitted from a room to another one passing through a wall as shown in Figure 1. This phenomenon occurs when the heat is exchanged from a fluid, such as air, to another one passing through a solid body, such as a window, a door or a wall. The wall is characterized by an own thermal capacity and a temperature that changes according to the quantity of heat that flows through it. Thus, by using (1), we obtain:

$$q_1 - q_2 = \Delta Q = C_w \frac{dT_w}{dt} \tag{2}$$

where q_1 [J] is the heat coming from the warmer room, q_2 [J] is the heat lost toward the colder room, T_i [K] and T_o [K] are the temperatures of the two rooms, T_{s1} [K] and T_{s2} [K] are the temperatures of the two wall surfaces, T_w [K] is the temperature and C_w $\left[\frac{J}{K}\right]$ is the thermal capacity of the wall. Considering the heat balance on the surface between the wall and the room 1 it is possible to determine q_1 and q_2 as:

$$\begin{aligned} q_1 &= h_i A(T_i - T_{s1}) = \frac{KA}{\frac{d}{2}}(T_{s1} - T_w) \\ q_2 &= h_o A(T_{s2} - T_o) = \frac{KA}{\frac{d}{2}}(T_w - T_{s2}) \end{aligned} \tag{3}$$

where h_i $\left[\frac{J}{m^2K}\right]$ and h_o $\left[\frac{J}{m^2K}\right]$ are the convective coefficient, K $\left[\frac{J}{mK}\right]$ is the thermal conductance of the wall, A [m^2] is the area of the surface and d [m] is the thickness of the wall. By combining these two equations and substituting them in (2), we obtain:

$$C_w \frac{dT_w}{dt} = \frac{1}{R_1}(T_i - T_w) - \frac{1}{R_2}(T_w - T_o) \tag{4}$$

where R_1 and R_2 are coefficients that represent the combination of convection and conduction resistances.
A single room is usually bordering with more different areas, for example with the ground through the floor, or with the external environment or with other rooms. Thus, it is important to take into account all the different heat flows to estimate the temperature evolution of a room. Moreover, heat sources like HVAC systems that can heat or cool the room are also usually present.
The whole system can now be written as follows:

$$\begin{cases} C_i\dot{T}_i = \sum_{j=1}^{k} q_j + \sum_{j=k+1}^{n} q_j = \sum_{j=1}^{k} \frac{1}{R_j}(T_i - T_{wj}) + \sum_{j=k+1}^{n} q_j \\ C_{w1}\dot{T}_{w1} = \frac{1}{R_1}(T_i - T_{w1}) - \frac{1}{R_{o1}}(T_{w1} - T_{o1}) \\ \dots \\ C_{wk}\dot{T}_{wk} = \frac{1}{R_k}(T_i - T_{wk}) - \frac{1}{R_{ok}}(T_{wk} - T_{ok}) \end{cases} \tag{5}$$

where T_{ok} are the temperatures of the $k-th$ adjacent room, $q_1 \dots q_k$ are the heat flows gained and lost to the external rooms and $q_{k+1} \dots q_n$ are the exogenous heat sources like HVAC systems, the direct irradiation from the sun, electrical devices, etc.
The aim is to write a thermal model of the room based on:

- the temperature of the adjacent areas;
- the measured heat flow, or an estimation of it, of the exogenous inputs.

By applying the Laplace transform to (5) the system can be rewritten as:

$$\begin{cases} sT_i(s)C_i = \sum_{j=1}^{k} \frac{1}{R_j}(T_i(s) - T_{wj}(s)) + \sum_{j=k+1}^{n} q_j(s) \\ sT_{w1}(s)C_{w1} = \frac{1}{R_1}(T_i(s) - T_{w1}(s)) - \frac{1}{R_{o1}}(T_{w1}(s) - T_{o1}(s)) \\ \dots \\ sT_{wk}(s)C_{wk} = \frac{1}{R_k}(T_i(s) - T_{wk}(s)) - \frac{1}{R_{ok}}(T_{wk}(s) - T_{ok}(s)) \end{cases} \tag{6}$$

As it is shown in (6), the temperature of the adjacent rooms appears only in the wall thermal balance equations, therefore the temperatures of the walls can be collected from all but the first equations and substituted into the first one. By assuming that the room has a parallelepiped shape ($k = 6$) and each wall border has a different area (four lateral walls, floor and ceiling), the transfer function between the temperature of a room and the temperature of the other rooms becomes a seventh order transfer function. Note that T_i depends on several inputs, like temperatures of the border rooms and exogenous inputs in general, and unknown parameters:

$$T_i(s) = f(T_{o1}(s), \dots, T_{o6}(s), q_7(s), \dots, q_n(s); R_1, \dots, R_6, R_{o1}, \dots, R_{o6}, C_i, C_{w1}, \dots, C_{w6}) \quad (7)$$

The second step is to transform this continuous transfer function in a discrete time transfer function by applying the Tustin bilinear transform [3] (for the sake of clarity we assume now that all the exogenous inputs are equal to zero):

$$s = \frac{2(z-1)}{\tau(z+1)} \quad (8)$$

and therefore, by substituting the s value into (7), the discrete time transfer function of the room is obtained:

$$T_i(z) = \frac{(a_{10}z^7 + \dots + a_{16}z + a_{17})T_{o1}(z) + \dots + (a_{60}z^7 + \dots + a_{66}z + a_{67})T_{o6}(z)}{b_{10}z^7 + \dots + b_{16}z + b_{17}} \quad (9)$$

By collecting the term z^7, we obtain:

$$T_i(z) = \frac{(a_{10} + \dots + a_{17}z^{-7})T_{o1}(z) + \dots + (a_{60} + \dots + a_{67}z^{-7})T_{o6}(z)}{b_{10} + \dots + b_{17}z^{-7}} \quad (10)$$

and therefore,

$$b_{10}T_i(z) = (a_{10}+\dots+a_{17}z^{-7})T_{o1}(z)+\dots+(a_{60}+\dots+a_{67}z^{-7})T_{o6}(z)-(b_{11}z^{-1}+\dots+b_{17}z^{-7})T_i(z) \quad (11)$$

which can be easily rewritten as follows:

$$T_i(k) = \mathbf{g_1} \cdot \mathbf{T}(k) + \dots + \mathbf{g_8} \cdot \mathbf{T}(k-7) \quad (12)$$

where,

$$\mathbf{g_1} = \begin{bmatrix} \frac{a_{10}}{b_{10}} \\ \vdots \\ \frac{a_{60}}{b_{10}} \\ 0 \end{bmatrix} \quad \dots \quad \mathbf{g_8} = \begin{bmatrix} \frac{a_{17}}{b_{10}} \\ \vdots \\ \frac{a_{67}}{b_{10}} \\ \frac{b_{17}}{b_{10}} \end{bmatrix} \quad \text{and} \quad \mathbf{T} = \begin{bmatrix} T_{o1} \\ \vdots \\ T_{o6} \\ T_i \end{bmatrix} \quad (13)$$

Therefore by applying the same consideration to the exogenous inputs, the complete black-box model is defined as follow:

$$T_i(k) = \sum_{j=1}^{8} \{\mathbf{g_j}\mathbf{T}(k-j+1) + \mathbf{p_j}\mathbf{q}(k-j+1)\} \quad (14)$$

where

$$\mathbf{p_1} = \begin{bmatrix} l_{11} \\ \vdots \\ l_{1m} \end{bmatrix}, \quad \dots \quad \mathbf{p_8} = \begin{bmatrix} l_{81} \\ \vdots \\ l_{8m} \end{bmatrix}, \quad \mathbf{q} = \begin{bmatrix} q_1 \\ \vdots \\ q_m \end{bmatrix} \quad (15)$$

Furthermore the system is also characterised from measurement noise, thus, the black-box model becomes an ARMAX model:

$$T_i(k) = \sum_{j=1}^{8} \{\mathbf{g_j}(\mathbf{T}(k-j+1) + \varepsilon(k-j+1)) + \mathbf{p_j}\mathbf{q}(k-j+1)\} \quad (16)$$

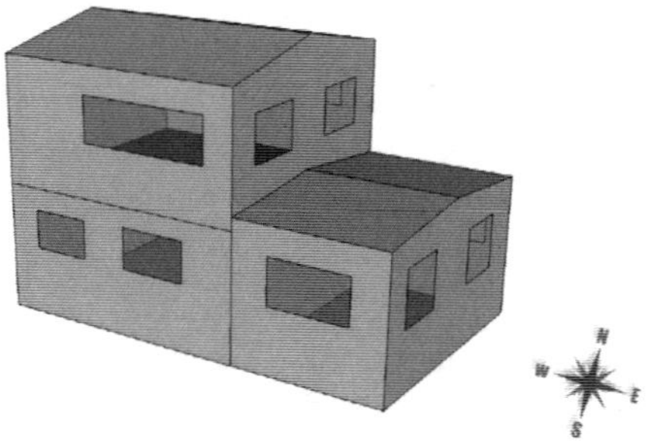

Figure 2. Model of test building implemented in SketchUp 2013.

where ε denotes the measurement noises.
The parameters estimation technique that is used to evaluate all the unknown parameters is the well-known least squares methodology [3]. This method is applied on (12) to determine all the $\mathbf{g_j}$ and $\mathbf{p_j}$ unknown parameters. Even though this equation is a direct consequence of the thermodynamical equations, during all the passages the parameters have lost their physical meaning, in fact such ARMAX models are defined black-box models.

3. Parameters identification methodology

In order for the parameters estimation methodology to be effective, it is essential to take into account that suitable exciting signals are necessary but, most of all, it has to be taken into account that all the nonlinearities (for example due to the solar irradiance effect) present in the true systems [4] have been neglected. For this reason, a self-calibrating methodology is proposed in order to automatically update the estimation when the operating conditions change.
The procedure consists therefore in performing an experiment during the first 24 hours (which can be any day of the year, when the fan-coil units are employed) and in estimating the ARMAX parameters based on the collected data. In this context, the choice of the set-point signals for the different rooms represents a crucial issue (see Section 5.1). Then, each x_1 hours, the estimation temperature error for the ith room, defined as

$$e_i = \frac{\sum_k |T_{oi}(k) - T_i(k)|}{N} \tag{17}$$

is evaluated, where N is the number of samples in the considered interval which is determined by taking into account the preceding x_2 hours. If the estimation error (17) is greater than a given threshold x_3, then the least squares procedure is applied again by using the data collected in the preceding x_4 hours. In this context a maximum variation of $\pm 5\%$ for each parameter is allowed in order to account for typically slow weather changes from one season to another.
Note that x_i, $i = 1, \ldots, 4$ are design parameters that have to be selected properly in order for the overall procedure to be effective. The choice of these parameters, as well of the other factors which might influence the results is discussed in Section 5.

4. Model of the test building

In order to discuss the practical issues involved in the method described in the previous section, a large number of simulations have been performed by using TRNSYS (using a sampling period of 4 min). In this paper we consider tests performed on the two floor building shown in Figure 2, which has been implemented in SketchUp 2013 [2]. The ground floor is composed by two rooms while there is just one room in the first floor. The characteristics of the three rooms are given in Table 1.

Room	Dimensions	Windows
East side ground floor	3x4x2.5	4
West side ground floor	5x4x2.5	5
First floor	5x4x2.5	4

Table 1. Area and number of windows of each room in the test building.

Element	u-value $\left[\frac{W}{m^2K}\right]$
external walls	0.510
ground floor	0.039
external roof	0.316
internal walls	0.508
internal ceiling	4.153
external windows	2.890

Table 2. Thermal conductivities of the test building.

The building is fictitiously placed in Milan Linate (Italy) and it is not surrounded by other houses. A file with the weather data of the whole year is available and therefore, by using the TRNSYS functionality to introduce it into the simulation algorithm, it is possible to take into account the real environmental conditions.
The most important parameters that characterize the building are the thermal conductivities (u-values) of the walls and of the windows. These parameters describe the thermal insulation coefficients of the main parts of the building and therefore they are responsible of the heat exchange rate between the rooms. They have been set as written in Table 2. The employed HVAC system consists of one fan-coil unit (Type 600 block in the TRNSYS HVAC library) and one temperature sensor for each room to keep the system as simple as possible. The temperature is controlled by means of PID low level controllers (see subsection 5.3).

5. Practical issues

As already mentioned in Section 3, for an effective implementation of the estimation methodology it is essential to appropriately choose the design parameters x_i, $i = 1, \dots, 4$ as well as considering other relevant issues as the set-point signals to be employed during the first 24 hours and the tuning of the PID controllers for the regulation of the temperature of each room.
In order to simplify the analysis (and the design in practical cases), a reasonable selection can be done for the parameters x_1, x_2 and x_3. In particular, the value x_1 can be fixed equal to 8 hours in order to provide an accurate model without increasing too much the computational burden. By applying a similar reasoning, x_2 is selected equal to 12 hours and x_3 equal to 1 K. The other design choices are discussed in the following subsections (where simulation results are obtained with the previous values of x_1, x_2 and x_3).

5.1. Signals for the initial estimation

Since the first day estimation plays obviously a key role in the overall method, it is essential to accurately select the set-point signals to be employed in this context. From a practical point of view, it would be very nice to employ routine operating set-points, in order to avoid a possible uncomfortable situation for the user. However, the result provided in this case depend on the user's choice and they are in general not accurate, that is, the self-calibrating procedure will be applied many times in the subsequent days, as it will be shown hereafter. In order to solve this issue we propose to use a pseudo-random signal, which is built by following some restrictions related to the comfort of the residents:

Journal of Physics: Conference Series **570** (2014) 062004 doi:10.1088/1742-6596/570/6/062004

- the upper acceptable temperature is 22°C;
- the lower acceptable temperature is 17°C;
- the set-point of each room is changed each hour and it can increase or decrease only with a $\Delta T = 1$.

Obviously, the first two constraints are related to the case where the first 24 hours are in the cold season but the same rationale can be easily applied to other periods. Actually, a difference of five degrees is necessary to provide a good estimation and it has to be stressed that the values are in any case in a liveable temperature range.
The advantage of using a pseudo-random signal is illustrated in Figures 3 and 4 where the proposed method is compared with the use of routine set-points. They have been selected as shown in Tables 3 and 4. Two cases are considered where different pseudo-random signals and different routine set-point signals are chosen (obviously, if the set-point signals are selected by the user, it should be taken into account that the choice is entirely subjective and this also represents a drawback with respect to the pseudo-random signal approach. All the simulations have been performed by using well-tuned PID controllers (see subsection 5.3) and a value $x_4 = 48$ (see subsection 5.2). They are related to the first days of November (however, the considerations done hereafter can be done also for other periods of the year). From this illustrative examples it appears that a more accurate model is obtained by using the pseudo-random set-point signals than by using the routine set-point signals. Indeed, the model parameters need to be updated much less times during the following 19 days (in both cases a maximum of three times per room with the pseudo-random signals and seven times with the user-chosen signals).

Room	Time intervals (hours of the day)					
	0-6	6-8	8-12	12-16	16-22	22-24
East side ground floor	17	17	21	21	21	18
West side ground floor	17	18	20	20	20	17
First floor	21	21	21	17	20	20

Table 3. Case 1 of the user-chosen set-point signals (in Celsius degrees) for the first 24 hours.

Room	Time intervals (hours of the day)	
	0-16	16-24
East side first floor	19	18
West side first floor	20	21
Second floor	18	17

Table 4. Case 2 of the user-chosen set-point signals (in Celsius degrees) for the first 24 hours.

5.2. Size of recalibration dataset array

Regarding the selection of the parameter x_4, namely, of the size of the array of data to be used for a new least-squares estimation of the model parameters when the error exceeds the given threshold, the following reasoning can be done. From one point of view, in order to reduce the computational burden and to consider the change of climate along the year it is sensible to use a small amount of data, for example 24 hours. However, this choice can introduce errors because the less samples are used the less filtered is an unusual behaviour of the system. For instance, in winter season the model is tuned to work with low environmental temperature and if a day hotter than usual occurs, the model error can exceed the 1^oC threshold and the algorithm will estimate

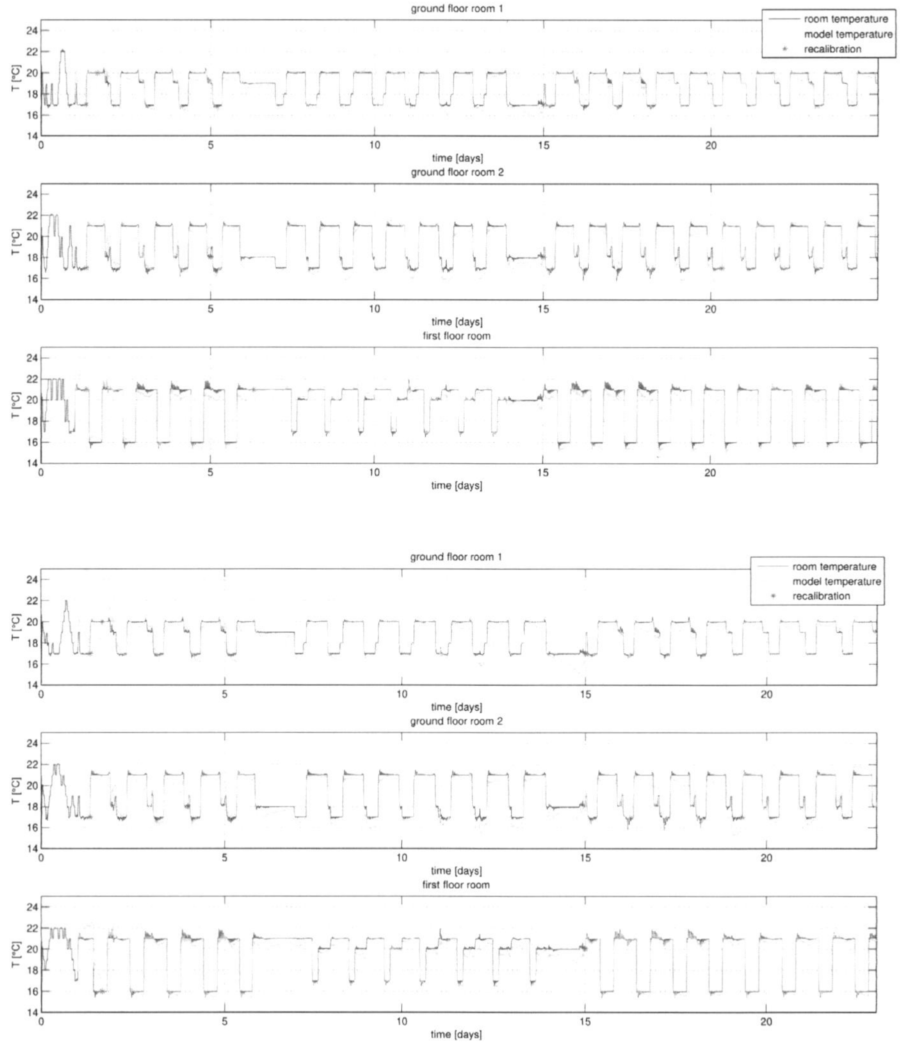

Figure 3. Results related to the use of pseudo-random set-point signals. Top: case 1. Bottom: case 2. Blue line: measured temperature. Green line: model temperature. The application of the recalibration strategy is indicated by '*'.

again the parameters. However, if only last 24 hours are taken into account, the parameters will be estimated based on a uncommon winter day and this may provoke another error the day after. On the other hand if 48 hours data are used, the model will be recalibrated based on two days data and the winter day will filter the contribution of the unusual one. For this reason, a suitable choice is to select $x_4 = 48$ hours. Illustrative examples with $x_4 = 24$ hours choice are shown in Figure 5 (two examples with a different pseudo-random initial set-point signals are considered) and it can be compared with the 48 hours choice employed in Figure 3.
It appears that the longer time interval provides a more accurate modelling with less recalibration procedures applied in the subsequent days.

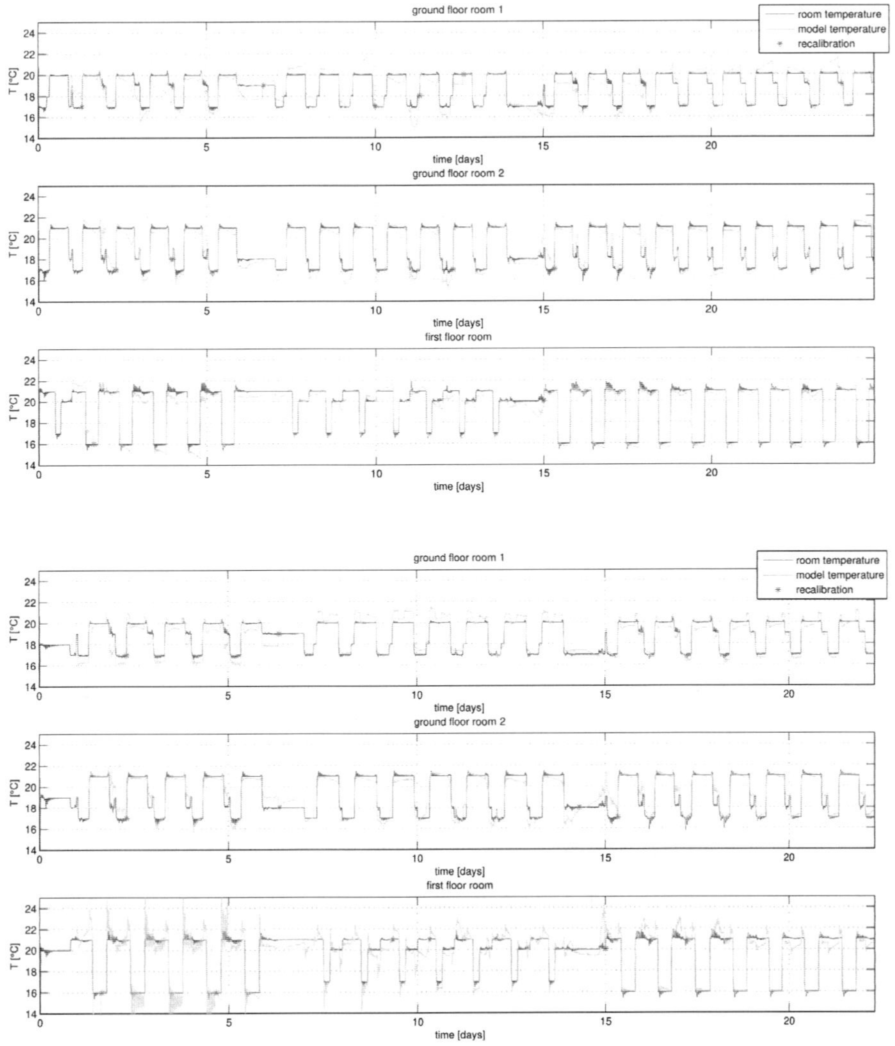

Figure 4. Results related to the use of user-chosen set-point signals. Top: case 1 (Table 3). Bottom: case 2 (Table 4). Blue line: measured temperature. Green line: model temperature. The application of the recalibration strategy is indicated by '*'.

5.3. Tuning of the PID controllers

As a final design choice, the influence of the tuning of the PID controller on the estimation result is discussed. We consider two different kinds of PID tuning. The first one is performed in such a way the set-point step response (for each room) has a fast settling time but without overshoots (the related results are those previously shown in Figure 3). On the contrary, in the second case a more aggressive strategy is selected, that is, (small) overshoots are allowed in order to reduce the rise time. Results related to this latter case are shown in Figure 6. Once again, for the sake of completeness, two different pseudo-random set-point signals have been considered (note that also $x_4 = 48$ hours has been fixed). By comparing the results it appears that, overall, the less

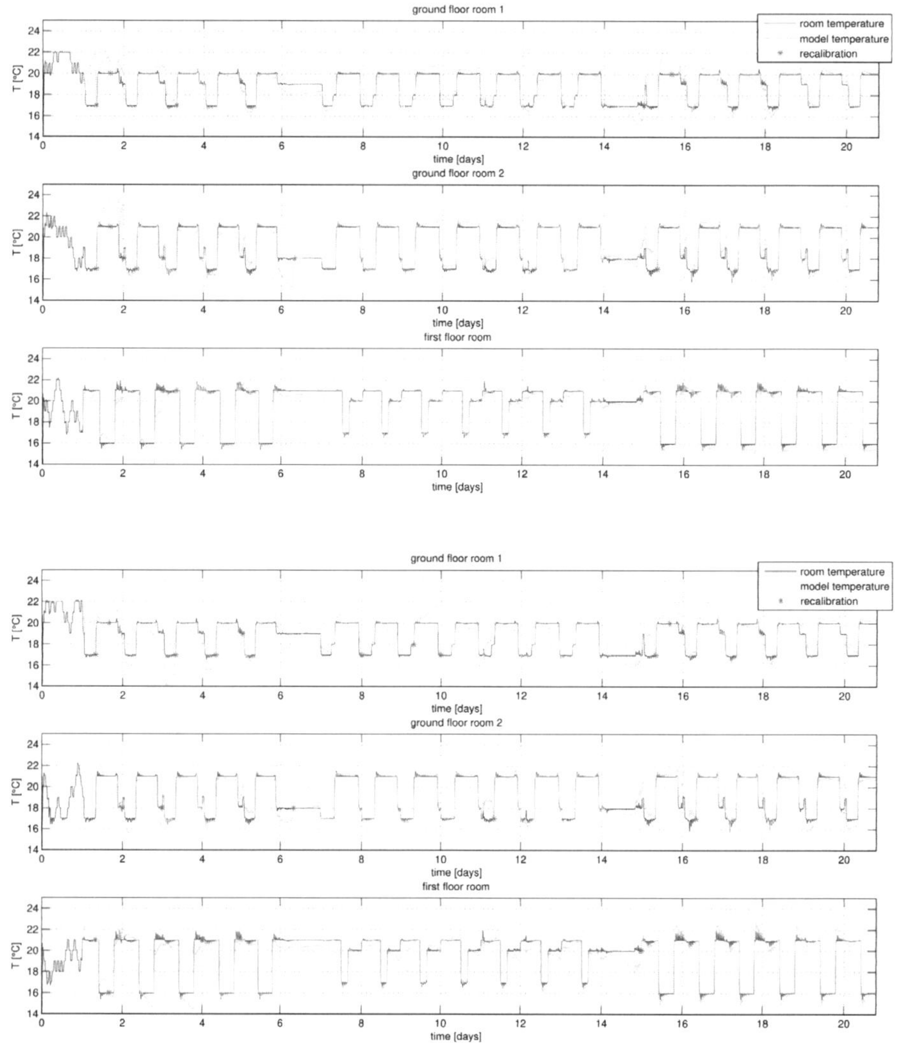

Figure 5. Results related to the use of $x_4 = 24$ hours and different pseudo-random set-point signals (top and bottom). Blue line: measured temperature. Green line: model temperature. The application of the recalibration strategy is indicated by '*'.

aggressive tuning provides the best performance in terms of modelling errors, as witnessed by the reduced number of new estimations of the model parameters.

Table 5 shows the mean absolute temperature error in all the performed tests. The error is defined as in (17) where N represents the number of instants of the whole test time interval.

6. Conclusions

In this paper we have proposed a methodology for the determination of an ARMAX model of the temperature of the rooms of a smart building. The technique is based on a first initial estimation of the model parameters (during the first 24 hours, by applying pseudo-random set-point signals)

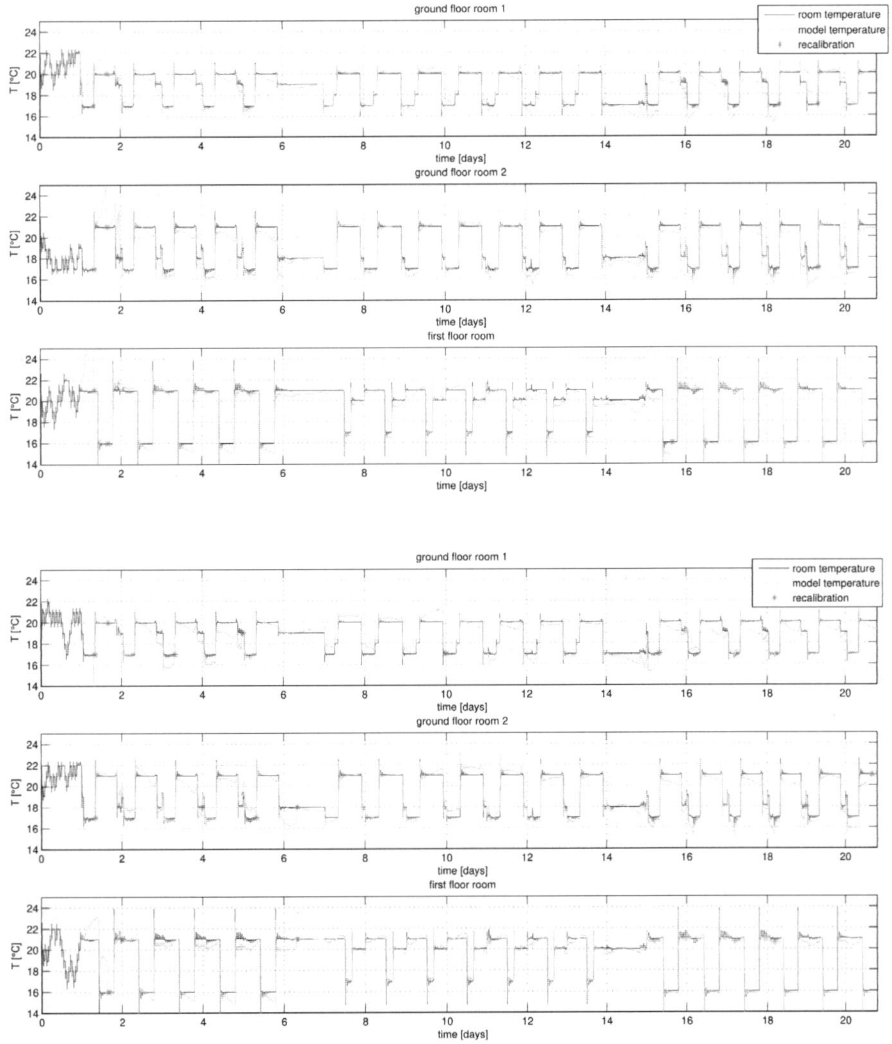

Figure 6. Results related to the use of a more aggressive PID tuning and different pseudo-random set-point signals (top and bottom). Blue line: measured temperature. Green line: model temperature. The application of the recalibration strategy is indicated by '*'.

and then on the application of a self-calibration procedure. Practical issues related to the main design choices have been discussed and it has been shown that the proposed solution are capable to provide the required balancing between the need to follow the system changes along the year (mainly due to the weather change) and to avoid too many recalibration procedures.
Future work will include the comparison between the presented modelling approach with others proposed in the literature and the use of the obtained model for (optimal) control purposes.

Test	West side ground floor	East side ground floor	First Floor
Fig. 3 Top	0.3419	0.4265	0.4111
Fig. 3 Bottom	0.4392	0.4661	0.3755
Fig. 4 Top	0.5428	0.4104	0.3411
Fig. 4 Bottom	0.4431	0.4519	0.5815
Fig. 5 Top	0.4281	0.4060	0.3622
Fig. 5 Bottom	0.5244	0.3776	0.4848
Fig. 6 Top	0.4198	0.4333	0.3843
Fig. 6 Bottom	0.7697	0.4474	0.3473

Table 5. Mean absolute temperature error in all the performed tests.

References

[1] http://www.environmentalleader.com/2009/04/27/building-sector-needs-to-reduce-energy-use-60-by-2050.
[2] http://www.sketchup.com/it/products/sketchup-pro.
[3] K. J. Åström and B. Wittenmark. *Computer-Controlled Systems.* Dover Publications, Englewood cliffs (NJ), 2011.
[4] M. Castilla. *Advanced comfort control techniques for energy efficient buildings.* PhD thesis, University de Almeria, Almeria, E, 2013.
[5] M. Castilla, J. Alvarez, M. Berenguel, M. Perez, J. Guzman, and F. Rodriguez. Comfort optimization in a solar energy research center. In *Proceedings of IFAC Conference on Control Methodologies and Technology for Energy Efficiency*, pages 36–41, Portugal, 2010.
[6] M. Castilla, J. Alvarez, M. Berenguel, F. Rodriguez, J. Guzman, and M. Perez. A comparison of thermal comfort predictive control strategies. *Energy and Buildings*, 43:2737–2746, 2011.
[7] M. Castilla, J. Alvarez, M. Ortega, and M. Arahal. Neural network and polynomial approximated thermal comfort models for HVAC systems. *Building and Environment*, 59:107–115, 2013.
[8] P. M. Ferreira, A. E. Ruano, S. Silva, and E. Z. E. Conceico. Neural networks based predictive control for thermal comfort and energy savings in public buildings. *Energy and Buildings*, 55:238–251, 2014.
[9] A. Kelman, Y. Ma, and F. Borrelli. Analysis of local optima in predictive control for energy efficient buildings. In *Proceedings of 50th IEEE Conference on Decision and Control and European Control Conference (CDC-ECC)*, pages 5125–5130, Orlando,FL,USA, 2011.
[10] Z. O'Neill, S. Narayanan, and R. Brahme. Model based thermal load estimation in buildings. In *Proceedings of Fourth National Conference of IBPSA-USA*, pages 474–481, New York City (NY), 2010.
[11] S. Privara, Z.Vana, D. Gyalistras, J. Cigler, C. Sagerschnig, M. Morari, and L. Ferkl. Modeling and identification of a large multi-zone office building. In *Proceedings of IEEE International Conference on Control Applications (CCA)*, pages 55–60, Denver (USA), 2011.
[12] P. Radecki and B. Hencey. Online thermal estimation, control, and self-excitation of buildings. In *Proceedings of 52th IEEE Conference on Decision and Control*, pages 4802–4807, Florence (I), December 2013.
[13] M. Trcka and J. Hensen. Overview of HVAC system simulation. *Automation in Construction*, 2:93–99, 2010.
[14] S. Wang and X. Xu. Simplified building model for transient thermal performance estimation using ga-based parameter identification. *International Journal of Thermal Sciences*, 45:419–432, 2006.
[15] X. Xu and S. Wang. A simplified dynamic model for existing buildings using ctf and thermal network models. *International Journal of Thermal Sciences*, 47:1249–1262, 2008.
[16] R. Yedra, F. Rodriguez, M. Castilla, and M. Arahal. A neural network model for energy consumption prediction of ciesol bioclimatic building. In *Proceedings of the* 8th *International Conference on Soft Computing Models in Industrial and Environmental Applications*, Salamanca (S), 2013.
[17] B. Yu and A. H. C. van Paassen. Simulink and bond graph modeling of an air-conditioned room. *Simulation Modelling Practice and Theory*, 12:61–76, 2004.
[18] X. Zhang, G. Schildbach, D. Sturzenegger, and M. Morari. Scenario-based MPC for energy efficient building climate control under weather and occupancy uncertainty. In *Proceedings of the European Control Conference (ECC)*, pages 1029–1034, Zurich (SW), 2013.

Optimising base station location for UMTS cellular networks

G Kalata, I Pozniak-Koszalka, L Koszalka[1] and A Kasprzak
Department of Systems and Computer Networks
Wroclaw University of Technology, Wroclaw, Poland

E-mail: leszek.koszalka@pwr.edu.pl

Abstract. Rapid development of universal mobile telecommunication systems put demands on tools for assisting planning of cellular network infrastructure. The tools need to focus on critical issues in modern cellular networks and techniques used for previous generation system no longer serve useful. In this paper, an algorithm based on Branch & Bound approach is proposed for solving base station location problem, covering interference levels, traffic demands and power control mechanism. The efficiency of the algorithm is evaluated with respect to existing approaches for solving this problem – using the designed and implemented experimentation system.

1. Introduction

The problem of choosing base station location is one of the basic problems that arise in 3G network planning, such as *Universal Mobile Telecommunication System* (UMTS). Scarcity of radio resources and high cost of infrastructure and resources put high demands on the quality of network planning. Since 3G systems are based on wideband code-division multiple access (W-CDMA), all connections share the same bandwidth. The capacity of the system depends on achievable signal-to-interference ratio (SIR) values. Because 3G systems tend to be very sensitive to interference, it should be ensured that any base station (BS) does not generate or receive too much of it [1]. Thus, the power control techniques are employed. This mechanism ensures that each device in the network transmits and receives just enough power to allow communication [2]. The mobile stations emission power is limited; therefore a minimal SIR level needed for communication may be unreachable in presence of high interference. Also, because traffic distribution can vary with time (and with them the interference levels), the area covered by a single BS can vary significantly. Usually, in planning phase of cellular network, a set of candidate sites for base stations is considered. Also traffic distribution needs to be supplied; usually it is estimated using empirical prediction models. In this work, the problem defined in [3] is taken into consideration. The model focuses on up-link direction (i.e. mobile station to BS), quality requirements in terms of SIR minimal levels and power limits.

The rest of work is organised as follows. In section 2 the problem is formulated. In section 3 two known algorithms based on randomized greedy procedures and Tabu Search are described. Then, in section 4 the algorithm based on Branch & Bound approach is presented. Section 5 and section 6 describe the created simulation environment and the propagation model, respectively. In section 7 the designed experiment setup is considered and the obtained computational results are presented. The conclusions appear in section 8.

[1] Corresponding author.

Journal of Physics: Conference Series **570** (2014) 062005 doi:10.1088/1742-6596/570/6/062005

2. Problem statement

The problem of the base station location may be interpreted as finding such a subset of a given set of base station candidate sites *S*, which results in the lowest cost and lowest minimal power of transceivers. The solution must cover all given test points and guarantee at least minimal connection quality. The problem can be formulated (basing on [3]) as follows:

Given

- set of *candidate sites* (CS) $S = \{1, \dots, m\}$ with corresponding installation cost $c_j, j \in S$,
- set of *test points* (TP), $I = \{1, \dots, n\}$ with corresponding required number of simultaneously active connections $u_i, i \in I$,
- propagation gain matrix

$$G = \{g_{ij}\}_{1 \le i \le n, 1 \le j \le m}, 0 < g_{ij} \le 1,$$

To find

- base station design $Y = \{y_j\}, j \in S$,

$$y_j = \begin{cases} 1 & \text{if a BS is installed in j} \\ 0 & \text{otherwise} \end{cases},$$

- test points assignment $X = \{x_{ij}\}, i \in I, j \in S$,

$$x_{ij} = \begin{cases} 1 & \text{test point i is assigned to BS j} \\ 0 & \text{otherwise} \end{cases},$$

Such that

$$\sum_{j=1}^{m} c_j y_j + \lambda \sum_{i=1}^{n} \sum_{j=1}^{m} u_i \frac{1}{g_{ij}} x_{ij} \text{ is minimal} \quad (1)$$

Subject to the constraints

$$\sum_{j=1}^{m} x_{ij} = 1, i \in I \quad (2)$$

$$x_{ij} \le min\left\{1, \frac{g_{ij} P_{max}}{P_{target}}\right\}, i \in I, j \in S \quad (3)$$

$$y_j \left(\sum_{h=1}^{n} \sum_{t=1}^{m} u_h \frac{g_{hj}}{g_{ht}} x_{ht} - 1 \right) \le \frac{1}{SIR_{min}}, j \in S \quad (4)$$

The first term of cost function (1) refers to the total cost of base station design, the second one to the total needed power. The constraint (2) implies that all test points should be covered, the constraint (3) ensures that TP can be assigned only to BS with enough signal strength, and the constraint (4) implies that for each BS used, the signal-to-interference ratio (SIR) is high enough to allow communication. As can be seen, after choosing candidate sites, one must also assign each TP to one of the available BS. Solution to this assignment sub-problem is trivial. As was shown in [3] each traffic point i has to be assigned to *j*-th BS with the highest propagation gain $\boldsymbol{g_{ij}}$. The problem is NP-hard.

3. Related work – Reference algorithms

To make some reference, two greedy randomized procedures and Tabu Search algorithm were adopted from [3]. In following subsection this algorithm will be described. The common procedure used in all three algorithms is the construction of X matrix. Because all of these algorithms can generate intermediate solutions, that are unfeasible. As has been stated in Section 2, each i-th TP should be assigned to available j-th BS that has the highest g_{ij} to ensure lowest interference and power demands. So, the X matrix construction procedure first assigns each TP to BS according to above rule. For every base station SIR is computed, and if it is lower than SIR_{min}, then all test points associated with it are considered for elimination. Next, the test point with the highest emission power is unassigned and the procedure repeats, until all BS fulfils SIR demands. Due to elimination of test point that cannot be handled, one can infer how much of the traffic can be served using current subset of BS. The drawback of this approach is that it quickly becomes computationally expensive.

3.1. Randomized greedy add procedure

This procedure starts from empty set of BS and in each iteration evaluates set with added BS from remaining candidate sites. The candidate sites are scored according to greedy function which captures the current total cost and the amount of currently covered traffic. This greedy function incorporates trade-off parameter between cost and amount of covered traffic. Then, from few best candidates according to greedy function one is chosen randomly, and added to the current solution. This procedure is repeated, until adding new base station no longer improves the greedy function.

3.2. Randomized greedy remove procedure

This procedure is similar to the previous one. It starts from an initial solution consisting of all the base stations, and then iteratively removes BS according to greedy function in analogous manner that previous procedure. However, the greedy function is somewhat different. It includes an additional term, which measures the number of additional connections that could be served by the solution. The advantage of this procedure is that, if a feasible solution exists, it always finds it - since solution including all candidates must fulfil the constraints (if instance is solvable).

3.3. Tabu Search

The Tabu Search algorithm [4], [5] uses swap moves and greedy goal function from randomized greedy remove procedure. The swap moves remove from the solution currently included BS and add one BS from currently unused sites. Because the number of all possible swap moves is too large to enumerate and evaluate them all, only some "best" candidates for each currently used BS are taken into consideration. The "best" refers to candidates that have higher propagation gains with respect to currently used BS. Some fraction of "best" candidates are systematically evaluated, the other are evaluated with a given probability. The algorithm was used in multi-run configuration, that is, the algorithm was invoked 10 times, and the best solution from 10 runs was returned. The initial solution for each run was generated using one of the randomized greedy procedures, and then 200 iterations of Tabu Search was computed.

4. Branch & Bound algorithm

The Branch & Bound is a general algorithm used for solving optimisation problems. The method was first proposed in [6] and widely developed [7]. The algorithm enumerates possible solutions by means of solution tree traversal. It uses lower bound and other criteria to eliminate branches of solution tree that represents fruitless solutions. In [4] Branch & Bound algorithm was used to solve differently modelled problem of base station optimisation, which is also NP-hard. They employed greedy solution tree traversing to ensure that optimal solution is found as early as possible. For smaller instances the algorithm could be used to provide the exact solution, and for larger it can be simply turn into heuristic by limiting the number of iterations. This concept was adopted into this paper.

Journal of Physics: Conference Series **570** (2014) 062005 doi:10.1088/1742-6596/570/6/062005

In the beginning of the activity of the proposed algorithm, for each i-th TP, a list of feasible BS is created. This list is then sorted using signal strength g_{ij} from higher to lower. Additionally, the BS candidate sites are sorted according to their scores, which are evaluated as follows. For each i-th TP the score of every feasible BS that this TP can be assigned to, is increased by 1. Thus, BS that can cover many test points will be placed on the beginning of the list. These two kinds of lists are used in the next actions of the algorithm.

4.1. Initial solution
At first, an initial solution is evaluated in order to get the upper bound of optimal solution's objective function. The lower bound for the number of BS needed to serve all traffic points is evaluated. It is easily to show that the maximum number of TP that can be served by any BS, while fulfilling demand for SIR, can be computed using the equation (5):

$$max_{TP} = \left\lfloor \frac{1}{SIR_{min}} + 1 \right\rfloor \tag{5}$$

what implies that for n TP the lower bound of number of BS min_{BS} can be expressed by (6):

$$min_{BS} = \left\lceil \frac{n}{max_{TP}} \right\rceil \tag{6}$$

Thus, the first best min_{BS} BS is included into initial solution. Then, every TP is assigned to best available BS (using per TP list). After that in any iteration one of the remaining BS is added, and the solution is checked for feasibility and objective function evaluated. If solution is feasible, and it is better (according to objective function defined by (1)) than a previous solution, it becomes current initial solution and the next BS is added. If current solution is not better than a previous one, the initial solution deriving procedure is interrupted.

4.2. Branching
The solution tree is constructed in the following way. At each node of the tree the next BS from best BS list (evaluated at the beginning) is taken into consideration, and two new nodes produced. One is considered to not use this BS, and the other to do so. For each of the nodes the lower bound is evaluated. If lower bound is higher than current upper bound, the node is removed. Remaining nodes are checked for feasibility, and if they pass they are added to active nodes priority queue. The priority of the nodes is its greedy score. In any iteration of Branch & Bound algorithm, one node of the solution tree is popped out from the queue. Its lower bound is compared with current upper bound (which could change meanwhile) and further expanded, or if it is at maximum depth of the tree it is evaluated using objective function. If such solution is feasible and has the lowest objective function value so far, it becomes the best solution and sets the new upper bound. The algorithm ends after given number of iterations (acts like heuristic) or when the whole tree is traversed (the obtained solution is known to be optimal).

4.3. Lower bound of objective function
In order to compute lower bound of the objective function the one must assign each TP to some of the available BS. Because only a part of solution is known at that time, the following strategy is used. For each TP, the BSs from feasible list are considered in order from best to worst. If given BS is included in partial solution and it is installed, then TP is assigned to it. If in partial solution it is not installed, the next BS from list is checked. If given BS is not included in partial solution, then TP is assigned to it. Then objective function is computed, including cost only for BS that partial solution uses, and the total power is calculated for assignment of BS described above.

4.4. Feasibility checking

Each partial solution is checked against feasibility. First it is checked (using assignment of TP from lower bound computing procedure) that every test point has been served. If this requirement is fulfilled, it could be expected that solutions in this branch will be able to serve all TPs. The next check that is performed, is number of TP assigned to each BS used by partial solution. If some BS has more than max_{TP} assigned, it is considered as invalid solution. Then for each of the used BS the SIR is calculated, and if it is lower than SIR_{min} then the partial solution is considered as invalid too.

4.5. Greedy score

For computing greedy score, the TP that can be served using BS included in partial solution need to be determined. Total served traffic T_{max} is calculated, as the sum of simultaneous connection demand u_i of TP that can be served. Then greedy score is simply calculated by (7):

$$greedyScore = lowerBound - T_{max} \quad (7)$$

5. Experimentation environment

As part of this work a new experimentation environment was developed following ideas proposed in [8]. The implemented simulator works with 2.5D and 3D vector data that can be imported into application from two different formats:

- Text format (used by Munich Test Site Building Data,
- CityGML (using *libcitygml* library).

When building data is imported, other information such as test points, BS candidate sites, frequency used etc. can be added and whole data saved into simulator's own binary file format.
Simulator user interface consists of four tabs:

- *Scenario setup.* Here user can setup change test points and BS candidate sites. New sites/test points can be added manually, or generated randomly. Main parameters of simulation can be changed here as well.
- *Solver.* Here user can calculate G matrix and run experiments with implemented algorithms.
- *Propagation prediction.* Here user can compute signal strength prediction for regular grid of points. This can be useful to get general idea about range of specific BS on whole city area. The precision (number of grid points) can be set by user. The Signal strength information is visualised using colours.
- *Full view.* This tab is intended to explore scenario in 3D view.

In each of the first three tabs there is smaller scenario preview window. Every view of scenario can be manipulated using mouse and keyboard to change camera direction, zoom, and centre point. The experimentation environment allows changing the following parameters:

- Used frequency,
- Minimum signal-to-interference ratio (SIR) SIR_{min},
- Maximum power of mobile equipment P_{max},
- Minimum signal power requested at base station P_{target},
- Ground level - useful for data which includes underground part,
- Criterion trade-off parameter λ.

Internally, binary space partitioning (BSP) tree is used to speed up geometric look-ups. For each transmitter and receiver position the faces intersecting on vertical plane between transmitter and receiver are found using BSP tree, then cross section between that plane and each of the faces is computed, and cross section of buildings is reconstructed. It is then used to find roofs, and estimate the

Journal of Physics: Conference Series **570** (2014) 062005 doi:10.1088/1742-6596/570/6/062005

parameters of the empirical model. Then according to the equations given in section 6 the loss in dBm is computed. It can be then converted to normalised propagation gain using the following formula (8)

$$g = \frac{1}{10^{\frac{l_p}{10}}} \quad (8)$$

6. Empirical COST-Walfish-Ikegami Model

There are many models created for prediction of radio wave propagation and signal strength. For the purposes of this work, the Empirical COST-Walfisch-Ikegami Model [9] was used. The motivation for choosing this model was simplicity and high prediction quality in urban environments. The considered model is a statistical model. The prediction is based on statistics of buildings in vertical plane between receiver and transmitter. The model captures propagation over the rooftops, which is main source of signal propagation in urban environments. The main parameters of the model are:

- Frequency f (800 . . . 2000 MHz),
- Height of the transmitter h_{TX} (4 . . . 50 m),
- Height of the receiver h_{RX}(1 . . . 3 m),
- Distance d between transmitter and receiver (20 . . . 5000 m.

The following statistical parameters are estimated:

- Mean building height h_{ROOF},
- Mean width of street w,
- Mean building separation b.

The fourth parameter - the orientation of road φ - is omitted here because the orientation of the road cannot be easily determined for every point, especially on crossings.

The model distinguishes between two situations: the "line of sight" (LOS) and "no line of sight" (NLOS). In the LOS case, the propagation loss l_p between transmitter and receiver can be simply expressed by the formula (9):

$$l_p = 42.6 + 26 \cdot \log\frac{d}{\text{km}} + 20 \cdot \log\frac{f}{\text{MHz}} \quad (9)$$

In case of NLOS, the expression (10) is used:

$$l_p = \begin{cases} l_0 + l_{rts} + l_{msd}, & l_{rts} + l_{msd} > 0 \\ l_0, & l_{rts} + l_{msd} \le 0 \end{cases} \quad (10)$$

where l_0 is free space loss, l_{msd} is multiple screen diffraction loss, and l_{rts} is the rooftop-to-street diffraction loss. The free space loss is calculated from formula (11):

$$l_0 = 32.44 + 20 \cdot \log\frac{d}{\text{km}} + 20 \cdot \log\frac{f}{\text{MHz}} \quad (11)$$

The rooftop-to-street diffraction loss term denoted by l_{rts} (12) describes the loss which occurs on the wave coupling into the street where the receiver is located.

$$l_{rts} = -16.9 - 10 \cdot \log\frac{w}{\text{m}} + 10 \cdot \log\frac{f}{\text{MHz}} + 20 \cdot \log\frac{h_{ROOF} - h_{RX}}{\text{m}} + l_{ori} \quad (12)$$

where l_{ori} is omitted, because it depends on street orientation φ. It could be replaced by an empirical correction obtained from the calibration with measurements. An approximation for the multi-screen diffraction loss l_{msd} is computed using the formula (13):

$$l_{msd} = l_{bsh} + k_a + k_d \cdot \log\frac{d}{\text{km}} + k_f \cdot \log\frac{f}{\text{MHz}} - 9 \cdot \log\frac{b}{\text{m}} \tag{13}$$

with

$$l_{bsh} = \begin{cases} -18 \cdot \left(1 + \frac{h_{TX} - h_{ROOF}}{\text{m}}\right), & h_{TX} > h_{ROOF} \\ 0, & h_{TX} \leq h_{ROOF} \end{cases}$$

$$k_a = \begin{cases} 54, & h_{TX} > h_{ROOF} \\ 54 - 0.8 \cdot \frac{h_{TX} - h_{ROOF}}{\text{m}}, & h_{TX} \leq h_{ROOF} \wedge d \geq 0.5\text{km} \\ 54 - 0.8 \cdot \frac{h_{TX} - h_{ROOF}}{\text{m}} \frac{d/\text{km}}{0.5}, & h_{TX} \leq h_{ROOF} \wedge d < 0.5\text{km} \end{cases}$$

$$k_d = \begin{cases} 18, & h_{TX} > h_{ROOF} \\ 18 - 15 \cdot \frac{h_{TX} - h_{ROOF}}{h_{ROOF} - h_{RX}}, & h_{TX} \leq h_{ROOF} \end{cases}$$

$$k_f = -4 + \left(\frac{f/\text{MHz}}{925} - 1\right) \cdot \begin{cases} 0.7, & \text{for medium sized city} \\ 1.5, & \text{for metropolitan centers} \end{cases}$$

7. Investigation

The tests were conducted on 5 small instances characterized by 22 candidate sites with $u_i = 1$ (following data used in [3]). Additional two instances were generated, using 30 BS, and 80 TS (instance 6), and using 60 BS, and 130 TS (instance 7) – both instances uniformly distributed in {1, 2, 3}. The used parameters concerning frequency, interference ratio, and power are presented in table 1. The parameters of greedy procedures and Tabu Search were used as in [3].

Table 1. Parameters of instances used in experiment.

Parameter	Value
Frequency	2000 [MHz]
SIR_{min}	0.031250
P_{max}	30 [dBm]
P_{target}	-100 [dBm]
λ	1.0

Using the designed and implemented experimentation system described in section 5 – the simulation experiments were carried out. The five algorithms were compared:

- RGA (Randomized Greedy Add)
- RGR (Randomized Greedy Remove)
- MTSA (Multistart Tabu Search using RGA)
- MTSR (Multistart Tabu Search using RGR)
- B&B (Branch & Bound Algorithm)

The obtained results are shown in table 2, where in columns are presented: 'cost', i.e., the value of cost function defined by (1), and '#BS' - the number of base stations in the solution, and in table 3 where in column is presented 'time', i.e., time of finding the solution. Note that the results for RGA for instances 6 and 7 are removed because a solution found was not feasible.

Table 2. Cost comparison between algorithms.

	RGA		**RGR**		**MTSA**		**MTSR**		**B&B**	
	Cost	#BS	Cost	#BS	Cost	#BS	Cost	#BS	Cost	#BS
Inst 1	2.89	7	2.21	7	1.17	10	1.19	10	0.36	21
Inst 2	2.54	5	3.99	8	1.18	11	1.79	8	0.36	20
Inst 3	2.49	8	2.04	7	1.02	10	1.05	10	0.33	20
Inst 4	1.21	10	1.44	14	1.11	11	1.20	13	0.65	19
Inst 5	1.31	8	1.76	9	0.84	10	0.80	10	0.22	22
Inst 6	-	-	1.41	20	1.08	16	1.05	19	0.64	24
Inst 7	-	-	1.64	40	1.28	32	1.24	28	0.80	50

Table 3. Time comparison between algorithms.

	RGA		**RGR**	**MTSA**	**MTSR**	**B&B**
Inst 1	198 [ms]		200 [ms]	160 [s]	154 [s]	17 [ms]
Inst 2	119 [ms]		182 [ms]	128 [s]	125 [s]	83 [ms]
Inst 3	204 [ms]		288 [ms]	184 [s]	187 [s]	97 [ms]
Inst 4	285 [ms]		180 [ms]	197 [s]	197 [s]	138 [ms]
Inst 5	199 [ms]		173 [ms]	224 [s]	228 [s]	17 [ms]
Inst 6	-	-	635 [ms]	616 [s]	465 [s]	2.8 [s]
Inst 7	-	-	635 [ms]	4.5 [h]	3.07 [h]	302 [s]

It may be observed that RGA was unable to find feasible solution for bigger instances. The remaining algorithms were always able to find feasible solution. For RGR the time needed to obtain solution was not growing significantly with size of the instance, but the solutions found were worse that those obtained by more time consuming algorithms. The both variants of Tabu Search algorithms, MTSA and MTSR, delivered better solutions that randomised greedy RGA and RGR, but the computational time was growing significantly with larger size of instance.

The Branch & Bound algorithm was the only one delivering the exact solution. The solutions found were usually one order of magnitude better in terms of the cost function. For small instances the computational time was the smallest, however, it grows significantly with instance size, but not as much as for Tabu Search. One could also observe that Branch & Bound and also RGR for bigger instances, was using significantly more base stations that Tabu Search variants. This can be attributed to the fact, that for the cost function with $\lambda = 1$, the main objective was to minimise power emitted and the installation cost was negligible.

8. Conclusion

The Randomized greedy add procedure turned out to serve good trade-off between time spent on finding solution and quality of the solution obtained. The Tabu Search variants were far more computationally expensive and still were obtaining solution order of magnitude worse than optimal ones. The proposed Branch & Bound algorithm could be considered as promising – the obtained results of simulation experiments showed remarkable advantages. However, rapid increase of computational time with the size of problem instance indicates that further improvement and extensions are needed. We are planning an improvement of the algorithm by applying hybrid meta-heuristics techniques, following ideas presented in [10] and [11]. Moreover, the problem defined and considered in this paper is one of the simpler ones and future work could focus on more detailed base station location problem models.

Acknowledgement
This work was supported by the statutory funds of the Faculty of Electronics, Department of Systems and Computer Networks, Wroclaw University of Technology, Wroclaw, Poland.

References

[1] Wacker A, Laiho J, Novosad T, Rautianinen T and Reava K 2007 *WCDMA Radio Network Planning* (John Wiley & Sons) 93-196
[2] Novosad T, Soldani D, Sipila K, Kola T and Wacker A 2007 *Introduction to WCDMA for UMTS* (John Wiley & Sons) 19-92
[3] Amaldi E, Capone A and Malucelli F 2003 Planning UMTS base station location: optimization models with power control and algorithms *IEEE Trans. on Wireless Communications* **2**(5) 939-952
[4] Mathar R and Schmeink M 2003 Optimal base station positioning and channel assignment for 3G mobile networks by integer programming *Annals of Oper. Res.* **107**(1-4) 225-236
[5] Gendreau M and Potvin J Y 2005 *Search methodologies – introductory tutorials in optimization and decision support techniques* (Springer – New York) 165-186
[6] Land H and Doig A G 1960 An automatic method of solving discrete programming problems *Econometrica* **28**(3) 497-520
[7] Liu G S and Zhang J Z 2002 A new branch and bound algorithm for solving quadratic programs with linear complementarity constraints *J. Comp. and Appl. Math.* **146**(10) 77-87
[8] Koszalka L, Lisowski D and Pozniak-Koszalka I 2006 *Lecture Notes in Computer Science* Springer vol. 3984 58-67
[9] Hemant K S, Sharma S and Pandey K K 2011 Survey of propagation model in wireless network *Intern. J. of Comp. Sc. Issues* 468-472
[10] Borkowski J and Kasprzak A 2014 Algorithms for base station location in wireless networks with cost minimization *Computer Systems Engineering: Theory and Applications Proc. of 10-13 Polish British Workshops* OW Wroclaw 22-34
[11] Regula P, Pozniak-Koszalka I and Koszalka L 2011 Evolutionary algorithms for base station placement in mobile networks *Lecture Notes in Computer Science* Springer vol. 6592 1-10

Journal of Physics: Conference Series **570** (2014) 072001 doi:10.1088/1742-6596/570/7/072001

Comparative study of the multiple fault isolation approaches based on the structures of residual sets

Michał Bartyś

Institute of Automatic Control and Robotics, Warsaw University of Technology, św. Boboli 8, 02-525 Warsaw, Poland

E-mail: bartys@mchtr.pw.edu.pl

Abstract. This paper presents a comparative study of the applicability features of the three chosen multiple fault isolation approaches based on the binary evaluated structures of residual sets. Firstly, all considered approaches are briefly described. Then, their computational effectiveness is discussed. Next, the low-level implementation of the compared approaches is characterised and the set of experimental results is presented. Finally, the results of experiments are discussed and the appropriate recommendations for diagnostic engineering practice are formulated.

1. Introduction

The goal of this paper is quite practical. It is aimed to give a theoretical background for a well documented rules of how to make reasonable choice of multiple fault isolation approach for real implementations. The scope of this paper covers three chosen multiple fault isolation approaches based on a binary evaluated structures of residual sets.

According to the definition given in [1], the fault isolation is the determination of the kind, location and time of detection of a fault. In this paper, we will principally refer to the multiple fault location abstracting from their detection time and considering a fault kind as a linguistic conjunction of a single fault labels.

The problem of multiple (simultaneous faults) arises particularly when there is no possibility or there is not enough time or capacities to fix all previously presented single faults. Intuitively, the multiple faults are that the more probable the more complex the process is. As far as the multiple faults are observed, the symptoms of single faults are overlapping each other. As long as combined symptoms of the single faults do not match any single fault signature, the hypothesis regarding multiple faults is reasonable. Otherwise, the multiple faults may introduce the effect referred to as fault cancellation. This rare phenomenon relies on the cancellation of a single fault residual as the result of the mutual overlapping of the truly existing faults. The cancellation of residual of a fault makes fault isolation process unreliable.

The multiple fault isolation problems have been studied mainly from the Fault Detection and Isolation (FDI) and Artificial Intelligence (AI) perspectives [2, 3, 4, 5]. The FDI methods are principally focused on generation and processing of residuals sensitive to particular faults and insensitive to others. The Fault Detection (FD) deals with the methods of residual generation while the Fault Isolation (FI) deals with location of faults, time of their origin and their kind. FI approaches make use from residuals that are transformed to the form of diagnostic signals in the process referred to as residual evaluation. Different threshold approaches (constant, adaptive, fuzzy, statistic) are applied on this stage [6, 7, 8, 9] in order to yield the diagnostic signals. The particular diagnostic signals which are indicative for the given faults are referred to as the fault symptoms. The Artificial Intelligence (AI) or consistency based methods are principally focussed on fault isolation based on reasoning from the first principles i.e., using knowledge of structure and behavior of the system. Some of AI multiple fault approaches [4, 5, 10, 11] are searching for the minimal sets of faults which are consistent with the observed symptoms according to the parsimony principle. Some of them are searching for all probable multiple faults despite their probability [12, 13].

In this paper, we will provide a deep-in comparative study of the properties of the three chosen approaches belonging to the FDI group and based on the binary evaluated residuals.

2. Fault isolation system structure

Consider a model based diagnostic system. Let a finite set F of the faults f_i in this system be defined as:

$$F = \{f_i : i = 1..n\} \tag{1}$$

and a finite set of diagnostic signals S:

$$S = \{s_j : j = 1..m\}. \tag{2}$$

Let us now discuss the relation between both sets in the form of Cartesian product $R_{FS} \subset F \times S$. Here, the relation R_{FS} constitutes a set of $n \cdot m$ ordered pairs (bi-element relations) $\langle f_i, s_j \rangle$. According to the geometrical interpretation of the Cartesian product, the relation is the set of $n \cdot m$ points of the plane defined in F and S coordinates. It is possible to spread out a three-dimensional mesh over the $F \times S$ plane by attributing diagnostic signal values $v_{j,i}$ of all diagnostic signals s_j for all f_i faults. The three-dimensional mesh is easily transformable into the form of a two-dimensional $m \cdot n$ matrix V of the diagnostic signal values $v_{j,i}$.

$$V = \begin{bmatrix} v_{1,1} & \dots & v_{1,n} \\ \dots & v_{j,i} & \dots \\ v_{m,1} & \dots & v_{m,n} \end{bmatrix} = [v_{j,i}]_{m \times n} \tag{3}$$

The matrix V forms a fault isolation system structure [12]. More comprehensively, this matrix is referred to as a diagnostic matrix. If the diagnostic signals are bi-valued then the matrix V is called as a Binary Diagnostic Matrix (BDM) [9]. For convenience, the diagnostic matrix V will be further represented as a block matrix of n vectors V_i:

$$V = [V_i]_{n \times 1}, \tag{4}$$

such that each vector V_i contains the set of diagnostic signals values associated exclusively with the single i-th fault:

$$V_i = [v_{1,i}, v_{2,i}, ... v_{j,i}]^T. \tag{5}$$

The specific vector V_i that contains all diagnostic signals associated with a particular fault builds up a numerical or symbolic signature (pattern) of this fault.

Consider a diagnostic system for which a binary diagnostic matrix is known. Let the output of this system in a time instant t be a set of residual values $R^t = \{r_j^t : j = 1..m\}$. Let now the set of residuals R^t be transformed onto the set of values of diagnostic signals $V^t = \{v_j^t : j = 1..m\}$ in result of binary evaluation of residuals. The ordered set of diagnostic values V^t in the time instant t will be referred to as the vector of actual diagnostic signal values.

Journal of Physics: Conference Series **570** (2014) 072001 doi:10.1088/1742-6596/570/7/072001

The vector V^t can match one or even more single fault signatures in the diagnostic matrix. If this is the case, than this specific vector may simply indicate single faults. However, please pay attention that the vector of actual diagnostic signals values V^t may be understood as a vector that points out multiple faults too. All of the multiple fault isolation approaches considered in this paper [12, 13, 14] assume that probability of the multiple faults compared to the single faults is negligible. Hence, if the vector of actual diagnostic signals values V^t matches the single faults then the multiple faults are not considered and searched.

However, the question arises, what happens if the vector V^t does not match any fault signature in a fault isolation structure. The possible answers are as follows:

(i) there is no fault;

(ii) there are unknown faults;

(iii) there are multiple faults.

Case (i) is trivial. It takes place when all values of the actual vector of diagnostic signals are equal to 0. This however, does not exclude a case of unknown faults. Case(ii) indicates poor design of the diagnostic system and will not be further discussed. Case (iii) is the main concern of this paper.

3. The classic approach to multiple fault isolation

The approach to multiple fault isolation proposed by Gertler in [12] will be further referred to as a classic approach. Let us take into consideration the following set of assumptions:

1° a bi-valued diagnostic matrix of the diagnosed system is known;
2° all single faults are defined;
3° the signatures of all single faults are known;
4° the residual cancellation effects do not take place;
5° the signature of a multiple fault is the logical alternative of single fault signatures.

The above given assumptions are acceptable to some extent in practice because the residual cancellation effects rather have marginal meaning. Moreover, the diagnosed systems having duly prepared diagnostic matrices are not exceptional. Obviously, if the assumptions are not met than the diagnostic system should notify users that fault isolation results may be false.

The starting point of this approach is quite a trivial statement that the maximal fault multiplicity does not exceed the number n of single faults. Therefore, searching for the multiple faults might be arranged as an iterative process of searching for single, double, triple, ... , and n–multiple faults. In fact, each search needs to identify all combinations of single fault signatures which are consistent with the vector V^t of actual diagnostic signal values. Therefore, the total number of all possible combinations of all single fault signatures equals:

$$N = 2^n - 1. \tag{6}$$

From (6), it follows that the number of searches for multiple faults grows exponentially with the cardinality of the set of faults F. Therefore, the substantial drawback of this approach is a huge computational effort that must be paid to find out all multiple faults. On the other hand, its most important advantage is an extremely easy implementation. It is obvious, that this approach may not be considered as a fast if the number of fault signatures is significant.

Conclusion 1: The efficiency of a classic multiple fault isolation approach depends exclusively on the number of columns of the BDM matrix.

The classic approach is based on matching logical alternatives of signatures of single faults with the V^t pattern. Clearly, the zero values of a diagnostic signals do not influence the logic alternatives of signatures. Therefore, the binary diagnostic matrix may be compressed dynamically in horizontal direction by removing this V_i columns for which holds:

$$\bigvee_{j=1}^{m} \left(v_{j,i} \wedge \neg v_j^t\right) = 1; \quad \forall i = \{1,..,n\}. \tag{7}$$

This, in general, decreases the number of signatures of the diagnostic matrix. Let the symbol c denotes the number of rejected columns. Hence, the number of combinations of all single fault signatures in relation to the number N from (6) will drop down approximately 2^c times:

$$\eta = \frac{2^c - 1}{2^{n-c} - 1} \approx 2^c. \tag{8}$$

Observation 1: The horizontal compression of the binary diagnostic matrix is beneficial in that sense that may speed up the multiple fault isolation.

Observation 2: The binary diagnostic matrix may be further compressed in vertical direction. It is easy to note that irrelevant are all those rows of the diagnostic matrix for which holds:

$$v_j^t = 0; \quad \forall j = \{1,..,m\} \tag{9}$$

Observation 3: As results from (6), the vertical compression of the binary diagnostic matrix does not influence the total count of all possible combinations of single fault signatures.

4. The *MUFIA* approach

Generally speaking, the classical algorithm of multiple fault isolation is ineffective. Much more effective is Multiple Fault Isolation Approach ($MUFIA$) algorithm[14]. Principally, the $MUFIA$ approach consists of two parts. In the first part, the set of possible faults is determined. In the second part, the classic approach to multiple fault isolation is applied with the use of horizontal and vertical squeezing of the BDM matrix which are described in Sect. 3.

4.1. The first part of the approach: identification of possible faults

The following assumptions are adopted for the $MUFIA$ approach:

1° all signatures of all single faults are known;
2° if a fault occurs then all symptoms related to this particular fault will appear;
3° multiple faults cause a mismatch in the set of diagnostic signal values;
4° there is no match between vector of diagnostic signals and signatures of single faults.

At the beginning, the presence of single faults is checked based on a classic fault isolation algorithm described in Sect. 3. Here, the computational effort is relatively low and moreover straight proportional to the total number of the BDM columns. In case if the single faults are localised then the $MUFIA$ algorithm is considered as completed. Otherwise, according to assumption (3°) a search for a multiple faults starts-up.

The search for multiple faults begins with the decomposition of the diagnostic matrix into dynamic subsystems according to Dynamic Table of States approach (DTS) proposed by Kościelny in [15]. Each dynamic subsystem is defined by a set of possible faults and a set of diagnostic signals sensitive to these faults. The subsystem is created and updated immediately after the appearance of any subsequent fault symptom.

Let us assume that the first fault symptom $s_j^1 = 1$ will trigger the procedure that creates a subset F^1 consisting of all faults for which $s_j^1 = 1$:

$$(s_j^1 = 1) \Rightarrow F^1 \ ; \ j \in [1..m]. \tag{10}$$

Next, a specific subset S^1 of diagnostic signals is created. This subset contains all diagnostic signals that are necessary for the isolation of the faults belonging to the subset F^1:

$$S^1 = \{s_j \in S : \forall f_i \in F^1\} \ ; \ j \in [1..m]. \tag{11}$$

Clearly, the Cartesian product $F^1 \times S^1$ finally defines a dynamic binary diagnostic subsystem. Usually, the size of this subsystem is significantly reduced in comparison to the primary BDM. A reduced diagnostic matrix contains only $|S^1|$ rows and $|F^1|$ columns. The next fault symptom will trigger the next procedure that similarly assigns faults and diagnostic signals to the appropriate subsets F^2 and S^2. If the subsets S^1 and S^2 of the diagnostic signals are disjunctive:

$$S^1 \cap S^2 = \varnothing \tag{12}$$

then multiple faults can be isolated correctly in separate fault isolation processes. In this case, the subsets of possible faults F^1 and F^2 are also disjunctive. This allows for significant acceleration of multiple fault isolation because of two reasons: lower size of diagnostic subsystems and possibility of application of parallel processing of multiple fault isolation in each subsystem.

However, truly, there is no reason to assume that dynamically created subsets of diagnostic signals must always be disjunctive. Moreover, the benefits of application of DDS approach are disputable in case of some applications, where the size of BDM matrices is low. This applies for example for the embedded diagnostics of smart final control devices.

In this case it is still possible to apply the horizontal and vertical compression of the size of the BDM, according to (7) and Obs. 2. As results from (8), particularly valuable are those actions, that reduce the maximal number of columns of BDM without the loss in the set of multiple faults. Originally there is the horizontal as well as vertical compression applied in the $MUFIA$ approach. According to (7), the total count c of removed columns is equal to the cardinal number of the set C consisting of indices of excluded columns.

$$c = |C| \ ; \ C = \{i : i \in \{1, \dots, n\} : \bigvee_{j=1}^{m} (v_{j,i} \wedge \neg v_j^t) = 1\} \tag{13}$$

Conclusion 2: The number of columns which may be removed from BDM matrix without the loss of a multiple fault information depends on the distribution of its entries equalling 1 ($v_{j,i} = 1$) and on the distribution of the entries equalling 0 ($v_j^t = 0$) of the vector of actual diagnostic signal values V^t.

Conclusion 3: Because the number of combinations of all single fault signatures in relation to the number N (8) drop down approximately 2^c times, then the $MUFIA$ approach promises faster handling of multiple faults compared to conventional approach presented in Sect. 3.

Conclusion 4: The $MUFIA$ approach is more effective the more rich the BDM matrix is and the more sparse the vector V^t is.

Observation 5: It is easy to see that in the case of absence of faults i.e. ($v_j^t = 0; \ \forall j \in \{1 \, .. \, m\}$) the number of rejected columns is maximal and $c = n$ because:

$$\exists! v_{j,i} = 1; \ \ \forall j \in \{1 \, .. \, m\}, \forall i \in \{1 \, .. \, n\} \tag{14}$$

In order to meet requirements of mathematical formalism let us now define conversion ($v \Rightarrow d$) of the binary number v into integer number d:

$$\begin{cases} v_{j,i} = 0 \Rightarrow d_{j,i} = 0 \ ; \ \ v_j^t = 0 \Rightarrow d_j^t = 0 \\ v_{j,i} = 1 \Rightarrow d_{j,i} = 1 \ ; \ \ v_j^t = 1 \Rightarrow d_j^t = 1 \end{cases} \tag{15}$$

Defined conversion allows to rewrite formula (13) in alternative and practical form:

$$c = \sum_{i=1}^{n} \prod_{j=1}^{m} d_{j,i}(1 - d_j^t). \tag{16}$$

The horizontal and vertical compression of the BDM matrix results in its transformation to a matrix referred to as binary diagnostic submatrix (BDS) [13]. The subset F^* of faults that may constitute multiple faults and a subset S^* of diagnostic signals suitable for its isolation in the BDS is to obtained from the below given formulae (17) and (18):

$$F^* = \{f_i \in F\} \backslash \{f_i \in F : \forall i : \bigvee_{j=1}^{m} (v_{j,i} \wedge \neg v_j^t) = 1; \quad \forall j \in \{1 .. m\}\} \tag{17}$$

$$S^* = \{s_j \in S : \forall j : v_j^t = 1)\}. \tag{18}$$

The total number of all faults $f_i \in F^*$ in BDS that may constitute multiple faults in accordance with (16) is equal to:

$$n^* = n - c \tag{19}$$

and the total number m^* of all diagnostic signals in BDS equals to:

$$m^* = \sum_{j=1}^{m} d_j^t. \tag{20}$$

Usually not all v_j^t are equal 1. Therefore, the maximal number of rows $m^* \leq m$. Analogously, the number n^* of columns of this matrix is less or equal to n. Usually holds ($n^* \leq n$). Hence, according to (6), the submatrix BDS gives hope to achieve faster multiple fault isolation.

Observation 6. Despite the underlaying theory, presented in [14] and its extensions given above, it is obvious that BDS is at least a compressed dynamically BDM matrix in horizontal (7) as well as in vertical (9) direction.

Example 1. *The first part of an approach.*
Let us consider a BDM matrix and a vector V^t as in Tab. 1. Firstly, we apply the dynamic decomposition approach. The following subsets of faults and diagnostic signals are obtained according to (10–11):

$F^2 = \{f_1, f_3, f_4, f_5\}$, $S^2 = \{s_1, s_2, s_3, s_4, s_6\}$,
$F^3 = \{f_1, f_3, f_4\}$, $S^3 = \{s_1, s_2, s_3, s_4, s_6\}$,
$F^4 = \{f_2, f_3, f_5\}$, $S^4 = \{s_1, s_2, s_3, s_4, s_5, s_6\}$,
$F^6 = \{f_1, f_3, f_4\}$, $S^6 = \{s_1, s_2, s_3, s_4, s_6\}$.

As it can be easily seen, all subsets of diagnostic signals S^j are not disjunctive. Therefore, according to condition (12), the application of the dynamic decomposition approach is not possible. Let us now create a fault isolation subsystem based on formulas (7 and 9). Firstly, the columns are removed in accordance with (7). The result is depicted in the intermediate matrix in Tab. 1. Next, the rows of intermediate matrix are rejected in accordance with (9). The result is depicted in the right hand matrix in Tab. 1. Please note that V^* is a column vector for which holds: $\bigwedge_{j=1}^{m^*} v_j^{*t} = 1$. □

4.2. The second part of the approach: isolation of multiple faults
This part of the $MUFIA$ relies on searching for multiple faults in the BDS submatrix with the use of classic approach described in Sect. 3. Theoretically, all possible multiple faults can

Table 1. An example of a formation of a BDS.

S/F	f_1	f_2	f_3	f_4	f_5	V^t	S/F	f_1	f_3	f_4	V^t	S/F	f_1	f_3	f_4	V^*
s_1	0	1	0	0	1	0	s_1	0	0	0	0					
s_2	1	0	1	1	1	1	s_2	1	1	1	1	s_2	1	1	1	1
s_3	1	0	1	1	0	1	s_3	1	1	1	1	s_3	1	1	1	1
s_4	0	1	1	0	1	1	s_4	0	1	0	1	s_4	0	1	0	1
s_5	0	1	0	0	0	0	s_5	0	0	0	0	s_6	1	1	1	1
s_6	1	0	1	1	0	1	s_6	1	1	1	1					

be searched. However, as stated in [14] the searching may be stopped by any predefined fault multiplicity $K \leqslant n^*$. In this case, the number $N^*$of possible multiple faults equals:

$$N^* = \sum_{k=1}^{K} \binom{n^*}{k}. \tag{21}$$

If $K = n^*$ then the total number of possible multiple faults equals $N^* = 2^{n^*} - 1$.

Example 2. *The second part of the MUFIA approach.*
Let us continue reasoning regarding the multiple faults from Ex. 1. The set of possible faults $F^* = \{f_1, f_3, f_4\}$ and a set of diagnostic signals $S^* = \{s_2, s_3, s_4, s_6\}$ were obtained in the first part of the $MUFIA$ approach. The possible multiple faults together with their hypothetic signatures are shown in Tab. 2. The hypothetic signatures were created in accordance with assumption (2°) in Sect. 4.1. As can be seen from the Tab. 2, vector V^* matches one single fault $\{f_3\}$, two double faults $\{f_1f_3\}$ and $\{f_3f_4\}$ and one triple fault $\{f_1f_3f_4\}$. □

Table 2. Matrix of all possible combinations of single fault signatures in BDS.

S^*/F^*	f_3	f_1f_3	f_1f_4	f_3f_4	$f_1f_3f_4$	V^*
s_2	1	1	1	1	1	1
s_3	1	1	1	1	1	1
s_4	1	1	0	1	1	1
s_6	1	1	1	1	1	1

5. The *MFI* approach

The $MUFIA$ algorithm is very effective, particularly if condition (12) is met. If not, it is still effective by implementation of a powerful workbench of the BDM matrix compressing tools. Particularly effective is the horizontal compression which significantly cuts down computational effort (16). But the $MUFIA$ approach has at least one serious drawback. Still, its numerical complexity depends on a combinatory manner on the number of columns compressed in the BDM (19). This allows to think that the $MUFIA$ approach is particularly useful in cases where a relatively low fault multiplicities are of concern or where the number of the BDM matrix columns is low. This tends to be an indication of its usefulness for applications in large scale systems, where higher fault multiplicities and rich sets of faults are not exceptional.

In this section, we will briefly introduce and discuss a multiple fault isolation approach MFI [13] which takes inspiration from the $MUFIA$ approach and propose an alternative solution. The main objective of this section is to present, prove and define conditions of the computational efficiency of the MFI approach.

Firstly, the main savings in computational effort have to be achieved in the MFI by the application of an approach in which checking of all hypothetic multiple faults is not necessary at all, as in $MUFIA$ or classic approaches. Instead of, the MFI simply generates the sets of multiple faults.

Secondly, the MFI approach is based on an idea of composition (synthesis) of multiple faults instead of analysis of possible multiple faults as in $MUFIA$ and classic approaches.

Thirdly, the MFI approach is based on a stepwise composition of the multiple faults in vertical instead of horizontal direction of the BDS submatrix.

5.1. Basics of MFI approach

In fact, the MFI rises from the same root as the $MUFIA$ approach. The assumptions presented in Sect 4.1 as well as in the first part of the approach are identical. Completely different is the second part of the approach. Even the sketchy analysis indicates that this part has dominant influence on the effectiveness of the approach. Therefore, we will only refer to this part of the MFI approach.

The MFI generates multiple fault tracks making use of the features of the matrix of paths introduced in [13]. Each path of this matrix points out exactly one multiple fault. Therefore, it is sufficient to define this matrix and determine all its paths.

Let path p_p be any set of non-zero elements of the matrix $BDS[m^*; n^*]$ leading through its m^* rows and such that each element of this path belongs to each separate row of the BDS .

$$p_p = \{v_{j,i} \in BDS : \; v_{j,i} \neq 0 \; ; \; \forall i \in \{1, n^*\}; \; \forall j \in \{1, m^*\}\} \; ; \; |p_p| = m^* \tag{22}$$

The total number of all paths p_p in the BDS is equal to:

$$N_p = \prod_{j=1}^{m^*} \sum_{i=1}^{n^*} d_{j,i} \tag{23}$$

In order to minimize the number of searched paths, the non-zero elements of the BDS matrix may shift left. Thus in general, the number of elements in each row of a squeezed matrix might be different and is equal to the number of ones in each row of the BDS. Let us denote by n_p the maximal number of ones in any row of the squeezed BDS matrix:

$$n_p = max_{j=1}^{m^*} \left(\sum_{i=1}^{n^*} d_{j,i} \right) \tag{24}$$

The theoretical maximal number N_p of all paths is easy to obtain from (23) by substitution $\sum_{i=1}^{n^*} d_{j,i} = n_p$. Hence:

$$N_p = n_p^{m^*} \tag{25}$$

At least for a part of BDS matrices holds: $n_p < n^*$. Those matrices might be further squeezed. This allows for additional horizontal compression of the previously compressed BDS matrix and thus, promises further reduction of computational effort. The only condition is, that squeezing operation cannot loose information about multiple faults which is, in fact, encoded in the structure of the BDS. Therefore, before being squeezed, all non-zero values from each column of the BDS will be replaced by their corresponding fault labels. Each sequence of m^*

single fault labels belonging to different rows forms a multiple fault label. Clearly, the order of single fault labels in the multiple fault label is irrelevant. This enables a significant reduction of the maximum number of relevant paths in comparison to (25). The maximal number of all multiple fault labels equals:

$$N_p^r = n_p m^* \tag{26}$$

The formula (26) indicates proportional dependence of the maximal number of paths to be searched to the product of the number of columns n_p and the number of rows m^* of the squeezed matrix of paths. In contrast, the number of possible paths to be searched, that is indicated by the same BDS matrix rise exponentially in respect to the number of maximal theoretical fault multiplicity n^* as indicate (21) and (25). Please keep in mind that, the fault symbol which is repeated in the same path does not increase fault multiplicity i.e. $\{f_1 f_1 f_2 f_2\} \equiv \{f_1 f_2\}$. On the contrary, each repeated fault symbol reduce fault multiplicity in particular path by one. In other words, the real count of multiple faults may be significantly lower then indicated in (26).

In the worst case the BDS matrix cannot be squeezed. But, if the BDS matrix is not squeezable, i.e. $n_p = n^*$, then it does not give an assumption to state that the $MUFIA$ approach is more effective. It is obvious, that in this case:

$$N_p^r \leqslant N^*. \tag{27}$$

Therefore, as can be derived from (26), it is sufficient to hold condition:

$$n^* m^* \leqslant 2^{n^*} - 1. \tag{28}$$

to expect better effectiveness of the MFI over the $MUFIA$ approach.

Conclusion 5: As results from (28), the MFI in comparison to the $MUFIA$ approach, is particularly effective if the fault multiplicity is greater or equal to 3 because usually $m^* \approx n^*$.

Example 3. *An example of the MFI approach.*
Firstly let us consider the BDS matrix as in Tab. 1 (right hand matrix). The operation of horizontal squeezing is not possible for this matrix because $n_p = n^* = 3$. The maximal theoretical number of paths according to (26) equals $N_p^r = 3 \cdot 4 = 12$. The maximal number of searches of the multiple faults (21) for the $MUFIA$ approach equals $N^* = 2^3 - 1 = 7$. In this case, the MFI approach is expected to have worse effectiveness than the $MUFIA$.

Let us now apply the MFI approach. Firstly, a matrix of fault paths is generated by simple replacement of the non-zero entries of the BDS matrix by the appropriate fault labels. All non-zero BDS entries $v_{j,i}$ are replaced by the labels f_i as shown below and all zero elements are removed. This matrix is referred as to matrix of fault paths.

S/F	f_1	f_3	f_4
s_2	1	1	1
s_3	1	1	1
s_4	0	1	0
s_6	1	1	1

$\Rightarrow$

f_1	f_3	f_4
f_1	f_3	f_4
	f_3	
f_1	f_3	f_4

The process of generation of multiple faults from matrix of fault paths rely on searching for all possible combinations of all fault labels originating from different rows of this matrix. Appropriate rules are given in [13]. These rules define the principles of the combination and reduction of the rows of the matrix of paths. In fact, the rules allow for transformation of the matrix of paths into the one row matrix of multiple fault labels. In order to exemplify this process, let us consider the first two rows of the right hand matrix shown above. Incidentally, both rows are identical. Possible combinations without repetitions of the fault labels from these both rows are: $\{f_1f_1\}$, $\{f_1f_3\}$, $\{f_1f_4\}$, $\{f_3f_3\}$, $\{f_3f_4\}$, $\{f_4f_4\}$. Thus, first two rows of the matrix of fault paths may be replaced by one row containing: $\{f_1\}$, $\{f_1f_3\}$, $\{f_1f_4\}$, $\{f_3\}$, $\{f_3f_4\}$, $\{f_4\}$. Analogously, third and fourth row of this matrix may be replaced by the one row containing: $\{f_1f_3\}$, $\{f_3\}$, $\{f_3f_4\}$. Result of combinations of rows (1, 2) and (3, 4) of the matrix of fault paths are depicted below.

$\{f_1\}$	$\{f_1f_3\}$	$\{f_1f_4\}$	$\{f_3\}$	$\{f_3f_4\}$	$\{f_4\}$
	$\{f_1f_3\}$		$\{f_3\}$	$\{f_3f_4\}$	

The row reduction process is repeated until the matrix will be reduced to the single row. This row contains sequenced labels of all single and multiple faults.

$\{f_3\}$	$\{f_1f_3\}$	$\{f_3f_4\}$	$\{f_1f_3f_4\}$

6. Implementation

All three multiple fault isolation approaches were considered and implemented in a 16-bit micro-controller based platform (MSP430). The software was coded by means of a low level symbolic address programming language. The best programming practices were used to guarantee the highest performance of the code. Wherever it was possible, the same common procedures were applied for all these implementations. The execution time of each individual implementation was captured by means of the machine cycle counter. In the MSP430 family of micro-controllers, each machine cycle lasts reverse proportionally to the machine clock frequency. The counting of machine cycles allows direct comparison of the computational effectiveness of the investigated algorithms despite the frequency of the oscillator foreseen for the final application.

The diverse binary diagnostic matrices were examined in the experimental investigations. The outputs of all implementations were standardised. A standardised output record contain: total number of isolated multiple faults, index of isolated faults, distribution of fault multiplicities and the number of machine cycles used. In fact, the number of the machine cycles expresses indirectly the effectiveness of the implementation. In this study, the relative execution time indices t_r are defined as the ratio of the number of machine cycles used by the currently tested implementation to the number of either minimum or maximum number of machine cycles used by the reference implementation. The reference number of counts is updated with each change of BDM matrix. The analysis of the values of the efficiency indices allows the decision making regarding reasonability of the final choice of the multiple fault approach. These indices are particularly indicative for embedded applications in the smart field devices. As a rule, these devices have strongly limited computational power. Obviously, the results of the experimental tests are BDM matrix case sensitive. Chosen results of experimental tests are presented in Figs. 1–4.

Fig. 1 depicts the relative execution time t_r versus fault multiplicity k for MFI approach implementation. In this case, a diagonal $BDM_{[k\times k]}$ matrix was tested. As can be seen, the relative execution time rise quadratic with the number of the fault multiplicity. Hopefully, this is not in contradiction to quadratic relationship laid down by the formula (26).

Fig. 2. refers also to the implementation of the MFI approach. It depicts the relative execution time versus the number of rows and columns of a diagonal BDM matrix. As can

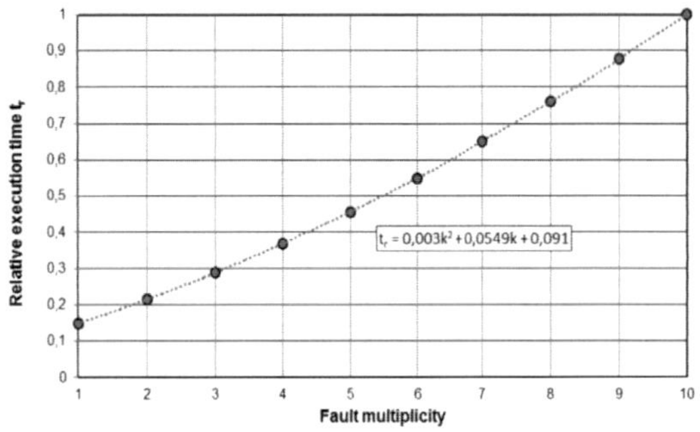

Figure 1. The relative execution time versus fault multiplicity k for MFI implementation.

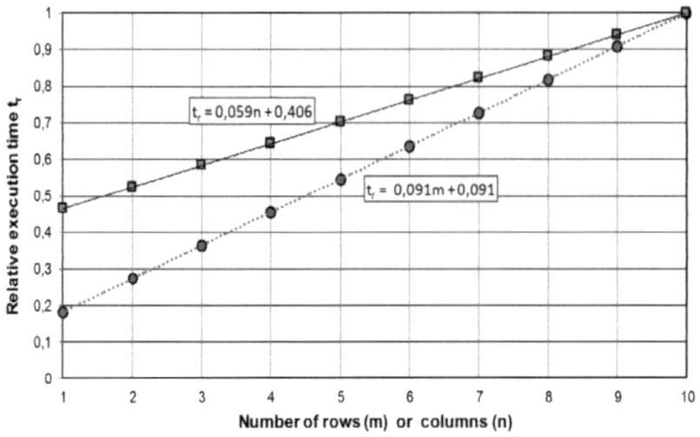

Figure 2. The relative execution time versus the number of rows or columns.

be seen, the relative execution time is proportional to the number of rows as well as to the number of columns. Hence, the expected execution time is proportional to the product $n \cdot m$. This observation further strengthens the credibility of the formula (26). The execution time is increasingly dependent on the number of rows than the number of columns. Fig. 3 depicts the relative execution time t_r versus maximal declared fault multiplicity k for all studied multiple fault isolation approaches. In this case, a diagonal, constant size [10;10] BDM matrix was tested. As can be seen, the relative execution time values for diagonal matrix are almost equal for the classic and the $MUFIA$ approaches. It comes from the adopted assumption that the condition (12) regarding disjunctive sets of diagnostic signals is not considered in this test. In this case the MFI approach is much faster beginning from fault multiplicities higher than 1. This is even more optimistic result than expected in conclusion 5.

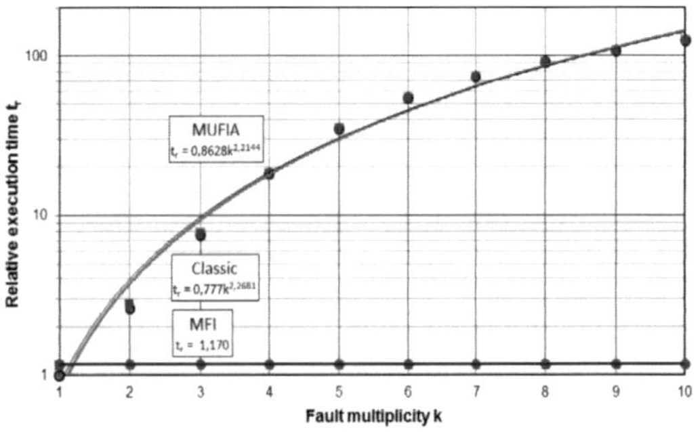

Figure 3. The t_r time versus fault multiplicity k for all studied fault isolation approaches.

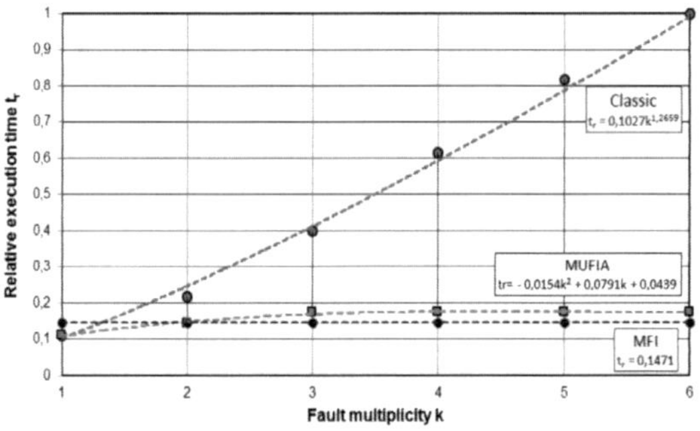

Figure 4. The t_r time versus fault multiplicity k in case of diagnostics of steam-water line.

An example of application to the diagnostic case of the steam-water line of the power station [14] based on the [9:6] BDM matrix is shown in Fig. 4. The condition (12) fails for this matrix. The only two multiple faults are present in the diagnosed system (one double and one triple). Therefore, the relative execution time of the $MUFIA$ is flat for all $k \geqslant 3$. The relative execution time of the MFI is also flat for all $k \geqslant 1$, because the number of rows and columns of the BDM matrix is constant. Here, the MFI approach is slower for the low fault multiplicities and at least ca. 20% faster starting from the fault multiplicities higher than 2. This observation further strengthens conclusion 5.

7. Summary

A theoretical analysis dealing with the applicability features of the three chosen multiple fault isolation approaches based on a binary diagnostic matrix is presented in this paper. The classic, $MUFIA$ and MFI approaches are discussed in order to elaborate practical hints and recommendations. All discussed approaches were implemented and examined in order to verify the correctness of the theoretical analysis. The following general practical recommendations can be withdrawn from presented above discussion:

(i) the classic approach should be used only in exceptional cases, where the number of all single faults in the system is very low (e.g. $n \leqslant 3$) and the expected implementation burden would be minimal;

(ii) the $MUFIA$ approach might be used, where the number of columns in each squeezed BDS matrix for each dynamically decomposed subsystem is very low (e.g. $n^* \leqslant 3$);

(iii) otherwise, the application of the MFI approach is strongly recommended.

Potentially, it is beneficial to use above given sets of recommendations for the development of the smart multiple fault isolation systems making dynamic decisions concerning the choice of the currently called multiple fault isolation procedure.

Acknowledgments

This work was supported by the Polish National Science Centre under the grant No DEC-2011/01/B/ST7/06183 entitled: *"Methods of solution of fundamental problems of fault diagnostics in complex dynamic systems"*.

References

[1] Isermann R and Ballé P 1997 *Trends in the Application of Model Based Fault Detection and Diagnosis of Technical Processes* (Control Engineering Practice) **5**(5) pp 709–719
[2] Travé–Massuyés L 2014 *Bridges between diagnosis theories from control and AI perspectives.* In: Korbicz J and Kowal M (eds.) *Intell. Syst. in Technical and Medical Diagnostics* (Heidelberg: Springer) **230** pp 3–28
[3] Venkatasubramanian V and Rengaswamy R and Yin K and Kavuri S N 2003 *A review of process fault detection and diagnosis: Part I: Quantitative model-based methods* (Computers & Chemical Engineering) **27** pp 293–311
[4] Nyberg M and Krysander M 2003 *Combining AI, FDI, and statistical hypothesis–testing in a framework for diagnosis.* Proc. of 5th IFAC Symp. Safeprocess, (Washington DC) pp 891–896
[5] de Kleer J and Williams B 1987 *Diagnosing multiple faults* (Artificial Intelligence) **32**(1) pp 97–130
[6] Isermann R 2006 *Fault Diagnosis Systems: An Introduction from Fault Detection to Fault Tolerance* (Berlin, Heidelberg: Springer) p 475
[7] Patton R and Frank P and Clark R. (Eds) 2000 *Issues of Fault Diagnosis for Dynamic Systems* (New York: Springer-Verlag) p 599
[8] Blanke M and Kinnaert M and Lunze J and Staroswiecki M 2006 *Diagnosis and Fault-Tolerant Control* (New York: Springer-Verlag) p 684
[9] Korbicz J and Kościelny J M and Kowalczuk Z and Cholewa W. (eds.) (2004). *Fault Diagnosis. Models, Artificial Intelligence, Applications* (Berlin: Springer) p 920
[10] Reiter R A 1987 *Theory of Diagnosis from First Principles* (Artificial Intelligence) **32**(1) pp 57–95
[11] Nyberg M 2006 *A Fault Isolation Algorithm for the Case of Multiple Faults and Multiple Fault Types* (Proc. of IFAC Safeprocess. Beijing) pp 679–689
[12] Gertler J 1998 *Fault Detection and Diagnosis in Engineering Systems* (New York: Marcel Dekker Inc) p 484
[13] Bartyś M 2014 *Multiple Fault Isolation Algorithm Based on Binary Diagnostic Matrix* in Korbicz J and Kowal M, (eds.) *Intell. Syst. in Technical and Medical Diagnosis* (Heilderberg: Springer) **230** pp 441–452
[14] Kościelny J M and Bartyś M and Syfert M 2012 *Methods of Multiple Fault Isolation in Large Scale Systems IEEE Transactions On Control Systems Technology*, **20**(5) pp 1302–1310
[15] Kościelny J M 1995 *Fault Isolation in Industrial Processes by Dynamic Table of States Method, Automatica* **31**(5) pp 747–753

An enhanced approach to actuator fault estimation design for linear continuous-time systems

Dušan Krokavec, Anna Filasová, Vladimír Serbák, and Pavol Liščinský
Department of Cybernetics and Artificial Intelligence, Faculty of Electrical Engineering and Informatics, Technical University of Košice, Letná 9, 042 00 Košice, Slovakia
E-mail: dusan.krokavec@tuke.sk; anna.filasova@tuke.sk; vladimir.serbak@tuke.sk; pavol.liscinsky@tuke.sk

Abstract. An enhanced approach to fault estimation systems design, adjusted for linear continuous-time systems, is proposed in the paper. Based on LMI principle the method exploits state-space observer principle in an adaptive fault estimation scheme for single actuator faults. A simulation example, subject to different type of failures, demonstrates the effectiveness of the proposed form of the fault estimation technique.

1. Introduction

Operating conditions in modern engineering systems are still exposed to possibility of system failure. Any failure of the sensors, actuators or other system components can drastically change the system behavior. Fault tolerant control (FTC) allows a strategy to improve reliability of the whole system and so many techniques have been proposed especially for sensor and actuator failures with application to a wide range of engineering fields. In some cases, fault estimation strategies only are needed to carry on controlling the faulty system and, respecting this fact, many sophisticated modifications have been developed, e.g., sliding mode observers, neural network based approaches and adaptive observer technique.

To estimate actuator faults for the linear time invariant systems without external disturbance the principles based on adaptive observers are frequently used, which make estimation of actuator faults by integrating the system output errors. First introduced in [11], this principle was applied also for descriptor systems [10], linear systems with time delays [3], [13], system with nonlinear dynamics [1], a class of nonlinear systems described by Takagi-Sugeno fuzzy models [6], [7] and linear stochastic Markovian jumping systems [5]. Some generalizations can be found in [15].

Following the results presented in [12], two methods for actuator fault estimation based on the adaptive observer technique are proposed in the paper. Following examination of the model-based fault estimation schemes, an enhanced algorithm using H_∞ approach is provided. Applied enhanced conditions in the scheme increase rapidity and develop a general framework for fast fault estimation in adaptive observer structures for linear deterministic systems. The approach utilizes the measurable input and output vector variables, design conditions are based on linear matrix inequality (LMI) technique giving an effective way to calculate the observer parameters.

The paper is organized as follows. Ensuing the introduction given in Sec. 1, Sec. 2 presents the problem formulation focusing on assumptions about the system and actuator fault properties.

In Sec 3 a short description of the main properties of methods exploiting the adaptive observer technique for actuator faults estimation in linear systems is presented and an enhanced condition of the H_∞ observer existence are analyzed and proven in Sec. 4. In Sec. 5, simulation results are presented and, finally, some concluding remarks are reached in Sec. 6.

Throughout the paper, the following notation was used: $\boldsymbol{x}^T$, $\boldsymbol{X}^T$ denotes the transpose of the vector $\boldsymbol{x}$ and the matrix $\boldsymbol{X}$, respectively, diag$[\cdot]$ enters up a block diagonal matrix, rank$(\cdot)$ remits the rank of a matrix, for a square matrix $\boldsymbol{X} < 0$ means that $\boldsymbol{X}$ is a symmetric negative definite matrix, the symbol $\boldsymbol{I}_n$ indicates the n-th order unit matrix, $I\!R$ notes the set of real numbers, and $I\!R^n$, $I\!R^{n\times r}$ refer to the set of all n-dimensional real vectors and $n \times r$ real matrices, respectively.

2. Problem formulation

A linear dynamic multi-input, multi-output (MIMO) system in presence of an unknown fault can be described by the state-space equations in the following form

$$\dot{\boldsymbol{q}}(t) = \boldsymbol{A}\boldsymbol{q}(t) + \boldsymbol{B}\boldsymbol{u}(t) + \boldsymbol{E}\boldsymbol{f}(t), \tag{1}$$

$$\boldsymbol{y}(t) = \boldsymbol{C}\boldsymbol{q}(t), \tag{2}$$

where $\boldsymbol{q}(t) \in I\!R^n$, $\boldsymbol{u}(t) \in I\!R^r$, and $\boldsymbol{y}(t) \in I\!R^p$ are vectors of the system, input and output variables, respectively, $\boldsymbol{f}(t) \in I\!R^s$ is the unknown fault vector, $\boldsymbol{A} \in I\!R^{n\times n}$ is the system dynamic matrix, $\boldsymbol{E} \in I\!R^{n\times s}$ is the fault input matrix and $\boldsymbol{B} \in I\!R^{n\times r}$ and $\boldsymbol{C} \in I\!R^{p\times n}$ are the system input and output matrices.

To estimate actuator faults and the system states simultaneously, the following adaptive state estimator is proposed [2], [8]

$$\dot{\boldsymbol{q}}_e(t) = \boldsymbol{A}\boldsymbol{q}_e(t) + \boldsymbol{B}\boldsymbol{u}(t) + \boldsymbol{E}\boldsymbol{f}_e(t) + \boldsymbol{J}(\boldsymbol{y}(t) - \boldsymbol{y}_e(t)), \tag{3}$$

$$\boldsymbol{y}_e(t) = \boldsymbol{C}\boldsymbol{q}_e(t), \tag{4}$$

where $\boldsymbol{q}_e(t) \in I\!R^n$ is the state observer vector, $\boldsymbol{f}_e(t)$ is an estimation of the fault $\boldsymbol{f}(t)$, $\boldsymbol{y}_e(t) \in I\!R^m$ is the vector of estimated output variables, $\boldsymbol{J} \in I\!R^{n\times p}$ is the estimator gain matrix while $n > p$.

The task is to design the matrix $\boldsymbol{J}$ in such a way that the observer dynamics matrix $\boldsymbol{A}_e = \boldsymbol{A} - \boldsymbol{J}\boldsymbol{C}$ is stable and $\boldsymbol{f}_e(t)$ approximates a slowly varying actuator fault $\boldsymbol{f}(t)$. Actuator faults are represented as a exchange of the matrix $\boldsymbol{E}$ to $\boldsymbol{B}$ for $s = r$.

The state observer (3), (4) is combined with the law for the fault estimation updating of the form [12]

$$\dot{\boldsymbol{f}}_e(t) = \boldsymbol{G}\boldsymbol{H}^T\boldsymbol{e}_y(t), \tag{5}$$

where

$$\boldsymbol{e}_y(t) = \boldsymbol{y}(t) - \boldsymbol{y}_e(t), \tag{6}$$

$\boldsymbol{H} \in I\!R^{p\times s}$ is the gain matrix and $\boldsymbol{G} = \boldsymbol{G}^T > 0$, $\boldsymbol{G} \in I\!R^{s\times s}$ is a learning weight matrix that has to be set interactively in the design step.

It has to be noted that a modifications of (5) were proposed in [14] for time varying $\boldsymbol{f}(t)$ in the form

$$\dot{\boldsymbol{f}}_e(t) = \boldsymbol{G}\boldsymbol{H}^T(\sigma\boldsymbol{e}_y(t) + \dot{\boldsymbol{e}}_y(t)), \tag{7}$$

where $\sigma \in I\!R$ is a positive scalar, determining together with $\boldsymbol{G}$ the learning rate. A generalization of the adaptive state estimator is given in [10], where

$$\dot{\boldsymbol{q}}_e(t) = \boldsymbol{A}\boldsymbol{q}_e(t) + \boldsymbol{B}\boldsymbol{u}(t) + \boldsymbol{E}\boldsymbol{f}_e(t) + \boldsymbol{J}(\boldsymbol{y}(t) - \boldsymbol{y}_e(t)) + \boldsymbol{J}_d(\dot{\boldsymbol{y}}(t) - \dot{\boldsymbol{y}}_e(t)), \tag{8}$$

$$\boldsymbol{y}_e(t) = \boldsymbol{C}\boldsymbol{q}_e(t), \tag{9}$$

$$\dot{\boldsymbol{f}}_e(t) = \boldsymbol{L}\boldsymbol{e}_y(t) + \boldsymbol{L}_d\dot{\boldsymbol{e}}_y(t) \tag{10}$$

It is obvious that if $\boldsymbol{J}_d = \boldsymbol{0}$ and if it can be set $\boldsymbol{L}$ as $\sigma\boldsymbol{L}_d$ with $\boldsymbol{L}_d = \boldsymbol{G}\boldsymbol{H}^T$ then this estimator reduces to the so called fast adaptive fault estimator (3), (4), (7). Analogously, if $\boldsymbol{J}_d = \boldsymbol{0}$, $\boldsymbol{L}_d = \boldsymbol{0}$ and $\boldsymbol{L} = \boldsymbol{G}\boldsymbol{H}^T$ the estimator reduces to the zero-integral estimator (3), (4), (5). In this work the zero-integral estimator is used for slowly-varying faults.

Assumption 1 *The couple $(\boldsymbol{A}, \boldsymbol{C})$ is observable and $rank(\boldsymbol{C}\boldsymbol{E}) = rank(\boldsymbol{E})$.*

Assumption 2 *The unknown fault vector, changing unexpectedly when a fault occurs, is piecewise constant, differentiable and bounded, i.e., $|\boldsymbol{f}(t)| \leq \boldsymbol{f}_{max} < \infty$, the upper bound $\boldsymbol{f}_{max}$ of the fault magnitude is known, and the value of $\boldsymbol{f}(t)$ is set to zero until a fault occurs.*

Under these assumptions design of the matrix parameters of the observers has to ensure asymptotic convergence of the estimation errors (6) and

$$\boldsymbol{e}_f(t) = \boldsymbol{f}(t) - \boldsymbol{f}_e(t) \tag{11}$$

to zero values. Note, Assumption 2 implies that the derivative $\boldsymbol{e}_f(t)$ with respect to time can be considered as

$$\dot{\boldsymbol{e}}_f(t) = -\dot{\boldsymbol{f}}_e(t). \tag{12}$$

3. The adaptive observer principle in actuator faults estimation

If single actuator faults influence the system through different input vectors (columns of the matrix $\boldsymbol{B}$), it is possible to avoid design of estimators with the tuning matrix parameter $\boldsymbol{G} > 0$ and formulate the design task through the set of LMIs and a linear matrix equality.

Proposition 1 *The actuator fault estimator is stable if there exist symmetric positive definite matrix $\boldsymbol{P} \in I\!R^{n \times n}$ and matrices $\boldsymbol{H} \in I\!R^{p \times s}$, $\boldsymbol{Y} \in I\!R^{n \times p}$ such that*

$$\boldsymbol{P} = \boldsymbol{P}^T > 0, \tag{13}$$

$$\boldsymbol{P}\boldsymbol{A} + \boldsymbol{A}^T\boldsymbol{P} - \boldsymbol{Y}\boldsymbol{C} - \boldsymbol{C}^T\boldsymbol{Y}^T < 0, \tag{14}$$

$$\boldsymbol{P}\boldsymbol{E} = \boldsymbol{C}^T\boldsymbol{H}. \tag{15}$$

When the above conditions hold, the observer gain matrix is given by

$$\boldsymbol{J} = \boldsymbol{P}^{-1}\boldsymbol{Y} \tag{16}$$

and the adaptive fault estimation algorithm is

$$\dot{\boldsymbol{f}}_e(t) = \boldsymbol{G}\boldsymbol{H}^T\boldsymbol{C}\boldsymbol{e}_q(t). \tag{17}$$

where

$$\boldsymbol{e}_q(t) = \boldsymbol{q}(t) - \boldsymbol{q}_e(t). \tag{18}$$

matrix $\boldsymbol{H}$ is obtained as an LMI variable and $\boldsymbol{G} \in I\!R^{s \times s}$ is an interactive setting symmetric positive definite matrix.

Proof: (compare, e.g., [3], [14]) From the system model (1), (2) and the observer model (3), (4) it can be seen that

$$\begin{aligned}\dot{\boldsymbol{e}}_q(t) &= \dot{\boldsymbol{q}}(t) - \dot{\boldsymbol{q}}_e(t)\\ &= \boldsymbol{A}\boldsymbol{q}(t) + \boldsymbol{B}\boldsymbol{u}(t) + \boldsymbol{E}\boldsymbol{f}(t) - \boldsymbol{A}\boldsymbol{q}_e(t) - \boldsymbol{B}\boldsymbol{u}(t) - \boldsymbol{E}\boldsymbol{f}_e(t) - \boldsymbol{J}(\boldsymbol{y}(t) - \boldsymbol{y}_e(t))\\ &= (\boldsymbol{A} - \boldsymbol{J}\boldsymbol{C})\boldsymbol{e}_q(t) + \boldsymbol{E}\boldsymbol{e}_f(t),\end{aligned} \tag{19}$$

where the observer error is

$$\boldsymbol{e}_q(t) = \boldsymbol{q}(t) - \boldsymbol{q}_e(t). \tag{20}$$

Since $\boldsymbol{e}_q(t)$ is linear with respect to system parameters, it is possible to consider the Lyapunov function candidate in the following form

$$v(\boldsymbol{e}_q(t)) = \boldsymbol{e}_q^T(t)\boldsymbol{P}\boldsymbol{e}_q(t) + \boldsymbol{e}_f^T(t)\boldsymbol{G}^{-1}\boldsymbol{e}_f(t) \tag{21}$$

where $\boldsymbol{P} > 0$, $\boldsymbol{G} > 0$ are symmetric positive definite matrices and an actuator fault satisfied the condition

$$\dot{\boldsymbol{f}}_e(t) = \boldsymbol{G}\boldsymbol{H}^T\boldsymbol{e}_y(t) = \boldsymbol{G}\boldsymbol{H}^T\boldsymbol{C}\boldsymbol{e}_q(t), \tag{22}$$

as well as

$$\boldsymbol{e}_f(t) = \boldsymbol{f}_0 - \boldsymbol{f}_e(t) \;\Rightarrow\; \dot{\boldsymbol{e}}_f(t) = -\dot{\boldsymbol{f}}_e(t). \tag{23}$$

Then, the derivative of $v(\boldsymbol{e}_q(t))$ with respect to t is

$$\dot{v}(\boldsymbol{e}_q(t)) = \dot{v}_0(\boldsymbol{e}_q(t)) + \dot{v}_1(\boldsymbol{e}_q(t)) < 0, \tag{24}$$

where

$$\begin{aligned}
\dot{v}_0(\boldsymbol{e}_q(t)) &= \dot{\boldsymbol{e}}_q^T(t)\boldsymbol{P}\boldsymbol{e}_q(t) + \boldsymbol{e}_q^T(t)\boldsymbol{P}\dot{\boldsymbol{e}}_q(t)\\
&= ((\boldsymbol{A}-\boldsymbol{J}\boldsymbol{C})\boldsymbol{e}_q(t) + \boldsymbol{E}\boldsymbol{e}_f(t))^T\boldsymbol{P}\boldsymbol{e}_q(t) + \boldsymbol{e}_q^T(t)\boldsymbol{P}((\boldsymbol{A}-\boldsymbol{J}\boldsymbol{C})\boldsymbol{e}_q(t) + \boldsymbol{E}\boldsymbol{e}_f(t))\\
&= \boldsymbol{e}_q^T(t)\left((\boldsymbol{A}-\boldsymbol{J}\boldsymbol{C})^T\boldsymbol{P} + \boldsymbol{P}(\boldsymbol{A}-\boldsymbol{J}\boldsymbol{C})\right)\boldsymbol{e}_q(t)\\
&+ \boldsymbol{e}_q^T(t)\boldsymbol{P}\boldsymbol{E}\boldsymbol{e}_f(t) + \boldsymbol{e}_f^T(t)\boldsymbol{E}^T\boldsymbol{P}\boldsymbol{e}_q(t)
\end{aligned} \tag{25}$$

and

$$\dot{v}_1(\boldsymbol{e}_q(t)) = \dot{\boldsymbol{e}}_f^T(t)\boldsymbol{G}^{-1}\boldsymbol{e}_f(t) + \boldsymbol{e}_f^T(t)\boldsymbol{G}^{-1}\dot{\boldsymbol{e}}_f(t) = -\dot{\boldsymbol{f}}_e^T(t)\boldsymbol{G}^{-1}\boldsymbol{e}_f(t) - \boldsymbol{e}_f^T(t)\boldsymbol{G}^{-1}\dot{\boldsymbol{f}}_e(t). \tag{26}$$

Substituting (22) into (26) leads to

$$\begin{aligned}
\dot{v}_1(\boldsymbol{e}_q(t)) &= -\boldsymbol{e}_q^T(t)\boldsymbol{C}^T\boldsymbol{H}\boldsymbol{G}\boldsymbol{G}^{-1}\boldsymbol{e}_f(t) - \boldsymbol{e}_f^T(t)\boldsymbol{G}^{-1}\boldsymbol{G}\boldsymbol{H}^T\boldsymbol{C}\boldsymbol{e}_q(t)\\
&= -\left(\boldsymbol{e}_q^T(t)\boldsymbol{C}^T\boldsymbol{H}\boldsymbol{e}_f(t) + \boldsymbol{e}_f^T(t)\boldsymbol{H}^T\boldsymbol{C}\boldsymbol{e}_q(t)\right).
\end{aligned} \tag{27}$$

Thus, substituting (25) and (27) into (19), the following inequality is obtained

$$\begin{aligned}
\dot{v}(\boldsymbol{e}_q(t)) &= \boldsymbol{e}_q^T(t)\left((\boldsymbol{A}-\boldsymbol{J}\boldsymbol{C})^T\boldsymbol{P} + \boldsymbol{P}(\boldsymbol{A}-\boldsymbol{J}\boldsymbol{C})\right)\boldsymbol{e}_q(t)\\
&+\boldsymbol{e}_q^T(t)\left(\boldsymbol{P}\boldsymbol{E} - \boldsymbol{C}^T\boldsymbol{H}\right)\boldsymbol{e}_f(t) + \boldsymbol{e}_f^T(t)\left(\boldsymbol{E}^T\boldsymbol{P} - \boldsymbol{H}^T\boldsymbol{C}\right)\boldsymbol{e}_q(t) < 0.
\end{aligned} \tag{28}$$

If there is set the condition

$$\boldsymbol{e}_q^T(t)\left(\boldsymbol{P}\boldsymbol{E} - \boldsymbol{C}^T\boldsymbol{H}\right)\boldsymbol{e}_f(t) + \boldsymbol{e}_f^T(t)\left(\boldsymbol{E}^T\boldsymbol{P} - \boldsymbol{H}^T\boldsymbol{C}\right)\boldsymbol{e}_q(t) = 0, \tag{29}$$

then the last equality gives

$$\boldsymbol{P}\boldsymbol{E} - \boldsymbol{C}^T\boldsymbol{H} = 0 \tag{30}$$

and, evidently, (30) implies (15).

Using the above given condition, the resulting formula for $\dot{v}(\boldsymbol{e}_q(t))$ takes the form

$$\dot{v}(\boldsymbol{e}_q(t)) = \boldsymbol{e}_q^T(t)\left((\boldsymbol{A}-\boldsymbol{J}\boldsymbol{C})^T\boldsymbol{P} + \boldsymbol{P}(\boldsymbol{A}-\boldsymbol{J}\boldsymbol{C})\right)\boldsymbol{e}_q(t) < 0, \tag{31}$$

where

$$\boldsymbol{P}(\boldsymbol{A}-\boldsymbol{J}\boldsymbol{C}) + (\boldsymbol{A}-\boldsymbol{J}\boldsymbol{C})^T\boldsymbol{P} < 0. \tag{32}$$

Since the inequality (32) can be written as

$$\boldsymbol{P}\boldsymbol{A} - \boldsymbol{P}\boldsymbol{J}\boldsymbol{C} + \boldsymbol{A}^T\boldsymbol{P} - \boldsymbol{C}^T\boldsymbol{J}^T\boldsymbol{P} < 0, \tag{33}$$

by introducing the substitution

$$\boldsymbol{P}\boldsymbol{J} = \boldsymbol{Y}, \tag{34}$$

it is possible to express (33) as (14). This concludes the proof. ■

4. Enhanced design conditions

Analyzing (19), i.e., the differential equation of the form

$$\dot{\boldsymbol{e}}_q(t) = (\boldsymbol{A} - \boldsymbol{JC})\boldsymbol{e}_q(t) + \boldsymbol{E}\boldsymbol{e}_f(t), \tag{35}$$

it is evident that $\boldsymbol{e}_f(t)$ acts on the state error dynamics as an unknown disturbance and, evidently, this differential equation is so not autonomous after a fault occurrence. Reflecting this fact, the enhanced approach is proposed to decouple Lyapunov matrix $\boldsymbol{P}$ from the system matrices $\boldsymbol{A}$, $\boldsymbol{C}$ by introducing a slack matrix $\boldsymbol{Q}$ in the observer stability condition, as well as to decouple the tuning parameter δ from the matrix $\boldsymbol{G}$ in the learning rate setting and to use it to tune the observer dynamic properties. Since the design principle for unknown input observer can not be used in the case if a fault is to be estimated, the impact of faults on observer dynamics is moreover minimized with respect to the H_∞ norm of the transfer functions matrix of $\boldsymbol{e}_y$ and $\boldsymbol{e}_f$, while a reduce in the fault amplitude estimate is easily countervailable using the matrix $\boldsymbol{G}$. In this sense it is proposed to formulate the enhanced design conditions as follows:

Theorem 1 *The actuator fault estimator is stable if for given positive $\delta \in I\!R$ there exist symmetric positive definite matrices $\boldsymbol{P} \in I\!R^{n\times n}$, $\boldsymbol{Q} \in I\!R^{n\times n}$, matrices $\boldsymbol{H} \in I\!R^{p\times s}$, $\boldsymbol{Y} \in I\!R^{n\times p}$ and a positive scalar $\gamma \in I\!R$ such that*

$$\boldsymbol{P} = \boldsymbol{P}^T > 0, \quad \boldsymbol{Q} = \boldsymbol{Q}^T > 0, \quad \gamma > 0, \tag{36}$$

$$\begin{bmatrix} \boldsymbol{QA} + \boldsymbol{A}^T\boldsymbol{Q} - \boldsymbol{YC} - \boldsymbol{C}^T\boldsymbol{Y}^T + \boldsymbol{C}^T\boldsymbol{C} & * & * \\ \boldsymbol{P} - \boldsymbol{Q} + \delta\boldsymbol{QA} - \delta\boldsymbol{YC} & -2\delta\boldsymbol{Q} & * \\ \boldsymbol{0} & \delta\boldsymbol{E}^T\boldsymbol{Q} & -\gamma\boldsymbol{I}_{rf} \end{bmatrix} < 0, \tag{37}$$

$$\boldsymbol{C}^T\boldsymbol{H} = \boldsymbol{QE}. \tag{38}$$

When the above conditions are affirmative the estimator gain matrix is given by the relation

$$\boldsymbol{J} = \boldsymbol{Q}^{-1}\boldsymbol{Y} \tag{39}$$

and the adaptive fault estimation algorithm is given by (17).

Here and hereafter, $*$ denotes the symmetric item in a symmetric matrix.

Proof: Using Krasovskii theorem [4], the Lyapunov function candidate can be considered as follows

$$v(\boldsymbol{e}_q(t)) = \boldsymbol{e}_q^T(t)\boldsymbol{P}\boldsymbol{e}_q(t) + \boldsymbol{e}_f^T(t)\boldsymbol{G}^{-1}\boldsymbol{e}_f(t) + \int_0^t \left(\boldsymbol{e}_y^T(r)\boldsymbol{e}_y(r) - \gamma\boldsymbol{e}_f^T(r)\boldsymbol{e}_f(r)\right) \mathrm{d}r, \tag{40}$$

where $\boldsymbol{P} = \boldsymbol{P}^T > 0$, $\boldsymbol{G} = \boldsymbol{G}^T > 0$, $\gamma > 0$ and γ is an upper bound of square of H_∞ norm of the transfer function matrix $\boldsymbol{e}_f \asymp \boldsymbol{e}_y$. Then the derivative of $v(\boldsymbol{e}_q(t))$ with respect to t has to be negative, i.e.,

$$\begin{aligned} \dot{v}(\boldsymbol{e}_q(t)) \; &= \dot{\boldsymbol{e}}_q^T(t)\boldsymbol{P}\boldsymbol{e}_q(t) + \boldsymbol{e}_q^T(t)\boldsymbol{P}\dot{\boldsymbol{e}}_q(t) + \dot{\boldsymbol{e}}_f^T(t)\boldsymbol{G}^{-1}\boldsymbol{e}_f(t) \\ &+ \boldsymbol{e}_f^T(t)\boldsymbol{G}^{-1}\dot{\boldsymbol{e}}_f(t) + \boldsymbol{e}_y^T(t)\boldsymbol{e}_y(t) - \gamma\boldsymbol{e}_f^T(t)\boldsymbol{e}_f(t) < 0. \end{aligned} \tag{41}$$

If it is assumed that the statements (22), (23) hold, then substitution of (22) into (41) leads to

$$\begin{aligned} \dot{v}(\boldsymbol{e}_q(t)) \; &= \dot{\boldsymbol{e}}_q^T(t)\boldsymbol{P}\boldsymbol{e}_q(t) + \boldsymbol{e}_q^T(t)\boldsymbol{P}\dot{\boldsymbol{e}}_q(t) \\ &- \boldsymbol{e}_q^T(t)\boldsymbol{C}^T\boldsymbol{H}\boldsymbol{G}\boldsymbol{G}^{-1}\boldsymbol{e}_f(t) - \boldsymbol{e}_f^T(t)\boldsymbol{G}^{-1}\boldsymbol{G}\boldsymbol{H}^T\boldsymbol{C}\boldsymbol{e}_q(t) \\ &+ \boldsymbol{e}_y^T(t)\boldsymbol{e}_y(t) - \gamma\boldsymbol{e}_f^T(t)\boldsymbol{e}_f(t) < 0, \end{aligned} \tag{42}$$

$$\begin{aligned}\dot{v}(\boldsymbol{e}_q(t)) &= \dot{\boldsymbol{e}}_q^T(t)\boldsymbol{P}\boldsymbol{e}_q(t) + \boldsymbol{e}_q^T(t)\boldsymbol{P}\dot{\boldsymbol{e}}_q(t) - \boldsymbol{e}_q^T(t)\boldsymbol{C}^T\boldsymbol{H}\boldsymbol{e}_f(t) - \boldsymbol{e}_f^T(t)\boldsymbol{H}^T\boldsymbol{C}\boldsymbol{e}_q(t) \\ &+ \boldsymbol{e}_y^T(t)\boldsymbol{e}_y(t) - \gamma\boldsymbol{e}_f^T(t)\boldsymbol{e}_f(t) < 0,\end{aligned} \tag{43}$$

respectively.

Since (35) implies

$$(\boldsymbol{A} - \boldsymbol{J}\boldsymbol{C})\boldsymbol{e}_q(t) + \boldsymbol{E}\boldsymbol{e}_f(t) - \dot{\boldsymbol{e}}_q(t) = 0, \tag{44}$$

it is possible to define the following condition based on the equality (44)

$$\left(\boldsymbol{e}_q^T(t)\boldsymbol{Q} + \dot{\boldsymbol{e}}_q^T(t)\delta\boldsymbol{Q}\right)((\boldsymbol{A} - \boldsymbol{J}\boldsymbol{C})\boldsymbol{e}_q(t) + \boldsymbol{E}\boldsymbol{e}_f(t) - \dot{\boldsymbol{e}}_q(t)) = 0, \tag{45}$$

where $\boldsymbol{Q} \in I\!R^{n\times n}$ is a symmetric positive definite matrix and $\delta \in I\!R$ is a positive scalar. Then, inserting (45) and its transposition into (43), the following has to be satisfied

$$\begin{aligned}\dot{v}(\boldsymbol{e}_q(t)) &= \dot{\boldsymbol{e}}_q^T(t)\boldsymbol{P}\boldsymbol{e}_q(t) + \boldsymbol{e}_q^T(t)\boldsymbol{P}\dot{\boldsymbol{e}}_q(t) - \boldsymbol{e}_q^T(t)\boldsymbol{C}^T\boldsymbol{H}\boldsymbol{e}_f(t) - \boldsymbol{e}_f^T(t)\boldsymbol{H}^T\boldsymbol{C}\boldsymbol{e}_q(t) \\ &+ (\boldsymbol{e}_q^T(t)\boldsymbol{Q} + \dot{\boldsymbol{e}}_q^T(t)\delta\boldsymbol{Q})((\boldsymbol{A} - \boldsymbol{J}\boldsymbol{C})\boldsymbol{e}_q(t) - \dot{\boldsymbol{e}}_q(t)) \\ &+ ((\boldsymbol{A} - \boldsymbol{J}\boldsymbol{C})\boldsymbol{e}_q(t) - \dot{\boldsymbol{e}}_q(t))^T(\boldsymbol{Q}\boldsymbol{e}_q(t) + \delta\boldsymbol{Q}\dot{\boldsymbol{e}}_q(t)) \\ &+ (\boldsymbol{e}_q^T(t)\boldsymbol{Q} + \dot{\boldsymbol{e}}_q^T(t)\delta\boldsymbol{Q})\boldsymbol{E}\boldsymbol{e}_f(t) + \boldsymbol{e}_f^T(t)\boldsymbol{E}^T(\boldsymbol{Q}\boldsymbol{e}_q(t) + \delta\boldsymbol{Q}\dot{\boldsymbol{e}}_q(t)) \\ &+ \boldsymbol{e}_y^T(t)\boldsymbol{e}_y(t) - \gamma\boldsymbol{e}_f^T(t)\boldsymbol{e}_f(t) < 0.\end{aligned} \tag{46}$$

If the following condition is introduced

$$\boldsymbol{e}_f^T(t)\left(\boldsymbol{E}^T\boldsymbol{Q} - \boldsymbol{H}^T\boldsymbol{C}\right)\boldsymbol{e}_q(t) + \boldsymbol{e}_q^T(t)\left(\boldsymbol{Q}\boldsymbol{E} - \boldsymbol{C}^T\boldsymbol{H}\right)\boldsymbol{e}_f(t) = 0, \tag{47}$$

this gives the equality

$$\boldsymbol{Q}\boldsymbol{E} - \boldsymbol{C}^T\boldsymbol{H} = 0, \tag{48}$$

which implies (38).

The condition (47) allows to write (46) as

$$\begin{aligned}\dot{v}(\boldsymbol{e}_q(t)) &= \dot{\boldsymbol{e}}_q^T(t)\boldsymbol{P}\boldsymbol{e}_q(t) + \boldsymbol{e}_q^T(t)\boldsymbol{P}\dot{\boldsymbol{e}}_q(t) \\ &+ (\boldsymbol{e}_q^T(t)\boldsymbol{Q} + \dot{\boldsymbol{e}}_q^T(t)\delta\boldsymbol{Q})((\boldsymbol{A} - \boldsymbol{J}\boldsymbol{C})\boldsymbol{e}_q(t) - \dot{\boldsymbol{e}}_q(t)) \\ &+ (\boldsymbol{e}_q^T(t)(\boldsymbol{A} - \boldsymbol{J}\boldsymbol{C})^T - \dot{\boldsymbol{e}}_q^T(t))(\boldsymbol{Q}\boldsymbol{e}_q(t) + \delta\boldsymbol{Q}\dot{\boldsymbol{e}}_q(t)) \\ &+ \delta\dot{\boldsymbol{e}}_q^T(t)\boldsymbol{Q}\boldsymbol{E}\boldsymbol{e}_f(t) + \delta\boldsymbol{e}_f^T(t)\boldsymbol{E}^T\boldsymbol{Q}\dot{\boldsymbol{e}}_q(t) + \boldsymbol{e}_y^T(t)\boldsymbol{e}_y(t) - \gamma\boldsymbol{e}_f^T(t)\boldsymbol{e}_f(t)\end{aligned} \tag{49}$$

and to prescribe the stability condition as

$$\dot{v}(\boldsymbol{e}_q(t)) = \boldsymbol{e}_q^{\bullet T}(t)\boldsymbol{P}^{\bullet}\boldsymbol{e}_q^{\bullet}(t) < 0, \tag{50}$$

where

$$\boldsymbol{P}^{\bullet} = \begin{bmatrix} \boldsymbol{Q}(\boldsymbol{A} - \boldsymbol{J}\boldsymbol{C}) + \boldsymbol{C}\boldsymbol{C}^T + (\boldsymbol{A} - \boldsymbol{J}\boldsymbol{C})^T\boldsymbol{Q} & \boldsymbol{P} - \boldsymbol{Q} + \delta(\boldsymbol{A} - \boldsymbol{J}\boldsymbol{C})^T\boldsymbol{Q} & \boldsymbol{0} \\ \boldsymbol{P} - \boldsymbol{Q} + \delta\boldsymbol{Q}(\boldsymbol{A} - \boldsymbol{J}\boldsymbol{C}) & -2\delta\boldsymbol{Q} & \delta\boldsymbol{Q}\boldsymbol{E} \\ \boldsymbol{0} & \delta\boldsymbol{E}^T\boldsymbol{Q} & -\gamma\boldsymbol{I}_{rf} \end{bmatrix} < 0. \tag{51}$$

Introducing the notation

$$\boldsymbol{Q}\boldsymbol{J} = \boldsymbol{Y} \tag{52}$$

then (51) implies (37). This concludes the proof. ■

5. Illustrative example

In the example, there is considered the system (1), (2) in the state-space representation, where the system matrices are

$$\boldsymbol{A} = \begin{bmatrix} -3.2341 & -0.0356 & 0.0200 & 0.0267 \\ -0.0356 & -3.1883 & -0.0502 & 0.0189 \\ 0.0200 & -0.0502 & -3.2575 & 0.0100 \\ 0.0267 & 0.0189 & 0.0100 & -3.1537 \end{bmatrix}, \quad \boldsymbol{B} = \boldsymbol{E} = \begin{bmatrix} -0.3802 & 1.4370 \\ 0.6959 & -0.5393 \\ 2.3171 & -0.4301 \\ 0.0000 & 0.6507 \end{bmatrix},$$

$$\boldsymbol{C} = \begin{bmatrix} 1 & 2 & 1 & 1 \\ 1 & 1 & 0 & 2 \end{bmatrix}.$$

Solving (13)-(15) with respect to the LMI matrix variables $\boldsymbol{P}$, $\boldsymbol{H}$, and $\boldsymbol{Y}$ using Self-Dual-Minimization (SeDuMi) package [9] for Matlab, the estimator parameter design problem was solved as feasible and

$$\boldsymbol{P} = \begin{bmatrix} 0.1770 & 0.2107 & 0.0321 & 0.0457 \\ 0.2107 & 0.6888 & 0.0678 & 0.2060 \\ 0.0321 & 0.0678 & 0.1586 & -0.0955 \\ 0.0457 & 0.2060 & -0.0955 & 0.6740 \end{bmatrix}, \quad \boldsymbol{Y} = \begin{bmatrix} -0.3691 & 0.3090 \\ -0.7451 & -0.1267 \\ 0.0251 & 0.1647 \\ 0.2357 & -0.8632 \end{bmatrix},$$

$$\boldsymbol{H} = \begin{bmatrix} 0.4026 & -0.1207 \\ -0.2489 & 0.2774 \end{bmatrix}.$$

The estimator gain matrix was computed using (16) as follows

$$\boldsymbol{J} = \begin{bmatrix} -1.2534 & 2.8937 \\ -1.1442 & -0.6872 \\ 1.5003 & -0.0182 \\ 0.9967 & -1.2693 \end{bmatrix},$$

and guaranties the stable actuator fault observer, where the system matrix eigenvalues spectrum is

$$\rho(\boldsymbol{A}_e) = \rho(\boldsymbol{A} - \boldsymbol{J}\boldsymbol{C}) = \{-2.1745 \ \ -2.6067 \ \ -3.3377 \pm 0.1566\,\mathrm{i}\}.$$

Setting the tuning parameter $\boldsymbol{G}$ as follows

$$\boldsymbol{G} = \begin{bmatrix} 1.7 & 1.0 \\ 1.0 & 5.5 \end{bmatrix}$$

the observer fault response is given in Fig. 1. This figure presents the fault signal, as well as its estimation, reflecting at first a single actuator fault in the the second actuator. This actuator fault starts at the time instant $t = 30s$ and is applied for $40s$. The learning parameter $\boldsymbol{G}$ has been set experimentally, considering the maximal value of the fault signal amplitude. Then, at the time instant $t = 100s$ a fault of the first actuator is applied for $40s$.

From the simulation results in Fig. 1 it can be observed that the differences between the signals reflecting single actuator faults and the observer approximated ones tends to zero.

Setting the tuning parameters $\delta = 5$ and solving (36)-(38) with respect the LMI matrix variables $\boldsymbol{P}$, $\boldsymbol{Q}$, $\boldsymbol{H}$, and $\boldsymbol{Y}$, γ, the estimator parameter design problem was solved as feasible and the design parameters were

$$\boldsymbol{P} = \begin{bmatrix} 25.7515 & 18.1103 & 3.3556 & -4.3147 \\ 18.1103 & 61.2780 & 4.9193 & 4.9418 \\ 3.3556 & 4.9193 & 20.7111 & -7.5051 \\ -4.3147 & 4.9418 & -7.5051 & 62.0896 \end{bmatrix}, \quad \boldsymbol{Y} = \begin{bmatrix} -0.6530 & -0.7548 \\ -1.6303 & -1.6135 \\ -0.2333 & 0.5191 \\ 0.4721 & -4.5613 \end{bmatrix},$$

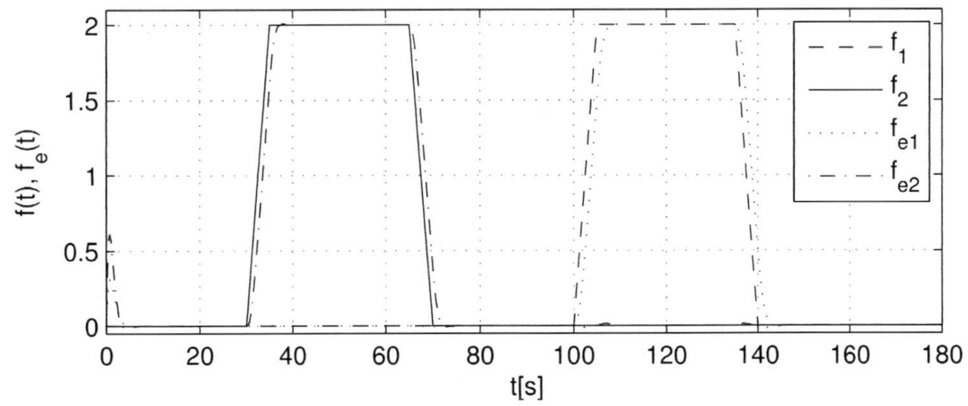

Figure 1. The actuator faults and its estimation using fault observer designed pursuant to Proposition 1

$$\boldsymbol{Q} = \begin{bmatrix} 1.9347 & 2.1134 & 0.1955 & 0.5689 \\ 2.1134 & 5.9730 & 0.3846 & 1.9950 \\ 0.1955 & 0.3846 & 1.2356 & -0.7994 \\ 0.5689 & 1.9950 & -0.7994 & 7.2932 \end{bmatrix}, \quad \boldsymbol{H} = \begin{bmatrix} 3.0563 & -0.9781 \\ -1.8683 & 2.9047 \end{bmatrix}.$$

The obtained estimator gain matrix $\boldsymbol{J}$ from (39) is

$$\boldsymbol{J} = \begin{bmatrix} -0.0057 & -0.0353 \\ -0.0256 & -0.0104 \\ -0.0010 & 0.0063 \\ 0.0091 & -0.0743 \end{bmatrix}.$$

which gives that all eigenvalues of the estimator dynamic matrix spectrum $\rho(\boldsymbol{A}_e)$ are stable, where

$$\rho(\boldsymbol{A}_e) = \rho(\boldsymbol{A} - \boldsymbol{J}\boldsymbol{C}) = \{\ -2.9469\ \ -3.1321\ \ -3.2411\ \ -3.2704\ \}.$$

Setting the learning parameter $\boldsymbol{G}$ as follows

$$\boldsymbol{G} = \begin{bmatrix} 1.3 & 0.5 \\ 0.5 & 2.1 \end{bmatrix}$$

the observed actuator faults are illustrated in Fig. 2. This figure presents the fault signal, as well as its estimation, reflecting at first a single actuator fault in the the second actuator, starting at the time instant $t = 30s$ and applied for $40s$, then the fault of the first actuator is introduced in the time instant $t = 100s$ and lasts for $40s$. The learning parameter $\boldsymbol{G}$ and the tuning parameter δ were set experimentally considering the maximal value of fault signal amplitude. Similarly to the previous case, the faults considered in simulation do not cause closed-loop instability. In the simulation results in Fig. 2 it can be seen that the exists very small differences between the signals reflecting single actuator faults and the observer approximate ones for piecewise constant actuator faults.

It has to be noted that now the matrix norms of $\boldsymbol{G}$ and $\boldsymbol{J}$ are substantially smaller then the matrix norms of these matrices computed using the standard design conditions (13)-(15). Moreover, since in this case matrix norm of $\boldsymbol{H}$ is significantly higher, LMI solution was carried out with higher precision.

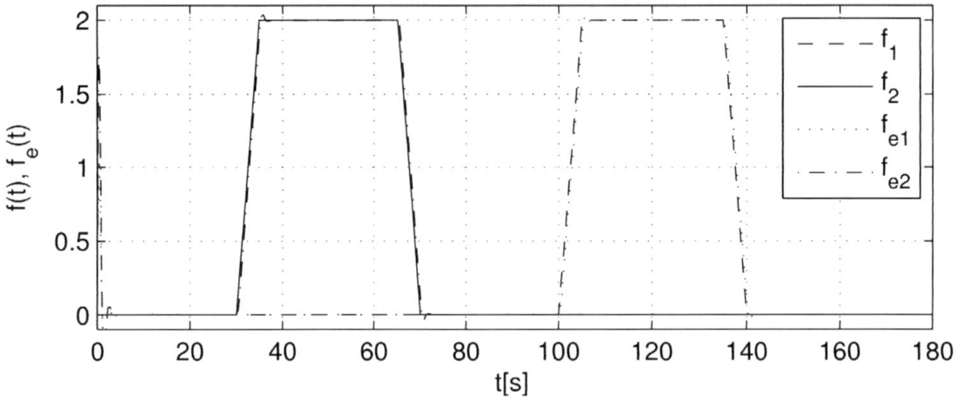

Figure 2. The actuator faults and its estimation with enhanced condition

6. Concluding remarks

Presented fault estimation method for linear continuous-time systems provides useful and easily implementable structure in process of fault detection, isolation and identification. Proposed approach to fault estimation design utilizing enhanced design conditions allows even better results, where the occurred actuator faults are estimated sooner and more precise as can be seen in simulation results. Tuning parameters $\boldsymbol{G}$ and δ were set interactively, incorrect values of these parameters would result in unstable or noisy response of the fault estimation signals. A simulation example, subject to given type of failures, demonstrates the effectiveness of the proposed form of the fault estimation design technique.

Acknowledgment

The authors thank an anonymous reviewer, who has made several pertinent comments that have enhanced the quality of this paper.

The work presented in the paper was supported by VEGA, the Grant Agency of the Ministry of Education and the Academy of Science of Slovak Republic, under Grant No. 1/0348/14. This support is very gratefully acknowledged.

References

[1] Ding X and Frank P 1993 An adaptive observer-based fault detection scheme for nonlinear dynamics systems *Prepr. 12th IFAC World Congres* Sydney Australia vol 1 307-10

[2] Fan J Zhang Y and Zheng Z 2011 Integrated adaptive fault diagnosis and state-feedback control for systems with constant actuator faults and control input constraints *Proc ASME 2011 Int. Design Engineering Technical Conferences & Computers and Information in Engineering Conference IDETC/CIE 2011* Washington DC USA 1-8

[3] Filasová A Gontkovič D and Krokavec D 2013 Observer-based fault estimation for linear systems with distributed time delay *Archives of Control Sciences* **23** 169-86

[4] Haddad W.M and Chellaboina V 2008 *Nonlinear Dynamical Systems and Control. A Lyapunov-Based Approach* (Princeton:Princeton University Press)

[5] He S and Liu F 2012 Adaptive observer-based fault estimation for stochastic Markovian jumping systems *Abstract and Appl. Analysis* **2012** 176419 11

[6] Khedher A Benothman K Benrejeb M Maquin D 2010 Adaptive observer for fault estimation in nonlinear systems described by a Takagi-Sugeno model *Proc. 18th Mediterranean Conf. Control and Automation MED'10* Marrakech Morocco 261-6

[7] Krokavec D and Filasová A 2013 Actuator faults reconstruction using reduced-order fuzzy observer structures *Proc. 12th EUCA European Control Conference ECC'13* Zurich Switzerland 4299-304

[8] Patton R.J and Klinkhieo S 2009 Actuator fault estimation and compensation based on an augmented state observer approach *Proc. Joint 48th IEEE Conference on Decision and Control and 28th Chinese Control Conference* Shanghai P.R. China 8482-7
[9] Peaucelle D Henrion D Labit Y and Taitz K 2002 *User's Guide for SeDuMi Interface 1.04* (Toulouse: LAAS-CNRS)
[10] Shi F and Patton R.J 2014 Simultaneous state and fault estimation for descriptor systems using an augmented PD observer *Prepr. 19th IFAC World Congres* Cape Town South Africa 8006-11
[11] Wang H and Daley S 1996 Actuator fault diagnosis. An adaptive observer-based technique *IEEE Tran. Automatic Control* **41** 1073-78
[12] Zhang F Liu G and Fang L 2009 Actuator fault estimation based on adaptive H_∞ observer technique *Proc. IEEE Int. Conf. Mechatronics and Automation* Changchun, China 352-57
[13] Zhang K and Jiang B 2008 Analysis and design of adaptive fault estimation for time-varying delay systems *Proc. 27th Chinese Control Conference* Kunming China 38-42
[14] Zhang K Jiang B and Cocquempot V 2008 Adaptive observer-based fast fault estimation *Int. J. Control Automation and Systems* **6** 320-26
[15] Zhang K Jiang B and Shi P 2013 *Observer-Based Fault Estimation and Accomodation for Dynamic Systems* (Berlin:Springer-Verlag)

Approximate active fault detection and control

Jan Škach, Ivo Punčochář and Miroslav Šimandl

Department of Cybernetics and NTIS - New Technologies for the Information Society, Faculty of Applied Sciences, University of West Bohemia, Univerzitní 22, 30614 Plzeň, Czech Republic

E-mail: janskach@kky.zcu.cz, ivop@kky.zcu.cz, simandl@kky.zcu.cz

Abstract. This paper deals with approximate active fault detection and control for nonlinear discrete-time stochastic systems over an infinite time horizon. Multiple model framework is used to represent fault-free and finitely many faulty models. An imperfect state information problem is reformulated using a hyper-state and dynamic programming is applied to solve the problem numerically. The proposed active fault detector and controller is illustrated in a numerical example of an air handling unit.

1. Introduction

Automatic control is a well-established research field that aims at automatic process control without direct human intervention. In consequence energy can be saved, costs can be reduced, and quality can be improved. In recent decades, the problem of automatic fault detection (FD) has become a topical issue in many technical and nontechnical areas where increased reliability and efficiency is demanded. Safety critical applications in aircraft, trains, cars, or power plants must be essentially accurate, reliable, robust, etc. FD in such applications prevents casualties and unwanted damage. The purpose of FD may be different for home and commercial appliances. One of the contemporary trends aims at reducing energy consumption of buildings.

Two main approaches to FD exist: passive FD and active FD. Passive FD is the most widely used [1, 2]. A passive detector generates decisions about possible faults based on input-output data obtained from a system. However, unreliable decisions may be generated due to insufficient information in the input-output data. To alleviate this issue active fault detection utilizes a feedback to the system to improve the quality of decisions [3]. In addition, active fault detection and control was considered in [4] to simultaneously fulfill detection and control aims. A unified formulation of active fault detection and control was presented in [5]. A designer of an active fault detector and controller (AFDC) can compromise between controlling the system and exciting it to improve the quality of fault detection.

The existing literature [6–8] considers active FD mainly for linear systems and a finite time horizon. An active fault detector that generates a sufficiently informative input signal for a nonlinear system over an infinite time horizon was proposed in [9]. Although this active fault detector is powerful for detecting faults in a system, potential control aims must be considered separately. A goal of this paper is to design an AFDC for a class of nonlinear stochastic systems over an infinite time horizon, where the optimal compromise between fault detection and control should be achieved by considering detection and control aims simultaneously.

The paper is organized as follows. In Section 2 the problem of active fault detection and control is formulated. A multiple model approach is employed to represent a fault-free and faulty behaviors of a system. Section 3 presents a solution to the stated problem which employs various approximations. The optimization problem is reformulated such that numerical methods of dynamic programming can be used. The proposed approach is demonstrated in a numerical example of an air-handling unit in Section 4.

2. Problem formulation

In control systems, there are typically several types of possible faults such as actuator, sensor, and component faults. A common problem is comprised of abrupt change detection. A suitable model of the system behavior is achieved by a multiple model framework. Subsequently, the detection and control aims are expressed by a design criterion that evaluates costs of states, inputs, and decisions.

Let us assume an observed and controlled system that can be described at a time step $k \in \mathcal{T} = \{0, 1, \ldots\}$ by the following discrete-time nonlinear model

$$\mathbf{x}_{k+1} = \mathbf{f}\left(\mathbf{x}_k, \mu_k, \mathbf{u}_k\right) + \mathbf{w}_k, \tag{1}$$

where $\mathbf{x}_k^{\mathrm{a}} = \left[\mathbf{x}_k^{\mathrm{T}}, \mu_k\right]^{\mathrm{T}} \in \mathbb{R}^{n_x} \times \mathcal{M}$ represents a hybrid state of the system, $\mathbf{x}_k \in \mathbb{R}^{n_x}$ is a fully measurable common part of the state, $\mu_k \in \mathcal{M} = \{1, 2, \ldots, N\}$ is an unknown index of the fault-free or faulty model, $\mathbf{u}_k \in \mathcal{U} \subset \mathbb{R}^{n_u}$ is the input, $\mathcal{U} = \{\bar{\mathbf{u}}^1, \bar{\mathbf{u}}^2, \ldots, \bar{\mathbf{u}}^M\}$ is a discrete set of admissible inputs, and $\mathbf{w}_k \in \mathbb{R}^{n_x}$ is the state noise with the known conditional probability density function (pdf) $p_{\mathbf{w}}\left(\mathbf{w}_k | \mathbf{x}_k^{\mathrm{a}}\right)$. The mean value $\mathbf{m}_{\mathbf{w}}(\mathbf{x}_k^{\mathrm{a}})$ and the covariance matrix $\mathbf{P}_{\mathbf{w}}(\mathbf{x}_k^{\mathrm{a}})$ of the state noise are known and depend on the hybrid state $\mathbf{x}_k^{\mathrm{a}}$. The system is characterized by a known vector function $\mathbf{f} : \mathbb{R}^{n_x} \times \mathcal{M} \times \mathcal{U} \mapsto \mathbb{R}^{n_x}$. The behavior of the system depends on μ_k which determines the nonlinear function $\mathbf{f}_{\mu_k} : \mathbb{R}^{n_x} \times \mathcal{U} \mapsto \mathbb{R}^{n_x}$

$$\mathbf{f}\left(\mathbf{x}_k, \mu_k, \mathbf{u}_k\right) = \begin{cases} \mathbf{f}_1\left(\mathbf{x}_k, \mathbf{u}_k\right) & \text{if } \mu_k = 1, \\ \mathbf{f}_2\left(\mathbf{x}_k, \mathbf{u}_k\right) & \text{if } \mu_k = 2, \\ \vdots & \\ \mathbf{f}_N\left(\mathbf{x}_k, \mathbf{u}_k\right) & \text{if } \mu_k = N. \end{cases} \tag{2}$$

It is assumed that $\mathbf{f}_1$ represents the fault-free behavior of the system and the faulty behaviors are defined by functions $\mathbf{f}_i$ for $i \in \{2, 3, \ldots, N\}$. The switching between the fault-free and faulty models is defined by a stationary finite-state Markov chain with the known transition probabilities $P_{i,j} = P\left(\mu_{k+1} = j | \mu_k = i\right)$. The pdf $p(\mathbf{x}_0)$ and the probability $P(\mu_0)$ of mutually independent initial conditions $\mathbf{x}_0$ and μ_0 are known. Note that the fault occurrence in real systems is always unknown and the transition probabilities are to be estimated.

The active fault detector and controller generates the decision $d_k \in \mathcal{M}$ about the possible faults and the input signal $\mathbf{u}_k$ that is injected into the system. The AFDC can be described as

$$\begin{bmatrix} d_k \\ \mathbf{u}_k \end{bmatrix} = \boldsymbol{\rho}_k \left(\begin{bmatrix} \mathbf{x}_0^k \\ \mathbf{u}_0^{k-1} \end{bmatrix} \right), \tag{3}$$

where $\boldsymbol{\rho}_k : \mathbb{R}^{n_x(k+1)} \times \mathcal{U}^k \mapsto \mathcal{M} \times \mathcal{U}$ is an unknown nonlinear vector function. The sequence $\boldsymbol{\rho} = \{\boldsymbol{\rho}_0, \boldsymbol{\rho}_1, \ldots\}$ is called a policy. Note that $\mathbf{x}_i^j = \left[\mathbf{x}_i^{\mathrm{T}}, \mathbf{x}_{i+1}^{\mathrm{T}}, \ldots, \mathbf{x}_j^{\mathrm{T}}\right]^{\mathrm{T}}$ denotes the vector of stacked variables. It is supposed that for all μ_k exists a stabilizing policy $\boldsymbol{\rho}$.

The goal is to find a suitable detection and control law represented by a policy $\boldsymbol{\rho}$ such that a chosen design criterion $J(\boldsymbol{\rho})$ respecting detection and control aims is minimized. The design criterion has the following form

$$J\left(\boldsymbol{\rho}\right) = \lim_{F \to +\infty} \mathbb{E}\left\{ \sum_{k=0}^{F} \lambda^k L\left(\mu_k, d_k, \mathbf{x}_k, \mathbf{u}_k\right) \right\}, \tag{4}$$

where $\lambda \in (0,1)$ is a discount factor, $\mathbb{E}\{\cdot\}$ is the expectation operator. The cost function $L : \mathcal{M} \times \mathcal{M} \times \mathbb{R}^{n_x} \times \mathcal{U} \mapsto \mathbb{R}^+$ is defined as

$$L(\mu_k, d_k, \mathbf{x}_k, \mathbf{u}_k) = \alpha L^{\mathrm{d}}(\mu_k, d_k) + (1-\alpha) L^{\mathrm{c}}(\mathbf{x}_k, \mathbf{u}_k), \tag{5}$$

where $\alpha \in [0,1]$ is a weighting factor, $L^{\mathrm{d}} : \mathcal{M} \times \mathcal{M} \mapsto \mathbb{R}^+$ is a detection cost function, $L^{\mathrm{c}} : \mathbb{R}^{n_x} \times \mathcal{U} \mapsto \mathbb{R}^+$ is a control cost function and $\mathbb{R}^+$ denotes a set of non-negative real numbers. It is assumed that the cost functions L^{d}, L^{c}, and L are bounded which makes the criterion well defined. The cost functions L^{d} together with L^{c} are chosen individually to satisfy requirements of a given problem.

Note that the problem formulation allows a designer to choose between three special cases. For $\alpha = 0$, the problem can be seen as optimal stochastic control, whereas $\alpha = 1$ represents a problem of active fault detection. Finally, $\alpha \in (0,1)$ compromises between the detection and control aims.

3. Active fault detector and controller

The design of an AFDC is divided into several steps. The presented problem formulation is a dynamic optimization problem that belongs to a class of imperfect state information problems. To use standard methods of dynamic programming, a reformulation as a perfect state information problem is employed. The optimal AFDC is obtained as a function of a state by solving the Bellman equation [10]. The Bellman equation is solved numerically by various approximation methods [11, 12]. An approximate solution to the Bellman equation is obtained by a quantization of the hyper-state space and using an iterative algorithm known as the value iteration (VI). The approximate AFDC is designed by yielding all partial results.

3.1. Problem reformulation

The problem reformulation is achieved by introducing a hyper-state $\mathbf{s}_k$ that includes a common state $\mathbf{x}_k$ of the system and a sufficient statistics for the unknown index μ_k. Since details of the problem reformulation using the hyper-state are presented in [9], only the result is provided in the following text.

The system, AFDC, and design criterion are reformulated using the hyper-state $\mathbf{s}_k$ which is defined as

$$\mathbf{s}_k = \begin{bmatrix} \mathbf{x}_k \\ \mathbf{b}_k \end{bmatrix} \in \mathcal{S} = (\mathbb{R}^{n_x} \times \mathcal{B}) \subset \mathbb{R}^{n_s}, \tag{6}$$

where $\mathbf{b}_k \in \mathcal{B}$, $\mathcal{B} = \{\mathbf{b} \in \mathbb{R}^{N-1} : \mathbf{b} \geq 0,\ \mathbf{1}^{\mathrm{T}}\mathbf{b} \leq 1\}$ is a belief state with components $b_{k,i} = P\left(\mu_k = i | \mathbf{x}_0^k, \mathbf{u}_0^{k-1}\right)$ referring to the unknown scalar variable μ_k. The belief state $\mathbf{b}_k$ can be understood as a conditional distribution of μ_k. The original model (1) of the system is replaced by a new time-invariant model

$$\mathbf{s}_{k+1} = \boldsymbol{\varphi}(\mathbf{s}_k, \mathbf{u}_k, \mathbf{x}_{k+1}), \tag{7}$$

with a nonlinear vector function $\boldsymbol{\varphi} : \mathcal{S} \times \mathcal{U} \times \mathbb{R}^{n_x} \mapsto \mathcal{S}$. Note that besides random variable $\mathbf{x}_{k+1}$ all arguments of the function $\boldsymbol{\varphi}$ are known or can be computed. The time-variant AFDC (3) is reformulated using stationary policies $\sigma : \mathcal{S} \mapsto \mathcal{M}$ and $\boldsymbol{\gamma} : \mathcal{S} \mapsto \mathcal{U}$ as follows

$$\begin{bmatrix} d_k \\ \mathbf{u}_k \end{bmatrix} = \begin{bmatrix} \sigma(\mathbf{s}_k) \\ \boldsymbol{\gamma}(\mathbf{s}_k) \end{bmatrix} = \bar{\boldsymbol{\rho}}(\mathbf{s}_k). \tag{8}$$

Lastly, the optimality discounted criterion (4) is reformulated as

$$\bar{J}(\bar{\rho}) = \lim_{F\to+\infty} \mathbb{E}\left\{\sum_{k=0}^{F} \lambda^k \bar{L}(d_k, \mathbf{s}_k, \mathbf{u}_k)\,|\mathbf{s}_0\right\}, \tag{9}$$

where $\bar{L}(d_k, \mathbf{s}_k, \mathbf{u}_k) = \alpha\bar{L}^{\mathrm{d}}(d_k, \mathbf{s}_k) + (1-\alpha)\bar{L}^{\mathrm{c}}(\mathbf{s}_k, \mathbf{u}_k)$, $\bar{L}^{\mathrm{d}}(d_k, \mathbf{s}_k)$ is a detection cost function equivalent to $L^{\mathrm{d}}(\mu_k, d_k)$ and $\bar{L}^{\mathrm{c}}(\mathbf{s}_k, \mathbf{u}_k)$ is a control cost function equivalent to $L^{\mathrm{c}}(\mathbf{x}_k, \mathbf{u}_k)$. These functions are given as

$$\bar{L}^{\mathrm{d}}(d_k, \mathbf{s}_k) = \mathbb{E}\left\{L^{\mathrm{d}}(\mu_k, d_k)\,|d_k, \mathbf{x}_0^k, \mathbf{u}_0^{k-1}\right\}, \tag{10}$$

$$\bar{L}^{\mathrm{c}}(\mathbf{s}_k, \mathbf{u}_k) = L^{\mathrm{c}}\left([s_{k,1}, \ldots, s_{k,n_x}]^{\mathrm{T}}, \mathbf{u}_k\right). \tag{11}$$

3.2. Approximate active fault detector and controller design

The optimal AFDC is based on a solution to the Bellman equation. The Bellman function V^* represents the optimal expected costs incurred from given time step to infinity and satisfies the Bellman equation

$$V^*(\mathbf{s}_k) = \min_{d_k\in\mathcal{M}, \mathbf{u}_k\in\mathcal{U}} \mathbb{E}\left\{\bar{L}(d_k, \mathbf{s}_k, \mathbf{u}_k) + \lambda V^*(\mathbf{s}_{k+1})\,|d_k, \mathbf{s}_k, \mathbf{u}_k\right\}. \tag{12}$$

The optimal detector σ^* and the optimal controller γ^* are given as

$$d_k^* = \sigma^*(\mathbf{s}_k) = \arg\min_{d_k\in\mathcal{M}} \alpha\bar{L}^{\mathrm{d}}(d_k, \mathbf{s}_k), \tag{13}$$

$$\mathbf{u}_k^* = \gamma^*(\mathbf{s}_k) = \arg\min_{\mathbf{u}_k\in\mathcal{U}} \mathbb{E}\left\{(1-\alpha)\bar{L}^{\mathrm{c}}(\mathbf{s}_k, \mathbf{u}_k) + \lambda V^*(\mathbf{s}_{k+1})\,|\mathbf{s}_k, \mathbf{u}_k\right\}. \tag{14}$$

The Bellman function V^* is computed off-line by solving (12) and then the AFDC is implemented to provide on-line detection and control by means of (13) and (14). However, the analytical solution to the Bellman equation (12) is almost impossible to find. Since the AFDC is designed for a nonlinear system, numerical methods are employed to find an approximate solution to the Bellman equation.

It is necessary to reduce an infinite number of states. The approximate solution is based on hyper-state space quantization. A uniform grid of points $\bar{\mathbf{s}} \in \mathbb{R}^{n_s}$ defined by the Cartesian product $\mathcal{S}^{\mathrm{g}} \equiv \mathcal{S}_1^{\mathrm{g}} \times \mathcal{S}_2^{\mathrm{g}} \times \ldots \times \mathcal{S}_{n_s}^{\mathrm{g}}$ of discrete sets $\mathcal{S}_j^{\mathrm{g}}$ is specified over a region with non-negligible probabilities of hyper-state trajectories. A non-grid hyper-state is projected to the grid using an aggregation function such that the Euclidean distance between the non-grid hyper-state and the grid point is minimized. In the paper it is supposed that the Bellman function V^* is approximated by a piecewise constant function $\bar{V}$. Note that the control space is assumed to be discrete with a reasonable number of controls.

An approximation of the unknown Bellman function V^* can be determined by the VI algorithm [13]. Following the general theoretical results, the VI algorithm consists in iterative computation of the contraction mapping

$$\bar{V}^{(i+1)}(\bar{\mathbf{s}}_k) = \min_{d_k\in\mathcal{M}, \mathbf{u}_k\in\mathcal{U}} \mathbb{E}\left\{\bar{L}(d_k, \bar{\mathbf{s}}_k, \mathbf{u}_k) + \lambda\bar{V}^{(i)}(\bar{\mathbf{s}}_{k+1})\,|d_k, \bar{\mathbf{s}}_k, \mathbf{u}_k\right\}, \tag{15}$$

where $i = 0, 1, 2, \ldots$ is an iteration index. Given any initial Bellman function $\bar{V}^{(0)}$ the VI algorithm converges to the fixed point. In practice, the iteration process is terminated after a fixed number of iterations.

In the iterative computation (15), it is necessary to compute the conditional expected value of the Bellman function. Due to nonlinearity of the system, the conditional expected value is computed approximately using the unscented transform with a parameter κ [14].

4. Numerical example

Control of temperature in a lecture hall and active fault detection in a corresponding air handling unit (AHU) are considered in this numerical example. The AHU mixes the ambient air and indoor air together with ratio proportional to the damper position. The mixed air is heated or cooled by a heating or cooling coil before it is supplied to the lecture hall. The AFDC can be focused on active fault detection of a stuck damper or it can be aimed at the temperature control of the lecture hall. The AHU and lecture hall temperature model can be described by the following continuous-time nonlinear model [15]

$$\begin{aligned} C^{\mathrm{i}}\dot{T}^{\mathrm{i}}(t) &= \dot{m}^{\mathrm{s}}c^{\mathrm{s}}\left(T^{\mathrm{s}}(t) - T^{\mathrm{i}}(t)\right) + \frac{1}{R^{\mathrm{ow}}}\left(T^{\mathrm{a}}(t) - T^{\mathrm{i}}(t)\right) + P^{w}(t), \\ T^{\mathrm{s}}(t) &= \Delta(t)T^{\mathrm{i}}(t) + (1 - \Delta(t))\,T^{\mathrm{a}}(t) + T^{\mathrm{hc}}(t), \\ T^{\mathrm{hc}}(t) &= s^{\mathrm{hc}}(t)\frac{P^{\mathrm{hc}}\eta^{\mathrm{hc}}}{\dot{m}^{\mathrm{s}}c^{\mathrm{s}}}, \end{aligned} \tag{16}$$

where $T^{\mathrm{i}}(t)$ [°C] is the indoor air temperature, $T^{\mathrm{a}}(t)$ [°C] is the ambient air temperature, $T^{\mathrm{s}}(t)$ [°C] refers to the supply air temperature, $\Delta(t) \in \mathcal{U}^{\mathrm{N}} = \{0, 0.1, \ldots, 0.9\}$ is the damper position and $P^{w}(t)$ represents the disturbance power load generated for example by occupancy of the lecture hall or its equipment. Note that the damper is fully opened to the ambient air when $\Delta(t) = 0$ and it cannot be fully closed due to fresh air circulation, therefore $\max(\Delta(t)) = 0.9$. The thermal capacity of the lecture hall is $C^{\mathrm{i}} = 7.8 \cdot 10^{6}$ [$\mathrm{J} \cdot {}^{\circ}\mathrm{C}^{-1}$], the specific thermal capacity of supply air is $c^{\mathrm{s}} = 1012$ [$\mathrm{J} \cdot \mathrm{kg}^{-1} \cdot {}^{\circ}\mathrm{C}^{-1}$], mass flow rate of the supply air is $\dot{m}^{\mathrm{s}} = 1$ [$\mathrm{kg} \cdot \mathrm{s}^{-1}$] and the external wall thermal resistance is $R^{\mathrm{ow}} = 7 \cdot 10^{-4}$ [${}^{\circ}\mathrm{C} \cdot \mathrm{W}^{-1}$]. The increase or decrease in temperature of the supply air caused by heating or cooling $T^{\mathrm{hc}}(t)$ [°C] is defined by the coil switching $s^{\mathrm{hc}}(t) \in \mathcal{U}^{\mathrm{L}} = \{-1, 0, 1\}$, the heating power $P^{\mathrm{hc}} = 4 \cdot 10^{4}$ [W] of the AHU, and the coil efficiency $\eta^{\mathrm{hc}} = 0.6$. The system input is $s^{\mathrm{hc}}(t)$ together with $\Delta(t)$ and the set of admissible inputs is defined as $\mathcal{U} = \mathcal{U}^{\mathrm{L}} \times \mathcal{U}^{\mathrm{N}}$. In the paper, it is assumed that the ambient air temperature remains constant, i.e. $\dot{T}^{\mathrm{a}}(t) = 0$, and the considered fault is represented by the damper stuck at the fully open position. A possible fault of the system is represented by damper stuck at the position fully opened to the ambient air.

The nonlinear model (16) discretized by the forward Euler method with the sampling period $T_{\mathrm{s}} = 300$ [s] has the following form

$$f_i\left(\mathbf{x}_k, \mathbf{u}_k\right) = \mathbf{A}\mathbf{x}_k + \mathbf{B}^{\mathrm{L}}\mathbf{u}_{k,1} + \mathbf{B}_i^{\mathrm{N}}\left(\mathbf{x}_{k,1} - \mathbf{x}_{k,2}\right)\mathbf{u}_{k,2} + \mathbf{w}_k, \tag{17}$$

where $i \in \mathcal{M} = \{1, 2\}$ and the following definitions hold

$$\begin{aligned} &\mathbf{A} = \begin{bmatrix} 1 - K_1T_{\mathrm{s}} - K_2T_{\mathrm{s}} & K_1T_{\mathrm{s}} + K_2T_{\mathrm{s}} \\ 0 & 1 \end{bmatrix}, \mathbf{B}^{\mathrm{L}} = \begin{bmatrix} K_1K_3T_{\mathrm{s}} \\ 0 \end{bmatrix}, \mathbf{B}_1^{\mathrm{N}} = \begin{bmatrix} K_1T_{\mathrm{s}} \\ 0 \end{bmatrix}, \mathbf{B}_2^{\mathrm{N}} = \begin{bmatrix} 0 \\ 0 \end{bmatrix}, \\ &\mathbf{x}_k = \begin{bmatrix} \mathbf{x}_{k,1} \\ \mathbf{x}_{k,2} \end{bmatrix} = \begin{bmatrix} T_k^{\mathrm{i}} \\ T_k^{\mathrm{a}} \end{bmatrix}, \mathbf{u}_k = \begin{bmatrix} \mathbf{u}_{k,1} \\ \mathbf{u}_{k,2} \end{bmatrix} = \begin{bmatrix} s_k^{\mathrm{hc}} \\ \Delta_k \end{bmatrix}, \mathbf{w}_k = \begin{bmatrix} \mathbf{w}_{k,1} \\ \mathbf{w}_{k,2} \end{bmatrix}, \\ &K_1 = \frac{\dot{m}^{\mathrm{s}}c^{\mathrm{s}}}{C^{\mathrm{i}}}, K_2 = \frac{1}{C^{\mathrm{i}}R^{\mathrm{ow}}}, K_3 = \frac{P^{\mathrm{hc}}\eta^{\mathrm{hc}}}{\dot{m}^{\mathrm{s}}c^{\mathrm{s}}}. \end{aligned} \tag{18}$$

The state noise $\mathbf{w}_k$ is independent of the system state. The noise $\mathbf{w}_{k,1}$ represents disturbance power load and has the Laplace distribution with the location parameter $\varpi = 0$ and the scale parameter $\beta = 0.08$. The noise $\mathbf{w}_{k,2}$ is described by Dirac delta function $\delta\left(\mathbf{w}_{k,2}\right)$. The system initially starts as fault-free, i.e. $P\left(\mu_0 = 1\right) = 1$. The switching between fault-free and faulty models is defined by the transition probabilities $P\left(\mu_{k+1} = i | \mu_k = j\right) = 0.02$ for $i, j \in \mathcal{M}, i \neq j$.

The detection cost function $L^{\mathrm{d}}(\mu_k, d_k)$ and control cost function $L^{\mathrm{c}}(\mathbf{x}_k, \mathbf{u}_k)$ are chosen as

$$L^{\mathrm{d}}(\mu_k, d_k) = \begin{cases} 0 & \text{if } d_k = \mu_k, \\ 1 & \text{otherwise,} \end{cases}$$
$$L^{\mathrm{c}}(\mathbf{x}_k, \mathbf{u}_k) = \sum_{i=1}^{n_u} |\mathbf{p}_i^{\mathrm{hc}} \mathbf{u}_{k,i}| + \mathbf{q}_1 \left(1 - e^{-\mathbf{q}_2\left(\mathbf{x}_{k,1} - x^{\mathrm{ref}}\right)^2}\right), \tag{19}$$

where $\mathbf{p}^{\mathrm{hc}} = \left[\mathbf{p}_1^{\mathrm{hc}}, \mathbf{p}_2^{\mathrm{hc}}\right]^{\mathrm{T}} = [1, 0]^{\mathrm{T}}$ and $\mathbf{q} = [\mathbf{q}_1, \mathbf{q}_2]^{\mathrm{T}} = [60, 1]^{\mathrm{T}}$ are parameters, and $x^{\mathrm{ref}} = 23$ [°C] is the reference indoor temperature. From (19) it follows that the detection aim is to minimize the probability of making a wrong decision about an actual system model. The control aim is designed to minimize costs of a compromise between a weighted price of control actions and following the reference temperature. It can be derived that the detector is given by the following decision rule

$$d_k = \begin{cases} 1 & \text{if } P(\mu_k = 1 | \mathbf{x}_0^k, \mathbf{u}_0^{k-1}) \geq 0.5, \\ 2 & \text{otherwise.} \end{cases} \tag{20}$$

The design parameter of the UT is set to $\kappa = 3$, the discount factor $\lambda = 0.98$, and the weighting factor is $\alpha = 0.99$. For the constant ambient air temperature $T^{\mathrm{a}} = 21$ [°C], the uniform grid is defined by $\mathcal{S}_1^{\mathrm{g}} = \{5, 5.1, \ldots, 30\}$, $\mathcal{S}_2^{\mathrm{g}} = \{21\}$, and $\mathcal{S}_3^{\mathrm{g}} = \{0, 0.01, \ldots, 1\}$.

The VI algorithm was terminated after performing $n_{\mathrm{iter}} = 30$ iterations. The corresponding approximate Bellman function $\bar{V}$ for fixed $\mathbf{s}_{k,2} = 21$ [°C] is shown in Figure 1. The global minimum of $\bar{V}$ is around the value of the indoor reference temperature which is caused by the control cost function L^{c}. It can also be seen that approximate Bellman function $\bar{V}$ attains a local maximum when the uncertainty of the model is high, i.e. $\mathbf{s}_{k,3}$ is close to 0.5. When the indoor air temperature is close to the ambient air temperature, $\bar{V}$ slightly increases as well. This is caused by the definition (17) of the fault-free and faulty models which differ just in $\mathbf{B}^{\mathrm{N}}$.

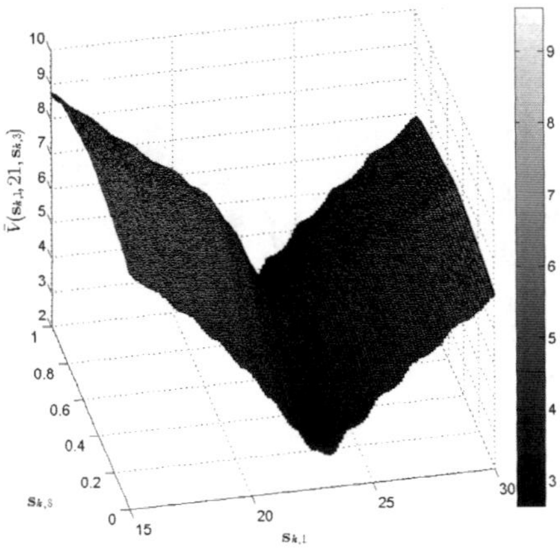

Figure 1. The approximate Bellman function for fixed $\mathbf{s}_{k,2} = 21$ [°C].

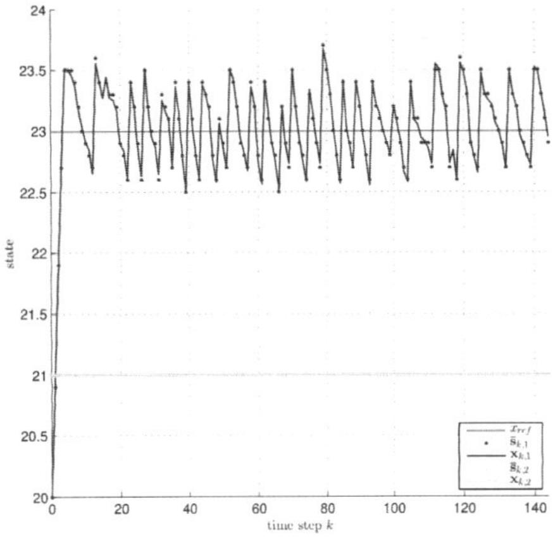

Figure 2. Typical state trajectories of the system with the AFDC for a time horizon of 12 hours.

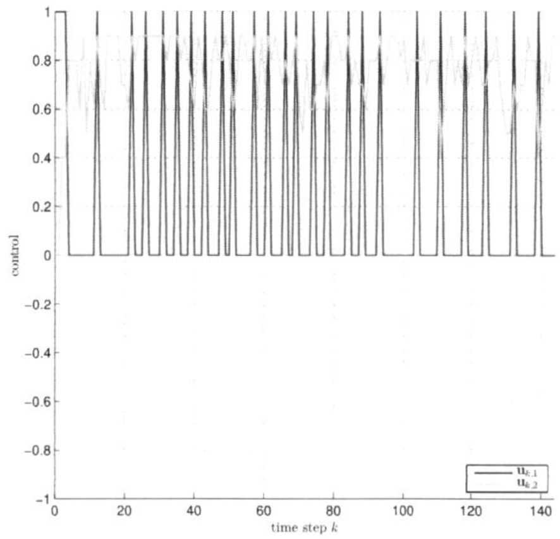

Figure 3. Typical control actions of the AFDC for a time horizon of 12 hours.

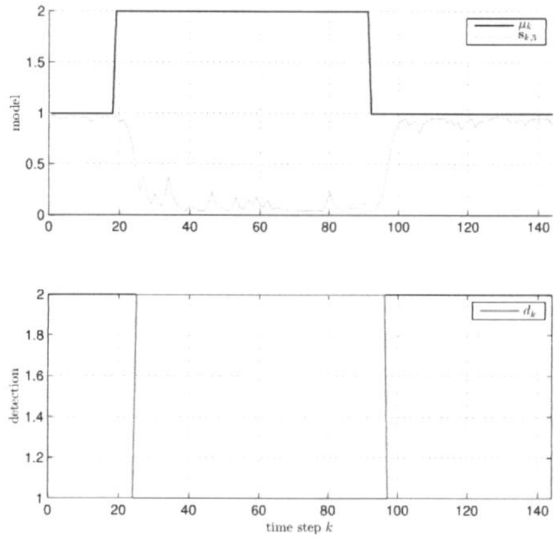

Figure 4. Typical true model, probability of the fault-free model, and a decision of the AFDC for a time horizon of 12 hours.

Next, a simulation of system trajectories was performed for a time horizon of 12 hours. Typical state trajectories of the system with the AFDC are depicted in Figure 2. A development of the state trajectories indicates that the indoor air temperature follows the reference value with oscillations caused by a discrete amount of power delivered during the sampling period. The trajectories of $\mathbf{u}_{k,1}$ and $\mathbf{u}_{k,2}$ are depicted in Figure 3. Since the ambient air temperature is lower than the reference air temperature, the AFDC keeps the damper closed to the ambient air. Figure 4 shows that the true model is correctly detected with a delay of approximately 5 steps. It was shown that the AFDC fulfilled the detection aim as well as the control aim. The designed AFDC is able to stabilize the system both for fault-free and faulty model as depicted in the Figures 2, 3, and 4. However, in general a stability is not guaranteed.

Each action of the approximate control policy $\hat{\gamma}(\bar{\mathbf{s}}_k) = [\mathbf{u}_{k,1}, \mathbf{u}_{k,2}]^{\mathrm{T}}$ of the AFDC is shown in Figures 5 and 6. These two figures represent the control policy functions for coil switching and damper position. A value of control is defined by color hue. Neither heating nor cooling is needed when the actual indoor temperature is close to the reference indoor temperature as shown in Figure 5. Heating or cooling is demanded when the actual indoor temperature is lower or higher than the reference indoor temperature, respectively. Figure 6 shows that when the indoor and ambient air temperatures attain approximately the same values it may be difficult to decide between fault-free and faulty model. In such a case the damper fully opens to ambient air to differentiate the temperatures. The damper is closed $\mathbf{u}_{k,2} = 0.9$ for majority of states to prevent unwanted cooling of the indoor air.

In the numerical example of the AHU, the weighting factor α that compromises detection and control aims was set to $\alpha = 0.99$. To illustrate an influence of the parameter α on detection and control aims, the following set of α values $\mathcal{A} = \{0, 0.01, 0.5, 0.99, 0.999, 1\}$ was chosen. For each of the $\mathcal{A}$ element an estimation of detection J^{d} and control J^{c} part of the design criterion (4)

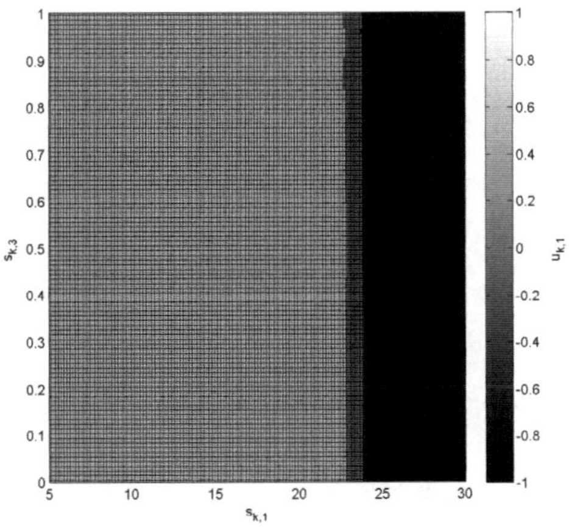

Figure 5. The component $\mathbf{u}_{k,1}$ of the approximate control policy $\hat{\gamma}$ for the AHU.

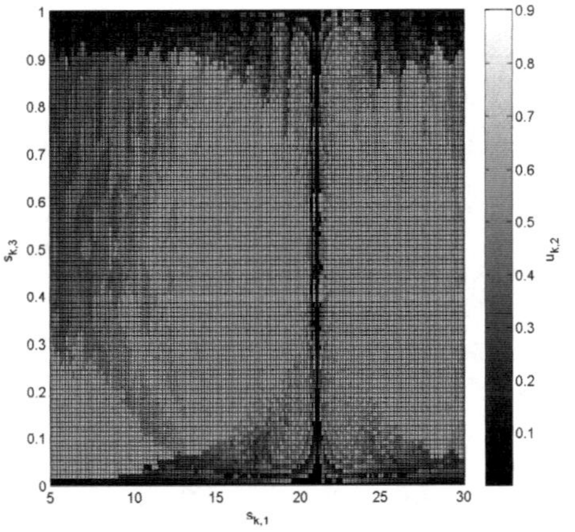

Figure 6. The component $\mathbf{u}_{k,2}$ of the approximate control policy $\hat{\gamma}$ for the AHU.

was performed. The detection and the control parts are expressed as

$$J^{\mathrm{d}} = \lim_{F\to+\infty} \mathbb{E}\left\{\sum_{k=0}^{F} \lambda^k L^{\mathrm{d}}(\mu_k, d_k)\right\},$$
$$J^{\mathrm{c}} = \lim_{F\to+\infty} \mathbb{E}\left\{\sum_{k=0}^{F} \lambda^k L^{\mathrm{c}}(\mathbf{x}_k, \mathbf{u}_k)\right\}. \quad (21)$$

The values J^{d} and J^{c} were computed approximately using 10 000 Monte Carlo simulations of state trajectories for a finite time horizon $F = 500$. The Figure 7 shows that the estimates $\hat{J}^{\mathrm{d}}$ and $\hat{J}^{\mathrm{c}}$ attain their lowest values for $\alpha = 1$ and $\alpha = 0$, respectively. Note that the solutions on the Pareto front are not uniformly distributed with respect to the weighting factor α. The value $\hat{J}^{\mathrm{c}}$ remains approximately the same for $\alpha \in [0, 0.99]$ and only quality of detection changes. Thus one might consider only $\alpha \geq 0.99$ when a compromise between active fault detection and control should be considered in this case.

5. Conclusion

This paper is one of the first to present a design of an approximate active fault detector and controller (AFDC) for a class of nonlinear stochastic systems over an infinite time horizon. The problem formulation allows a compromise to be made between detection and control aims. The approximate AFDC was designed by solving the Bellman equation over a discrete grid of hyper-states and successfully demonstrated in a numerical example of an air handling unit.

Acknowledgments

This work was supported by the European Regional Development Fund (ERDF), project NTIS New Technologies for the Information Society, European Centre of Excellence, CZ.1.05/1.1.00/02.0090 and by the Czech Science Foundation, project GAP103/13/07058J.

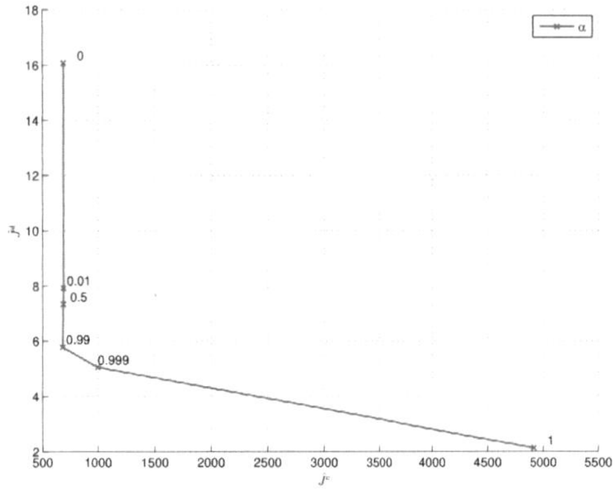

Figure 7. Pareto front for the AHU optimization problem indicating an influence of parameter α on estimates of detection and control parts of criterion (4).

References

[1] Basseville M and Nikiforov I V 1993 *Detection of Abrupt Changes - Theory and Application* (New Jersey, USA: Prentice Hall)

[2] Isermann R 2011 *Fault-Diagnosis Applications* (Heidelberg, Germany: Springer)

[3] Kerestecioglu F and Zarrop M B 1994 *International Journal of Control* **59** 1063–1084

[4] Blackmore L and Williams B 2006 Finite Horizon Control Design for Optimal Discrimination between Several Models *Proceedings of the 45th IEEE Conference on Decision and Control* (San Diefo, California: Ieee) pp 1147–1152

[5] Šimandl M and Punčochář I 2009 *Automatica* **45** 2052–2059

[6] Kerestecioglu F 1993 *Change detection and input design in dynamical systems* (Taunton: Research Studies Press)

[7] Niemann H 2006 *IEEE Transactions on Automatic Control* **51** 1572–1578

[8] Šimandl M, Punčochář I and Královec J 2005 Rolling horizon for active fault detection *Proceedings of the 44th IEEE Conference on Decision and Control* (Seville, Spain: Ieee) pp 3789–3794

[9] Šimandl M, Škach J and Punčochář I 2014 Approximation Methods for Optimal Active Fault Detection *Proceedings of the 22nd Mediterranean Conference on Control and Automation* (Palermo, Italy) pp 103–108

[10] Bertsekas D P 2000 *Dynamic Programming and Optimal Control (Volume I)* 2nd ed (Belmont, Massachusetts: Athena Scientific)

[11] Bertsekas D P and Tsitsiklis J N 1996 *Neuro-Dynamic Programming* (Belmont, Massachusetts: Athena Scientific)

[12] Powell W B 2007 *Approximate Dynamic Programming: Solving the Curses of Dimensionality* 1st ed (Hoboken, New Jersey: John Wiley & Sons, Inc.)

[13] Lewis F L and Vrabie D 2009 *Circuits and Systems Magazine, IEEE* 40–58

[14] Julier S J and Uhlmann J K 1997 A New Extension of the Kalman Filter to Nonlinear Systems *Proceedings of the SPIE 3068, Signal Processing, Sensor Fusion, and Target Recognition VI* (Orlando, FL, USA) pp 182–193

[15] Ma Y, Kelman A, Daly A and Borelli F 2012 *IEEE Control Systems* 44–64

Diagnosis of nonlinear systems using kernel principal component analysis

M. Kallas, G. Mourot, D. Maquin and J. Ragot

Centre de Recherche en Automatique de Nancy (CRAN), CNRS UMR 7039, Université de Lorraine, France

E-mail: {maya.kallas, gilles.mourot, didier.maquin, jose.ragot}@univ-lorraine.fr

Abstract. Technological advances in the process industries during the past decade have resulted in increasingly complicated processes, systems and products. Therefore, recent researches consider the challenges in their design and management for successful operation. While principal component analysis (PCA) technique is widely used for diagnosis, its structure cannot describe nonlinear related variables. Thus, an extension to the case of nonlinear systems is presented in a feature space for process monitoring. Working in a high-dimensional feature space, it is necessary to get back to the original space. Hence, an iterative pre-image technique is derived to provide a solution for fault diagnosis. The relevance of the proposed technique is illustrated on artificial and real dataset.

1. Introduction

The last century has seen an increasingly emergence of manufacturing and process industries, with the rise of energy costs and environmental regulations. In these environments, processes are highly automated. Therefore, the monitoring algorithms are important to detect any fault that might occur.

Principal component analysis (PCA) has been widely used for sensors' fault detection and isolation [8, 18] and more generally for process monitoring [1, 15, 10]. Taking for example the detection methods using a model known a priori, one of the most successful PCA technique is based on the comparison between the value of a variable at a given time and the reconstructed value of this variable from the other variables. The PCA considers the most relevant eigenvectors of the data covariance matrix. Therefore, this reconstruction uses the eigenvectors that span the subspace called secondary or residual corresponding to the smallest eigenvalues. As shown in [8], it is possible to extend this reconstruction to n variables, subject to observability, where we can reconstruct simultaneously p variables out of the $n - p$ remaining ones.

Despite the proven performances of this technique, it is necessary to mention that it is applied on a system whose variables are mainly linearly related. In fact, the PCA is a linear projection and identifies only linear structures in a given dataset. Different techniques have been introduced to learn the nonlinearities leading to the nonlinear PCA or more generally using the kernel machines, the so-called kernel PCA (KPCA). In process monitoring or diagnosis, KPCA has been used for fault identification [3, 9, 6] and more generally for process monitoring [7, 14].

KPCA has been widely used for specific applications such as missing data for partially observed systems, like the case of image processing [2, 16]. Therefore, the association of KPCA and discriminant analysis, for fault identification, was proposed in data processing in order to increase the performance of each technique [4, 11].

It is important to mention that KPCA approach considers an overall system, since it attempts to explain the maximum variance of the data taken all together, but without taking into account any existing local relationships between neighboring data. Therefore, the authors of [5] propose the Local Kernel Principal Component Analysis (LKPCA) technique for preserving to some extent the local structure of data, then apply it on the benchmark TE [1]. In the same spirit, in [12] Orthogonal Neighborhood Preserving Embedding (ONPE) technique is also supposed to preserve the proximity of points of the original space, during data processing in the feature space.

In order to isolate the fault affecting some given data, diagnosis techniques developed in a linear framework rely heavily on the idea of structuring indicators of failure. The extension of this structure to the nonlinear case remains difficult because of the pre-image problem, but has been the subject of several studies. Fu *et al.* in [6] applied the KPCA on several subsets of data, each using only a part of the original variables, the difficulty is obviously to build these subsets. In [5], the contribution evaluation, widely used in PCA, is extended to the nonlinear case to provide fault isolation.

Multiscale approaches have also been proposed to try to take into account the existence of nonlinear correlations between variables at different scales. This resulted in MultiScale KPCA (MSKPCA) techniques with applications for fault detection [20, 1].

While few results making use of KPCA on industrial applications have been published, some studies about sets of simulated data or relative to international benchmarks or dealing with measures from pilot laboratory have been presented in the field of air [8], water [13, 14], combustion [17], batch type process [10], and propulsion system [19].

Regarding the three main phases of diagnosis, namely fault presence detection, fault isolation and estimation of its amplitude, the first point has been the subject of various studies. The other two points are less addressed justifying our proposed approach.

The rest of the paper is organized as follows: while section 2 introduces the KPCA technique, section 3 presents a detailed method for fault detection, isolation and estimation of its magnitude. Finally, section 4 illustrates the efficiency of the proposed technique on real and artificial examples, where the sensitivity of the square prediction error (SPE) due to the presence of the fault, and the number of retained principal components (PCs) is detailed and a discussion for choosing the parameters is discussed.

2. Diagnosis using KPCA

Principal component analysis has been widely used for fault detection. However, such technique is limited by its linear nature. Many studies in the past two decades take into consideration the nonlinear models, and introduce the kernel machines due, on the one hand to the development of the statistical learning theory, and on the other hand to the computational efficiency of the corresponding algorithms. This is illustrated with the kernel PCA, the nonlinear version of PCA.

2.1. Kernel PCA and fault detection

Let the kernel $\kappa\colon \mathcal{X} \times \mathcal{X} \mapsto \mathbb{R}$ be positive semi-definite, where $\mathcal{X}$ is an input space, namely $\sum_{i,j} \alpha_i\, \alpha_j\, \kappa(\boldsymbol{x}_i,\, \boldsymbol{x}_j) \geq 0$ for all $\alpha_i, \alpha_j \in \mathbb{R}$ and all $\boldsymbol{x}_i, \boldsymbol{x}_j \in \mathcal{X}$. Let $\Phi(\cdot)$ denotes the mapping function from the input space $\mathcal{X}$ into a high dimensional feature space $\mathcal{H}$, where the inner product is done. The kernel is given by $\kappa(\boldsymbol{x}_i, \boldsymbol{x}_j) = \Phi^\top(\boldsymbol{x}_i)\Phi(\boldsymbol{x}_j)$.

Using this concept, a given data matrix, defined by n variables describing m measures,

$$X = \begin{bmatrix} \boldsymbol{x}_1^\top \\ \vdots \\ \boldsymbol{x}_m^\top \end{bmatrix} \quad \boldsymbol{x}_i \in \mathbb{R}^n, \quad X \in \mathbb{R}^{m \times n}$$

is mapped to $\mathcal{H}$, more precisely $\boldsymbol{x}_i \underset{\Phi}{\rightarrow} \boldsymbol{\phi}_i = \Phi(\boldsymbol{x}_i), \boldsymbol{\phi}_i \in \mathbb{R}^h$. In a matrix form, we can write

$$\mathfrak{X} = \begin{bmatrix} \boldsymbol{\phi}_1^\top \\ \vdots \\ \boldsymbol{\phi}_m^\top \end{bmatrix}, \quad \mathfrak{X} \in \mathbb{R}^{m \times h},$$

where $h >> n$ is the dimension of the feature space.

We evaluate the covariance matrix, in the feature space $\mathcal{H}$ using:

$$S = \frac{1}{m-1}\sum_{i=1}^{m}\boldsymbol{\phi}_i\boldsymbol{\phi}_i^\top \tag{1}$$

$$= \frac{1}{m-1}\mathfrak{X}^\top\mathfrak{X}. \tag{2}$$

It is important to mention that $\Phi(\cdot)$ is not explicitly defined, one can evaluate the Gram matrix $\mathfrak{X}^\top\mathfrak{X}$ using the kernel function $\kappa(\boldsymbol{x}_i, \boldsymbol{x}_j) = \Phi^\top(\boldsymbol{x}_i)\Phi(\boldsymbol{x}_j)$. For the rest of the paper, the radial type of kernels is studied, in particular the Gaussian kernel defined by:

$$\kappa(\boldsymbol{x}_i, \boldsymbol{x}_j) = \exp\left(-\frac{(\boldsymbol{x}_i - \boldsymbol{x}_j)^T(\boldsymbol{x}_i - \boldsymbol{x}_j)}{c}\right), \tag{3}$$

where the influence of c is studied lately. This type of kernels induces infinite-dimensional feature space, namely $h \to \infty$, and it is represented by m dimensions. Let K be the matrix with $\kappa(\boldsymbol{x}_i, \boldsymbol{x}_j)$ elements, we obtain:

$$K = \mathfrak{X}\mathfrak{X}^\top = \begin{bmatrix} \boldsymbol{\phi}_1^\top\boldsymbol{\phi}_1 & \cdots & \boldsymbol{\phi}_1^\top\boldsymbol{\phi}_m \\ \vdots & \ddots & \vdots \\ \boldsymbol{\phi}_m^\top\boldsymbol{\phi}_1 & \cdots & \boldsymbol{\phi}_m^\top\boldsymbol{\phi}_m \end{bmatrix} \tag{4}$$

As in the case of the conventional PCA, KPCA seeks to resolve the eigenvector equation in the feature space, by evaluating the eigenvectors $\boldsymbol{\nu}_i$ and eigenvalues λ_i of the covariance matrix S. Let $\boldsymbol{\alpha}$ and λ be an eigenvector and its corresponding eigenvalue of the matrix K:

$$\boldsymbol{\nu} = \lambda^{-1}\mathfrak{X}^\top\boldsymbol{\alpha} \tag{5}$$

In the feature space, the eigenvectors $\boldsymbol{\nu}_i$ form a matrix $P = [\boldsymbol{\nu}_1\cdots\boldsymbol{\nu}_\ell, \boldsymbol{\nu}_{\ell+1}\cdots\boldsymbol{\nu}_m]$, where ℓ is the number of principal components (PC). The first part $[\boldsymbol{\nu}_1\cdots\boldsymbol{\nu}_\ell]$ will be denoted as $P_f \in \mathbb{R}^{m\times\ell}$ and the second $[\boldsymbol{\nu}_{\ell+1}\cdots\boldsymbol{\nu}_m]$ as $\tilde{P}_f \in \mathbb{R}^{m\times(m-\ell)}$ representing the eigenvectors defining respectively the principal and residual spaces, where the two are complementary subspaces. We can write

$$P_f = \left[\frac{1}{\lambda_1}\mathfrak{X}^\top\boldsymbol{\alpha}_1 \cdots \frac{1}{\lambda_\ell}\mathfrak{X}^\top\boldsymbol{\alpha}_\ell\right], \tag{6}$$

where $\boldsymbol{\alpha}_1 > \cdots > \boldsymbol{\alpha}_\ell$. The choice of the number ℓ of PCs has been the subject of many studies where [21] describes some of them.

2.2. Analysis of a new observation

During the learning phase of the KPCA model, we seek the ℓ value, and the different hyper-parameters of the model. However, it is important to validate the model through new observations. Let us denoted by $\boldsymbol{x}$ such new observation, its image in the feature space is

$$\boldsymbol{\phi} = \Phi(\boldsymbol{x}), \quad \boldsymbol{\phi} \in \mathbb{R}^m,$$

for which, the projection on the principal and residual spaces are given by:

$$\begin{cases} \boldsymbol{t} = P_f^\top\boldsymbol{\phi}, & \in \mathbb{R}^\ell \\ \tilde{\boldsymbol{t}} = \tilde{P}_f^\top\boldsymbol{\phi}, & \in \mathbb{R}^{m-\ell} \end{cases} \tag{7}$$

Journal of Physics: Conference Series **570** (2014) 072004 doi:10.1088/1742-6596/570/7/072004

We evaluate the statistics SPE, where the dot product in $\mathcal{H}$ is replaced by the kernel function as given

$$\begin{cases} SPE & = \tilde{\boldsymbol{t}}^\top \tilde{\boldsymbol{t}} \\ & = \boldsymbol{\phi}^\top \tilde{P}_f \tilde{P}_f^\top \boldsymbol{\phi} \\ & = \boldsymbol{\phi}^\top (I - P_f P_f^\top) \boldsymbol{\phi} \\ & = \boldsymbol{\phi}^\top \boldsymbol{\phi} - \boldsymbol{\phi}^\top P_f P_f^\top \boldsymbol{\phi} \\ & = \kappa(\boldsymbol{x}, \boldsymbol{x}) - \boldsymbol{\kappa}^\top(\boldsymbol{x})\, C\, \boldsymbol{\kappa}(\boldsymbol{x}) \end{cases} \tag{8}$$

where

$$\begin{cases} \boldsymbol{\kappa}(\boldsymbol{x}) & = [\kappa(\boldsymbol{x}_1, \boldsymbol{x}), \cdots, \kappa(\boldsymbol{x}_m, \boldsymbol{x})]^\top \\ C & = P\, \Lambda^{-1}\, P^\top \end{cases} \tag{9}$$

It is preferred to have centered data in the feature space. Thus, the statistics SPE will be transformed as

$$\begin{cases} \overline{SPE} & = \kappa_c(\boldsymbol{x}, \boldsymbol{x}) - \boldsymbol{\kappa}_c{}^\top(\boldsymbol{x})\, C\, \boldsymbol{\kappa}_c(\boldsymbol{x}) \\ \boldsymbol{\kappa}_c(\boldsymbol{x}) & = (I - E)(\boldsymbol{\kappa}(\boldsymbol{x}) - K\, \boldsymbol{u}_m) \\ \kappa_c(\boldsymbol{x}, \boldsymbol{x}) & = 1 - 2\boldsymbol{u}_m{}^\top \boldsymbol{\kappa}(\boldsymbol{x}) + \boldsymbol{u}_m{}^\top K \boldsymbol{u}_m \\ \boldsymbol{u}_m & = \frac{1}{m} [1 \ \ldots \ 1]^\top \\ E & = m\, \boldsymbol{u}_m\, \boldsymbol{u}_m^\top \end{cases} \tag{10}$$

The analysis of $\overline{SPE}$ indicates the presence of a fault in the measure $\boldsymbol{x}$, based on a detection threshold defined during the learning stage.

3. Fault isolation and estimation with KPCA

Any new observation $\boldsymbol{x}$ to be analyzed, subject to a possible additive fault, can be expressed in terms of the true value $\boldsymbol{x}^*$ as:

$$\boldsymbol{x} = \boldsymbol{x}^* + \boldsymbol{\xi}\, f, \quad \boldsymbol{x} \in \mathbb{R}^n, \boldsymbol{\xi} \in \mathbb{R}^{n \times d}, \tag{11}$$

where $\boldsymbol{\xi}$ is the fault's direction and f its amplitude. The true value to be estimated is given by:

$$\boldsymbol{x}^* = \boldsymbol{x} - \boldsymbol{\xi}\, f \tag{12}$$

3.1. Analysis of the validity of an observation

By substituting any observation $\boldsymbol{x}$ with its real value defined by (12), the $\overline{SPE}$ in (10) is

$$\overline{SPE} = \kappa_c(\boldsymbol{x} - \boldsymbol{\xi}\, f, \boldsymbol{x} - \boldsymbol{\xi}\, f) - \boldsymbol{\kappa}_c{}^\top(\boldsymbol{x} - \boldsymbol{\xi}\, f) C\, \boldsymbol{\kappa}_c(\boldsymbol{x} - \boldsymbol{\xi}\, f). \tag{13}$$

If $\overline{SPE}$ is greater than a predefined threshold, a fault is detected. However in order to isolate and estimate its value, we seek a direction $\boldsymbol{\xi}$ and a value of f that make the $\overline{SPE}$ less than the threshold. In (11) and (13), only the value $\boldsymbol{x}$ of the measure is known. The fault direction $\boldsymbol{\xi}$ and the amplitude of the fault f must be determined. Therefore, we seek $\boldsymbol{\xi}$ and f that minimize the index $\overline{SPE}$. To this end, for each $\boldsymbol{\xi}$ of the possible directions, the fixed point iterative technique is used to resolve the optimization problem with

$$\widehat{f} = \frac{\boldsymbol{\xi}^\top B^\top(\boldsymbol{x} - \boldsymbol{\xi}\widehat{f}) \left[\boldsymbol{u}_m + (I - E)\, C\, \boldsymbol{\kappa}_c(\boldsymbol{x} - \boldsymbol{\xi}\widehat{f})\right]}{\boldsymbol{\kappa}^\top(\boldsymbol{x} - \boldsymbol{\xi}\widehat{f}) \left[\boldsymbol{u}_m + (I - E)\, C\, \boldsymbol{\kappa}_c(\boldsymbol{x} - \boldsymbol{\xi}\widehat{f})\right]} \tag{14}$$

where

$$\begin{cases} B(\boldsymbol{x} - \boldsymbol{\xi}\widehat{f}) & = \begin{bmatrix} \kappa(\boldsymbol{x} - \boldsymbol{\xi}\, \hat{f}, \boldsymbol{x}_1)\, (\boldsymbol{x}_1 - \boldsymbol{x})^\top \\ \vdots \\ \kappa(\boldsymbol{x} - \boldsymbol{\xi}\, \hat{f}, \boldsymbol{x}_m)\, (\boldsymbol{x}_m - \boldsymbol{x})^\top \end{bmatrix} \\ \boldsymbol{\kappa}(\boldsymbol{x} - \boldsymbol{\xi}\widehat{f}) & = \left[\kappa(\boldsymbol{x}_1, \boldsymbol{x} - \boldsymbol{\xi}\, \hat{f}), \cdots, \kappa(\boldsymbol{x}_m, \boldsymbol{x} - \boldsymbol{\xi}\, \hat{f})\right]^\top \\ \boldsymbol{\kappa}_c(\boldsymbol{x} - \boldsymbol{\xi}\widehat{f}) & = (I - E)(\boldsymbol{\kappa}(\boldsymbol{x} - \boldsymbol{\xi}\, \hat{f}) - K\, \boldsymbol{u}_m) \\ \kappa(\boldsymbol{x}_i, \boldsymbol{x}_j) & = \exp\left(-\frac{(\boldsymbol{x}_i - \boldsymbol{x}_j)^\top (\boldsymbol{x}_i - \boldsymbol{x}_j)}{c}\right) \end{cases} \tag{15}$$

When the magnitude f is estimated, the measure $\boldsymbol{x}$ can be corrected with (12) and $\overline{SPE}$ is recalculated with the corrected measure using (13). A meaningful assessment of the fault direction and amplitude results in a significant decrease of $\overline{SPE}$.

This is done for any given direction $\boldsymbol{\xi}$. However, it is worth nothing that the direction in not defined, since the diagnosis aims to localize the fault among the component of $\boldsymbol{x}$. KPCA technique, proposed in [6], operates on different subsets of variables, here, we propose an alternative in order to identify the fault's direction. In case of a single fault, affecting the jth component of the observation $\boldsymbol{x}$, the direction is represented by $\boldsymbol{\xi}_j = \begin{bmatrix}0 & \dots & 1 & \dots & 0\end{bmatrix}$ where 1 is at the jth position. Therefore, for all $j = 1 \cdots n$, the optimal amplitude $\hat{f}_j$ is the one that minimizes $\overline{SPE}$, and makes it below a given threshold.

Remark 1. *Interpretation of the equation* (14)
Let $\boldsymbol{w}$ *of elements* w_i *be a weighting vector:*

$$\boldsymbol{w} = \boldsymbol{u}_m + (I - E)\, C\, \boldsymbol{\kappa}_c(\boldsymbol{x} - \boldsymbol{\xi}\widehat{f}) \tag{16}$$

Thus, we have

$$\boldsymbol{\kappa}^\top(\boldsymbol{x} - \boldsymbol{\xi}\widehat{f})\boldsymbol{w} = \sum_{i=1}^{m} \kappa(\boldsymbol{x}_i, \boldsymbol{x} - \boldsymbol{\xi}\widehat{f}) w_i \tag{17}$$

By considering a fault direction $\boldsymbol{\xi}_j$, *the term* $\boldsymbol{\xi}_j^\top B^\top(\boldsymbol{x} - \boldsymbol{\xi}\widehat{f})$ *of* (14) *is used to select the* j^{th} *element of* $\boldsymbol{x}_i - \boldsymbol{x}$. *Therefore, we can write:*

$$\boldsymbol{\xi}_j^\top B^\top(\boldsymbol{x} - \boldsymbol{\xi}\widehat{f})\, \boldsymbol{w} = \sum_{i=1}^{m} \kappa(\boldsymbol{x}_i, \boldsymbol{x} - \boldsymbol{\xi}\widehat{f})(\boldsymbol{x}_i - \boldsymbol{x})^\top \boldsymbol{\xi}_j\, w_i \tag{18}$$

Therefore, the equation (14) *is written as*

$$\hat{f} = \frac{\displaystyle\sum_{i=1}^{m} \kappa(\boldsymbol{x}_i, \boldsymbol{x} - \boldsymbol{\xi}\widehat{f})\, w_i\, (\boldsymbol{x}_i - \boldsymbol{x})^\top \boldsymbol{\xi}_j}{\displaystyle\sum_{i=1}^{m} \kappa(\boldsymbol{x}_i, \boldsymbol{x} - \boldsymbol{\xi}\widehat{f}) w_i} \tag{19}$$

The term $\kappa(\boldsymbol{x}_i, \boldsymbol{x} - \boldsymbol{\xi}\widehat{f})\, w_i$ *provides a weight for the scalar* $(\boldsymbol{x}_i - \boldsymbol{x})^\top \boldsymbol{\xi}_j$, *leading to the* j^{th} *element of the distance* $\boldsymbol{x}_i - \boldsymbol{x}$ *between the new observation and the* i^{th} *one of the dataset.*

3.2. Algorithm

As aforementioned, two parameters indicate the right fault detection. The first one is the amplitude estimate of the fault. As the direction of the fault is not known, it is necessary to test all possible directions $\boldsymbol{\xi}_j = (0, \dots, 1, 0, \dots, 0)$, where solving (14) provides an estimate of the fault $\hat{f}_j$. A first indicator $I_{f,j}$ is defined by the value of the estimated of the fault over the jth variable with respect to a predefined threshold δ_f.

The second parameter takes into consideration the $\overline{SPE}$. It must be either zero for fault-free measures or corrected ones, or more precisely for robustness issues with respect to uncertainties, must be less than a predefined threshold δ_s. In practice, this value is defined by the user and corresponds to the minimal magnitude needed to detect a fault. Effectively, the combination of the two tests is given in the following procedure steps:

(i) Define the detection thresholds δ_f and δ_s

(ii) Define the fault directions $\xi_j, j = 1, \dots, n$ when all the components are null except the jth.

(iii) Estimate the fault magnitude $\hat{f}_j$ at each instant for a specific direction $\xi_j, j = 1, \dots, n$, then evaluate $\overline{SPE}_j$ after correcting the faulty measure.

Table 1. Fault signatures' table

	δ_1	δ_2	δ_3	δ_4
r_1	$\times$	$\times$	$\times$	0
r_2	$\times$	0	0	$\times$

(iv) Define the amplitude validity parameter by

$$I_{f,j} = \frac{1}{2}\left(1 + sgn(\hat{f}_j > \delta_f)\right)$$

(v) Define the $\overline{SPE}_j$ validity parameter by:

$$I_{s,j} = \frac{1}{2}\left(1 - sgn(\overline{SPE}_j < \delta_s)\right)$$

(vi) Evaluate the fault presence indicator

$$I_j = I_{f,j} \,\&\, I_{s,j}$$

The adjustment of thresholds δ_s and δ_f affects the performance of fault detection and isolation. It can be chosen carefully at the learning stage. Based on different tests, we decided to use an automatic adjustment based on the comparison of different estimated amplitudes $\hat{f}_j$ and a comparison of different $\overline{SPE}_j$ criteria. Thus, these thresholds are defined as follows:

$$\left\{\begin{array}{lcl} \delta_s & = & \frac{1}{n}\sum_{j=1}^{n}\overline{SPE}_j \\ \delta_f & = & \frac{1}{n}\sum_{j=1}^{n}\hat{f}_j \end{array}\right. \tag{20}$$

4. Experiments

In this section, we study the relevance of the proposed technique by applying it on simulated and real data.

4.1. Example on simulated dataset

The data in this example were generated from a model with $n = 4$ variables $\boldsymbol{x}_i$, $i = 1, \ldots, 4$ linked by two nonlinear redundancy equations, namely:

$$\left\{\begin{array}{lcl} \boldsymbol{x}_3 - \boldsymbol{x}_1\boldsymbol{x}_2 & = & 0 \\ \boldsymbol{x}_4 - \boldsymbol{x}_1^2 & = & 0 \end{array}\right. \tag{21}$$

As we can see from table 1, variables $\boldsymbol{x}_1$ and $\boldsymbol{x}_4$ can be isolated due the difference of the signature, where r_j denotes the redundancy equations with respect to faults f_i affecting variable $\boldsymbol{x}_i$ represented by δ_i.

To construct the KPCA model, 50 measures were used to form the training data, while the testing set was limited to 25 samples, a fault of constant amplitude affects $\boldsymbol{x}_2$. Figure 1 illustrates the data, where the shaded area represents the fault's presence. In order to assess the relevance of estimating the fault's magnitude, $\overline{SPE}$ defined with (10) is evaluated.

Let us consider the case when a fault of magnitude -1.20 occurs on a variable, as for example the situation presented within figure 1.

As aforementioned, four fault estimates were performed according to directions ξ_i, where $i = 1, \ldots, 4$. Columns 3 and 4 of table 2 indicate the estimated fault and $\overline{SPE}$ when the

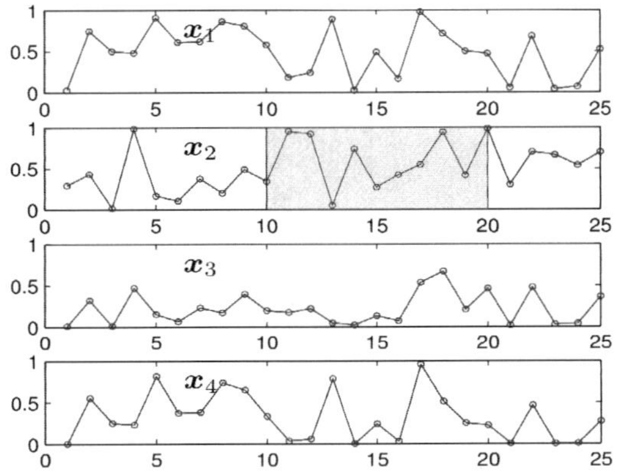

Figure 1. Values of the four variables with a fault on $\boldsymbol{x}_2$

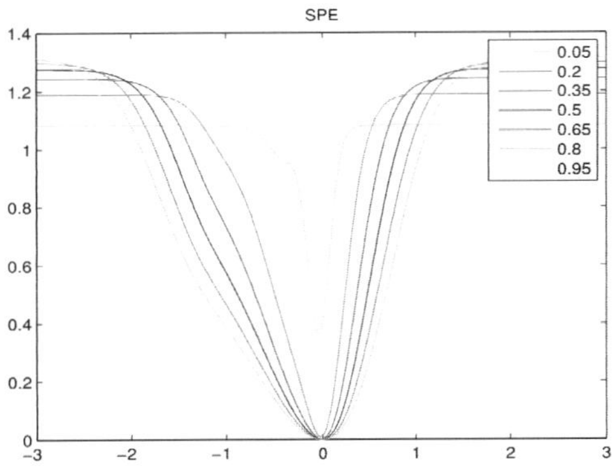

Figure 2. Illustration of the $\overline{SPE}$ with respect to f for different values of c.

measures are corrected. Data collected at $t = 5$ are therefore not affected by any fault, since $\overline{SPE}$ and $\hat{f}$ are null. At $t = 15$, we note that in directions ξ_i such that $i = 1, 3, 4$ a fault was found, whose value cannot be accepted as significant, because it does not make $\overline{SPE}$ null. The ξ_2 direction is the one whose fault is on variable $\boldsymbol{x}_2$, which after correction makes null the $\overline{SPE}$. Hence, the amplitude of the fault has been correctly estimated.

Table 2. Faults and criteria

Time	Direction ξ	Fault $\hat{f}$	Criterion $\overline{SPE}$
5	1	0.00	−0.01
	2	−0.01	0.01
	3	0.00	0.00
	4	0.00	0.00
15	1	0.05	1.14
	2	−1.20	0.00
	3	0.49	1.03
	4	0.11	1.13

The relevance of the proposed method is demonstrated by the preceding example. However, it requires the setting of some parameters. Therefore, it is important to study the influence, on the one hand, of the number of eigenvalues ℓ retained in the feature space (1), and on the other hand, the hyper-parameter c in (3).

Figures 3 and 4 visualize the influence of these two parameters for six values of $c \in \{0.025, 0.1, 0.25, 0.5, 1, 5\}$ and two values for PCs in $\{3, 12\}$. Each figure shows the evolution of $\overline{SPE}$ as a function of the fault amplitude affecting only the variable $\boldsymbol{x}_2$. On each sub-figure, the vertical dashed blue line recalls the fault amplitude on $\boldsymbol{x}_2$ which is -1.2. The vertical red line positions the estimated fault value for the considered direction ξ_2. These two lines are almost overlapping, reflecting the correct estimate of fault magnitude.

The first sub-figures in 3 and in 4 report an extreme case, as established with a very small hyper-parameter ($c = 0.025$), where the first one uses a reduced number of PCs in the feature space. Despite the particular shape of $\overline{SPE}$ and its values, the fault is therefore isolated on $\boldsymbol{x}_2$

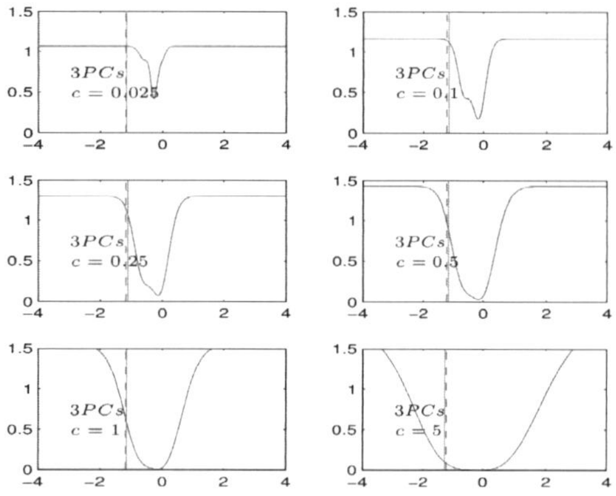

Figure 3. $\overline{SPE}$ with $c \in [0.025, 5]$ and 3 PCs with respect to f

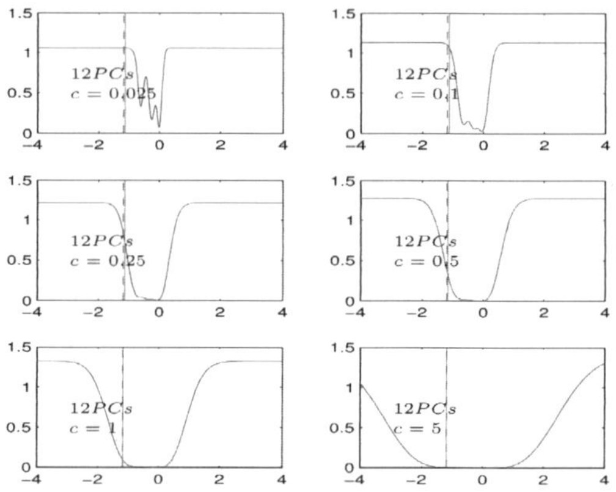

Figure 4. $\overline{SPE}$ with $c \in [0.025, 5]$ and 12 PCs with respect to f

and its amplitude is correctly estimated. However, the second one with a larger number of PCs presents a criterion near 0 when the fault amplitude tends to 0.

4.2. $\overline{SPE}$ sensitivity due to fault and choice of PCs

$\overline{SPE}$ is sensitive when a fault is present and it is affected by the number of PCs retained in feature space.

4.2.1. Sensitivity due to fault presence Let us start by an example in order to study the sensitivity of the $\overline{SPE}$ in presence of a fault. Figure 2 illustrates the evolution of the $\overline{SPE}$ with respect to f for different values of c, where $c \in [0.05, 1]$ with a step of 0.15. The first 15 PCs are considered, using the same configuration of the simulated data, from which we can conclude that

- If $f >> c$, the $\overline{SPE}$ reaches a constant.
- If $c \to 0$, the $\overline{SPE}$ is not 0 when there is no fault.

It is important to mention that for the second case, where c is very small, we must choose more PCs, and the value of the $\overline{SPE}$ decreases.

The $\overline{SPE}$ is thus affected by the fault f, the choice of c and the considered $\boldsymbol{x}^*$. We will study analytically this sensitivity. Let $\boldsymbol{x}$ be as defined in (11), namely $\boldsymbol{x} = \boldsymbol{x}^* + \boldsymbol{\xi}_j f$. In order to evaluate $\kappa(\boldsymbol{x}, \boldsymbol{x}_l)$, the expression of the exponential is developed using

$$
\begin{aligned}
(\boldsymbol{x} - \boldsymbol{x}_l)^\top (\boldsymbol{x} - \boldsymbol{x}_l) &= (\boldsymbol{x}^* - \boldsymbol{x}_l + \boldsymbol{\xi}_j f)^\top (\boldsymbol{x}^* - \boldsymbol{x}_l + \boldsymbol{\xi}_j f) \\
&= (\boldsymbol{x}^* - \boldsymbol{x}_l)^\top (\boldsymbol{x}^* - \boldsymbol{x}_l) + 2f(\boldsymbol{x}^* - \boldsymbol{x}_l)^\top \boldsymbol{\xi}_j + f^2 \\
&= \sum_{\substack{t=1 \\ t \neq j}}^{m} (\boldsymbol{x}_t^* - \boldsymbol{x}_{lt})^2 + (\boldsymbol{x}_j^* - \boldsymbol{x}_{lj})^2 + 2f(\boldsymbol{x}_j^* - \boldsymbol{x}_{lj}) + f^2
\end{aligned}
\tag{22}
$$

In a general manner, the kernel $\kappa(\boldsymbol{x}, \boldsymbol{x}_l)$ can be defined as

$$
\kappa(\boldsymbol{x}, \boldsymbol{x}_l) = \prod_{\substack{t=1 \\ t \neq j}}^{m} \exp\left\{ -\frac{(\boldsymbol{x}_t^* - \boldsymbol{x}_{lt})^2}{c} \right\} \exp\left\{ -\frac{f^2}{c} \left(1 + \frac{(\boldsymbol{x}_j^* - \boldsymbol{x}_{lj})}{f} \right)^2 \right\}
\tag{23}
$$

Let us study the influence of f and $\boldsymbol{x}_j^* - \boldsymbol{x}_{lj}$.

- If f and $\boldsymbol{x}_j - \boldsymbol{x}_{lj}$ are of the same sign :

$$1 < 1 + \frac{\boldsymbol{x}_j^* - \boldsymbol{x}_{lj}}{f} \Rightarrow \exp\left\{-\frac{f^2}{c}\left(1 + \frac{(\boldsymbol{x}_j^* - \boldsymbol{x}_{lj})}{f}\right)^2\right\} < \exp\left\{-\frac{f^2}{c}\right\} \tag{24}$$

Therefore

$$\exp\left\{-\frac{f^2}{c}\right\} \xrightarrow[f\to\infty]{} 0 \tag{25}$$

By taking the gaussian distribution, if $f > 3\sqrt{\frac{c}{2}}$, the $\exp\left\{-\frac{f^2}{c}\right\} < 0.01$. In this case, if $\boldsymbol{\kappa}(\boldsymbol{x}) \to 0$ then $\overline{SPE}$ in (10) is written as

$$\overline{SPE} = 1 + \boldsymbol{u}_m{}^\top K\boldsymbol{u}_m - (K\,\boldsymbol{u}_m)^\top (I-E)C(I-E)(K\,\boldsymbol{u}_m).$$

This expression does not depend on $\boldsymbol{\kappa}(\boldsymbol{x})$, thus it is a constant which is its maximum.

- If f and $\boldsymbol{x}_j - \boldsymbol{x}_{lj}$ are of opposite sign :
  - ◇ $\left|\frac{\boldsymbol{x}_j^* - \boldsymbol{x}_{lj}}{f}\right| \ll 1$, then if $|f| > \max_l(|\boldsymbol{x}_j^* - \boldsymbol{x}_{lj}|)$, we are in the same situation described in (25) and $\exp\{-\frac{f^2}{c}\} \xrightarrow[f\to\infty]{} 0$. In this case, we might say that c should have a small value. However from figure 2, $\overline{SPE}$ is not null in absence of fault. Therefore, in order to detect the presence of such fault, f should have an impact greater than the threshold obtained for $f = 0$.
  - ◇ $\left|\frac{\boldsymbol{x}_j^* - \boldsymbol{x}_{lj}}{f}\right| \gg 1$, then if $|f| < \min_l(|\boldsymbol{x}_j^* - \boldsymbol{x}_{lj}|)$, then

$$\exp\left\{-\frac{f^2}{c}\left(1 + \frac{(\boldsymbol{x}_j^* - \boldsymbol{x}_{lj})}{f}\right)^2\right\} < \exp\left\{-\frac{1}{c}(\boldsymbol{x}_j^* - \boldsymbol{x}_{lj})^2\right\} \qquad \forall l$$

In this case, $\boldsymbol{\kappa}(\boldsymbol{x}) \to \boldsymbol{\kappa}(\boldsymbol{x}^*)$, where $\boldsymbol{x}^*$ is the fault-free value. $\overline{SPE}$ is given by:

$$\overline{SPE} = 1 - 2\boldsymbol{u}_m{}^\top\boldsymbol{\kappa}(\boldsymbol{x}^*) + \boldsymbol{u}_m{}^\top K\boldsymbol{u}_m - (\boldsymbol{\kappa}(\boldsymbol{x}^*) - K\,\boldsymbol{u}_m)^\top (I-E)C(I-E)(\boldsymbol{\kappa}(\boldsymbol{x}^*) - K\,\boldsymbol{u}_m)$$

The obtained value is $\overline{SPE(\boldsymbol{x}^*)}$, therefore, the fault f will be considered as noise and cannot be detected for any value of c.

4.2.2. Sensitivity due to principal functional space dimension tuning If we choose ℓ PCs, the $\overline{SPE}$ criterion is given by:

$$\left\{\begin{array}{l} \overline{SPE_\ell} = \kappa_c(\boldsymbol{x} - \boldsymbol{\xi}\,f, \boldsymbol{x} - \boldsymbol{\xi}\,f) - \boldsymbol{\kappa}_c^\top(\boldsymbol{x} - \boldsymbol{\xi}\,f)C_\ell\,\boldsymbol{\kappa}_c(\boldsymbol{x} - \boldsymbol{\xi}\,f) \\ C_\ell = P_\ell\,\Lambda_\ell^{-1}\,P_\ell^\top \end{array}\right. \tag{26}$$

If we use $\ell + 1$ PCs, we obtain:

$$\left\{\begin{array}{l} C_{\ell+1} = P_{\ell+1}\,\Lambda_{\ell+1}^{-1}\,P_{\ell+1}^\top \\ P_{\ell+1} = \left[P_\ell \mid \boldsymbol{p}_{\ell+1}\right] \\ \Lambda_{\ell+1}^{-1} = \begin{bmatrix}\Lambda_\ell^{-1} & 0 \\ 0 & \lambda_{\ell+1}^{-1}\end{bmatrix} \end{array}\right. \tag{27}$$

Therefore, by combining (26) and (27), the expression of $\overline{SPE}$ is:

$$\left\{\begin{array}{lcl} C_{\ell+1} & = & C_{\ell+1} + \boldsymbol{p}_{\ell+1}\,\lambda_{\ell+1}^{-1}\,\boldsymbol{p}_{\ell+1}^\top \\ \overline{SPE_{\ell+1}} & = & \kappa_c - \boldsymbol{\kappa}_c^\top\,C_{\ell+1}\,\boldsymbol{\kappa}_c \\ & = & \kappa_c - \boldsymbol{\kappa}_c^\top\,(C_{\ell+1} + \boldsymbol{p}_{\ell+1}\,\lambda_{\ell+1}^{-1}\,\boldsymbol{p}_{\ell+1}^\top)\,\boldsymbol{\kappa}_c \\ & = & \overline{SPE_\ell} - \boldsymbol{\kappa}_c^\top\boldsymbol{p}_{\ell+1}\,\lambda_{\ell+1}^{-1}\,\boldsymbol{p}_{\ell+1}^\top\,\boldsymbol{\kappa}_c \end{array}\right. \tag{28}$$

Journal of Physics: Conference Series **570** (2014) 072004 doi:10.1088/1742-6596/570/7/072004

Thus, the variation of the $\overline{SPE}$ due to an increase of PCs is given by:

$$\Delta\overline{SPE_{\ell+1}} = -\frac{(\boldsymbol{\kappa}_c^\top \boldsymbol{p}_{\ell+1})^2}{\lambda_{\ell+1}} \tag{29}$$

This expression is used to adjust the contribution of different eigenvectors for $\overline{SPE}$, making it possible to obtain the number of PCs to retain. Figure 5 was established for a faulty observation, while figure 6 was established for fault-free observation. Both of them show the evolution of $\overline{SPE}$ and its variations as a function of the number of PCs. In both situations, 15 PCs are enough to obtain an acceptable $\overline{SPE}$.

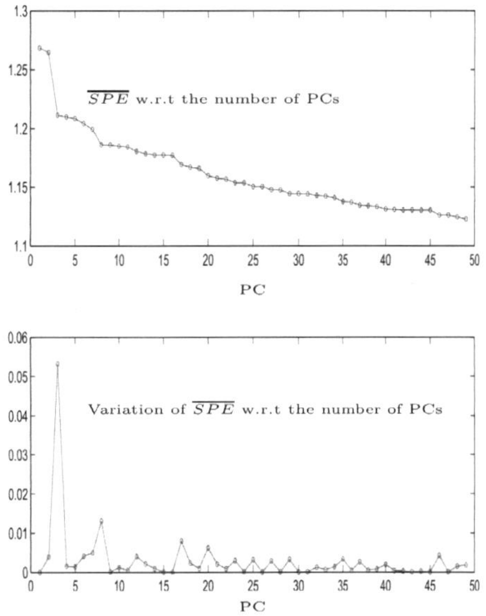

Figure 5. $\overline{SPE}$ with respect to the number of PCs, case of a faulty observation.

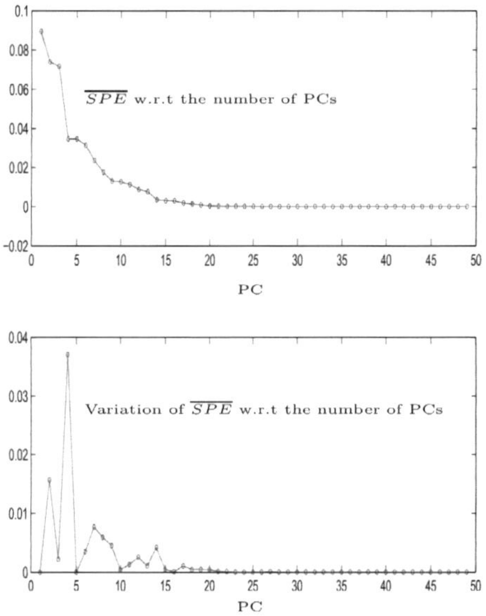

Figure 6. $\overline{SPE}$ with respect to the number of PCs, case of an observation without fault.

4.3. Chemical reactor

For the CSTR considered reactor, the two state variables have a concentration C_a of a specie A and a temperature T in the reactor. Data were generated by varying the concentration C_{af} of A and the volumetric flowrate q. The properties of the reactor are summarized in table 3.

For a steady state, and for a supposed homogeneous reactor system, conservations of the mass of A specie and the energy result in the following two relationships:

$$\begin{cases} 0 = q\,(C_{af} - C_a) - k_0\, V\, \exp(-E/T)\, C_a \\ 0 = q\rho\, C_p\, (T_f - T) + V\, H\, k_0\, \exp(-E/T)\, C_a + U_a\, (T_c - T) \end{cases} \tag{30}$$

Two fault sensors are considered, the first of amplitude -3 affecting the measurement of variable T_c between 10 and 20, the second of amplitude 2 affecting the measurement of variable T between 30 and 40. Figure 7 includes 50 collected measures for different steady states of the reactor. After identifying the KPCA model for these 50 measures, the procedure for fault detection and estimation of their amplitudes was made on 50 other measures. Recall that for each measure of 5 variables at a given time, the procedure examines five possible situations of fault presence, the latter may affect any of the 5 variables.

Figure 8 shows the evolution over 50 observations of the five $\overline{SPE_i}$ criteria assessed after measures correction of their respective estimated faults. For the case of fault absence or corrected

Table 3. CSTR. Variables, parameters and their nominal values

T_c	270	Temperature outside CSTR
q	100	Volumetric Flowrate (m^3/sec)
V	140	Volume of CSTR (m^3)
ρ	1000	Density of A-B Mixture (kg/m^3)
C_p	.239	Heat capacity of A-B Mixture $(J/kg - K)$
H	5e4	Heat of reaction for A->B (J/mol)
E	8750	Activation energy
k_0	7.2e10	Pre-exponential factor $(1/sec)$
U_a	5e4	Overall Heat Transfer Coefficient
C_{af}	1	Feed Concentration (mol/m^3)
T_f	370	Feed Temperature (K)
C_a	0.9	Concentration of A in CSTR (mol/m^3)
T	305	Temperature in CSTR (K)

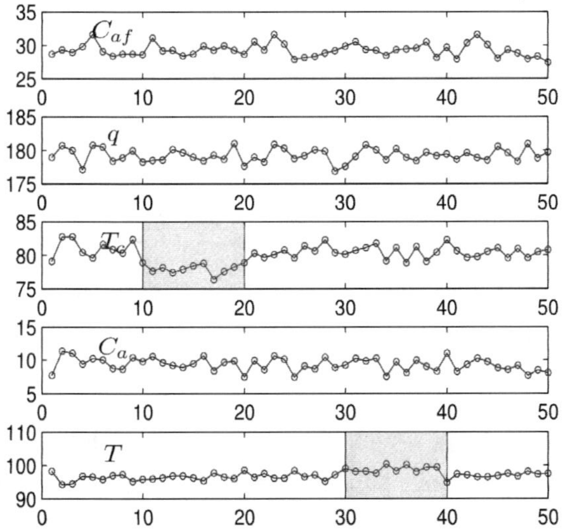

Figure 7. CSTR. Measures of five variables with faults on T_c between instances 10 and 20 and on T between 30 and 40

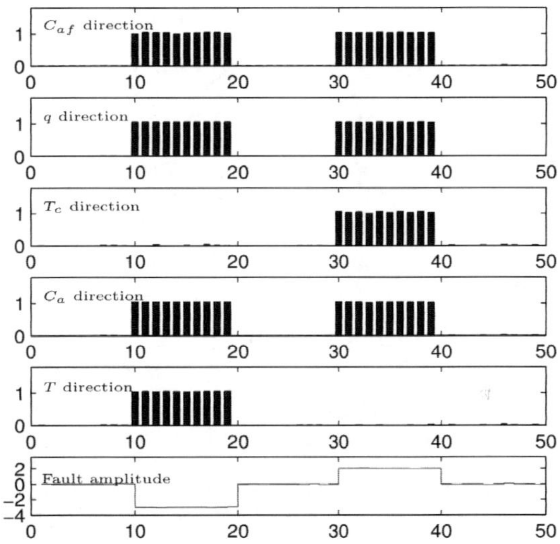

Figure 8. CSTR. Variation of $\overline{SPE}$ for the 5 fault directions and maximal fault amplitude

data, these criteria tend to zero if the number of PCs and the hyper-parameter were correctly chosen. Still, a $\overline{SPE}_i$, whose amplitude is significantly greater than zero, indicates the presence of wrongly corrected measures and therefore supposed to be false. Clearly, this figure shows the presence of a first fault on variable T_c between 10 and 20, and a second on variable T between 30 and 40. The last graph of this figure shows, for each one of the 50 observations, the maximal values of the 5 estimated faults magnitudes, which correspond to the ones used for the simulation.

5. Conclusion

In this paper, we propose to diagnose a nonlinear system using the KPCA technique. The latter can be described as the linear PCA but on transformed data using a nonlinear mapping function into a feature space.

The term diagnosis is defined as the detection of a fault, its isolation and the estimation of its amplitude. A technique used for diagnosis is the analysis of the $\overline{SPE}$. It should have a reduced value when evaluated on raw data or corrected ones. In order to estimate the magnitude of these errors, knowing that several faults can appear on multiple variables, the chosen method considers different fault directions. The performance of this approach has been demonstrated on real and artificial data, in terms of faults detection, isolation and estimation.

Many improvements can be addressed in particular to study the influence of some chosen learning database. The choice of the other parameters, the hyper-parameter and the number of PCs in the feature space is relative to each dataset.

References

[1] H.B. Aradhye, B.R. Bakshi, and R. Straus. Process monitoring by PCA, dynamic PCA, and multiscale PCA- theoretical analysis and disturbance detection in the tennessee eastman process. Technical report, AIChE Annual Meeting, 1999.

[2] M. Berar, M. Desvignes, G. Bailly, Y. Payan, and B. Romaniuk. Missing data estimation using polynomial kernels. In S. Singh, M. Singh, C. Apté, and P. Perner, editors, *International Conference on Advances in Pattern Recognition (1)*, volume 3686 of *Lecture Notes in Computer Science*, pages 390–399. Springer, 2005.

[3] M. Bin Shams. Fault identification using kernel principal component analysis. In *18th IFAC World Congress*, pages 4320–4325, Milano, Italy, 2011.

[4] P. Cui and J. Fang. KPCA plus FDA for fault detection. In D. Liu, S. Fei, Z. Hou, H. Zhang, and C. Sun, editors, *4th International Symposium on Neural Networks : Advances in Neural Networks*, volume 4493 of *Lecture Notes in Computer Science*, pages 597–606. Springer Berlin Heidelberg, 2007.

[5] X. Deng, X. Tian, and S. Chen. Modified kernel principal component analysis based on local structure analysis and its application to nonlinear process fault diagnosis. *Chemometrics and Intelligent Laboratory Systems*, 127:195–209, August 2013.

[6] K. Fu, L. Dai, T. Wu, and M. Zhu. Sensor fault diagnosis of nonlinear processes based on structured kernel principal component analysis. *Journal of Control Theory and Applications*, 7(3):264–270, 2009.

[7] Z. Ge, C. Yang, and Z. Song. Improved kernel PCA-based monitoring approach for nonlinear processes. *Chemical Engineering Science*, 64(9):2245 – 2255, 2009.

[8] M.F. Harkat, G. Mourot, and J. Ragot. An improved PCA scheme for sensor FDI: Application to an air quality monitoring network. *Journal of Process Control*, 16(6):625–634, July 2006.

[9] M. Hehe, H. Yi, and S. Hongbo. Statistics kernel principal component analysis for nonlinear process fault detection. In *9th World Congress on Intelligent Control and Automation*, 2011.

[10] J.M. Lee, C.K. Yoo, and I.B. Lee. Fault detection of batch processes using multiway kernel principal component analysis. *Computers & Chemical Engineering*, 28(9):1837–1847, 2004.

[11] Z. Li, G. Tan, and Y. Li. Fault diagnosis based on improved kernel fisher discriminant analysis. *Journal of Software*, 7(12):2657–2662, 2012.

[12] A. Miao, Z.H. Song, Q. Wen, and Z. Ge. Process monitoring based on generalized orthogonal neighborhood preserving embedding. In *8th IFAC Symposium on Advanced Control of Chemical Processes*, pages 148–153, Singapore, 2012.

[13] J. Ni, C. Zhang, L. Ren, and S.X. Yang. Abrupt event monitoring for water environment system based on KPCA and SVM. *IEEE Transactions on Instrumentation and Measurement*, 61(4):980–989, 2012.

[14] A. Nowicki, M. Grochowski, and K. Duzinkiewicz. Data-driven models for fault detection using kernel PCA: A water distribution system case study. *Applied Mathematics and Computer Science*, 22(4):939–949, 2012.

[15] S.J. Qin. Statistical process monitoring : basis and beyond. *Journal of Chemometrics*, 17(8-9):480–502, 2003.

[16] M. Scholz, F. Kaplan, C.L. Guy, J. Kopka, and J. Selbig. Non-linear PCA: a missing data approach. *Bioinformatics*, 21(20):3887–3895, 2005.

[17] D. Sun, G. Lu, H. Zhou, and Y. Yan. Condition monitoring of combustion processes through flame imaging and kernel principal component analysis. *Combustion Science and Technology*, 185(9):1400–1413, 2013.

[18] M. Tamura and S. Tsujita. A study on the number of principal components and sensitivity of fault detection using PCA. *Computers & Chemical Engineering*, 31(9):1035–1046, 2007.

[19] T. Xu. Sensor Fault Detection and Identification Method with KPCA in the Process of Aero-Engine Ground Testing. *Applied Mechanics and Materials*, 303:297–301, February 2013.

[20] T. Xuemin and D. Xiaogang. A fault detection method using multi-scale kernel principal component analysis. In *27th Chinese Control Conference*, pages 25–29, Kunming, Yunnan, China, 2008.

[21] Y. Zhang, S. Li, and Y. Teng. Dynamic processes monitoring using recursive kernel principal component analysis. *Chemical Engineering Science*, 72(0):78 – 86, 2012.

Implementation of a Fractional Model–Based Fault Detection Algorithm into a PLC Controller

Ryszard Kopka

Opole University of Technology, Sosnkowskiego 31, 45-272 Opole, POLAND

E-mail: r.kopka@po.opole.pl

Abstract. This paper presents results related to the implementation of model–based fault detection and diagnosis procedures into a typical PLC controller. To construct the mathematical model and to implement the PID regulator, a non–integer order differential/integral calculation was used. Such an approach allows for more exact control of the process and more precise modelling. This is very crucial in model–based diagnostic methods. The theoretical results were verified on a real object in the form of a supercapacitor connected to a PLC controller by a dedicated electronic circuit controlled directly from the PLC outputs.

1. Introduction

Improving the functionality of modern equipment and systems requires the use of very advanced fault detection mechanisms or procedures. Most of them are based on mathematical models of processes, built by differential equations, neural networks or fuzzy logic. But the efficiency of model–based fault detection methods are related to the precision of the model [7, 8]. On the other hand, technology development generates new materials, new equipment or systems for which the use of traditional mathematical calculus is insufficient. An example could be activated carbon technology used in supercapacitors. The differential equation model of non–integer order, so–called fractional order, allows for the described phenomena in porous materials to yield more detail than the integer one.

There are many examples in the use of fractional calculus [5, 10]. However, these are generally used as experimental or laboratory solutions. The paper presents results related to the implementation of a fractional model and fractional regulator into a typical, industrial PLC controller. The main constraints in such systems are computational speed and insufficient memory.

The paper presents the results of fault detection procedures based on a plant model implemented into a PLC controller. In a fractional plant simulator, the supercapacitor charged by voltage–controlled voltage source was used. The fractional order model of the process and two fractional $PI^{\lambda}D^{\mu}$ regulators were all implemented into a PLC controller. The minimal sampling time and maximal length of data blocks were determined.

This article is organized as follows. In Section 2, the basics of model-based fault detection mechanism implementation are presented. Section 3 describes the principles of fractional order calculus, its discretization scheme and problems with practical implementation. The experimental test setup, its configuration, software implementation into the PLC controller and model–based fault detection procedure implementation are described in detail in Section 4. Next,

Journal of Physics: Conference Series **570** (2014) 072005 doi:10.1088/1742-6596/570/7/072005

the results of model identification and $\mathrm{PI}^{\lambda}\mathrm{D}^{\mu}$ regulator configuration are shown. The last two sections are test results and a conclusion.

2. Model–Based Fault Detection

There are different approaches for fault detection using mathematical models. The goal is to detect faults in the process using a residual signal, expressed as the difference between measurable process signals and mathematical models. Figure 1 shows the basic structure of model–based fault detection. Based on the measured input signal $u(t)$ and the output signal $y(t)$, the detection module generates a difference signal $d(t)$. By comparison with nominal values, the analysis module detects changes leading to faults [2, 4, 9]. Model–based methods require

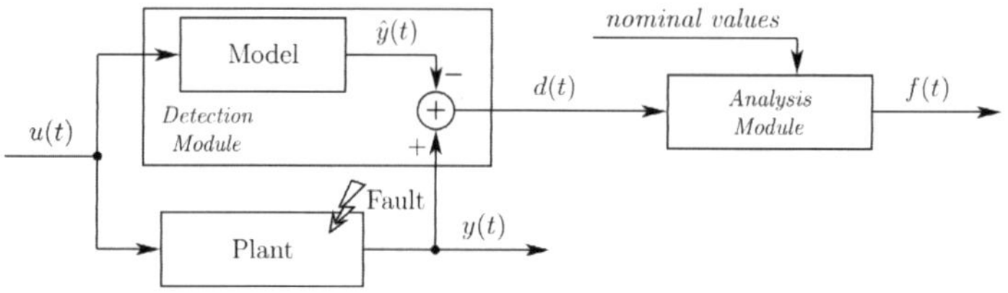

Figure 1. General scheme of model–based fault detection.

the knowledge of process dynamics in the form of mathematical structure and parameters. For process dynamics, the basic input/output models in the form of differential equations are used. But in most practical cases, the process parameters are not known. They must be determined with parameter estimation methods by measuring input and output signals.

3. Fractional Calculus

Fractional differential/integral calculus operator ${}_a\mathcal{D}_t^\alpha$ is a generalization of the integration and differentiation to the non–integer order, in which a and t denote the limits of the operation and α denotes the fractional order such that

$$ {}_a\mathcal{D}_t^\alpha = \begin{cases} \frac{\mathrm{d}^\alpha}{\mathrm{d}t^\alpha} & \text{for } \alpha > 0 \\ 1 & \text{for } \alpha = 0 \\ \int_a^t \mathrm{d}\tau^{-\alpha} & \text{for } \alpha < 0, \end{cases} \tag{1} $$

where it is assumed that $\alpha \in \mathbb{R}$. In the literature, there are a few definitions of the fractional order differ/integral operator that may be useful in a specific situation [1, 3]. One of the most widely used is the Grünwald–Letnikov. According to it, the fractional operator ${}_a\mathcal{D}_t^\alpha$ is defined as

$$ {}^{GL}{}_a\mathcal{D}_t^\alpha f(t) = \lim_{h\to 0} \frac{1}{h^\alpha} \sum_{j=0}^{\left[\frac{t-a}{h}\right]} (-1)^j \binom{\alpha}{j} f(t - jh), \tag{2} $$

where h is the time period, $m - 1 < \alpha \leq m$ while $m \in \mathbb{N}$, and $[\kappa]$ is the integer part of κ. The binomial $\binom{\alpha}{j}$ is calculated as

$$ \binom{\alpha}{j} = \begin{cases} 1 & \text{for } j = 0 \\ \frac{\alpha(\alpha-1)\ldots(\alpha-1+j)}{j!} & \text{for } j > 0. \end{cases} \tag{3} $$

To obtain the fractional model and controller in discrete time, the definition of Grünwald–Letnikov in discrete form was introduced as

$$\Delta_h^\alpha f(t)|_{t=kh} = \frac{1}{h^\alpha}\sum_{j=0}^{k}(-1)^j\binom{\alpha}{j}f(t-jh), \tag{4}$$

where $t = kh$ is the discrete sample in continuous time. Converting continuous time samples into a discrete domain $k = 0, 1, ...,$ equation (4) can be written as

$$\Delta_h^\alpha f(t) = \frac{1}{h^\alpha}\sum_{j=0}^{t}(-1)^j\binom{\alpha}{j}f(t-j). \tag{5}$$

There are a few discretization schemes of the Grünwald–Letnikov equation. The most commonly used are the backward difference (Euler), trapezoidal (Tustin) rule, and the Al–Alaoui operator. Taking the Euler rule, the fractional order difference in the discrete time domain for sequential time sample k can be calculated as

$$\Delta_h^\alpha x_k = \frac{1}{h^\alpha}\sum_{j=0}^{k}(-1)^j\binom{\alpha}{j}x_{k-j}. \tag{6}$$

The fractional order result according to (6) depends on all earlier samples. This is a major advantage of the fractional differ/integral calculus in relation to the integer order. But the infinite summation all previous measurements according to equation (6), in real devices must be limited to finite samples $L+1$. This is constrained by both the limited amount of memory and the calculation time. According to this, equation (6) takes the form

$$\Delta^\alpha x_k = \frac{1}{h^\alpha}\sum_{j=0}^{L}(-1)^j\binom{\alpha}{j}x_{k-j}. \tag{7}$$

The limitation of sample number reduces the accuracy expressed by equation (7). This is why the number of samples needs to be as large as possible. But in industrial controllers, which operate in real-time and have a limited memory, this is a serious problem.

4. Experimental Test Setup

The experimental test setup consists of an electronic circuit that simulates part of a real plant. A special electronic element known as a supercapacitor makes proper use of fractional calculus. The plant simulator is connected to a typical PLC controller by dedicated electronic circuits with voltage–controlled voltage and current sources. A fractional order PI$^\lambda$D$^\mu$ regulator and fractional plant model were both implemented into the PLC controller.

4.1. Fractional order PI$^\lambda$D$^\mu$ regulator

The scheme of the fractional PI$^\lambda$D$^\mu$ regulator is shown in Fig. 2. It has a typical PID configuration, except that integration and differentiation are of fractional order. The regulator transfer function is defined as

$$G_R(s) = K_p + \frac{K_i}{s^\lambda} + K_d s^\mu, \tag{8}$$

where integration and differentiation are of the orders λ and μ, respectively, with $\lambda, \mu \in \mathbb{R}$. To implement fractional calculus, according to equation (7), the arrays of the binomials $\binom{\lambda}{j}$ and $^{\mu}_{j}$ of length $L+1$ have to be calculated first.

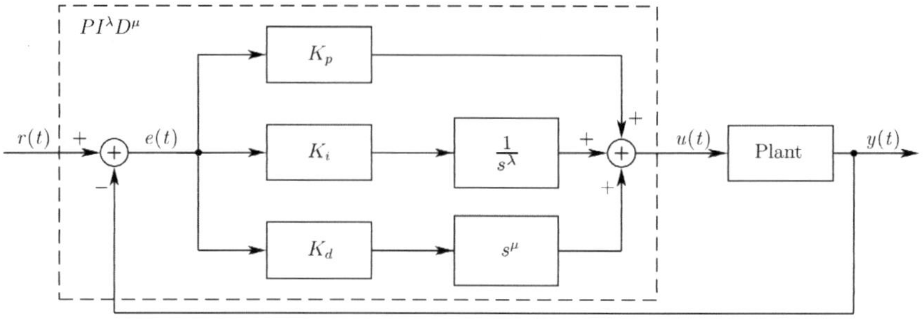

Figure 2. Configuration of fractional order $\mathrm{PI}^\lambda\mathrm{D}^\mu$ regulator.

4.2. Fractional order plant model

The plant was simulated by an electronic circuit built with resistors and a supercapacitor, as shown in Fig. 3(a). To implement the model into the PLC controller, the plant in the form of differential equations was defined as

$$y(t) + T_1 \frac{\mathrm{d}^\alpha y(t)}{\mathrm{d}t^\alpha} = u(t) + T_2 \frac{\mathrm{d}^\alpha u(t)}{\mathrm{d}t^\alpha}, \tag{9}$$

where $T_1 = (R_1 + R_2)C$ and $T_2 = R_2 C$. The resistance R_2 represents the series connection of R_{I1} and r_{ESR} (r_{ESR} represents the equivalent series resistance of the supercapacitor, see Fig. 3(b) and [6]). After simple calculations it can be presented as

$$\frac{\mathrm{d}^\alpha y(t)}{\mathrm{d}t^\alpha} = \frac{1}{(R_1 + R_2)C}(u(t) - y(t)) + \frac{R_2}{R_1 + R_2} \frac{\mathrm{d}^\alpha u(t)}{\mathrm{d}t^\alpha} \tag{10}$$

Based on (10), the mathematical fractional model of the plant was implemented into the PLC controller.

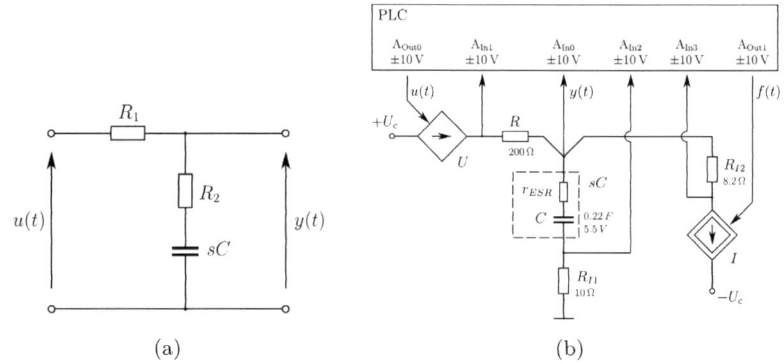

Figure 3. Electronic diagram of fractional plant simulator (a) and connection scheme of plant simulator to PLC controller (b).

4.3. Realization of fractional calculus into PLC controller

The fractional calculus was implemented into the PLC controller according to the scheme shown in Fig. 4. In the main memory, seven real-number arrays of length $L + 1$ were defined, where

$L = 900$. All binomials and sign arrays were calculated in the Startup Block. After that, the Main Block was started. The *Time Interrupt* was generated with sampling time $t_s = 250\,\text{ms}$, independent of the Main Block [11]. During every *Time Interrupt* procedure, the control signals for both plant and model output were calculated. According to equation (7), the calculations of the fractional $\text{PI}^\lambda\text{D}^\mu$ regulator and the fractional plant model outputs need to provide six times the $L+1$ summation: two for each regulator and two for the model output. In every *Time Interrupt* the actual value of the plant output was measured. Next, the error signals for plant and model $\text{PI}^\lambda\text{D}^\mu$ regulators were calculated. These values were introduced into two error arrays (DB6 and DB7). Based on them, the integral and differential calculations for $\text{PI}^\lambda\text{D}^\mu$ regulators were provided. The control signal for plant steering was written to the analog output (A_{Out0}), while the control signal for the model was used to calculate the model output. All calculations were made during $t_C \approx 230\,\text{ms}$. The analog output signal from the PLC was steering the voltage

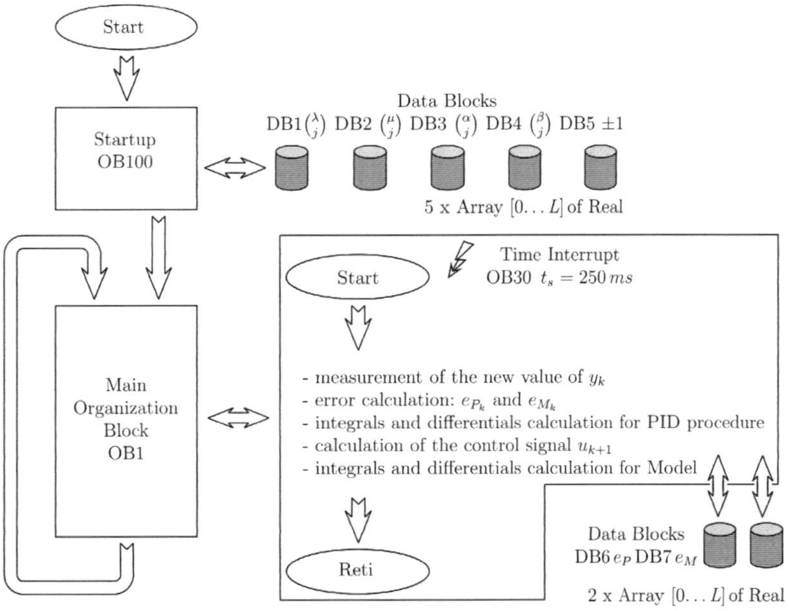

Figure 4. Algorithm of the PLC program.

source with a current intensity high enough to charge the supercapacitor. The actual voltage level was measured on the analog input (A_{In0}), while the supercapacitor current was measured by the voltage drop on the R_2 resistor (A_{In1}). The fault was simulated by a voltage-controlled current source discharging supercapacitor. The current source intensity was steered by voltage changes on the analog output (A_{Out1}). The detailed wiring diagram is shown in Fig. 3(b).

4.4. Configuration of model–based fault detecting structure

Based on the scheme in Fig. 1, a detailed configuration of the implemented model–based fault detection mechanism is shown in Fig. 5. As a fault indicator, the difference signal between the $\text{PI}^\lambda\text{D}^\mu$ regulators was taken. This solution allows one to keep the plant and model output signals at the same levels regardless of the faults. This guarantees the same dynamic behaviour and set point.

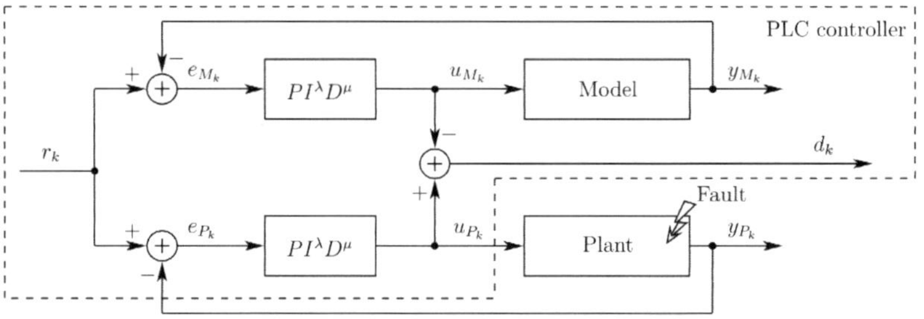

Figure 5. Implementation of the fault detection algorithm based on plant and model – control difference signals.

5. Model Parameters Identification and PID Configuration

There are several kinds of identification methods of plant model parameters. The goal of the time–domain identification process used in this work, was to obtain a fractional model parameters vector in the form of (10). An output error method was used with a non–linear least–square approach. The optimization criterion was the output error norm $\|\epsilon_k\|_2^2$ given by

$$\epsilon_k = y_k - \hat{y}_k, \tag{11}$$

where y_k was the plant output signal and $\hat{y}_k$ was obtained by simulation of the identified model under the input signal u_k. The best method to estimate the parameters so as to minimize the output error is numerical optimization by means of iterative procedures. θ denotes the vector of the identified parameters and $y(t)$ and $\hat{y}(t)$ represent the vectors of measured and calculated data, during simulations. The identification problem is to find a vector $\theta \in \Theta_{ad}$ that minimizes the quadratic criterion J, that is

$$\min_{\theta\in\Theta_{ad}} J, \tag{12}$$

where

$$J = \int_0^t \epsilon(\tau)^{\mathrm{T}} \mathrm{W} \epsilon(\tau) d(\tau). \tag{13}$$

Θ_{ad} stands for the set of admissible parameters, $\mathrm{W} = \mathrm{W}^{\mathrm{T}}$ is the positive definite weighting matrix and t denotes the simulation time. The problem (12) can be solved using many optimization algorithms. In this paper, the identification procedure was implemented in MATLAB® environment using the Nelder–Mead simplex method. The results of the model identification are presented in Tab. 1.

Several methods of tuning fractional $\mathrm{PI}^\lambda\mathrm{D}^\mu$ controllers are proposed. However, there is still no general one. Some authors propose optimization techniques applied to the tuning problem instead. But there are several problems in designing a fractional $\mathrm{PI}^\lambda\mathrm{D}^\mu$ controller via optimization. These include the type of plant to be controlled, optimization critera, parameters to optimize or methods to obtain initial parameters. The results of the $\mathrm{PI}^\lambda\mathrm{D}^\mu$ regulator tuning procedure are presented in Tab. 2. Figure 6(a) shows the step response output signals of the plant and model for different step values. One can see the identity quite well. Figure 6(b), however, shows the closed-loop plant and model-output signals with regulator parameters presented in Tab. 2.

6. Test Results

The tests were carried out with and without fault events. The faults had abrupt or incipient form. Fig. 7 shows the plant y_{P_k} and model y_{M_k} output signals under the reference signal r_k.

Table 1. Model identification parameters.

Parameter	Identified Value
C	0.31 F
R_1	85.82 Ω
R_2	71.07 Ω
α	−0.96

Table 2. $\mathrm{PI}^{\lambda}\mathrm{D}^{\mu}$ configuration parameters.

Parameter	Value
K_p	0.01
K_i	8.3
K_d	0.1
λ	−0.9
μ	0.05

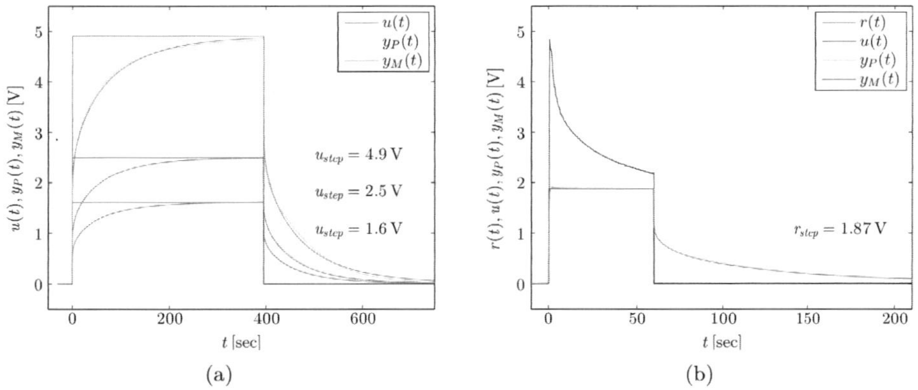

Figure 6. Process and model step responses for different reference values (a) and closed–loop output signals with $\mathrm{PI}^{\lambda}\mathrm{D}^{\mu}$ regulators (b).

The difference signal d_k was still contained within the area bounded by two threshold levels. Figure 8 shows the same signals but the plant has been damaged by an abrupt and incipient fault signal I_{f_k}. Although the control process was still provided (y_{P_k} followed the reference signal r_k), faults can be clearly seen on the difference signal d_k, which crossed the threshold line.

7. Conclusions

The growth in reliability demands result in developing new methods and procedures to more precisely and effectively diagnose and detect faults. In model–based methods, the crucial element is the precision of the mathematical model of the protected plant. It seems that fractional calculus meets these expectations. Related works show that scientific solutions developed in research laboratories can be implemented even in typical industrial controllers, although there are some limitations, such as calculation speed and memory size.

References

[1] Barbosa S. R., Tenreiro Machado J. A., Implementation of Discrete-Time Fractional-Order Controllers based on LS Approximations, *Acta Polytechnica Hungarica*, Vol. 3, No. 4, pp. 5–22 (2006)
[2] Blanke M., Kinnaert M., Lunze J., Staroswiecki M., *Diagnosis and Fault-Tolerant Control*, 2nd Edition, Springer-Verlag, Berlin Heidelberg (2006)
[3] Das S., *Functional Fractional Calculus for System Identification and Controls*, Springer-Verlag, Berlin Heidelberg (2008)

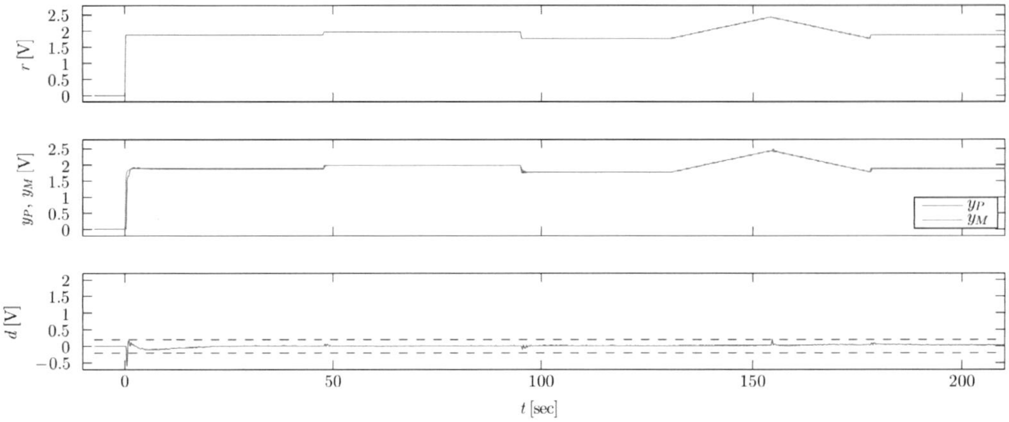

Figure 7. Difference signal $d(t)$ without faults.

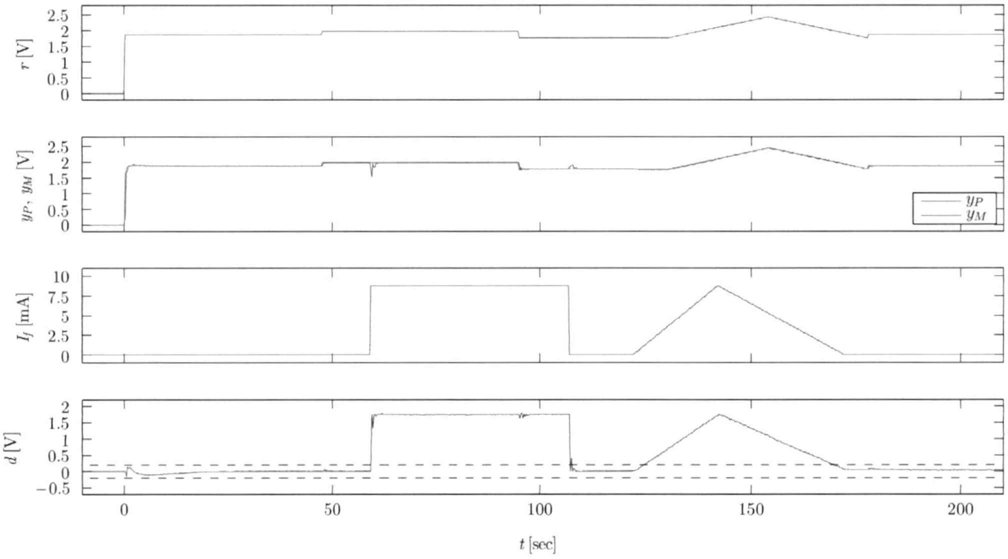

Figure 8. Difference signal $d(t)$ with abrupt and incipient faults.

[4] Isermann R., Model-based fault-detection and diagnosis status and applications, *Annual Reviews in Control*, 29, pp. 71–85 (2005)
[5] Kaczorek T., *Selected Problems of Fractional Systems Theory*, Springer-Verlag, Berlin Heidelberg (2011)
[6] Kopka R., Tarczynski W., Measurement system for supercapacitors equivalent parameters determination, *Metrology and Measurement Systems*, Vol. XX, No. 4, pp. 581–590, (2013).
[7] Korbicz J., Kościelny J.M., Kowalczuk Z., Cholewa W. (Eds.), *Fault Diagnosis*, Springer-Verlag, Berlin Heidelberg (2004)
[8] Korbicz J., Kościelny J.M. (Eds.), *Modeling, Diagnostics and Process Control*, Springer-Verlag, Berlin Heidelberg (2010)
[9] Noura H., Theilliol D., Ponsart J-Ch., Chamseddine A., *Fault-tolerant Control Systems*, Springer-Verlag, London (2009)
[10] Podlubny I., *Fractional-order Systems and Fractional-order Controllers*. Slovak Academy of Sciences Institute of Experimental Physics, UEF SAV, Kosice (1994)
[11] SIEMENS Manual, *Simatic S7-1200 Easy Book*, www.automatyka.siemens.pl, (2012)

Journal of Physics: Conference Series **570** (2014) 072006 doi:10.1088/1742-6596/570/7/072006

Development of model-based fault diagnosis algorithms for MASCOTTE cryogenic test bench

A. Iannetti[1,3], J. Marzat[2], H. Piet-Lahanier[2], G. Ordonneau[2], L. Vingert[2]
E-mail: alessandra.iannetti@cnes.fr

Abstract. This article describes the on-going results of a fault diagnosis benchmark for a cryogenic rocket engine demonstrator. The benchmark consists in the use of classical model-based fault diagnosis methods to monitor the status of the cooling circuit of the MASCOTTE cryogenic bench. The algorithms developed are validated on real data from the last 2014 firing campaign (ATAC campaign). The objective of this demonstration is to find practical diagnosis alternatives to classical redline providing more flexible means of data exploitation in real time and for post processing.

1 Introduction

Model-based fault detection methods are one of the most common diagnostic technics used in process monitoring for a wide range of applications (ex. nuclear, aeronautics, chemical etc.) [1], [2],[3]. A mathematical model represents the knowledge of the process and it is used to produce diagnosis flags based on deviations of parameters from their nominal expected values. These deviations (residuals) can be obtained from the evaluation of characteristic parameters of the system as in parameter estimation methods [4] or directly from sensor output estimation as in Kalman filtering approaches [5]. Finally statistical tests on the residuals are performed to obtain robust diagnosis flags (i.e. CUSUM test [6], [14]).

In the field of rocket engines, fault detection and diagnosis technics are called health monitoring systems (or health management systems if reconfiguration actions are possible). So far rocket engine diagnosis has been performed mostly via simple redlines over preselected sensors (fixed threshold). This approach has been preferred as it is simple and requires low computation effort. On the other hand its drawbacks are the high number of sensors and thresholds to be set for each targeted engine operating point leading to higher possibility of human errors and unjustified firing aborts. During the last decades, an important number of researches [7] have been dedicated to the improvement of the diagnosis methods for applications at test bench or during flight [8][9][10][11]. The difficulty of these systems is the complexity of the processes to monitor and the need for reliable diagnosis with short reaction times (from 10 milliseconds up to a few seconds).

The aim of this work is to find new possible alternatives to simple redline checking for the cryogenic rocket engine demonstrator MASCOTTE. Two diagnosis methods have been developed here, namely parameter identification and Kalman filtering. These methods have been implemented and tested on real data from the latest firing campaign of 2014 for the bi-dimensional nozzle ATAC [12].

1 CNES launchers directorate, 52 rue Jacques Hillairet, 75012 Paris, France
2 ONERA - The French Aerospace Lab, F-91123 Palaiseau, France
3 Author to whom any correspondence should be addressed

2 Mascotte description

MASCOTTE test bench was designed to study cryogenic rocket combustion and nozzle flow performances. It was built in 1994 from the cooperation of CNES and ONERA. Several version of the hardware were developed during the years for different purposes. The most recent version called ATAC is used to analyse rocket nozzle performances. The combustion chamber fed with gaseous hydrogen and oxygen is terminated by a 2D nozzle. The hardware has been equipped with a water cooling circuit in order to withstand high mixture ratio and pressure in the chamber and to operate for longer duration (around 60s). Figure 1 shows the chamber section and the bi-dimensional nozzle [12].

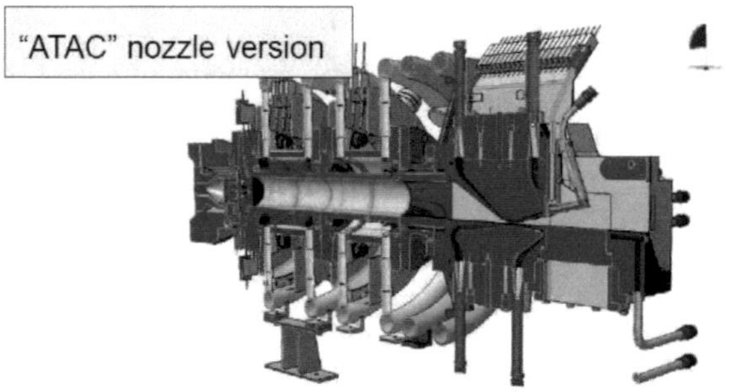

Figure 1 Combustion and nozzle section with water cooling system (yellow circuit for the chamber part and blue circuit for the nozzle part)

The bench is monitored and controlled via three main computers. The initial diagnosis system is based on independent redlines on relevant parameters. In particular for the water circuit all pressures, temperatures and mass flows are monitored and a test stop is issued whenever one or more of these parameters cross a redline for more than a minimum time. The redlines are set manually by the experts and validated via a dedicated bench firing before each test campaign. The bench is in open-loop control and only test abort over redlines or nominal test stop are possible. Figure 2 shows the water circuit architecture of the bench.

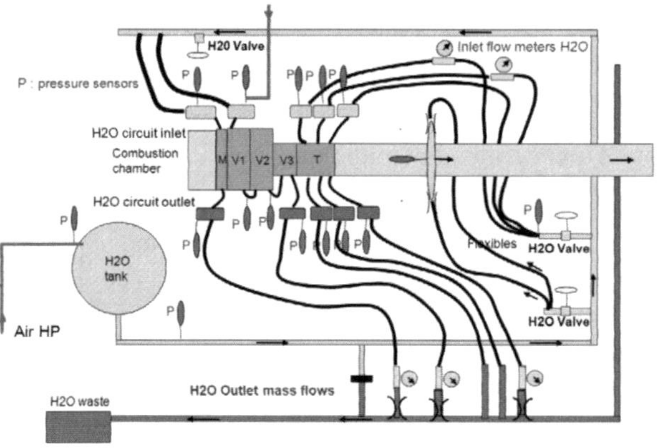

Figure 2 Test bench water cooling circuit: M = inlet chamber section, V1 to V3 = middle chamber sections, T = nozzle.

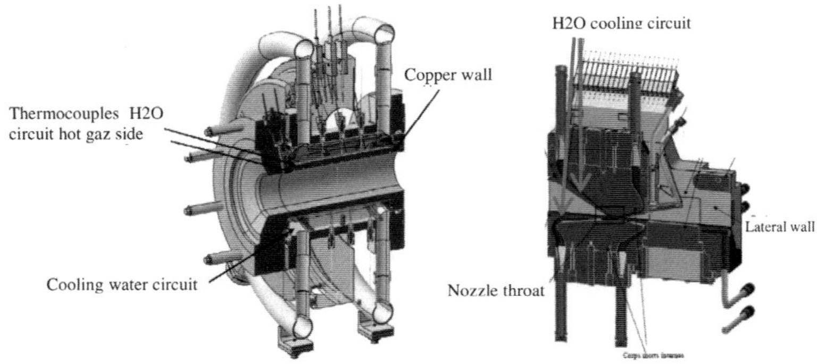

Figure 3 Chamber and nozzle section: water cooling detail.

3 Fault diagnosis methods

The selected methods for the diagnosis demonstration are classical model-based technics [1], [3]:

- Least-square parameter identification
- Kalman filter

An analysis of the possible modelling of the cooling system was performed and a simplified model was set up for describing the main processes in each branch of the water circuit. This model is the basis of the diagnosis strategy.

4 Water cooling system model

The chamber or nozzle section in Figure 3 is modelled through an inlet cavity, an orifice and an outlet cavity. Figure 4 shows the corresponding zones. Cavity 1 corresponds to the inlet volume of the cooling circuit, the orifice is the connecting tube and cavity 2 the water volume flowing on the chamber wall and heated up by the combustion process. This scheme can be applied to any other segment of the water circuit: the three chamber segments (V1 and V2, V3 in Figure 2) and the bi-dimensional nozzle part (T in figure 2). The only difference is in the parameter of the model depending on the geometry.

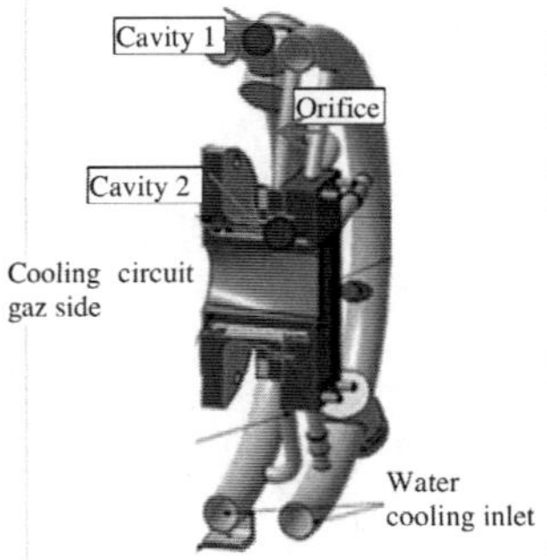

Figure 4 Water cooling modelling

This modelling approach is directly derived by the CNES software CARMEN [18] which is used for rocket engine systems modelling. It was extensively validated on different engines and propulsive systems. The idea here is to build a simple and generic model, but representative of the system dynamics and generic enough to be applicable to different system parts. The governing equations obtained from conservation laws applied to each element are given in equations (1), (2) and (3).

Cavity 1:

$$\frac{dP_1}{dt} = (q_1^e - q_1^s) \cdot \frac{a^2}{V_1}$$
$$\frac{dT_1}{dt} = \frac{1}{\rho \cdot V_1} \cdot (q_1^e T_1^e - q_1^s T_1) + \frac{\dot{Q}_1}{c_v \rho V_1} \quad (1)$$

Journal of Physics: Conference Series **570** (2014) 072006 doi:10.1088/1742-6596/570/7/072006

Liquid orifice:

$$P_1 - P_2 = \frac{k_p}{\rho S^2} \cdot q^2$$
$$q = \sqrt{\frac{P_1 - P_2}{k_p} \cdot \rho S^2} \tag{2}$$

Cavity 2:

$$\frac{dP_2}{dt} = (q_2^e - q_2^s) \cdot \frac{a^2}{V_2}$$
$$\frac{dT_2}{dt} = \frac{1}{\rho \cdot V_2} \cdot (q_2^e T_2^e - q_2^s T_2\ ) + \frac{\dot{Q}_2}{c_v \rho V_2} \tag{3}$$

List of variables:

P_1: pressure in cavity 1 (Pa)
P_2: pressure in cavity 2 (Pa)
k_p : pressure drop coefficient (non dimensional)
ρ : water density (kg/m^3)
a : speed of sound in water (m/s)
c_v: specific heat of water at constant volume
V_1 : volume of cavity 1 (m^3)
V_2 : volume of cavity 2 (m^3)
S : cross sectional area for the orifice element (m^2)
$q = q_1^s = q_2^e$: mass flow through the orifice element / outlet mass flow from cavity 1/ inlet mass flow to cavity 2 (kg/s)
q_1^e : cavity 1 inlet mass flow (kg/s)
q_2^s : cavity 2 outlet mass flow (kg/s)
T_1 : temperature in cavity 1 (K)
T_2 : temperature in cavity 2 (K)
$\dot{Q}_1$: heat flux on cavity 1 (W)
$\dot{Q}_2$: heat flux on cavity 2 (W)

Let consider first equations (1), (2) for establishing a model of pressure and mass flow evolution. Assuming $q_1^s = q_2^e$ the resulting equation, for each branch of the circuit is equation (4) below.

$$\frac{dP_2}{dt} = \left(\sqrt{\frac{P_1 - P_2}{k_p} \cdot \rho S^2} - q_2^s \right) \cdot \frac{a^2}{V_2} \tag{4}$$

4.1 Pressure drop coefficient

The pressure drop coefficient, k_p is calculated via a Blasius correlation [13] applicable for Reynolds number higher than 5000. This was verified on past nominal firing campaign as shown in Figure 5.

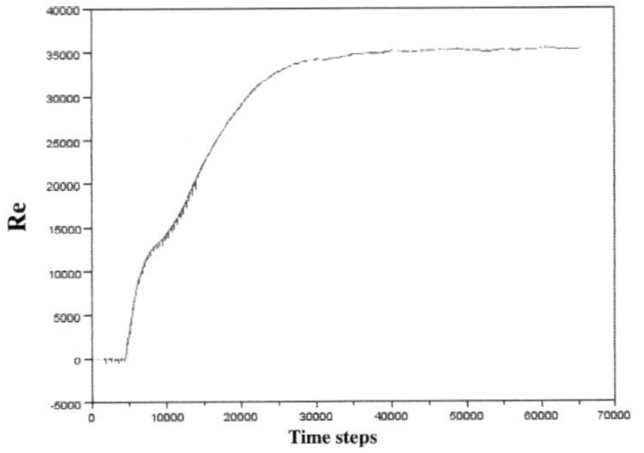

Figure 5 Reynolds number in water cooling chamber branch for a nominal Mascotte firing campaign (time in milliseconds)

The correlation is expressed in equation (5).

$$P_1 - P_2 = \Lambda \cdot \frac{L}{D_h} \cdot \frac{\rho \cdot v^2}{2}$$
$$\Lambda = 0.3164\, Re^{-0.25} \quad (5)$$

With:

$Re = \frac{D_h \cdot v \cdot \rho}{\mu}$ Reynold number (non-dimensional)
v = average speed in flow cross section (m/s)
μ = dynamic viscosity (kg/(s m))
D_h= characteristic dimension cross flow, hydraulic diameter
L = characteristic length of the flow

Rewriting the expression above with the mass flow rate expression $q = \rho \cdot v \cdot S$, we obtain:

$$K = 0.3164 \cdot \left(\frac{Q}{\frac{\pi D}{4} \cdot \mu} \right)^{-0.25} \cdot \frac{L}{D} \cdot \frac{1}{2} \quad (6)$$

Introducing the parameter $M = 0.3164 \cdot \left(\frac{1}{\frac{\pi D_h}{4} \cdot \mu} \right)^{-0.25} \cdot \frac{L}{D_h} \cdot \frac{1}{2}$ and the derivative of state variable $\dot{P}_2 = \frac{dP_2}{dt}$ in equation (4), we obtain equation (7).

$$\dot{P}_2 = \frac{a^2}{V_2} \cdot \left(-q_2\,(t) + q_2^{0.125}(t) \cdot \sqrt{\frac{\rho \cdot S^2}{M}} \cdot \sqrt{P_1(t) - P_2(t)} \right) \quad (7)$$

It should be noted that this model is only valid when P1>P2, otherwise, a flag could be raised.

4.2 *Final water cooling system model*

With additional constant parameters $b = \frac{a^2}{V}$, $c = \left(\frac{\rho \cdot S^2}{M}\right)^{\frac{1}{2}}$, the final model is expressed in equation (8).

$$\dot{P}_2 = b \cdot \left(-q_2(t) + c \cdot q_2^{0.125}(t) \cdot \sqrt{P_1(t) - P_2(t)} \right) \quad (8)$$

This equation will be applied to the chamber or the nozzle sections of the water cooling circuit with a measured mass flow either at outlet or inlet.

Journal of Physics: Conference Series **570** (2014) 072006 doi:10.1088/1742-6596/570/7/072006

5 Geometrical evaluation of equation parameters b, c and M

Geometrical data of each part of the water cooling circuit makes it possible to determine the numerical values for parameters c and b in equation (8). The obtained values are characteristic quantities of the given water circuit section supposed to be always constant.

The practical difficulty in their evaluation from their analytical expression is that the volumes of the cavities indicated in Section 4 cannot be precisely identified and depend also on the measurement location. The following evaluations constitute thus first guesses including measurement and modelling uncertainty: they will be verified on real data for nominal cases. For diagnosis purpose the reference value can also be identified through the method in Section 6.

Table 1. First evaluations of *b*, *c* and *M*

section	*b*	*c*	*M*
chamber	4.4 E-11	0.0045	0.12
nozzle	5E-11	0.0016	0.36

The above values are calculated with assumptions of constant values for the following physical data of water:

Table 2. Thermo physical data for water

Property	**value**	**unit**
Density, ρ	980	kg/m^3
Dynamic viscosity, μ	0.001	kg/m/s

6 Parameter identification algorithm

The model used in this section is derived from expression (8) with assumption of $\dot{P}_2 \sim 0$, and is thus applicable only in steady-state conditions.

$$q_2(t) = c \cdot q_2^{0.125}(t) \cdot \sqrt{P_1(t) - P_2(t)} \tag{9}$$

A recursive least-square identification algorithm is employed for estimating the value of *c* based on the measurements of q_2 , P_1 and P_2. Equation (9) can be written under the form $y = h \cdot c$ where $y = q_2(t)$ and $h = q_2^{0.125}(t) \cdot \sqrt{P_1(t) - P_2(t)}$.

The parameter is initialized with $c(0) = c_0$, where c_0 can be a geometrical guess from Section 5, and the parameter variance $v(0) = v_0$. The update at time step *k* is then performed as follows, to minimize $\|Y_N - H_N\|_2^2$ where Y_N and H_N are the sequences of *y* and *h* from time 0 to *N*.

$$\begin{aligned} \alpha(k) &= y(k) - h(k) \cdot c(k-1) \\ g(k) &= v(k-1)h(k)\big(\gamma + h(k) \cdot v(k-1) \cdot h(k)\big)^{-1} \\ c(k) &= c(k-1) + g(k) \cdot \alpha(k) \\ v(k) &= c(k-1) + g(k) \cdot \alpha(k) \end{aligned} \tag{10}$$

The parameter γ is an exponential forgetting factor: it allows giving more importance to recent data from sensors in the recursive scheme and accounting for unmodelled slow parameter variation. The residual considered can be the difference between *c(k)* and a known value c_{ref}, or the update residual $\alpha(k)$ [7]. Here, the first solution was retained (the second one is similar to the Kalman filter strategy developed in the next section).

7 Kalman filter fault diagnosis

The Kalman filter model allows to estimate at the same time the system characteristic parameters (just as seen with *b* and *c* in the model) and one or more state variables [5]. In the chosen example we decided to estimate the outlet (or inlet) pressure and the parameter $d = b \cdot c$. The inputs are the measured pressure and the mass flow (respectively inlet or outlet). There is no need to make assumptions on pressure derivatives to obtain the Kalman algorithm as needed to obtain equation (9) for the parameter identification approach. The filter is also efficient during transient phases. The state-space model can be written in discrete time form with time step k. The Kalman prediction is given in equation (11).

$$X(k+1) = f_d(U(k), X(k)) = X(k) + dt \cdot f(X(k), U(k)) \tag{11}$$

The input vector is $U = [q_2, P1]$ and the state variables vector is $X = [P2, d]$. The expression of $f(X(k), U(k))$ with $d = b \cdot c$ obtained from equation (8) is given in equation (12).

$$\begin{gathered} \dot{P}_2 = -bq_2\,(t) + d \cdot q_2^{0.125}(t) \cdot \sqrt{P_1(t) - P_2(t)} \\ \dot{d} = 0 \end{gathered} \tag{12}$$

Since this state-space model is nonlinear, an Extended Kalman filter (EKF) is considered (other extensions of Kalman filter, such as UKF, will be considered for future versions). The derivatives for equation (11) are $F_k = \left.\frac{\partial f_d}{\partial X}\right|_{X=X_k}$ and $L_k = \left.\frac{\partial f_d}{\partial U}\right|_{U=U_k}$, and the coefficients in equations (13) can be obtained from equations (11) and (12).

$$\begin{aligned} F_k(1,1) &= 1 + dt \cdot \left(\frac{X(k-1,2)}{2} \cdot U(k,1)^{0.125} \left(\frac{1}{U(k,2) - X(k-1,1)} \right)^{\frac{1}{2}} \right) \\ F_k(1,2) &= -dt \cdot \left(U(k,1)^{0.125} \sqrt{U(k,2) - X(k-1,1)} \right) \\ F_k(2,1) &= 0 \\ F_k(2,2) &= 1 \\ L_k(1,1) &= dt \cdot \left(-b + X(k-1,2) * 0.125 * U(k,1)^{0.125} \cdot \sqrt{U(k,2) - X(k-1,1)} \right) \\ L_k(1,2) &= -dt \cdot \left(\frac{X(k-1,2)}{2} \cdot U(k,1)^{0.125} \cdot \left(\frac{1}{U(k,2) - X(k-1,1)} \right)^{\frac{1}{2}} \right) \\ L_k(2,1) &= 0 \\ L_k(2,2) &= 0 \end{aligned} \tag{13}$$

The covariance matrix is predicted with the above coefficients as in equation (14):

$$Cov_{k|k-1} = F_{k-1}\, Cov_{k-1|k-1}\, F_{k-1}^T + L_{k-1}\, Q_{k-1}\, L_{k-1}^T + M_{k-1} \tag{14}$$

Where Q is the covariance matrix of the input noise for U and is based on the sensor noise and M is the covariance matrix of the state noise. Introducing the value $\widehat{Y_k}$ as the residual between the Kalman prediction and measured values Z_k (here is P1 or P2, the parameter d is only estimated, not measured), the residual obtained is given in equation (15).

$$\widehat{Y_k} = Z_k - C_k \hat{X}_{k|k-1} \tag{15}$$

The variance of the residual is expressed in equation (16) in terms of the state covariance matrix Cov and the measurement covariance matrix R_k (pressure measurement noise).

$$S_k = C_k Cov_{k|k-1} C_k^T + R_k \tag{16}$$

The Kalman filter gain expression is given in equation (17).

$$K_k = Cov_{k|k-1} C_k^T S_k^{-1} \tag{17}$$

The Kalman filter is used to update the X_k estimate and the covariance matrix Cov as described in equation (18).

$$\begin{aligned} \hat{X}_{k|k} &= \hat{X}_{k|k-1} + K_k \widehat{Y_k}\,, \\ Cov_{k|k} &= (I - K_k C_k) Cov_{k|k-1} \end{aligned} \tag{18}$$

The residual Y_k can be taken before or after the Kalman correction depending on the balance between the reliability of the model and of the available measurements. For this demonstration the residual is taken after Kalman correction at the end of each calculation step.

8 Residual analysis for decision

The outputs provided by the diagnosis algorithms are residuals that can be processed to raise detection flags. For the parameter identification approach, the residual is the difference between the estimated parameter *c* and its reference identified value. This value is constant during steady-state nominal operations (less than 1% noise as verified in the various firing tests used for validation). For the Kalman filter approach the value of the residual Y_k is directly the difference between the estimated state variable and its measured value. The diagnosis parameters are thus analyzed via the CUSUM test expressed in equations (19) [6],[14],[15] .

$$S_1(t) = \max(S_1(t-1) + r(t) - \mu_0 - \delta/2, 0)$$
$$S_2(t) = \max(S_2(t-1) - r(t) + \mu_0 - \delta/2, 0) \quad (19)$$

The parameter δ is the minimal size of the faulty variation to be detected. The decision rule is then expressed in equations (20).

if $(S_1 > \lambda)$ or $(S_2 > \lambda) \Rightarrow$ decide fault
else $\Rightarrow$ decide no fault
λ is a user threshold. (20)
$r(t)$ is the residual
μ_0= residual average

The value of δ can be set to the maximum allowable variation of the diagnostic parameter or the minimum detectable change. The value of threshold λ is based on the allowable detection delay.

9 Validation on real data

The developed tools have been at first validated using data from different firing campaigns of 2010 and 2014. The estimated value of parameter *c* has been compared to those obtained from test data and the maximum difference observed is lower than 10%. This parameter is observed to remain constant during nominal steady state operation regardless of the pressure and mass flow conditions in the system. It thus constitutes a viable reference parameter for the residual calculation.
Kalman estimation error for the pressure component proved to be less than 1% error.
In this section, one firing test is selected to present an example of application on a faulty behaviour. The algorithms are applied on the chamber section (V2 in Figure 2) and on the nozzle (section T in Figure 2). The monitored sensors are inlet /outlet pressure and mass-flow (P1, P2, q2).

9.1 ATAC campaign, May 14th 2014 results

During this test the MASCOTTE bench was fired with the bi-dimensional nozzle configuration ATAC. The test was nominally performed but some small water leakages appeared in the lateral cooling sections of the nozzle. This phenomenon was not detected in real time as the measurements still satisfied the larger redlines margins. Figure 6 and Figure 7 respectively show the behaviour of the water cooling for the chamber and the nozzle section. Some fluctuations on the outlet pressure signal of the nozzle can be observed while the behaviour on the chamber section is nominal.

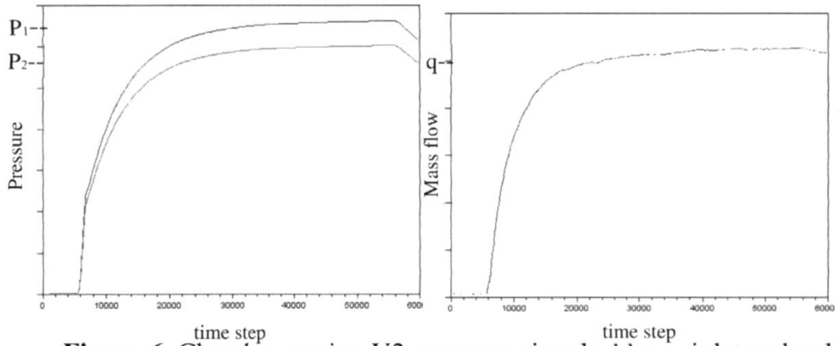

Figure 6. Chamber section V2: pressure signals, blue = inlet and red = outlet pressures, mass flow on the right

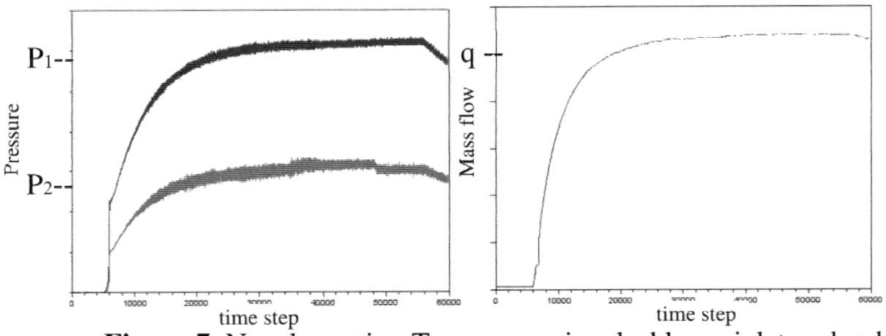

Figure 7. Nozzle section T: pressure signals, blue = inlet and red = outlet pressures, mass flow on the right

9.1.1 Parameter identification

The estimated value in steady state for the parameter c is 0.0042 for the chamber part and 0.0018 for the nozzle part. This result is close to the expected values of Section 5 and confirms the nominal behaviour of the system. On the nozzle parameter we can also identify a small fluctuation between 30000 and 50000 time steps.This oscillation is about 1% of the nominal parameter value c.

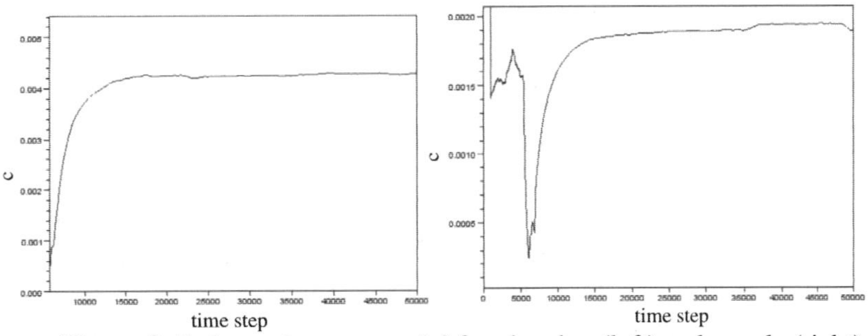

Figure 8. Estimated parameter 'c' for chamber (left) and nozzle (right)

The CUSUM test is set with δ=0.5% of c (that is 0.00001) and a threshold $\lambda = 3\delta$. The CUSUM and the detection flag for the nozzle part is shown below.No detection flag is raised for the chamber part, which allows isolation of the small fault on the nozzle.

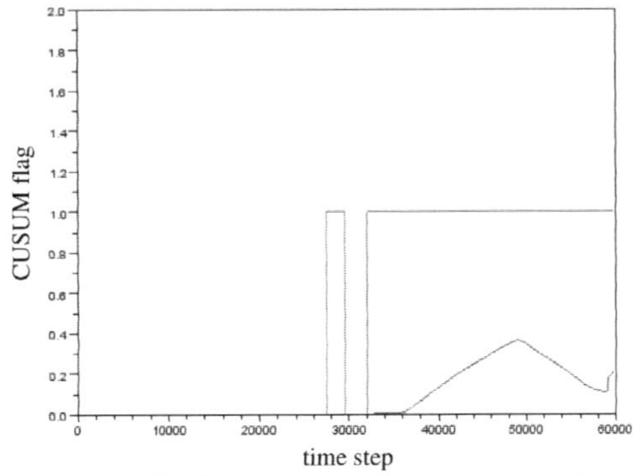

Figure 9. Nozzle CUSUM parameter (blue) and detection flag (red), δ =0.5% of c, λ= 3δ

Some other fluctuations are detected at around 25000 time steps but these cannot directly be associated to the pressure fluctuations. With less stringent CUSUM parameter such as δ=1% of c (δ=0.00002) and thresold λ=5δ it is possible to tune the test on the desired oscillations only.

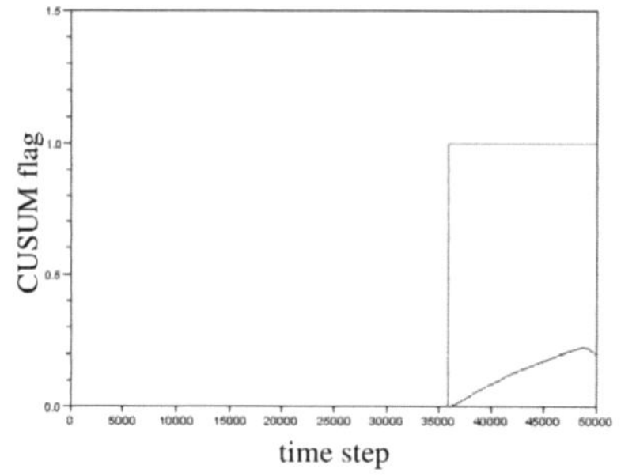

Figure 10. CUSUM parameter (blue) and detection flag (red), δ= 1% of c, $\lambda=5\delta$

The small variations of system variables, as those linked to the small water leakages, are visible on the identified parameter c but it is not possible to discriminate which variable produced them. The CUSUM test allows to raise a detection flag when the parameter has fluctuations and it needs to be tuned to the c parameter fluctuations, not those of the variables.

9.1.2 Kalman filter

The Kalman filter is able to follow very efficiently the pressure signal. The residual at each step is negligible.

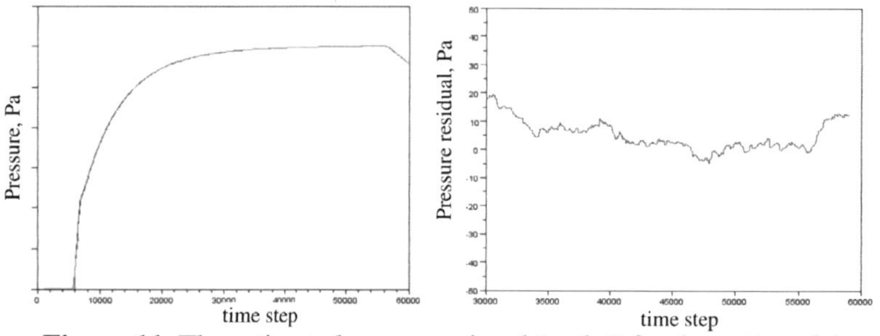

Figure 11. The estimated pressure signal (outlet) for the outlet of the chamber section and the measurement overlap (image on the left). The residual for the selected time frame of the faults (right) is negligible less than 0.1 % of the value.

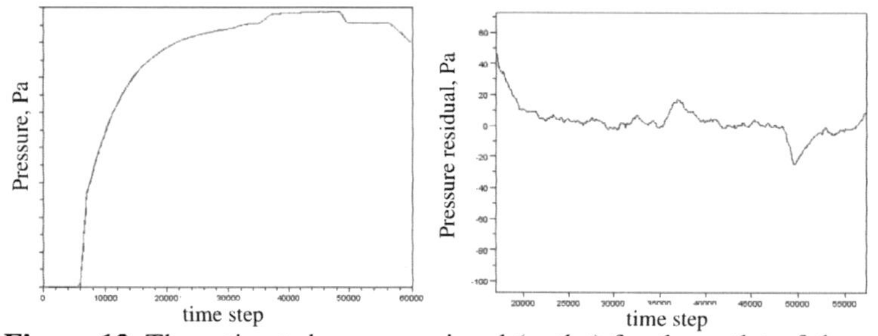

Figure 12. The estimated pressure signal (outlet) for the outlet of the nozzle section and the real measurement overlaps (image on the left). On the right is the residual for the selected time frame of the faults.

The residual curve for the nozzle part (Figure 12) shows two small variations at the time of the pressure measurement fluctuations: they correspond to the Kalman correction to follow the new trend of the pressure. The CUSUM test for the nozzle part is run during the steady state with δ=10 corresponding to the order of magnitude of the residual in the steady state nominal operation and the maximum allowed threshold is set to $\lambda = 3\delta$. A flag is raised during the transient phase of the fluctuations: when the pressure slightly increases and when it slightly decreases to set to its expected nominal value.

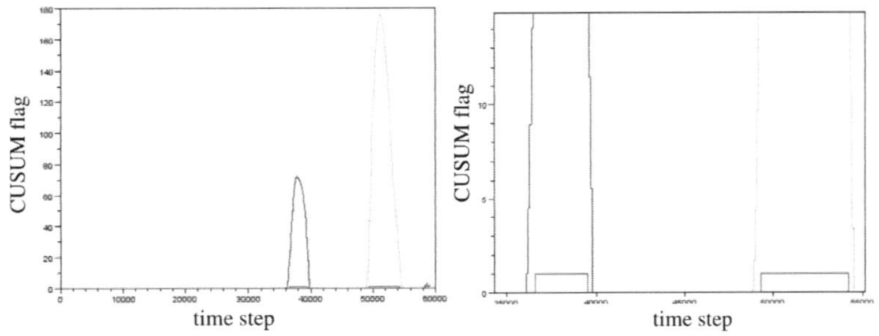

Figure 13. CUSUM flag for the outlet pressure measurement for the overall firing time frame on the left, and zoom for the time frame of the pressure fluctuations.

10 Synthesis of results and comparison with redlines

The aim of this work was to build up a set of tools to enhance currently used monitoring techniques for rocket engine and associated demonstrators. The results obtained so far showed multiple aspects for future developments:

- the parameter identification method coupled with the simplified model has been validated on the real data as the expected numerical values for the estimated parameter c has been obtained for the chamber and nozzle section during different test run;
- the Kalman filter approach allowed to produce a good estimate of the pressure measurements;
- the two algorithms are very simple to use and they are generic and applicable without modifications for different test run with different operating points;
- the CUSUM test provides a good tool for detection; it may require tuning on non-physical algorithm parameters such as c in case of parameter identification, or it may be based on the Kalman innovation variance. Via the CUSUM it was possible to detect small faults either on the identified parameter c or on the Kalman estimate of the pressure. The main difference in the two approaches is that in case of detection with the c parameter the fault cannot be associated to one measurement.

Some advantages of these technics compared to the classical redlines are the following:

- Parameter identification approach:
  - only one redline on parameter c is needed, compared to the three redlines for inlet/outlet pressures and mass flow monitoring,
  - the estimated parameter and its threshold do not depend on the operating point but only on geometrical data of the system and are always the same for different test runs. This is an important advantage over redlines which have to be set for each test run and can lead to human errors and unjustified tests stops.
- Kalman filtering approach:
  - the filter allows to estimate one or multiple variables and allows to detect unwanted changes;
  - coupled with the CUSUM test it can help detecting small variations into a measurement;
  - it can be used to validate a measurement.

11 Conclusions

This work is a first step towards an automatic diagnosis system for the MASCOTTE rocket engine demonstrator: the first step is to set up efficient alternatives to redlines for detection and diagnosis. This work will be completed with the improvement of the water circuit model, the analysis of other statistical tests for detection and with the application of the algorithms to other branches of the MASCOTTE test bench.

References

[1] S.X. Ding, „Model-based fault diagnosis techniques“, Springer Verlag, Berlin Heidelberg, 2008
[2] M. Blanke, M. Kinnaert, J. Lunze, M. Staroswiecki, „Diagnosis and Fault-Tolerant Control“, Springer Verlag, Berlin Heidelberg, 2003
[3] V. Venkatasubramanian, R. Rengaswamy, K. Yin, S.N. Kavuri,"A review of process fault detection and diagnosis: Part I: Quantitative model-based methods". Computers & chemical engineering, 27(3), 293-311, 2003
[4] R. Isermann,"Process fault detection based on modeling and estimation methods—a survey". Automatica, 20(4), 387-404, 1984
[5] E. Chow, A.S. Willsky,"Analytical redundancy and the design of robust failure detection systems". IEEE Transactions on Automatic Control, 29(7), 603-614, 1984
[6] J. Marzat, E. Walter, H. Piet-Lahanier, F. Damongeot, "Automatic tuning via kriging-based optimization of methods for fault detection and isolation", 2010 Conference on control and fault tolerant systems, Nice, France, October 6-8 2010
[7] A. Iannetti, "Overview on European efforts on health monitoring/management systems for rocket engines", Space Propulsion Conference, Köln, Germany, 2014
[8] S. Benoit, P.Bonert, S. Legonidec, P. Supié, „A diagnostic demonstrator: a platform for the evaluation of real time diagnostic data dedicated to space engines“, MFPT 2009
[9] H. Cicanek, "Space Shuttle main engine failure detection", American control conference, Boston MA, June 19-21 1985
[10] J. Wu, "Liquid propellant rocket engines health monitoring- a survey", ACTA Astronautica 56 (2005) 347-356
[11] IAEA Nuclear Energy Series, "On line Monitoring for improving Performance of Nuclear power plants, Part 2: Process and component condition monitoring and diagnostics", No. NP-T-1.2, 2013
[12] G. Ordonneau, P. Hervat, L. Vingert, S. Petitot, B. Pouffary, "First results of heat transfer measurements in a new water-cooled combustor on the Mascotte facility", 4th European conference for aerospace sciences (EUCASS), 2013
[13] I.E. Idel'cik,"Mémento des pertes de charge",3rd edition, éditions Eyrolles, 1986
[14] M. Basseville and I. Nikiforov, "Detection of abrupt changes: theory and application", Prentice Hall, Englewood Cliffs NJ, 1993
[15] J. Marzat, H. Piet-Lahanier, F. Damongeot, E. Walter, "Model based fault diagnosis for aerospace systems: a survey", Proceedings of the Institution of Mechanical Engineers, Part G: Journal of aerospace engineering, 226(10), 1329-1360, 2012
[16] R. Mehra, C. Rago, S. Seereeram, "Autonomous failure detection, identification and fault tolerant estimation with aerospace applications", IEEE Aerospace Conference, 2, 33-138, 1998
[17] C. Meyer, J. Zakrajsek, "Rocket engine failure detection using system identification techniques", AIAA-90-1993
[18] E. Cliquet, A. Iannetti, J. Masse, "Carmen, liquid propulsion systems simulation platform", EUCASS 2009

European Workshop on Advanced Control and Diagnosis IOP Publishing
Journal of Physics: Conference Series **570** (2014) 072007 doi:10.1088/1742-6596/570/7/072007

Diagnosis of an actuated seat in a constrained space via Takagi-Sugeno sliding mode observers

Kamel Bouibed[a], Lynda Seddiki[a], Kevin Guelton[b] and Herman Akdag[a]

[a]LIASD, EA 4383, University of Paris 8, Saint-Denis, France
[b]CReSTIC, EA 3804, University of Reims Champagne-Ardenne, Reims, France

E-mail: bouibed@ai.univ-paris8.fr

Abstract. This paper deals with actuator faults detection and isolation for an actuated seat described by Takagi-Sugeno multiple models. The goal is to ensure the comfort and the security of the users in simulator applications. Sliding mode observers based on T-S models are designed to estimate the system state vector. Residuals are generated by the comparison of measured and estimated outputs. In this work, a multi-observers technique is used. It consists of the construction of many observers such that each observer must be robust to noises and to other uncertainties but sensitive to one actuator fault. Simultaneous faults occurring on the actuated seat can be detected and isolated using this method. Simulation results are given to show the effectiveness of this approach.

1. Introduction

Nowadays, multimedia space simulators (flight simulator, automotive, video games ...) overgrown the field of new technologies. Users spend several hours a day on their chairs in front of the screens. For simulators, the user is in general installed on an articulated mobile platform equipped with an instrumented seat, a joystick and one or more wide-screen. For example, to be an efficient racing simulator, the user has to feel every sensation as in real situations, such as acceleration, braking or jumps.
In this study, which is part of an industrial project [1], we are interested in the diagnostic of the seat for such applications. Indeed, the seat is instrumented in a confined space and it has to ensure the comfort and the security of the users. Such systems integrate actuators, sensors and embedded intelligent systems (control, optimization, supervision, ...) whose aim is to increase their performances. This context is motivating the present diagnostic study with application on an actuated seat which can be seen as a robot with tree structure. Due to high reliability demands of such systems, recent supervision and fault diagnosis concepts are of particular importance [16]. There are three main approaches to deal with fault diagnosis [15], [12], [19], [10], [2]:

- Signal analysis-based approach,
- Knowledge-based approach,

[1] Confidentiality has been contracted with the industrial company, therefore, its name and the studied application can not be provided in this paper. All the parts of this contract has read this paper and agreed to its publication.

- Model-based approach.

Model-based approaches to fault diagnosis in dynamic processes have received considerable attention since the beginning of the 70's, both in the research context and in the domain of application on real processes [16], [9], [14], [11]. These model-based fault diagnosis systems are structured on two levels:

- Residual generation: Its purpose is to generate a residual signal which is sensible to a fault.
- Decision making: The residuals are examined for the likelihood of a fault and a decision rule is then applied to determine if some fault appears.

Most of model-based fault diagnosis methods are based on linear system models. For nonlinear dynamic systems, the fault diagnosis problem has been traditionally analyzed in two steps: First, the model is linearized around an operating point, and second, some techniques are applied to generate signal-residual, e.g. Kalman filter, observers, parity relations, parameter estimation [12]. Among nonlinear observers, T-S fuzzy model-based approaches have become popular, because such models provide universal approximations of nonlinear systems. Therefore in the past few decades, T-S fuzzy models have been the subject of many theoretical studies (e.g. [24] [22]) and applications (e.g. [13], [20], [21]). The major interest of such modelling approaches is that they allow extending some linear observer or controller design methodologies to nonlinear systems. In this work, we consider the model-based approach and the technique used in [1], in order to deal with the fault diagnosis problem. The strategy is based on obtaining a Takagi-Sugeno (T-S) fuzzy model and then using sliding mode observers to estimate the system state vector; the residual signal is then generated by the comparison of the measured and the estimated outputs. Observers based diagnosis is a technique that has been much discussed in the literature [17], [4], [25], [10], [3]. Without the occurrence of a faults, the residuals are close to zero while they deviate significantly from zero upon the occurrence of a faults on the system. The detection of the faults is generally quite easy; however, their localization is more delicate. Therefore, one frequently uses multi-observers to generate the residuals, whose are analysed through logical decision rules to find the location of the faults.
The paper is organized as follows: In section 2, the dynamical model of the seat and the corresponding Takagi-Sugeno fuzzy model are described. In section 3, we design a fuzzy sliding mode observer for the actuator fault detection and isolation. In section 4, simulation results show the effectiveness of the approach.

2. Description and modelling of the actuated seat

In this paper we will consider the diagnostic of an empty actuated seat, depicted in figure 1. That is to say, without considering the interaction with the user, which can be considered as modelling uncertainties and external disturbances. Therefore, in this preliminary study, one assumes that the robustness of the following designed observers will handle such uncertainties. The actuated seat contains three actuated joints r_1, θ_3, θ_4 and one unactuated joint θ_2, whose significations are given in Table 1 (r_1 is a prismatic joint, θ_i (j = 2,3,4) are rotary joints).

To synthesize the relevant observers, dynamical model of the seat is required. It is given as the following second order mechanical system :

$$J\left(q(t)\right)\ddot{q}(t) + D(q(t), \dot{q}(t))\dot{q}(t) + G(q(t)) = u(t) \tag{1}$$

where $q(t) = [r_1\ \theta_3\ \theta_4]^T$ is the vector of generalized coordinates, $J(q(t))$ is the inertia matrix (symmetric and positive definite), $D(q(t), \dot{q}(t))$ is the matrix of Coriolis and centrifugal forces, $G(q(t))$ is the vector of gravitation and $u(t)$ is the vector of the generalized forces produced by

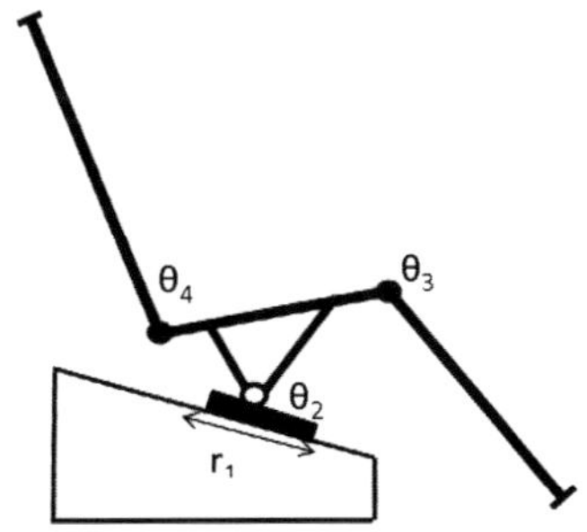

Figure 1. Different joints of the seat.

Table 1. Joints significations

r_1	Tracking position expressed in m
θ_2	Angle of the seating expressed in rad
θ_3	Angle of the legrest expressed in rad
θ_4	Angle of the seat-back expressed in rad

the actuators. The control input u is assumed to be given by some known feedback controllers. For lightening the mathematical expressions, the time t will be committed in the sequel when there is no ambiguity. The matrices of the model, obtained from the Lagrange formalism [18], are detailed as follows:

$$J(q) = \begin{bmatrix} M & (*) & (*) \\ m_3 L_3 \sin(\beta - \theta_{23}) & J_3 & 0 \\ -m_4 L_6 \sin(\beta - \theta_{24}) & 0 & J_4 \end{bmatrix}$$

$$D(q,\dot{q}) = \begin{bmatrix} 0 & -m_4 L_3 \cos(\beta - \theta_{23})\dot{\theta}_3 & m_6 l_6 \cos(\beta - \theta_{24})\dot{\theta}_4 \\ 0 & 0 & 0 \\ 0 & 0 & 0 \end{bmatrix}$$

$$G(q) = \begin{bmatrix} -Mg\cos\beta \\ -(gm_3 \frac{L_3}{2})\sin(\theta_{23}) \\ -(gm_4 \frac{L_4}{2})\sin(\theta_{24}) \end{bmatrix}$$

where $\theta_{23} = \theta_2 + \theta_3$; $\theta_{24} = \theta_2 + \theta_4$; M is the total mass of the seat; m_3 and m_4 are respectively the masses of the footrest and the headrest of the seat; L_3 and L_4 are respectively the lengths of the footrest and the headrest of the seat; β is the angle between the seat tracking axis and the horizontal plane; g is the gravitational acceleration.

Note that θ_2 is an unactuated joint, which is not considered as a generalized coordinate since it is geometrically linked to the prismatic joint r_1 by means of a guide (constrained regarding to the mechanical design of the seat). Therefore, the nonlinear relation between θ_2 and r_1 has been approximated from experimental measurements (see Figure 2) as a third order polynomial given by:

$$\theta_2 = ar_1^3 + br_1^2 + cr_1 + e \tag{2}$$

where $a = -36.54$, $b = 1.079$, $c = 1.49$ and $e = 0.1$.

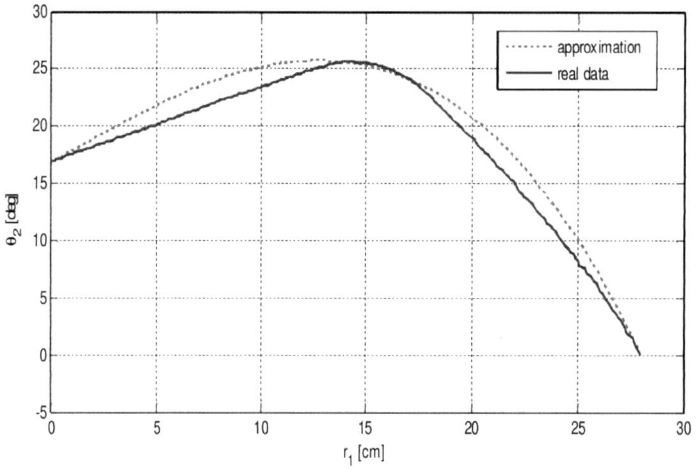

Figure 2. Approximation of θ_2 by a third order polynomial

2.1. State space dynamical model of the seat and Takagi-Sugeno modelling

Let us consider the state vector $x = [q\ \dot{q}]^T$ and the control input $u = [u_1 u_2 u_3]^T$, the model (1) can be rewritten as the following nonlinear state space model:

$$\begin{cases} \dot{x}(t) = A(q,\dot{q})x(t) + B(q)u(t) + Fd(t) \\ y(t) = Cx(t) \end{cases} \tag{3}$$

with

$$A(q,\dot{q}) = \begin{bmatrix} 0_3 & I_3 \\ -J(q)^{-1}\bar{G}(q) & -J^{-1}(q)D(q,\dot{q}) \end{bmatrix}$$

$$B = \begin{bmatrix} 0_3 \\ -J(q)^{-1} \end{bmatrix}$$

where $\bar{G}(q) \in R^{3\times3}$ is defined such that $G(q) = \bar{G}(q)q$; F and C are constant matrices of appropriate dimensions and $d(t)$ is an unknown input vector containing external disturbance or faults in the system.

Let us notice that all the nonlinear terms included in the matrices of the state space model (3) are bounded (since r_1, θ_3 and θ_4 are physically bounded and chosen such that they do not cross zero) and given by: $f_{J_1} = \sin(\beta - \theta_{23})$, $f_{J_2} = \sin(\beta - \theta_{24})$, $f_{G_1} = \frac{1}{r_1}$, $f_{G_2} = \frac{\sin(\theta_{23})}{\theta_3}$, $f_{G_3} = \frac{\sin(\theta_{24})}{\theta_4}$, $f_{D_1} = \cos(\beta - \theta_{23})\dot{\theta}_3$ and $f_{D_2} = \cos(\beta - \theta_{24})\dot{\theta}_4$. Therefore, a Takagi-Sugeno model

(4), which exactly matches the nonlinear model (3), is obtained by applying the well-known sector nonlinearity approach [23].

$$\dot{x}(t) = \sum_{i=1}^{r} h_i(z(t)) \left(A_i x(t) + B_i u(t) + F d(t)\right) \tag{4}$$

where $z(t)$ is the premise vector assumed to depend only on the state variables, r is the number of fuzzy sets for the right-hand side of the system (4), respectively; $h_i(z(t)) \geq 0$ are the membership functions that satisfy the convex sum property: $\sum_{i=1}^{r} h_i(z(t)) = 1$.

Note that, the above described T-S model contains 64 rules, therefore the matrices constituting the local subsystems and the membership functions are not detailed here for space reason. The reader may refer to [23] for more details on how to apply the well-known sector nonlinearity approach. Nevertheless, to illustrate the effectiveness of proposed T-S modelling approach, simulations of the seat have been performed in Matlab/Simulink to show that the T-S fuzzy model of the seat represents perfectly the nonlinear model, see Figures 3, 4 and 5.

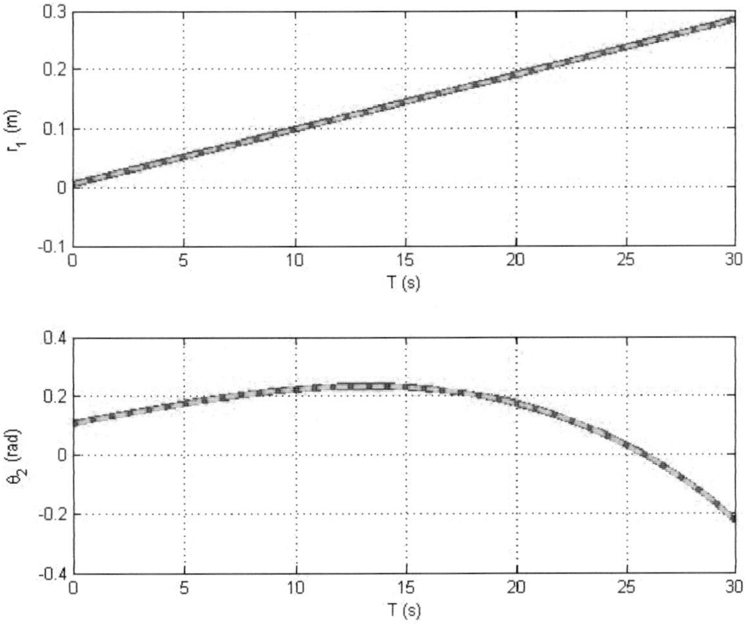

Figure 3. Comparison of the nonlinear model of the seat and its T-S model: tracking and seating position. – – – T-S Model, —— Nonlinear model

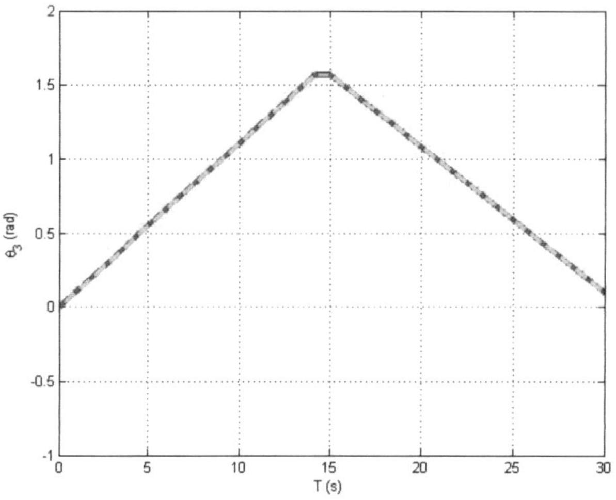

Figure 4. Comparison of the nonlinear model of the seat and its T-S model: legrest position. – – – T-S Model, —— Nonlinear model

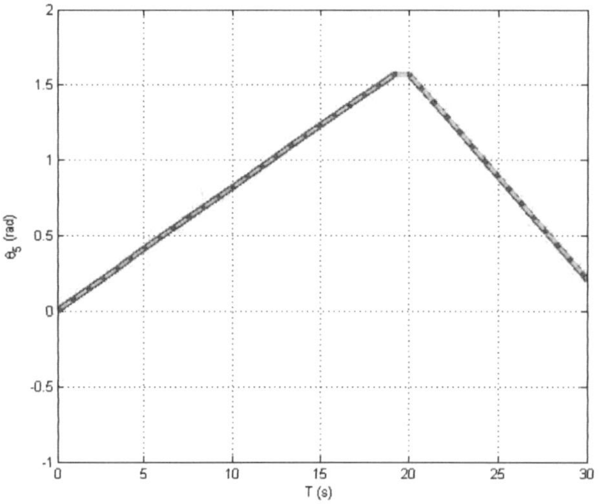

Figure 5. Comparison of the nonlinear model of the seat and its T-S model: seat-back position. – – – T-S Model, —— Nonlinear model

3. Synthesis of the sliding mode observer

The goal is now to design convenient observers dedicated to the diagnosis of the seat actuators faults. From the model-based approach, it is necessary to build the mathematical model of the

system that is to be observed. In the general case of nonlinear dynamic systems, the design of observers is not an easy task. Therefore, in this work we consider the Takagi-Sugeno fuzzy model described in (4) and derived from (3).
It is assumed that the T-S fuzzy model (4) is locally observable and our goal is to design an observer such that $x(t) - \hat{x}(t) \rightarrow 0$ as $t \rightarrow \infty$, where $\hat{x}(t)$ denotes the estimated state vector. In this work, one considers an adaptation of the fuzzy sliding mode observer design methodology proposed in [8] to the T-S model of the actuated seat. This method uses a number of local linear time-invariant observers; each local model represented by (4) is associated with a local observer given as follows:

$$\begin{cases} \dot{\hat{x}}(t) = A_i\hat{x}(t) + B_iu(t) + F_id(t) + L_i\left(y(t) - \hat{y}(t)\right) + \phi_i(t) \\ \hat{y}(t) = C_i\hat{x}(t) \end{cases} \tag{5}$$

where L_i is the observer gain and ϕ_i is the discontinuous vector of sliding mode which will be defined later.
By the use of the Parallel Distributed Compensation (PDC) technique [9], the total state estimation is the combination of local observer outputs.

$$\begin{cases} \dot{\hat{x}}(t) = \sum_{i=1}^{r} h_i(z(t))\left\{A_i\hat{x}(t) + B_iu(t) + Fd(t) + L_i\left(y(t) - \hat{y}(t)\right) + \phi_i(t)\right\} \\ \hat{y}(t) = \sum_{i=1}^{r} h_i(z(t))C_i\hat{x}(t) \end{cases} \tag{6}$$

To analyze the convergence of this observer, the state estimation error $e(t) = x(t) - \hat{x}(t)$ dynamics is examined:

$$\dot{e}(t) = \dot{x}(t) - \dot{\hat{x}}(t) \tag{7}$$

Using (4) and (6), one can find:

$$\dot{e}(t) = \sum_{i=1}^{r}\sum_{j=1}^{r} h_i(z(t))h_j(z(t))\left[\bar{A}_{ij}e(t) + F_id(t) - \phi_i(t)\right] \tag{8}$$

where $\bar{A}_{ij} = A_i - L_iC_j$.
One chooses the following Lyapunov function:

$$v(t) = e^T(t)Pe(t) \tag{9}$$

Deriving:

$$\dot{v}(t) = e^T(t)P\dot{e}(t) + \dot{e}^T(t)Pe(t) \tag{10}$$

Using (8), one finds:

$$\begin{aligned} \dot{v} = e^TP\textstyle\sum_{i=1}^{r}\sum_{j=1}^{r} h_i(z(t))h_j(z(t))\left(\bar{A}_{ij}e + F_id - \phi_i\right) + \\ \textstyle\sum_{i=1}^{r}\sum_{j=1}^{r} h_i(z(t))h_j(z(t))\left(\bar{A}_{ij}e + F_id - \phi_i\right)^T Pe \\ = \textstyle\sum_{i=1}^{r}\sum_{j=1}^{r} h_i(z(t))h_j(z(t))\left\{e^T(P\bar{A}_{ij} + \bar{A}_{ij}^TP)e + 2e^TPF_id - 2e^TP\phi\right\} \end{aligned} \tag{11}$$

Then:

$$\dot{v}(t) \leq -\gamma\left\|e(t)\right\|^2 + \sum_{i=1}^{r}\left\{2\left\|e^T(t)P\right\|\left\|F_id(t)\right\| - 2e^T(t)P\phi_i(t)\right\} \tag{12}$$

Journal of Physics: Conference Series **570** (2014) 072007 doi:10.1088/1742-6596/570/7/072007

In order to have $\dot{v}(t) < 0$, we propose that the discontinuous term has the following form:

$$\phi_i(t) = k_i \text{sign}\left(e^T(t)P\right) = k_i \frac{e^T(t)P}{\|e^T(t)P\|} \tag{13}$$

where $k_i > 0$ is constant and $P > 0$ so that the following Lyapunov inequality is satisfied:

$$P\bar{A}_{ij} + \bar{A}_{ij}^T P < 0 \tag{14}$$

From this, $lim_{t\to\infty}e(t) = 0$ if (12) and (13) are respected, therefore if:

$$2\left\|e^T(t)P\right\| \sum h_i(z(t)) \left\{\|F_i d(t)\| - k_i\right\} < 0 \tag{15}$$

then one obtains the following condition:

$$k_i > \|F_i d(t)\| \tag{16}$$

Furthermore, it should be mentioned that an observer with converging state estimation is considered as a stable observer. Nevertheless, its stability may be checked, for valisation purpose, from a well-known bounded real lemma (stability conditions) [23]. These conditions come from the inequalities (14), which can be efficiently solved via convex optimization algorithms within the linear matrix inequalities (LMI) framework [7]. Moreover, from the solution of such LMIs, the observer gains L_i can be obtained as in [1]. Hence, if a positive definite matrix P exists, the sliding mode fuzzy observer (6) is defined.

Finally, once the state or output is estimated, the residual signals are generated by the comparison of the measured and estimated output.

$$r(t) = y(t) - \hat{y}(t) \tag{17}$$

Therefore, analysing the residuals, one may detect and isolate actuator faults. Indeed, to detect and isolate the actuator faults, a set of three observers is used such that each observer is sensitive to one actuator fault [5], [6]. So the ith observer will have as inputs y and $\bar{u}_i$ which is the vector of inputs without the ith input u_i. This structure is presented in Figure 6 for the actuator fault detection and isolation. In the considered seat, a set of three observers for each actuator is sufficient. Each observer generates one residual which is sensitive to only one actuator fault.

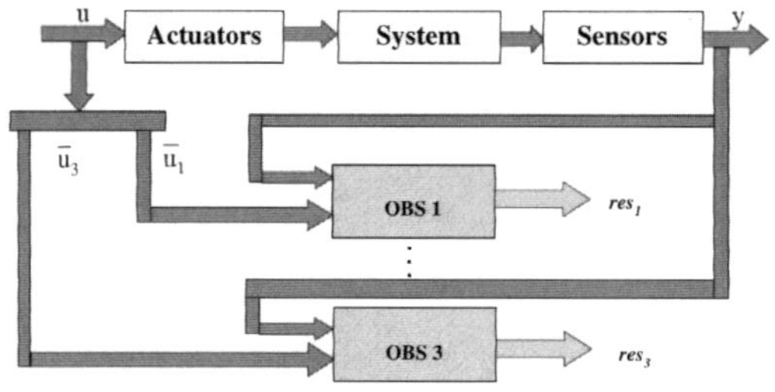

Figure 6. Multi-observers for actuator fault detection and isolation.

4. Simulation results

Simulations trials have been performed in the absence of faults, thresholds can be set on the basis of the maximum absolute values of the residuals . If some faults occur, they will cause deviations from zero of some residuals much greater than the differences caused by uncertainties and measurement noises. Thresholds should be chosen correctly in view to reduce the false alarms and the non detections. In our case, we assume that uncertainties and measurements noises do not have a great influence on the residuals. For our simulations, thresholds are chosen as $+/-0.5$ meters for the residual res_1, related to the prismatic joint r_1, and $+/-0.5$ radians for the residuals res_2 and res_3, respectively related to the rotary joints θ_3 and θ_4.

To show the effectiveness of the approach, some scenarios for some actuator faults are simulated using Matlab/Simulink. Figure 7 shows that the first residual res_1 is sensitive to the prismatic joint actuator fault where the other residuals are not sensitive. Figure 8 shows that the second residual res_2 is sensitive to the seat legrest rotary joint actuator fault while the others are not sensitive to this fault. Residuals res_3, res_4 and res_5 are sensitive to the seat footrest, the seat-back and the seat headrest actuator faults respectively. Figure 9 shows that we can detect the seat-back and the tracking actuator faults simultaneously.

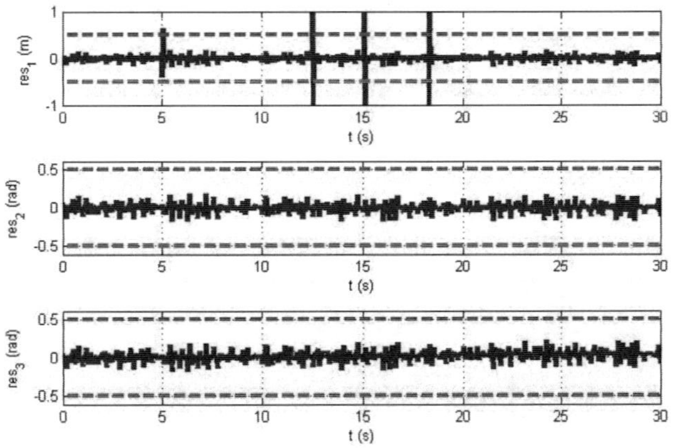

Figure 7. Detection of the tracking actuator fault.
– – – thresholds, —— residuals

Figure 7 shows the sensitivity of the first residual to the actuator of the first joint. This fault consists of blocking of the actuator several times at the instants 5s, 12.5s, 15s and 18.2s respectively. Similar fault is simulated for the actuator of the rotary joint 2 and detected by the second residual as it is shown by the Figure 8. Finally, simultaneous faults are simulated for the first and the third actuator and detected by the two residuals res_1 and res_3 as is shown in figure 9.

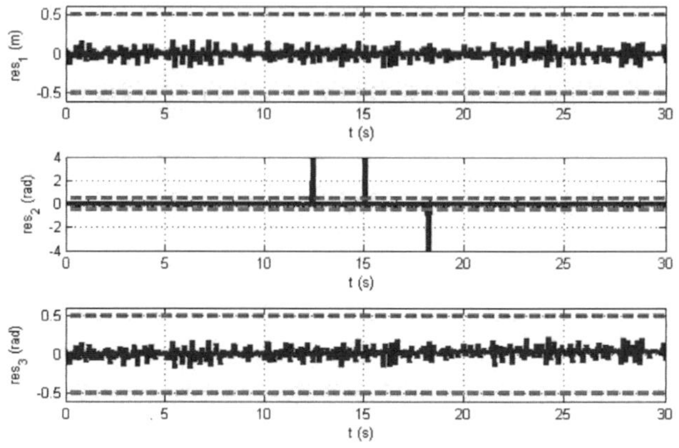

Figure 8. Detection of the legrest actuator fault.
– – – thresholds, —— residuals

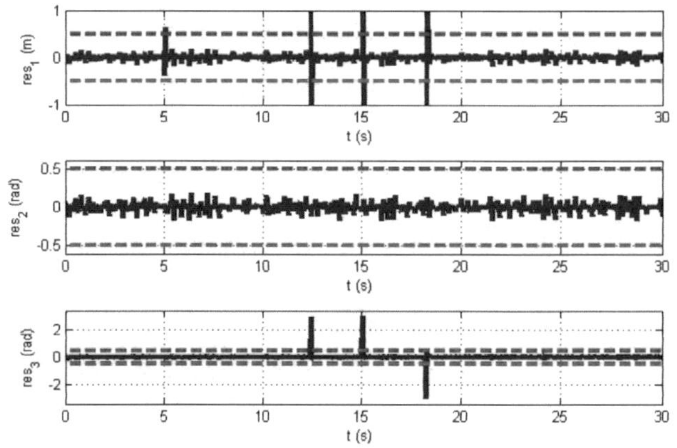

Figure 9. Detection of simultaneous faults : seat-back and tracking actuator. – – – thresholds, —— residuals

5. Conclusion

In this paper an analysis for fuzzy sliding mode observer based FDI applied to a Takagi-Sugeno fuzzy model of an actuated seat has been presented. A multi-observers approach is used to generate residuals that allow us to detect and to isolate actuator faults. In the actuated seat, considered as a robot with tree structure, it is important to detect the faulty actuator and then to reconfigure the control law. The effectiveness of the approach allows avoiding false alarms and non detections. Because of decentralization of actuators, we notice that each residual is sensitive to only one actuator fault. These residuals are presented as fault signature. Five observers are used and each of them is dedicated to only one actuator. Therefore, five residuals are obtained and each of them is sensitive to only one actuator fault. Simulation results show that each residual generated by the sliding mode observers is sensitive to only one actuator fault. All real systems have uncertain characteristics, and cannot be modelled perfectly. Therefore, one has considered some bounded modelling uncertainties. In the simulation results we can observe that robustness is achieved under parametric variations of the system. Residual only depend of external inputs or fault in the system. Therefore, the sliding mode technique provides a way to find a possible solution to this problem.

References

[1] E. Alcorta-Garcia and S. Saucedo-Flores. Fault detection and isolation based on takagi-sugeno modelling. In *IEEE International Symposium on Intelligent Control*, pages 673–678, 2003.
[2] M. Blanke, M. Kinnaert, J. Lunze, and M. Staroswiecki. *Diagnosis and Fault-Tolerant Control*. Springer, 2006.
[3] K. Bouibed, A. Aitouche, and M. Bayart. Sensor fault detection by sliding mode observer applied to an autonomous vehicle. In *International Conference on Advances in Computational Tools for Engineering Applications*, pages 621–626, Zouk Mosbeh, Lebanon, July 15-17 2009.
[4] K. Bouibed, A. Aitouche, and M. Bayart. Sensor and actuator fault detection and isolation using two model based approaches: Application to an autonomous electric vehicle. In *IEEE Mediterranean Conference on Control and Automation*, pages 1290 – 1295, Marrakech, Morocco, June 23-25 2010.
[5] K. Bouibed, L. Seddiki, K. Guelton, and H. Akdag. Actuator and sensor fault detection and isolation of an actuated seat via nonlinear multi-observers. *Systems Science & Control Engineering*, 2(1):150–160, 2014.
[6] K. Bouibed, L. Seddiki, K. Guelton, and H. Akdag. Actuator fault detection by nonlinear sliding mode observers: application to an actuated seat. 2nd International Confrence on Control and Fault Tolerant Control,739–743 , 2013.
[7] S.P. Boyd, L. El Ghaoui, E. Feron, and V. Balakrishnan. *Linear matrix inequalities in system and control theory*, volume 15. SIAM, 1994.
[8] B. Castillo-Toledo, and J. Anzurez-Marin. Model-based fault diagnosis using sliding mode observers to takagi-sugeno fuzzy model. In *IEEE International Symposium on, Mediterrean Conference on Control and Automation*, pages 652–657, Limassol, Cyprus, June 27-29 2005.
[9] J. Chen and R.J. Patton. *Robust model-based fault diagnosis for dynamic systems.* Kluwer Academic Publishers, Norwell, MA, USA, 1999.
[10] P.M. Frank, and X. Ding. Survey of robust residual generation and evaluation methods in observer-based fault detection systems. *Journal of Process Control*, 7(6):403 – 424, 1997.
[11] P. Gerland, D. Gros, H. Schulte, and A. Kroll. Robust adaptive fault detection using global state information and application to mobile working machines. In *Conference on Control and Fault-Tolerant Systems (SysTol)*, pages 813–818, 2010.
[12] J. Gertler. *Fault detection and diagnosis in engineering systems.* CRC, New York, 1998.
[13] K. Guelton, S. Delprat, and T.M. Guerra. An alternative to inverse dynamics joint torques estimation in human stance based on a takagi–sugeno unknown-inputs observer in the descriptor form. *Control Engineering Practice*, 16(12):1414–1426, 2008.
[14] D. Henry, S. Simani, and R.J. Patton. Fault detection and diagnosis for aeronautic and aerospace missions. In *Fault tolerant flight control*, pages 91–128. Springer, 2010.
[15] R. Isermann. Model-based fault-detection and diagnosis - status and applications. *Annual Reviews in Control*, 29(1):71 – 85, 2005.
[16] R. Isermann. *Fault-Diagnosis Applications: Model-Based Condition Monitoring: Actuators, drives, machinery, plants, sensors, and fault-tolerant systems.* Springerverlag Berlin Heidelberg, 2011.
[17] B. Jiang, and F.N. Chowdhury. Parameter fault detection and estimation of a class of nonlinear systems using observers. *Journal of the Franklin Institute*, 342(7):725 – 736, 2005.
[18] W. Khalil, and E. Dombre. *Modeling, identification and control of robots.* Butterworth-Heinemann, 2004.
[19] R.J. Patton, P.M. Frank, and R.N. Clarke. *Fault diagnosis in dynamic systems: theory and application.* Prentice-Hall, Inderscience Publishers, 1989.
[20] H. Schulte and K. Guelton. Descriptor modelling towards control of a two link pneumatic robot manipulator: At–s multimodel approach. *Nonlinear Analysis: Hybrid Systems*, 3(2):124–132, 2009.
[21] L. Seddiki, K. Guelton, and J. Zaytoon. Concept and takagi-sugeno descriptor tracking controller design of a

closed muscular chain lower-limb rehabilitation device. *IET Control Theory and Applications*, 4(8):1407–1420, 2010.

[22] K. Tanaka, T. Ikeda, and H.O. Wang. Fuzzy regulators and fuzzy observers: relaxed stability conditions and lmi-based designs. *Fuzzy Systems, IEEE Transactions on*, 6(2):250–265, 1998.

[23] K. Tanaka and H.O. Wang. *Fuzzy control systems design and analysis: a linear matrix inequality approach.* John Wiley & Sons, 2001.

[24] L. Xiaodong and Z. Qingling. New approaches to H_∞ controller designs based on fuzzy observers for ts fuzzy systems via lmi. *Automatica*, 39(9):1571–1582, 2003.

[25] M.H. Yang, S.C. Guo, Z.R. Xing, Y. Li, and J.Q. Qiu. Actuator fault detection and isolation for robot manipulators with the adaptive observer. *Advanced Materials Research*, 482:529–532, 2012.

Physically-based modeling of speed sensors for fault diagnosis and fault tolerant control in wind turbines

Wolfgang Weber[1], Jonas Jungjohann[2], Horst Schulte[2]

[1] Pepperl+Fuchs GmbH, 68307 Mannheim, Germany
[2] HTW Berlin, Department of Engineering I, Control Engineering, 12459 Berlin, Germany

E-mail: wweber@de.pepperl-fuchs.com, schulte@htw-berlin.de

Abstract.
In this paper, a generic physically-based modeling framework for encoder type speed sensors is derived. The consideration takes into account the nominal fault-free and two most relevant fault cases. The advantage of this approach is a reconstruction of the output waveforms in dependence of the internal physical parameter changes which enables a more accurate diagnosis and identification of faulty incremental encoders i.a. in wind turbines. The objectives are to describe the effect of the tilt and eccentric of the encoder disk on the digital output signals and the influence of the accuracy of the speed measurement in wind turbines. Simulation results show the applicability and effectiveness of the proposed approach.

1. Introduction

Due to the increasing dimensions and complexity of wind turbines, requirements regarding safety and availability continually rise. In particular, the increase of the equipment complexity caused an increase of probability of failure of individual components. Hence, in the recent years several approaches and methods have been proposed to detect and isolate faults in wind turbines on the system level which have been successfully applied in wind turbine simulation. A survey is given by the summary of a wind turbine FDI (fault detection and isolation) competition [1], where several actuator and sensor faults had to be detected within a pre-defined benchmark model [2]. In these studies, to prevent the loss of generality, a few sensor fault cases have been distinguished such as additive, scaling and fixed value faults. All of these are signal-based, which means the faults are modeled by event-triggered sensor signal adjustment in terms of signal addition, gain and compensation. Therefore the physical cause of faults in the sensor components are not previously considered. However, this information is useful to quantify the probability of occurrence and improved the fault reconstruction, both on the component and system level.

In this paper, a novel physically-based generic model of an incremental encoder, frequently used as rotor or generator speed sensor in wind turbines, is presented. In contrast to [2] a detailed mathematical model that takes into account the nominal case and systematic faults are proposed. Here, two relevant fault cases are studied in detail: First, the tilt of the encoder disk and second, the eccentric of the axis of rotation. The advantage of this approach is a reconstruction of waveforms in dependence of physical parameter changes which enables a more accurate fault identification.

This paper is organized as follows: In Section 2 the main components and control functions which have important influence on the performance of wind turbines, i.e. the energy production, are briefly described. Further, commonly used encoder types and their operating principles are presented. Sections 3 and 4 contain the main parts of this work. In Section 3 a generic white-box model framework for encoder type sensor systems is derived. The objective is to describe the effect of the tilt of the increment disk, the offset of the rotation axis on the digital output signals and the influence of the accuracy of the speed measurement. In Section 4, simulation studies including the nominal fault free and two common sensor fault cases are presented.

2. Sensors for wind turbine control

2.1. Functions of wind turbine control system

State of the art wind turbines in the 0.5 MW to 8 MW class have three major components which have important influence on the performance, i.e. the energy production of the system. First of all the nacelle including the rotor has to turn into the wind direction as good as possible. This function is called yaw and includes electrical drives which rotate the nacelle according to the signal of a wind measuring device (see Figure 2). The second function is called pitch and means the positioning of the rotor blades. Depending on the wind speed the rotor blade has to be adjusted in a way that a constant rotational speed is achieved (see Figure 1). Finally the rotor speed respectively the generator speed has to be determined (see Figure 3).

All these functions have performance as well as safety aspects. If the rotor speed exceeds a certain maximum value serious damage can happen, even the destruction of the wind turbine. Therefore, a safe shutdown function must be activated in case of exceeding the critical value. On the other hand the stand-still position of the rotor must be detected to enable safe access of service personnel.

If the nacelle turns with changing wind conditions there is a limit of revolutions into the same direction. A typical value is 3 to 4 turns into one direction. Exceeding this limit would cause the destruction of the cables which connect the power generation in the nacelle with the converters or other devices at the ground. Finally the determination of the blade position is crucial for the shutdown. Since up to date Mega Watt machines have rotors of huge size and weight like 20 tons or more, the full stop cannot be achieved by brakes any more. The rotor is controlled by turning the blades into a stall position to reduce the speed down to a value where the brakes could be safely applied.

Looking at all these functionalities it becomes obvious that a precise and reliable measuring of speeds and positions is a crucial aspect in the whole construction.

2.2. Encoder types for position and speed measurement

Traditionally there are two technical solutions to achieve this measuring task. In some cases inductive sensors are used to detect holes in a metal ring. By counting these incremental signals a control system could determine the actual speed but typically not a position. Therefore, incremental and absolute encoders have become state of the art, since they offer both types of signal and additionally a better resolution and precision. Because of the different requirements in the individual applications, there are a number of technologies and variants existing to provide solutions for each and every case:

- Incremental encoders with analogue or digital output (see Figure 4)
- Absolute encoders (single-turn; multi-turn)
- Absolute multi-turn encoders with
  - mechanical counter
  - battery buffered memory
- Absolute multi-turn encoders gearless, without battery by using Wiegand effect for counting

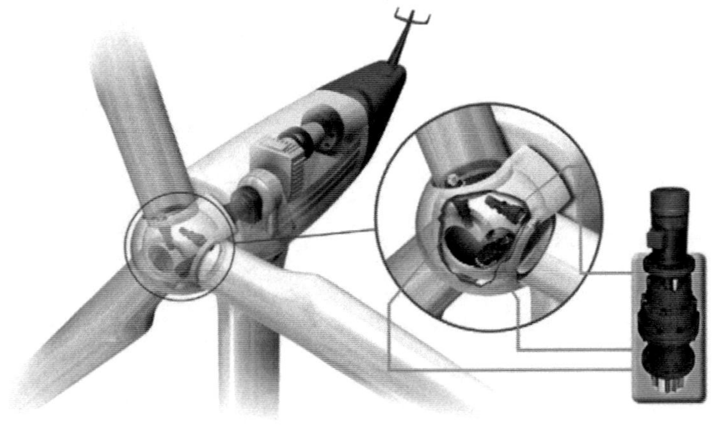

Figure 1. Shows the rotor of a wind turbine. The blade angle is adjusted according to the wind conditions to control the rotor speed.

Figure 2. The nacelle is turned into the wind by electrical drives. The position of the nacelles must be determined and also the number of turns in either direction to prevent cable destruction.

Figure 3. The rotor speed is measured directly at the low speed shaft and also at the fast moving generator shaft

- Absolute encoders with
  - fieldbus interfaces (CAN, Profibus, Profinet, EtherCAT, etc.)
  - magnetic, optical, capacitive or inductive working principle
  - non-contact magnetic or inductive encoder with external pole wheel.

Absolute encoders deliver position signals instead of incremental signals. The encoder disk contains a Gray Codes scale instead of simple one to one black and transparent transmissions. As an option absolute encoders can be executed as multi-turn versions, which keep counter values if the rotation exceeds 360 degrees several times. Nowadays optical systems are more and more replaced by magnetic encoders. The advantages of optical encoders were two-fold. Number one they could reach much higher rotational speed and number two they could deliver a much higher resolution. Newest developments have made at least the second point almost marginal. The resolution of magnetic encoders can reach up to 18 Bit, which lifts them to a

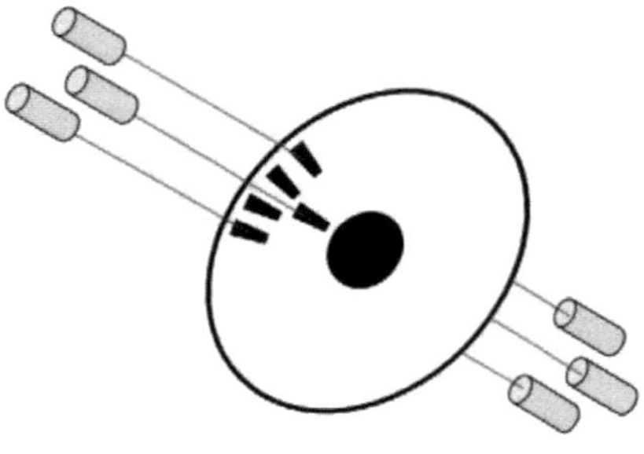

Figure 4. Typical arrangement of an incremental encoder. In case of an optical sensing system the encoder disk can be made of glass, plastic or metal. Two transmitter/receiver units detect the signal for the A and B signal. A third unit scans the zero position.

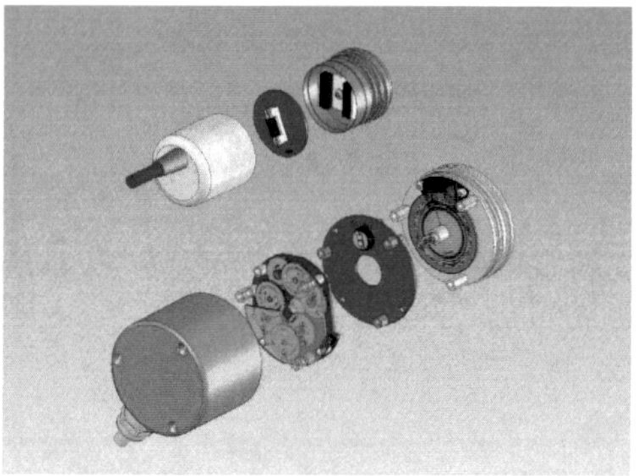

Figure 5. Comparison between two different encoder constructions. Both are multi-turn absolute encoders. One type shows an optical sensing system with a mechanical gear to count and store the multi-turn value. The other is a complete magnetic system with Wiegand wire for the multi-turn function.

comparable level as optical encoders. Since high revolutions do not take place in wind turbines but harsh environmental conditions became much more of an issue, magnetic encoders have become increasingly popular. The comparison in Figure 5 cearly shows that the magnetic version of a multi-turn encoder requires only a minimum of components and much less than the optical version with mechanical multi-turn unit. In addition to the increased robustness this has a significant influence on the life time calculation which shows in this specific case an advantage with a factor of 10.

Encoders are typically electro-mechanical devices (see Figure 6a). The mechanical parts are usually bearings. They are specifically subject to wear and tear. The calculation of the life-time of bearings is extremely difficult, because it depends heavily on external influences, such as temperature, kind of operation (low or high speed, long stand-still periods), type of grease or axial and radial axis loads. In most cases these environmental influences cannot clearly be specified. Therefore, there is a move in the direction of encoders without mechanical parts, respectively parts which are moving with mechanical contact.

In this case the magnet impulse unit is built in form of a ring which is placed outside the encoder detecting unit. This unit scans the magnetic field on a distance and is not touching anything (see Figure 6b). The detecting unit does not incorporate any moving mechanical parts, because it is sitting off-axis. One disadvantage of this construction is the small dimensional tolerance between the pole wheel and the sensor, which makes the installation in heavily vibrating machines impossible.

2.3. Accuracy of increments delivered by the output of an incremental encoder

There are various reasons why the signal of an incremental encoder is not 100% precise. In the state of the art products the electronic components are of sufficient quality and reliability. Usually limitations of the signal quality are caused by the mechanical construction. Especially manufacturing tolerances of different kind, such as eccentric or tilted mounting of the encoder disk or dimensional differences between the individual increments will cause limitations in the

Figure 6. A conventional encoder with ball bearings versus a non-contact magnetic encoder with pole wheel.

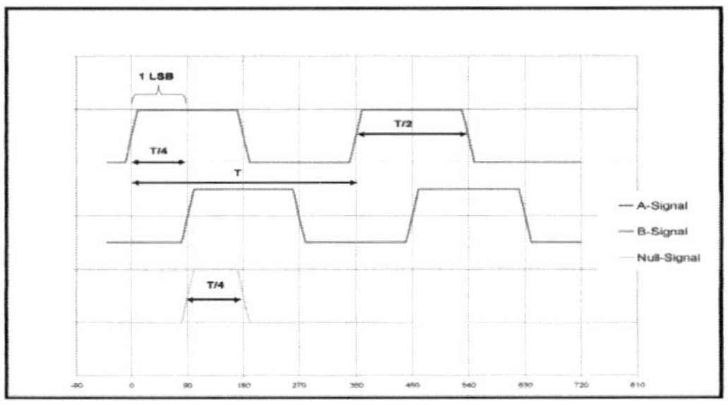

Figure 7. An encoder disk with 1024 increments and two channels (A + B) plus the zero track delivers 4 signal edges per increment. The $T/4$ measuring step is called the Least Significant Bit (LSB).

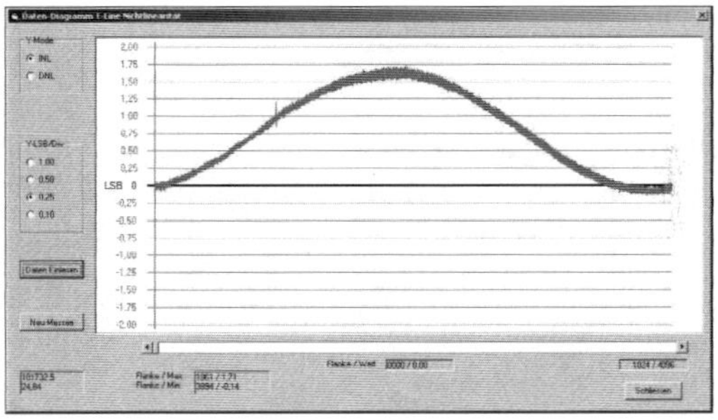

Figure 8. Measurement of the integral nonlinearity. The graph shows a maximum deviation of 1.71 LSB (see Figure 7)

achievable precision of the encoder signal.
There are two measurements which are typically executed to determine the accuracy of an incremental encoder. They are called differential and integral nonlinearity:

- The differential nonlinearity describes the dimensional deviation from each increment compared to the ideal model.
- The integral nonlinearity describes the summarized deviation over one revolution, which means the precision to reach exactly the same position after 360 degrees of rotation.

Taking the results shown in Figure 8 as an example the achievable accuracy would be 0.042%. Quiet often the encoder is coupled via a pinion wheel to the shaft or the drive. If such a pinion and the yaw bearing have a teeth ratio of 16 : 256 with a bearing diameter of 3 m, the positioning precision would be 0.063 m. If the same calculation would be applied on the revolution per minute of a rotor it would mean that at a typical value of 20 rpm the precision of the measurement would be 0.13 rpm.

Journal of Physics: Conference Series **570** (2014) 072008 doi:10.1088/1742-6596/570/7/072008

3. Physically-based modeling of rotor/generator speed sensors

3.1. Model assumptions

As shown in Figure 4 an incremental encoder consists of light sources as transmitter, a receiver such as photodiodes or phototransistors, a rotating encoder disk and a stationary aperture. It is essential to mention that the modeling framework based on following assumptions:

(i) The aperture size A is taken as identical to the photosensitive surface of the sensor.

(ii) A pure ray optics is assumed. That is the light propagates in straight lines from the transmitter through the aperture to the receiver.

(iii) A homogeneous intensity distribution of the radiation causes a voltage signal that is propotional to the current aperture size A

$$u = \frac{A}{A_{max}} \tag{1}$$

with $u \in [0, 1]$ is a normalized voltage signal.

3.2. Nominal fault-free case

The current voltage signal or aperture size for a fixed axis results from the intersection of the aperture and encoder increment and can be determined by

$$A = \int\limits_{\varphi=\varphi_1}^{\varphi=\varphi_2} \int\limits_{r=r_i}^{r=r_a} r \, \mathrm{d}r \, \mathrm{d}\varphi = \frac{1}{2} \left(r_a^2 - r_i^2\right) \left(\varphi_2 - \varphi_1\right) \tag{2a}$$

where r_i and r_a are the constant radii of a encoder increment of one track (Figure 9), φ is the variable of integration and is limited by the circular arc of each increment with the bounds φ_1 and φ_2. Because the increment is not always completely over the aperture, the following cases have to be distinguished:

(i) Increasing sectional area

$$A = \int\limits_{\varphi=\gamma_1}^{\varphi=\alpha_2} \int\limits_{r=r_i}^{r=r_a} r \, \mathrm{d}r \, \mathrm{d}\varphi \tag{2b}$$

(ii) Constant sectional area

$$A = \int\limits_{\varphi=\alpha_1}^{\varphi=\gamma_2} \int\limits_{r=r_i}^{r=r_a} r \, \mathrm{d}r \, \mathrm{d}\varphi \tag{2c}$$

(iii) Decreasing sectional area

$$A = \int\limits_{\varphi=\alpha_1}^{\varphi=\alpha_2} \int\limits_{r=r_i}^{r=r_a} r \, \mathrm{d}r \, \mathrm{d}\varphi \tag{2d}$$

(iv) No sectional area

$$A = 0 \tag{2e}$$

where α_1, α_2 are the fixed aperture angles and γ_1, γ_2 are the angle of orifice area of one increment. For illustrating, the encoder geometry, all coordinates and variables are shown in Figure 9.

In the next step, we have to take into account the rotation of the encoder disk for the nominal case. For this purpose, without loss of generality, it is shortly and conveniently to consider a

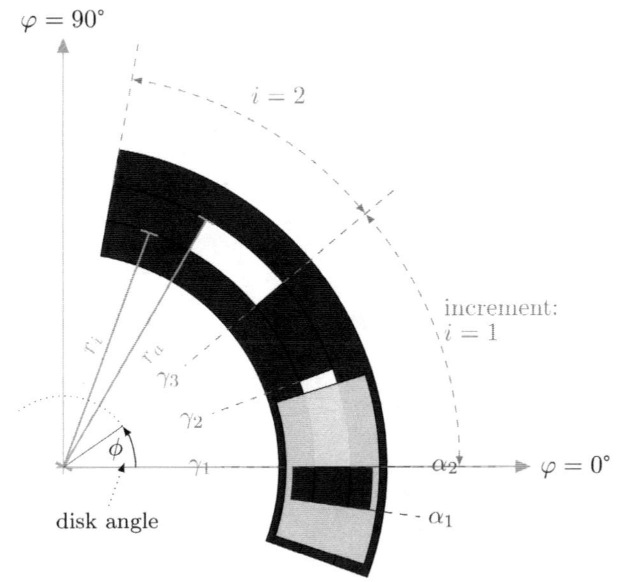

Figure 9. Geometry of aperture and increment disk for the fault-free case

rotating aperture instead of a rotating encoder disk. Using this approach, the aperture angle α_2 is equal to the current encoder disk angle ϕ. Therefore, the integral bounds can be determined by

$$\begin{aligned} \alpha_2 &= \phi \\ \alpha_1 &= \phi - \beta_B \ . \end{aligned} \tag{3}$$

where β_B is the orifice angle of the aperture. For the first increment this result to

$$\begin{aligned} \gamma_1 &= (i-1)\ \beta_I \\ \gamma_2 &= \gamma_1 + \beta_S \\ \gamma_3 &= \gamma_1 + \beta_I \end{aligned} \tag{4}$$

with β_S as angular width of increment orifice and β_I as angular width of increment $(2 \cdot \beta_S)$. The expression for γ_1 can be extended to $i \in \{2, 3, ...\}$ increments, see Figure 9. Further details can be found in [3]. Thus for each angular position the signal level is determined by previous equations (2) and (4).

3.3. Faulty case: Tilt encoder disk

Due to the manufacturing process the encoder disk can be tilt, so that disk and aperture are not parallel to each other. Then, the distance between disk and aperture is no longer constant and depends on the encoder disk angle ϕ. The tilt angle is denoted by ϑ and is related to the vertical disk position illustrated in Figure 10.

The starting point of the modeling approach is the parallel projection of an increment on the aperture plane. Instead of circles with constant radii r_i and r_a , a tilt encoder disk causes ellipse

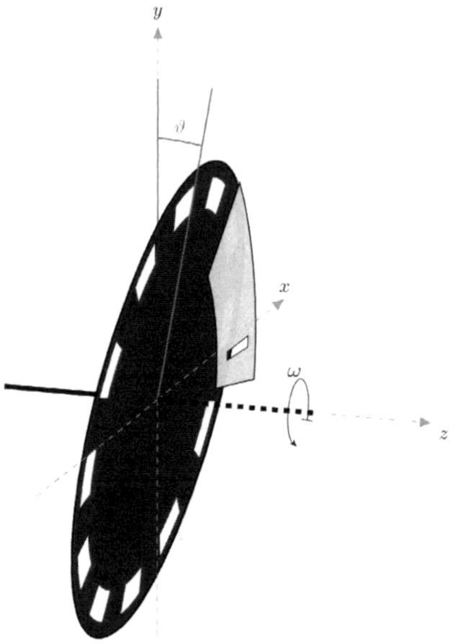

Figure 10. Tilt encoder disk

shaped projection of the increment edges on the aperture plane. This can be described by the following angle-dependent radii based on elliptic equation

$$r_i(\varphi) = \frac{r_{A_i} r_{B_i}}{\sqrt{(r_{A_i}\sin\varphi)^2 + (r_{B_i}\cos\varphi)^2}}, \quad r_a(\varphi) = \frac{r_{A_a} r_{B_a}}{\sqrt{(r_{A_a}\sin\varphi)^2 + (r_{B_a}\cos\varphi)^2}}$$

with

$$\begin{aligned} r_{A_i} &= r_i, & r_{B_i} &= r_i \cos\vartheta, \\ r_{A_a} &= r_a, & r_{B_a} &= r_a \cos\vartheta \end{aligned} \tag{5}$$

where r_A is the minor and r_B is the main axis of the ellipse. Taking this into account we get the integral representation to calculate the aperture size

$$A = \int\limits_{\varphi=\varphi_1}^{\varphi=\varphi_2} \int\limits_{r=r_i(\varphi)}^{r=r_a(\varphi)} r \, \mathrm{d}r \, \mathrm{d}\varphi = \int\limits_{\varphi=\varphi_1}^{\varphi=\varphi_2} \frac{1}{2} r^2 \Big|_{r_i(\varphi)}^{r_a(\varphi)} \mathrm{d}\varphi = \frac{1}{2} \int\limits_{\varphi=\varphi_1}^{\varphi=\varphi_2} r_a^2(\varphi) - r_i^2(\varphi) \, \mathrm{d}\varphi \, .$$

After analytical calculation of this integral we obtain the aperture size as a function of ϑ

$$A(\vartheta) = \frac{1}{2}\left(\arctan\left(\frac{1}{\cos\vartheta}\tan(\varphi_2)\right) - \arctan\left(\frac{1}{\cos\vartheta}\tan(\varphi_1)\right)\right)\left({r_a}^2\cos(\vartheta) - {r_i}^2\cos(\vartheta)\right) \, . \tag{6}$$

3.4. Faulty case: Eccentric axis of rotation

Due to manufacturing inaccuracy or age-related wear, it may be occur that the rotation axis is not exactly in the center of the increment disk. Then, similarly as in the previous case, r_i and r_a are not constant, and depending on the angle ϕ. In the following, it is assumed that

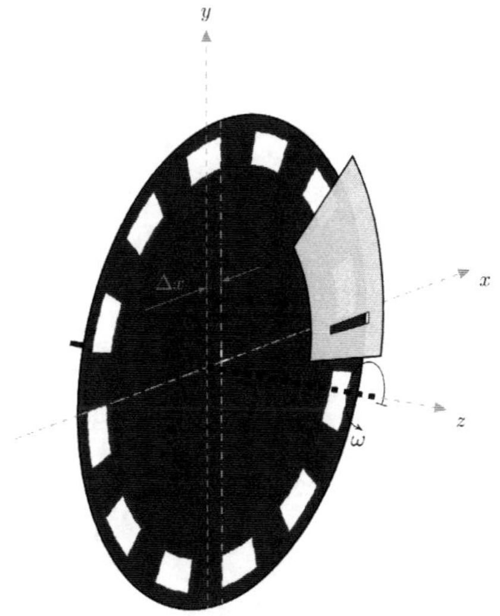

Figure 11. Eccentric axis of rotation

the displacement of the rotation axis can be described by Δx (see Figure 11) and the angle-dependent radii are denoted by $\tilde{r}_i$ and $\tilde{r}_a$. Applying the law of cosines to calculate $\tilde{r}_i$ and $\tilde{r}_a$, we get

$$\begin{aligned} {r_i}^2 &= \tilde{r}_i^2(\varphi) + \Delta x^2 - 2\,\tilde{r}_i(\varphi)\,\Delta x\,\cos(\varphi)\,, \\ {r_a}^2 &= \tilde{r}_a^2(\varphi)^2 + \Delta x^2 - 2\,\tilde{r}_a(\varphi)\,\Delta x\,\cos(\varphi)\,, \end{aligned}$$

and changed to the interesting quantities

$$\begin{aligned} \tilde{r}_i(\varphi) &= \Delta x\,\cos(\varphi) + \sqrt{{r_i}^2 - (\Delta x\,\sin(\varphi))^2}\,, \\ \tilde{r}_a(\varphi) &= \Delta x\,\cos(\varphi) + \sqrt{{r_a}^2 - (\Delta x\,\sin(\varphi))^2}\,. \end{aligned} \tag{7}$$

As for the previous case, the aperture size is

$$A = \int\limits_{\varphi=\varphi_1}^{\varphi=\varphi_2} \int\limits_{r=\tilde{r}_i(\varphi)}^{r=\tilde{r}_a(\varphi)} r\,\mathrm{d}r\,\mathrm{d}\varphi = \frac{1}{2}\int\limits_{\varphi=\varphi_1}^{\varphi=\varphi_2} \tilde{r}_a^2(\varphi) - \tilde{r}_i^2(\varphi)\,\mathrm{d}\varphi \tag{8}$$

After substituting (7) in (8) and some calculation we get the aperture size as a function of the displacement of the rotation axis

$$
\begin{aligned}
A(\Delta x) \;=\; & \frac{1}{2}\left(\Delta x^2 \sin(\varphi_2)\left(\sqrt{\left(\frac{r_a{}^2}{\Delta x^2}-\frac{1}{2}\sin^2(\varphi_2)\right)}+\sqrt{\left(\frac{r_i{}^2}{\Delta x^2}-\frac{1}{2}\sin^2(\varphi_2)\right)}\right)\right.\\
& - \;\Delta x^2 \sin(\varphi_1)\left(\sqrt{\left(\frac{r_a{}^2}{\Delta x^2}-\frac{1}{2}\sin^2(\varphi_1)\right)}+\sqrt{\left(\frac{r_i{}^2}{\Delta x^2}-\frac{1}{2}\sin^2(\varphi_1)\right)}\right)\\
& + r_a{}^2\left[\arctan\left(\frac{\Delta x\ \sin(\varphi_2)}{\sqrt{r_a{}^2\ 2\Delta x^2 \sin^2(\varphi_2)}}\right)-\arctan\left(\frac{\Delta x\cdot\sin(\varphi_1)}{\sqrt{r_a{}^2\ 2\Delta x^2 \sin^2(\varphi_1)}}\right)\right]\\
& - r_i{}^2\left[\arctan\left(\frac{\Delta x\ \sin(\varphi_2)}{\sqrt{r_i{}^2\cdot 2\Delta x^2 \sin^2(\varphi_2)}}\right)-\arctan\left(\frac{\Delta x\ \sin(\varphi_1)}{\sqrt{r_i{}^2\ 2\Delta x^2 \sin^2(\varphi_1)}}\right)\right]\\
& \left. + \;(\varphi_2-\varphi_1)\,(r_a{}^2-r_i{}^2)\right). \qquad (9)
\end{aligned}
$$

4. Impact of sensor faults on measurement accuracy

4.1. Signal processing

Finally, the impact of the fault cases on the sensor measurement accuracy are considered. First, the signal processing methods must be explained: In the first signal processing step, the raw analog signal A is transformed into a digital one by a comparator circuit. This digital signal is denoted by A_d. In order to ensure a high accuracy, sensor manufacturers use different circuit solutions for the same problem, which will not be discussed here. Then, in the second step, the current speed value n is calculated from the digital signal A_d. Two methods are commonly used:

(i) Counting the signal edges during a constant gate time T_g.

(ii) Measuring the time t_A that elapses between two successive signal edges.

The first method can be described by the expression

$$n = \frac{N_s}{T_g\,N} \qquad (10)$$

with n as the speed value, N as the number of increments on the encoder disk and N_s as the number of counted signal edges during the constant gate time T_g. The second method is given by

$$n = \frac{1}{t_A\,N} \qquad (11)$$

where t_A is the elapses time between two successive signal edges.

4.2. Impact of tilt encoder disk on measurement accuracy

The effects of a twisted encoder disk are illustrated by an example using a reasonable tilt angle $\vartheta = 0.75°$. For realistic results the two common methods for speed calculation (10) and (11) are taken into consideration. The speed variation over a sensor revolution with $N = 1024$ using (10) where $T_g = 10\,\mathrm{ms}$ is shown in Figure 12. For comparison, the same faulty case is considered with the second signal processing method (11) (see Figure 13). With both methods

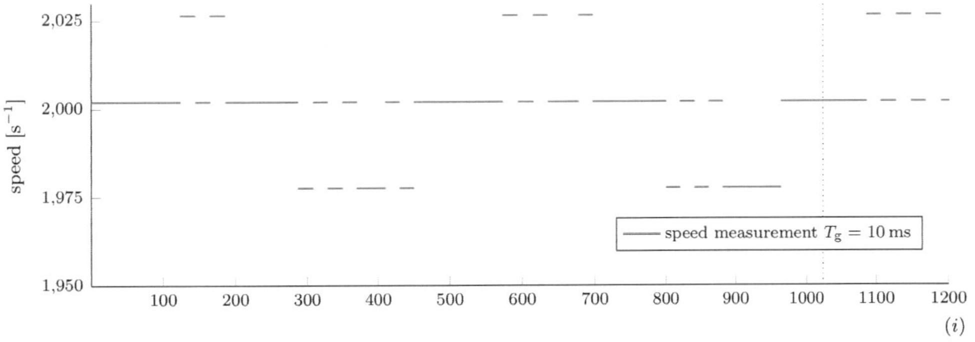

Figure 12. Effect on measured speed curve caused by a tilt encoder disk $\vartheta = 0.75°$ at true speed $n_0 = 2.0\,\mathrm{s}^{-1}$ using (10) where $T_g = 10\,\mathrm{ms}$ and $N = 1024$

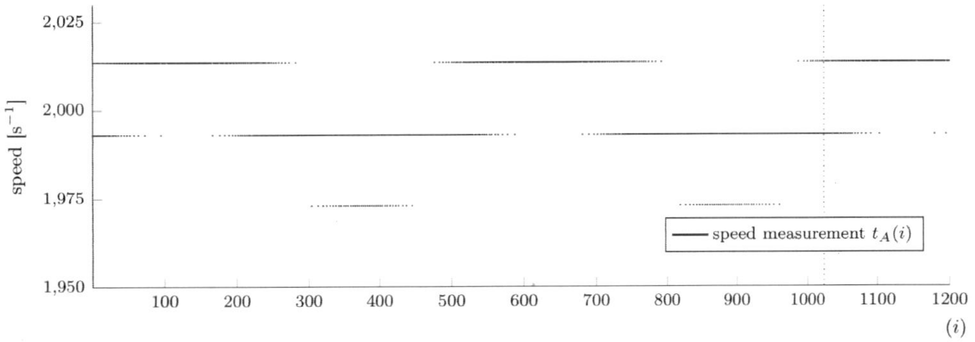

Figure 13. Effect on measured speed curve caused by a tilt encoder disk $\vartheta = 0.75°$ at true speed $n_0 = 2.0\,\mathrm{s}^{-1}$ using (11) where $N = 1024$

it is clearly illustrated, that the simulated measured curves vary significantly about the true value $n_0 = 2.0\,\mathrm{s}^{-1}$. The influence of the calculation methods are also clearly visible. Further simulations have shown, that the peak-to-peak value of n-curve increases with increasing tilt angle. But a purely analytical description of this relationship has not been derived yet.

4.3. Impact of eccentric rotation axis on measurement accuracy

Furthermore, the effects of a eccentric rotation axis are illustrated by a displacement of $\Delta x = 0.1\,\mathrm{mm}$. As before, the two different methods for speed calculation are taken into consideration. The variation of the measured speed curve caused by displacement of the rotation axis calculated with (10) is shown in Figure 14. For comparison, the same fault is considerd with the second signal processing method (11). The simulation result is shown in Figure 15. Also in this case, it is illustrated that the simulated measured curves vary significantly about the true value $n_0 = 2.0\,\mathrm{s}^{-1}$ influenced by a displacement of the rotation axis.

5. Conclusion and current work

In this paper, a generic physically-based model for encoder type speed sensors with an optical system was presented. Using the classical beam optics, the effect of tilt and eccentric of the

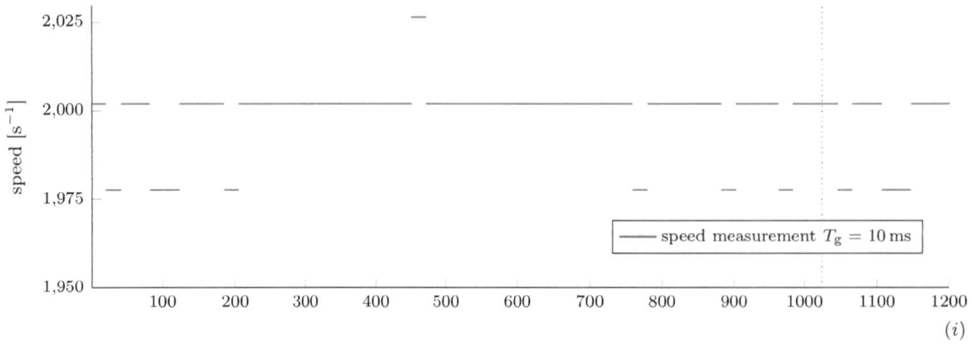

Figure 14. Effect on measured speed curve caused by displacement of the rotation axis $\Delta x = 0.1\,\text{mm}$ at true speed $n_0 = 2.0\,\text{s}^{-1}$ using (10) where $T_g = 10\,\text{ms}$ and $N = 1024$

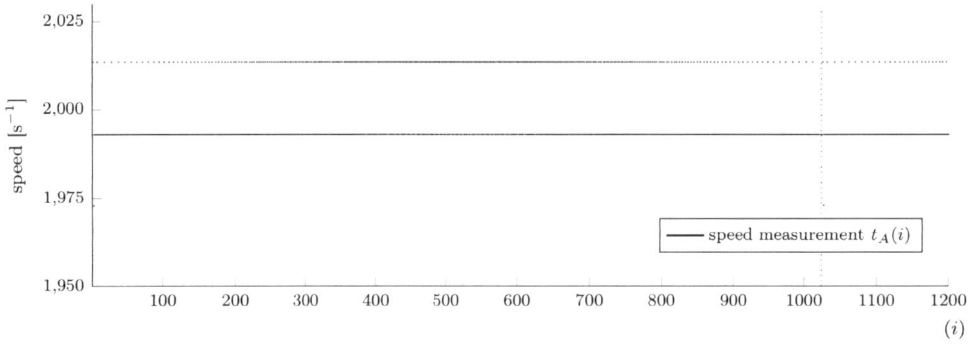

Figure 15. Effect on measured speed curve caused by displacement of the rotation axis $\Delta x = 0.1\,\text{mm}$ at true speed $n_0 = 2.0\,\text{s}^{-1}$ using (11) where $N = 1024$

encoder disk were analytically described as a function of the tilt angle and the displacement of the encoder axis. The influence of different methods of speed calculation was also investigated. On the basis of simulation studies using the analytical model the effects of non-nominal parameter values on the sensor speed signal could be made clearly visible. In current work, reduced models are derived with the objective to integrate these into the benchmark models [2] and [4] (based on [5],[6]) for a more realistic sensor failure scenarios for fault diagnosis of wind turbines.

References

[1] Odgaard P F and Stoustrup J 2012 *IFAC Symposium on Fault Detection, Supervision and Safety of Technical Processes* (Mexico City, Mexico) pp 102–107

[2] Odgaard P F, Stoustrup J and Kinnaert M 2009 *IFAC Symposium on Fault Detection, Supervision and Safety of Technical Processes* (Barcelona, Spain) pp 155–160

[3] Jungjohann J 2014 *Model-based fault identification for incremental encoders in wind turbines (in german)* Master Thesis, HTW Berlin, Control Engineering Group, (Berlin, Germany)

[4] Odgaard P F and Johnson K E 2013 *American Control Conference* (Washington D.C., USA), pp 4447–4452

[5] Jonkman, J (2010) NWTC Design Codes, FAST *http://wind.nrel.gov/designcodes/simulators/fast/* accessed 05-November-2010 (Colorado, USA)

[6] Jonkman, J, Butterfield S, Musial W, and Scott G (2009) *Definition of a 5-MW Reference Wind Turbine for Offshore Systems Development* NREL/TP-500-38060, Tech. Rep. (Colorado, USA)

A novel mathematical setup for fault tolerant control systems with state-dependent failure process

S Chitraganti[1], S Aberkane[1] and C Aubrun[1]

CRAN-CNRS, UMR 7039 CNRS, Faculté des Sciences et Techniques, Université de Lorraine, 54506 Vandoeuvre-les-Nancy, France.

E-mail: shaikshavali.chitraganti@univ-lorraine.fr

Abstract. In this paper, we consider a fault tolerant control system (FTCS) with state-dependent failures and provide a tractable mathematical model to handle the state-dependent failures. By assuming abrupt changes in system parameters, we use a jump process modelling of failure process and the fault detection and isolation (FDI) process. In particular, we assume that the failure rates of the failure process vary according to which set the state of the system belongs to.

1. Introduction

In fault tolerant control system (FTCS), apart from the actuator, plant, sensors, input, outputs, there are two major modules in general: a fault detection and identification (FDI) module and a reconfiguration controller. In the event of faulty behaviour reported by the FDI module, the reconfiguration controller provides a necessary feedback action such that over all control system is fault tolerant, which is the reason for naming it as FTCS. One can find myriads of contributions of FTCS in the literature for instance in [1], [4] etc.

In the presence of abrupt changes or multiple modes that involve randomness, FTCS with Markov processes have been investigated in [11], [6], [7], [2] and the references therein. In these works, the authors primarily assumed different state spaces for a failure process and an FDI process to incorporate fault detection delays, false alarms that may result from various errors in the FDI module. FTCS modelling in this situation can be given by

$$\left.\begin{aligned} \dot{x}(t) &= A_{\theta(t)}x(t) + B_{\theta(t)}u(\eta(t), x(t), t), \\ x(0) &= x_0, \end{aligned}\right\} \tag{1}$$

with the state vector $x(t) \in \mathbb{R}^n$, the initial state x_0, the control input $u(.,.,.) \in \mathbb{R}^m$, and the initial state $x_0 \in \mathbb{R}^n$. Let $A_{\theta(t)} \in \mathbb{R}^{n\times n}$, $B_{\theta(t)} \in \mathbb{R}^{n\times m}$, be the system matrices that depend on the mode $\theta(t)$, and are assumed to be known. Here $\{\theta(t), t \geq 0\} \in S$ and $\{\eta(t), t \geq 0\} \in T$ represent the failure process and the FDI process with finite state-spaces $S := \{1, 2, \cdots, M\}$ and $T := \{1, 2, \cdots, N\}$ respectively. An illustration of the overall FTCS is given in figure 1, where the controller reconfiguration depends on the FDI process at each time. In an ideal scenario of FDI, the FDI process $\{\eta(t), t \geq 0\}$ should match exactly with the failure process $\{\theta(t), t \geq 0\}$, which is not the case in practise, thus the reason for different state spaces S and T for $\{\theta(t), t \geq 0\}$ and $\{\eta(t), t \geq 0\}$.

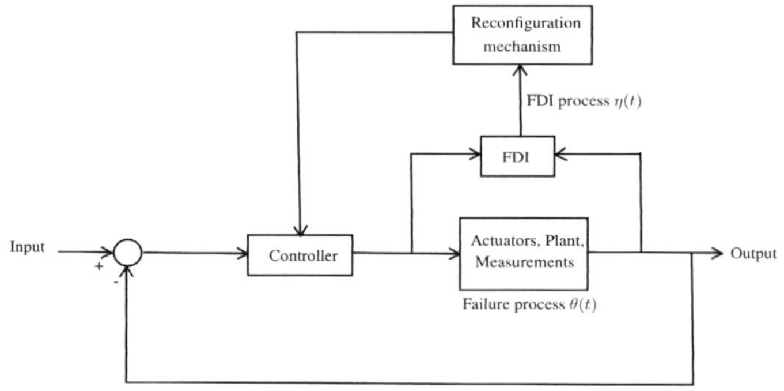

Figure 1. An illustration of the FTCS

We begin with the failure process.

In the presence of abrupt changes in system parameters that may cause failures in various components, the failure process $\{\theta(t), t \geq 0\}$ has been modelled as time-homogeneous Markov Process, which implies the constant failure rates, which is a quite restrictive assumption. In practice, the failure rates depend on the usage and/or the current state of the component. For example, it was examined in [8] that the failure rate of a component is dependent on its state at age t, where the state variable may be an amount of wear, stress etc. Also, for instance, in [9], a state-dependent Markov process was used to describe the random breakdown of cylinder lines in a heavy-duty marine diesel engine. Thus, we consider that the failure process $\theta(t)$ depends on the state variable. Next we move to the FDI process.

In an ideal case, by using various measurements of the FTCS, the FDI process should give accurate results whether a faulty behaviour has occurred in the system, which is used as a feedback to reconfigure the overall the FTCS to correct faulty behaviour. Thus, the FDI process depend directly on the failure process. But, in the presence of noisy measurements, it is difficult for the FDI process to provide accurate behaviour of the system. If the measurement process contain white noise, and memoryless tests are used to characterize the FDI process, then the FDI process can be modelled as a Markov jump process with finite space as examined in [11], [6] etc. In our case, for simplicity, we consider that the FDI process is Markovian. It is a reasonable assumption, because even if the FTCS contain state-dependent failure process, one can still employ memoryless tests for FDI process, which just depend on the measurements from the system. Thus, we consider the FDI process, which is Markovian.

Thus, we consider state-dependent failure process, which is also Markovian in some suitably defined sets, and the FDI process as a pure Markov process which just depend on the failure process. The presence of state-dependent failure process makes the FTCS modelling (1) far more complex. To the best of author's knowledge, a tractable the FTCS modelling (1) with state-dependent failure process has not been addressed, which forms a theme of this paper.

In the next section, we provide a tractable mathematical model of FTCS with state-dependent failure process.

2. Mathematical model of the FTCS with state-dependent failure process

Consider a mathematical model of FTCS:

$$\left.\begin{aligned} \dot{x}(t) &= A_{\theta(t)}x(t) + B_{\theta(t)}u(\eta(t), x(t), t), \\ x(0) &= x_0. \end{aligned}\right\} \tag{2}$$

Since we assume the Markovian modelling of the FDI process $\{\eta(t), t \geq 0\} \in T$ that depends on the failure process $\{\theta(t), t \geq 0\} \in S$, its evolutions for $m, n \in T$, $m \neq n$ and $z \in S$ can be considered as:

$$\Pr\{\eta(t+\triangle) = n|\eta(t) = m, \theta(t) = z\} = \delta_{mn,z}\triangle + o(\triangle), \tag{3}$$

where $\triangle > 0$ and $\lim_{\triangle \to 0} \frac{o(\triangle)}{\triangle} = 0$. Regarding the transition rates of $\eta(t)$, we make the following assumption:

Assumption 1. For each $z \in S$, $m, n \in T$, $\delta_{mn,z}$ is the transition rate of $\eta(t)$ from mode m to mode n with $\delta_{mn,z} \geq 0$ for $m \neq n$, and $\delta_{mm,z} = -\sum_{n=1, m\neq n}^{N} \delta_{mn,z}$. We further assume that $\delta_{mn,z}$ is bounded for all $z \in S$, $m, n \in T$.

Before proceeding with the state-dependent failure process $\{\theta(t), t \geq 0\}$, we make the following assumption regarding the different sets in $\mathbb{R}^n$.

Assumption 2. Let $\mathcal{K} = \{1, 2, \cdots, K\}$. We assume that the sets $\mathscr{C}_1, \mathscr{C}_2, \cdots \mathscr{C}_K \subseteq \mathbb{R}^n$, $\cup_{\iota=1}^{K} \mathscr{C}_\iota = \mathbb{R}^n$ and $\mathscr{C}_p \cap \mathscr{C}_q = \phi$ for any $p, q \in \mathcal{K}$, $p \neq q$. At any time $t \geq 0$, we assume that $x(t)$ belongs to exactly one of the sets $\mathscr{C}_l$, $l \in \mathcal{K}$. By its definition, $\mathscr{C}_1, \mathscr{C}_2, \cdots \mathscr{C}_K$ are Borel sets in $\mathbb{R}^n$.

For example, as an illustration, we can consider $\mathbb{R}^2$ space with three sets $\mathscr{C}_1, \mathscr{C}_2, \mathscr{C}_3$, which can be defined as: $\mathscr{C}_1 := \{x(t) \in \mathbb{R}^2, x_1^2(t) + x_2^2(t) \leq \xi_1\}$, $\mathscr{C}_2 := \{x(t) \in \mathbb{R}^2, \xi_1 < x_1^2(t) + x_2^2(t) \leq \xi_2\}$ and $\mathscr{C}_3 := \{x(t) \in \mathbb{R}^2, x_1^2(t) + x_2^2(t) > \xi_2\}$ for some $\xi_1 > 0$ and $\xi_2 > 0$. The illustration is given in figure 2, where the state variable $x(t)$ can belong to exactly one of the sets at each time $t \geq 0$.

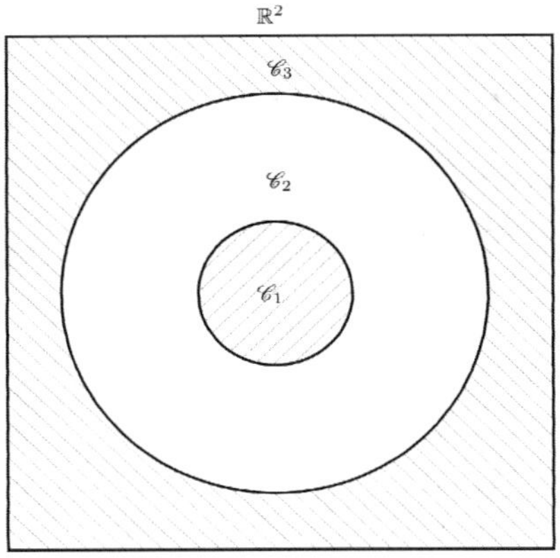

Figure 2. A schematic representation of the sets $\mathscr{C}_1, \mathscr{C}_2, \mathscr{C}_3$ in $\mathbb{R}^2$

Now, the evolutions of the state-dependent failure process $\{\theta(t), t \geq 0\}$ for $i, j \in S$, $i \neq j$ can be governed by:

$$\Pr\{\theta(t+\triangle) = j|\theta(t) = i, x(t)\} = \begin{cases} \lambda_{ij}^1\triangle + o(\triangle), & \text{if } x(t) \in \mathscr{C}_1, \\ \lambda_{ij}^2\triangle + o(\triangle), & \text{if } x(t) \in \mathscr{C}_2, \\ & \vdots \\ \lambda_{ij}^K\triangle + o(\triangle), & \text{if } x(t) \in \mathscr{C}_K. \end{cases} \tag{4}$$

Journal of Physics: Conference Series **570** (2014) 082001 doi:10.1088/1742-6596/570/8/082001

Regarding the transition rates of $\theta(t)$, we make the following assumption:

Assumption 3. Let $\mathcal{K} \triangleq \{1, 2, \cdots K\}$. For each $m \in \mathcal{K}$, λ_{ij}^m is the transition rate of $\theta(t)$ from mode i to mode j with $\lambda_{ij}^m \geq 0$ for $i \neq j$, and $\lambda_{ii}^m = -\sum_{j=1, j\neq i}^{N} \lambda_{ij}^m$. We further assume that λ_{ij}^m is bounded for all $m \in \mathcal{K}$.

Remark 1: Notice that the evolution of $\theta(t)$ in (4) depends on the state variable $x(t)$, hence $\theta(t)$ is called as a state-dependent failure process. Also observe that the failure rates are different depending on which set the state of the system belongs to. It is a valid assumption, because, in this setup, one can predefine suitable sets $\mathscr{C}_1, \mathscr{C}_2, \cdots \mathscr{C}_K$ and assign different failure rates in a suitable fashion, thus it is tractable. A more general jump system with state-dependent jump process without any other jump process (for example the FDI process in our case) is considered in [3] similar to (4), where sufficient conditions for stability and stabilizing controller in terms of linear matrix inequalities (LMIs) are obtained by utilizing techniques of probability theory and stochastic version of Lyapunov's second method.

3. Challenges and future work

We would like to highlight a couple of challenges in the FTCS with state-dependent failure setup given above.

1. Since the overall FTCS (2) is nonlinear, it is necessary to provide its flow as a function of time.
2. For a simple state-feedback reconfiguration control of the FTCS, usually one design state-feedback gains obtained from Lyapunov's second method, which requires the assumption of Markovian process of the joint variables of the Lyapunov candidate [5].

For point 1 above, one can describe the flows of the FTCS (2) when $x(t) \in \mathscr{C}_m$ for each $m \in \mathcal{K}$, which can be characterized in terms of suitably defined first exit times from different sets $\mathscr{C}_1, \mathscr{C}_2, \cdots \mathscr{C}_K$. For point 2 above, one can prove that the joint process formed by $(x(t), \theta(t), \eta(t))$ is Markov process by using arguments of probability theory.

Both of these issues has already been investigated in [3] for a more general system without FTCS configuration, which we would like to extend to the FTCS framework, which can provide answers to the reconfiguration control in FTCS with state-dependent failure process. This forms a future extension of this paper.

4. Conclusions

We provided a tractable mathematical model for FTCS with state-dependent failure process. For the underlying state-dependent failure process, we explicitly considered different failure rates depending on the set to which the state of the system belongs. Finally, we presented a future course of direction.

References

[1] Blanke M 1992 *Fault-tolerant control systems*, Springer.
[2] Huang J 2013 *Analysis and synthesis of semi-Markov jump linear systems and networked dynamic systems*, PhD thesis, University of Victoria.
[3] Chitraganti S 2014 *On stability and control of random jump linear systems*, PhD thesis, submitted, University of Lorraine.
[4] Zhang Y and Jin J 2008 *Bibliographical review on reconfigurable fault-tolerant control systems*, Annual reviews in control, **32.2**: 229-252.
[5] Boukas E K 2007 *Stochastic switching systems: analysis and design*, Springer.

[6] Mufeed M, Jin J and Zhang Y 2001 *Stochastic stability analysis of fault-tolerant control systems in the presence of noise,* IEEE Transactions on Automatic Control, **46.11**: 1810-1815.

[7] Tao F and Zhao Q 2007 *Synthesis of active fault-tolerant control based on Markovian jump system models,* IET Control Theory and Applications, **1.4**: 1160-1168.

[8] Bergman B 1978, *Optimal replacement under a general failure model,* Advances in Applied Probability, **10.2**: 431-451.

[9] Giorgio M, Guida M and Pulcini G 2010 *A state-dependent wear model with an application to marine engine cylinder liners*, Technometrics, **52.2**: 172-187.

[10] Yin G G and Zhu C 2009 *Hybrid switching diffusions: properties and applications*, **63**, Springer.

[11] Srichander R and Walker B 1993 *Stochastic stability analysis for continuous fault tolerant control systems*, International Journal of Control, **57.2**: 433-452.

Actuator fault diagnosis and fault-tolerant control: Application to the quadruple-tank process

Mariusz Buciakowski, Michał de Rozprza-Faygel, Joanna Ochałek, Marcin Witczak

Institute of Control and Computation Engineering, University of Zielona Góra, Poland

E-mail: {m.buciakowski, m.derozprza-faygel, j.ochalek, m.witczak}@issi.uz.zgora.pl

Abstract. The paper focuses on an important problem related to the modern control systems, which is the robust fault-tolerant control. In particular, the problem is oriented towards a practical application to quadruple-tank process. The proposed approach starts with a general description of the system and fault-tolerant strategy, which is composed of a suitably integrated fault estimator and robust controller. The subsequent part of the paper is concerned with the design of robust controller as well as the proposed fault-tolerant control scheme. To confirm the effectiveness of the proposed approach, the final part of the paper presents experimental results for considered the quadruple-tank process.

1. Introduction

Fault-Tolerant Control (FTC) systems can be divided into two of distinct classes [24], i.e., passive and active. Regarding the passive FTC [10, 11, 16, 3] systems, there is no need for fault diagnosis, owing to the fact that they are robust to a set of predefined faults. Although, the presented approach usually degrades the overall performance. Contrarily, the above-mentioned passive FTC scheme, active one, reacts to faults actively. To maintain the system stability and acceptable performance, controller reconfigures the control actions. The control system relies on Fault Detection and Isolation (FDI) [9, 18, 17, 20, 23, 14, 13, 19] as well as an accommodation technique [1] to achieve a challenging problem. As can be observed, the problems of FDI and FTC are treated separately by most of the works, that exist in the literature. According to the fact that perfect FDI and fault identification are impossible to achieve, there is always an inaccuracy related to this process. Hence that, there is a need for an integrated FDI and FTC schemes. In last decade, a number of books were focused on the emerging problem of the FTC. In particular, the work[7] is mainly concentrated on fault diagnosis. Moreover, it provided general rules for the hardware-redundancy-based FTC systems. The book [12] introduces the concepts of FTC divided into active and passive structures. It also investigates the problem of stability and performance of the FTC under imperfect (imprecise and delayed) fault diagnosis. In particular, the authors consider (under a chain of some, not necessarily easy to satisfy assumptions) the effect of an imperfect fault identification and a delayed fault detection. The proposed approach deals with the fault diagnosis scheme separately, during the design, excluding real integration of the fault diagnosis and the FTC. Plenty of practical case studies of FTC are treated in [15], i.e., a winding machine, a three-tank system, and an active suspension system. Unfortunately, the FTC integrated with the fault diagnosis is not studied, in spite of the incontestable appeal of the proposed approach. The paper is organised as follows. Section 2 introduces a general scheme of the proposed framework. Whilst section 3 describes the concept of the fault compensation mechanism. The subsequent section 4 presents the robust control framework, while the final part of the paper is concerned with an illustrative example.

2. A general description of system and fault-tolerant strategy

Let us consider a linear discrete-time system:

$$\boldsymbol{x}_{f,k+1} = \boldsymbol{A}\boldsymbol{x}_{f,k} + \boldsymbol{B}\boldsymbol{u}_{f,k} + \boldsymbol{B}\boldsymbol{f}_k + \boldsymbol{W}\boldsymbol{w}_k, \tag{1}$$

where $\boldsymbol{x}_{f,k} \in \mathbb{X} \subset \mathbb{R}^n$ is the state vector, $\boldsymbol{u}_{f,k} \in \mathbb{U} \subset \mathbb{R}^r$ stands for the input, $\boldsymbol{f}_k \in \mathbb{R}^r$ is the actuator fault, and $\boldsymbol{w}_k \in l_2$ denotes an exogenous disturbance vector, where:

$$l_2 = \{\mathbf{w} \in \mathbb{R}^n | \, \|\mathbf{w}\|_{l_2} < +\infty\}, \tag{2}$$

$$\|\mathbf{w}\|_{l_2} = \left(\sum_{k=0}^{\infty} \|\boldsymbol{w}_k\|^2\right)^{\frac{1}{2}}. \tag{3}$$

For the purpose of further deliberations, let us formulate the following condition [21]:

$$\text{rank}(\boldsymbol{B}) = r, \tag{4}$$

$$\boldsymbol{f}_k = \boldsymbol{f}_{k-1} + \bar{\boldsymbol{v}}_k, \quad \bar{\boldsymbol{v}}_k \in l_2. \tag{5}$$

The proposed control scheme is a combination of the fault compensation [22] and a classical control scheme:

$$\boldsymbol{u}_{f,k} = -\boldsymbol{K}\boldsymbol{x}_k - \hat{\boldsymbol{f}}_{k-1}, \tag{6}$$

where:

- $\boldsymbol{K}$ is the $\mathcal{H}_\infty$ controller designed to achieve robustness,
- $\hat{\boldsymbol{f}}_{k-1}$ is the fault estimate, which compensates the effect of a fault,

The control strategy design boils down to solving a two set of problems:

- to design a robust controller $\boldsymbol{K}$,
- to estimate the fault $\boldsymbol{f}_k$.

The proposed approach, described by (6) is presented in Fig. 1.

3. Fault estimation

The general objective of this section is to present the fault estimation technique. Thus, following [6, 18], by computing

$$\boldsymbol{H} = \boldsymbol{B}^+ = \left[\boldsymbol{B}^T\boldsymbol{B}\right]^{-1}\boldsymbol{B}^T, \tag{7}$$

and then multiplying (1) by $\boldsymbol{H}$, it can be shown that:

$$\boldsymbol{f}_k = \boldsymbol{H}\boldsymbol{x}_{f,k+1} - \boldsymbol{H}\boldsymbol{A}\boldsymbol{x}_{f,k} - \boldsymbol{u}_{f,k} - \boldsymbol{H}\boldsymbol{W}\boldsymbol{w}_k, \tag{8}$$

while its estimate can be given as:

$$\hat{\boldsymbol{f}}_k = \boldsymbol{H}\boldsymbol{x}_{f,k+1} - \boldsymbol{H}\boldsymbol{A}\boldsymbol{x}_{f,k} - \boldsymbol{u}_{f,k}, \tag{9}$$

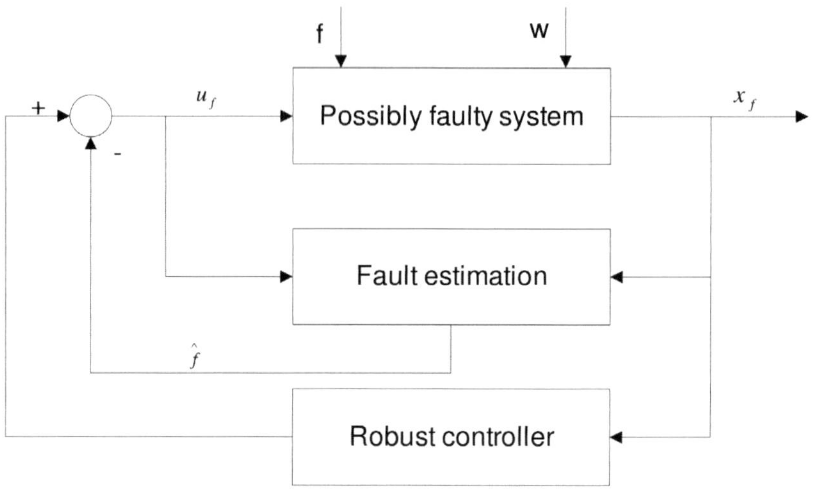

Figure 1. Scheme of the proposed robust predictive FTC

with the associated fault estimation error

$$\varepsilon_{f,k} = \boldsymbol{f}_k - \hat{\boldsymbol{f}}_k = -\boldsymbol{H}\boldsymbol{W}\boldsymbol{w}_k. \tag{10}$$

Since the general framework for computing the fault estimate (9) is given, then its computational feasibility can be verified. To obtain $\hat{\boldsymbol{f}}_k$, it is necessary to have $\boldsymbol{x}_{f,k+1}$. Thus, the only choice to compensate $\boldsymbol{f}_k$ in (1) is to use $\hat{\boldsymbol{f}}_{k-1}$. This determines the above-proposed control strategy:

$$\boldsymbol{u}_{f,k} = -\hat{\boldsymbol{f}}_{k-1} - \boldsymbol{K}\boldsymbol{x}_{f,k}. \tag{11}$$

Taking into account (10) and (5), it can be shown that

$$\hat{\boldsymbol{f}}_k = \hat{\boldsymbol{f}}_{k-1} + \boldsymbol{H}\boldsymbol{W}[\boldsymbol{w}_k - \boldsymbol{w}_{k-1}] + \bar{\boldsymbol{v}}_k. \tag{12}$$

Bearing in mind that all faults present in the real systems have a finite value and knowing that $\boldsymbol{w}_k, \bar{\boldsymbol{v}}_k \in l_2$ it is evident that there exists $\boldsymbol{v}_k$ such that:

$$\hat{\boldsymbol{f}}_k = \hat{\boldsymbol{f}}_{k-1} + \boldsymbol{v}_k, \quad \boldsymbol{v}_k \in l_2. \tag{13}$$

Thus, (11) can be written in an equivalent form, which will be used for further deliberations

$$\boldsymbol{u}_{f,k} = -\hat{\boldsymbol{f}}_k + \boldsymbol{v}_k - \boldsymbol{K}\boldsymbol{x}_{f,k}. \tag{14}$$

4. Controller design

The main objective of this section is to present the design procedure of the robust controller. Substituting (14) into (1) gives

$$\boldsymbol{x}_{f,k+1} = \boldsymbol{A}_1\boldsymbol{x}_{f,k} + [\boldsymbol{I} - \boldsymbol{B}\boldsymbol{H}]\boldsymbol{W}\boldsymbol{w}_k + \boldsymbol{B}\boldsymbol{v}_k, \tag{15}$$

with

$$\boldsymbol{A}_1 = \boldsymbol{A} - \boldsymbol{B}\boldsymbol{K}. \tag{16}$$

Analysing (15), and in particular

$$\left[\boldsymbol{I} - \boldsymbol{B}\left[\boldsymbol{B}^T\boldsymbol{B}\right]^{-1}\boldsymbol{B}^T\right]\boldsymbol{W}\boldsymbol{w}_k = \boldsymbol{W}\boldsymbol{w}_k - \boldsymbol{B}\left[\boldsymbol{B}^T\boldsymbol{B}\right]^{-1}\boldsymbol{B}^T\boldsymbol{W}\boldsymbol{w}_k, \tag{17}$$

along with the fact that any vector $\boldsymbol{W}\boldsymbol{w}_k \in \text{col}(\boldsymbol{B})$, where $\text{col}(\boldsymbol{B}) = \{\boldsymbol{\alpha} \in \mathbb{R}^n : \boldsymbol{\alpha} = \boldsymbol{B}\boldsymbol{\beta}\}$ for some $\boldsymbol{\beta} \in \mathbb{R}^r$ can be written as $\boldsymbol{W}\boldsymbol{w}_k = \boldsymbol{B}\bar{\boldsymbol{w}}_k$ for some non-zero $\bar{\boldsymbol{w}}_k$, leads (17) to

$$\boldsymbol{B}\bar{\boldsymbol{w}}_k - \boldsymbol{B}\left[\boldsymbol{B}^T\boldsymbol{B}\right]^{-1}\boldsymbol{B}^T\boldsymbol{B}\bar{\boldsymbol{w}}_k = \boldsymbol{0}. \tag{18}$$

This significant simplification of (15) yields its new form:

$$\boldsymbol{x}_{f,k+1} = \boldsymbol{A}_1\boldsymbol{x}_{f,k} + \boldsymbol{B}\boldsymbol{v}_k \tag{19}$$

The general framework for designing $\mathcal{H}_\infty$ robust controller is to determine the gain matrix $\boldsymbol{K}$ such that

$$\lim_{k\to\infty} \boldsymbol{x}_{f,k} = \boldsymbol{0} \quad \text{for } \boldsymbol{v}_k = \boldsymbol{0} \tag{20}$$

$$\|\boldsymbol{x}_f\|_{l_2} \leq \mu\|\boldsymbol{v}_k\|_{l_2} \quad \text{for } \boldsymbol{v}_k \neq \boldsymbol{0},\ \boldsymbol{x}_0 = \boldsymbol{0}. \tag{21}$$

Thus, the problem is to find a Lyapunov function V_k such that:

$$\Delta V_k + \boldsymbol{x}_{f,k}^T\boldsymbol{x}_{f,k} - \mu^2\boldsymbol{v}_k^T\boldsymbol{v}_k < 0,\ k = 0, \ldots \infty, \tag{22}$$

where

$$\Delta V_k = V_{k+1} - V_k, \tag{23}$$

and

$$V_k = \boldsymbol{x}_{f,k}^T\boldsymbol{P}\boldsymbol{x}_{f,k}. \tag{24}$$

Indeed, if $\boldsymbol{v}_k = \boldsymbol{0}$ then (22) boils down to

$$\Delta V_k + \boldsymbol{x}_{f,k}^T\boldsymbol{x}_{f,k} < 0,\ k = 0, \ldots \infty, \tag{25}$$

and hence $\Delta V_k < 0$, which leads to (20). If $\boldsymbol{v}_k \neq \boldsymbol{0}$ then (22) yields

$$J = \sum_{k=0}^{\infty}\left(\Delta V_k + \boldsymbol{x}_{f,k}^T\boldsymbol{x}_{f,k} - \mu^2\boldsymbol{v}_k^T\boldsymbol{v}_k\right) < 0, \tag{26}$$

which can be written as

$$J = -V_0 + \sum_{k=0}^{\infty}\boldsymbol{x}_{f,k}^T\boldsymbol{x}_{f,k} - \sum_{k=0}^{\infty}\mu^2\boldsymbol{v}_k^T\boldsymbol{v}_k < 0, \tag{27}$$

Knowing that $V_0 = 0$ for $\boldsymbol{x}_{f,0} = \boldsymbol{0}$, equation (27) leads to (21). For (24) the inequality (22) is

$$\Delta V + \boldsymbol{x}_{f,k}^T\boldsymbol{x}_{f,k} - \mu^2\boldsymbol{v}_k^T\boldsymbol{v}_k < 0, \tag{28}$$

with

$$\begin{aligned}\Delta V =& \boldsymbol{x}_{f,k}^T\left[\boldsymbol{A}_1^T\boldsymbol{P}\boldsymbol{A}_1 - \boldsymbol{P}\right]\boldsymbol{x}_{f,k} + \\ & \boldsymbol{x}_{f,k}^T\left[\boldsymbol{A}_1^T\boldsymbol{P}\boldsymbol{B}\right]\boldsymbol{v}_k + \\ & \boldsymbol{v}_k^T\left[\boldsymbol{B}^T\boldsymbol{P}\boldsymbol{A}_1\right]\boldsymbol{x}_{f,k} + \\ & \boldsymbol{v}_k^T\left[\boldsymbol{B}^T\boldsymbol{P}\boldsymbol{B}\right]\boldsymbol{v}_k.\end{aligned} \tag{29}$$

Thus, it can be shown that (28) is equivalent to

$$\begin{bmatrix} \boldsymbol{A}_1^T \boldsymbol{P} \boldsymbol{A}_1 + \boldsymbol{I} - \boldsymbol{P} & \boldsymbol{A}_1^T \boldsymbol{P} \boldsymbol{B} \\ \boldsymbol{B}^T \boldsymbol{P} \boldsymbol{A}_1 & \boldsymbol{B}^T \boldsymbol{P} \boldsymbol{B} - \mu^2 \boldsymbol{I} \end{bmatrix} \prec \boldsymbol{0}. \tag{30}$$

Lemma 1. *The following statements are equivalent [4, 2, 5]*

(i) There exists $\boldsymbol{X} \succ 0$ *such that*

$$\boldsymbol{V}^T \boldsymbol{X} \boldsymbol{V} - \boldsymbol{W} \prec 0 \tag{31}$$

(ii) There exists $\boldsymbol{X} \succ 0$ *such that*

$$\begin{bmatrix} -\boldsymbol{W} & \boldsymbol{V}^T \boldsymbol{U}^T \\ \boldsymbol{U} \boldsymbol{V} & \boldsymbol{X} - \boldsymbol{U} - \boldsymbol{U}^T \end{bmatrix} \prec 0. \tag{32}$$

The following theorem presents the main tool of this section

Theorem 1. *For a prescribed fault estimation uncertainty and disturbances attenuation level* $\mu > 0$ *for the* $\boldsymbol{x}_{f,k}$, *the* $\mathcal{H}_\infty$ *controller design problem for the system* (1) *is solvable if there exist* $\boldsymbol{U}$, $\boldsymbol{N}$ *and* $\boldsymbol{P} \succ \boldsymbol{0}$ *such that the following condition is satisfied:*

$$\begin{bmatrix} \boldsymbol{I} - \boldsymbol{P} & \boldsymbol{0} & \boldsymbol{A}\boldsymbol{U} - \boldsymbol{B}\boldsymbol{N} \\ \boldsymbol{0} & -\mu^2 \boldsymbol{I} & \boldsymbol{U}^T \boldsymbol{B}^T \\ \boldsymbol{U}^T \boldsymbol{A}^T - \boldsymbol{N}^T \boldsymbol{B}^T & \boldsymbol{U}\boldsymbol{B} & \boldsymbol{P} - \boldsymbol{U} - \boldsymbol{U}^T \end{bmatrix} \prec \boldsymbol{0}, \tag{33}$$

with $\boldsymbol{N} = \boldsymbol{K}\boldsymbol{U}$.

Finally, the design procedure boils down to solving (33) with respect to $\boldsymbol{U}$, $\boldsymbol{N}$ and $\boldsymbol{P}$, subsequently gain matrix of the controlled can be calculated as follow:

$$\boldsymbol{K} = \boldsymbol{N}\boldsymbol{U}^{-1}. \tag{34}$$

5. Case study

To verify the proposed approach, it was implemented for the quadruple-tank process. A schematic diagram of the process is presented in Fig. 2. The goal is to control the water level in tank 1, 2 and to estimate the faults for pump 1 and pump 2, respectively. The system can be described by following equations [8]:

$$\frac{dh_1}{dt} = -\frac{a_1}{A_1}\sqrt{2gh_1} + \frac{a_3}{A_1}\sqrt{2gh_3} + \frac{\gamma_1 k_1}{A_1} v_1 \tag{35}$$

$$\frac{dh_2}{dt} = -\frac{a_2}{A_2}\sqrt{2gh_2} + \frac{a_4}{A_2}\sqrt{2gh_4} + \frac{\gamma_2 k_2}{A_2} v_2 \tag{36}$$

$$\frac{dh_3}{dt} = -\frac{a_3}{A_3}\sqrt{2gh_3} + \frac{(1-\gamma_2) k_2}{A_3} v_2 \tag{37}$$

$$\frac{dh_4}{dt} = -\frac{a_4}{A_4}\sqrt{2gh_4} + \frac{(1-\gamma_1) k_1}{A_4} v_1 \tag{38}$$

$$\tag{39}$$

where A_i is cross–sectional area of the tank, a_i stands for cross–section of the outlet orifice, h_i is water level, v_i is voltage applied to pump, γ_i are determined from how the valves are set prior to an experiment, g is the acceleration of gravity. The numerical values of above parameters are as follows: $A_1 = A_3 = 28[cm^2]$, $A_2 = A_4 = 32[cm^2]$, $a_1 = a_3 = 0.071[cm^2]$, $a_2 = a_4 = 0.057[cm^2]$, $g = 981.0[cm/s^2]$. The model was linearised around the steady-state

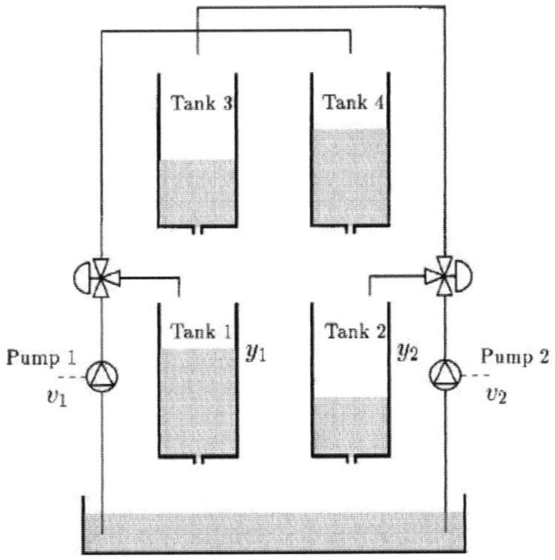

Figure 2. The quadruple-tank process [8]

$h_1^0 = 12.4[cm]$, $h_2^0 = 12.7cm$, $h_3^0 = 1.8cm$, $h_4^0 = 1.4cm$, $k_1 = 3.33[cm^3/Vs]$, $k_2 = 3.35[cm^3/Vs]$, $\gamma_1 = 0.7[cm^3/Vs]$, $\gamma_2 = 0.6[cm^3/Vs]$ and discretized by using Euler method with the sampling time $T_s = 0.1s$. The discrete time system is given by:

$$\boldsymbol{x}_{f,k+1} = \boldsymbol{A}\boldsymbol{x}_{f,k} + \boldsymbol{B}\boldsymbol{u}_{f,k} + \boldsymbol{B}\boldsymbol{f}_k + \boldsymbol{W}\boldsymbol{w}_k, \tag{40}$$

with

$$\boldsymbol{A} = \begin{bmatrix} 0.9984 & 0 & 0.0042 & 0 \\ 0 & 0.9989 & 0 & 0.0033 \\ 0 & 0 & 0.9958 & 0 \\ 0 & 0 & 0 & 0.9967 \end{bmatrix}, \quad \boldsymbol{B} = \begin{bmatrix} 0.0083 & 0 \\ 0 & 0.0062 \\ 0 & 0.0048 \\ 0.0031 & 0 \end{bmatrix},$$

$$\boldsymbol{W} = \begin{bmatrix} 0.0500 & 0 & 0 & 0 \\ 0 & 0.0500 & 0 & 0 \\ 0 & 0 & 0.0500 & 0 \\ 0 & 0 & 0 & 0.0500 \end{bmatrix},$$

and the exogenous disturbance input

$$\boldsymbol{w}_k \sim \mathcal{N}(\boldsymbol{0}, 0.1^2\boldsymbol{I}). \tag{41}$$

The robust controller was designed with (33) and $\mu = 0.9$. The obtained robust controller $\boldsymbol{K}$ is as follows:

$$\boldsymbol{K} = \begin{bmatrix} 106.2134 & 3.5115 & -4.1539 & 36.5174 \\ -0.0436 & 112.2489 & 62.4471 & 0.4812 \end{bmatrix}$$

Figure 3 presents the system evolution for the fault-free case. It can be seen that the state converges to the required setpoint for the following initial conditions $\boldsymbol{x}_{f,0} = [0.03, 0.04, 0.01, 0.05]$. For further comparative study, two control strategies are employed:

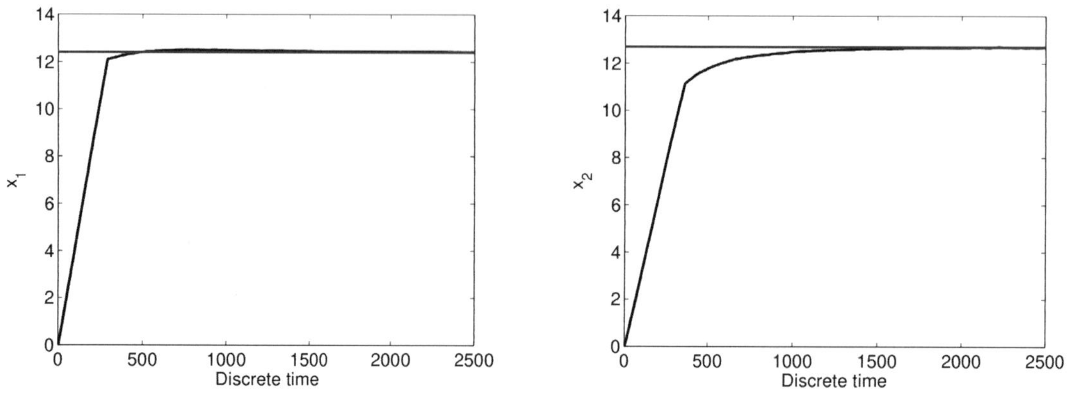

Figure 3. Set point, state $\boldsymbol{x}_1$ and $\boldsymbol{x}_2$ (dash line) (for $k = 0, \ldots, 2500$)

Without FTC

$$\boldsymbol{u}_{f,k} = -\boldsymbol{K}\boldsymbol{x}_{f,k}. \tag{42}$$

With FTC

$$\boldsymbol{u}_{f,k} = -\hat{\boldsymbol{f}}_{k-1} - \boldsymbol{K}\boldsymbol{x}_{f,k}. \tag{43}$$

Moreover, three fault scenarios are considered

S1: 5% performance decrease of the first actuator:

$$\boldsymbol{f}_{k,1} = \begin{cases} -0.05\boldsymbol{u}_{f,k} & 1500 \le k \le 1600, \\ 0 & \text{otherwise} \end{cases}$$
$$\boldsymbol{f}_{k,2} = 0.$$

S2: 10% performance decrease of the first actuator:

$$\boldsymbol{f}_{k,1} = \begin{cases} -0.1\boldsymbol{u}_{f,k} & 1500 \le k \le 1600, \\ 0 & \text{otherwise} \end{cases}$$
$$\boldsymbol{f}_{k,2} = 0.$$

S3: 5% and 10% performance decrease of the first and second actuator, respectively:

$$\boldsymbol{f}_{k,1} = \begin{cases} -0.05\boldsymbol{u}_{f,k,1} & 1500 \le k \le 1600, \\ 0 & \text{otherwise} \end{cases}$$
$$\boldsymbol{f}_{k,2} = \begin{cases} -0.1\boldsymbol{u}_{f,k,2} & 1500 \le k \le 1600, \\ 0 & \text{otherwise} \end{cases}$$

The results presented in Figs. 4–7, show the system performance with and without FTC strategy. Fault scenarios S1–S3 show the operation of the system for different actuator fault values. The control strategy without FTC provides the worst results for all considered fault scenarios S1–S3. The control strategy without FTC, which being a robust control only, does not consider any information about faults. Thus it is impossible to be realise the suitable recovery actions, while the control strategy with FTC gives better significantly better performance of the system.

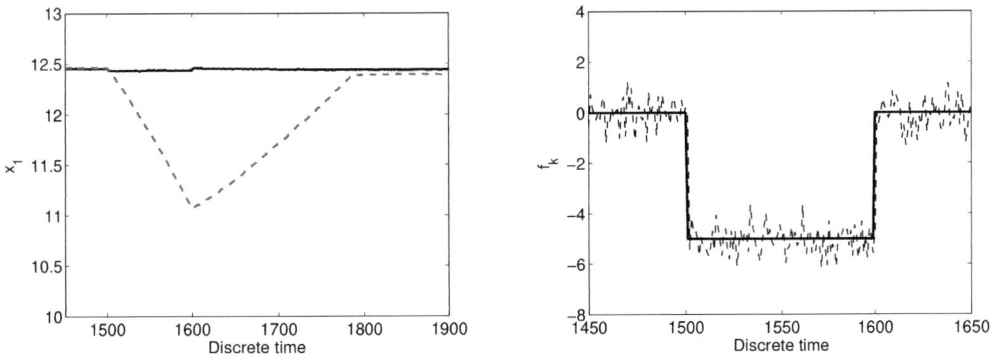

Figure 4. Performance of the system with (solid line) and without FTC (dashed line) (left) and fault (solid line) and its estimate (dashed line) (right) for scenario S1

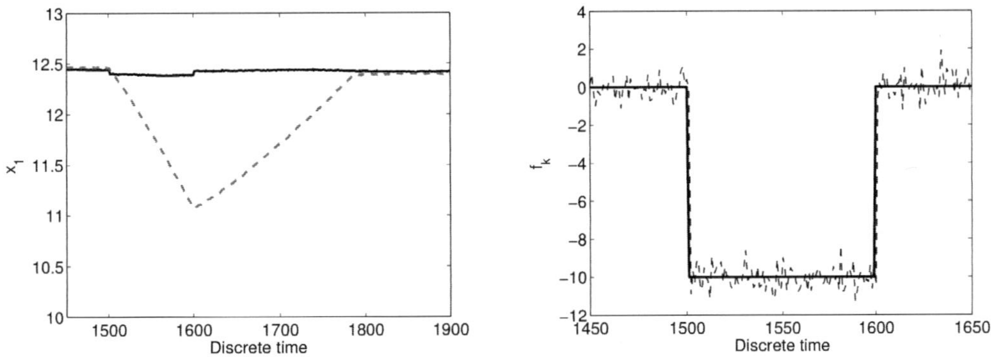

Figure 5. Performance of the system with (solid line) and without FTC (dashed line) (left) and fault (solid line) and its estimate (dashed line) (right) for scenario S2

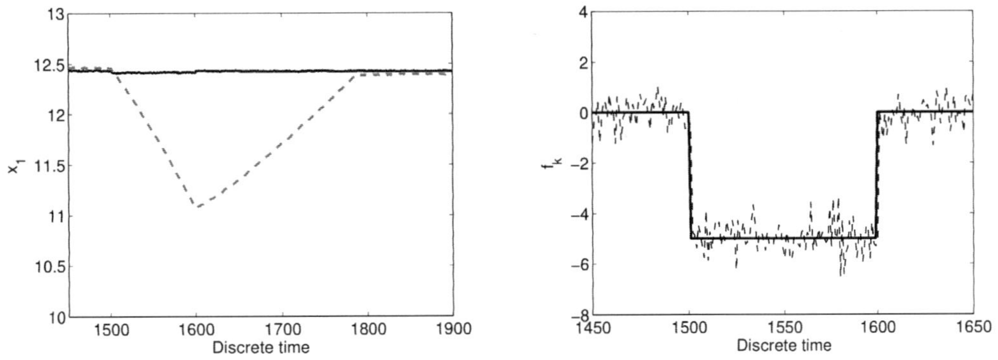

Figure 6. Performance of the system with (solid line) and without FTC (dashed line) (left) and fault (solid line) and its estimate (dashed line) (right) for scenario S3 (first actuator)

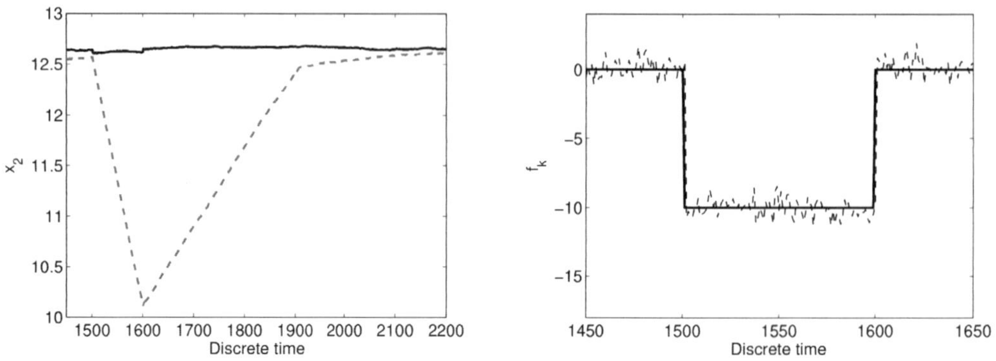

Figure 7. Performance of the system with (solid line) and without FTC (dashed line) (left) and fault (solid line) and its estimate (dashed line) (right) for scenario S3 (second actuator)

6. Conclusions

The main objective of the paper is to propose a robust fault-tolerant control scheme for a non-linear discrete-time systems. The achieved results were divide into the following points:

- minimisation the effects of exogenous disturbances and present fault estimation strategy, through the $\mathcal{H}_\infty$ approach,
- development the procedure for: fault estimation, fault compensation with robust controller,
- verification the proposed approach on the quadruple-tank process.

The proposed approach can be efficiently implemented to real-time system. The offline computations boil down to solving linear matrix inequalities. While the on-line computation boils down to calculate the fault and control signal. The proposed approach was applied to the quadruple-tank process. The achieved results clearly exhibit the high performance of the proposed scheme.

Acknowledgments

The work was supported by the National Science Centre of Poland under grant: 2014-2017

[1] M. Blanke, M. Kinnaert, J. Lunze, and M. Staroswiecki. *Diagnosis and Fault-Tolerant Control.* Springer-Verlag, New York, 2003.

[2] J. Daafouz, P. Riedinger, and C. Iung. Stability analysis and control synthesis for switched systems: a switched lyapunov function approach. *IEEE Transactions on Automatic Control*, 47(11):1883–1887, 2002.

[3] S. de Oca, V. Puig, M. Witczak, and L. Dziekan. Fault-tolerant control strategy for actuator faults using lpv techniques: application to a two degree of freedom helicopter. *International Journal of Applied Mathematics and Computer Science*, 22(1):161–171, 2012.

[4] M.C. de Oliveira, J. Bernussou, and J.C. Geromel. A new discrete-time robust stability condition. *Systems and Control Letters*, 37(4):261–265, 1999.

[5] F. Delmotte, T.M. Guerra, and M. Ksantini. Continuous takagi–sugeno's models: reduction of the number of lmi conditions in various fuzzy control design technics. *IEEE Transactions on Fuzzy Systems*, 15(3):426–438, 2007.

[6] S. Gillijns and B. De Moor. Unbiased minimum-variance input and state estimation for linear discrete-time systems. *Automatica*, 43:111–116, 2007.

[7] R. Isermann. *Fault Diagnosis Applications: Model Based Condition Monitoring, Actuators, Drives, Machinery, Plants, Sensors, and Fault-tolerant Systems.* Springer-Verlag, Berlin, 2011.

[8] Johansson K.H. The quadruple-tank process: A multivariable laboratory process with an adjustable zero. *IEEE Transactions on Control Systems Technolofy*, 8(2):456–465, 2000.

[9] J. Korbicz, J. Kościelny, Z. Kowalczuk, and W. Cholewa (Eds.). *Fault diagnosis. Models, Artificial Intelligence, Applications.* Springer-Verlag, Berlin, 2004.

Journal of Physics: Conference Series **570** (2014) 082002 doi:10.1088/1742-6596/570/8/082002

[10] Y. Liang, D. Liaw, and T. Lee. Reliable control of nonlinear systems. *IEEE Trans. Automatic Control*, 45(4):706–710, 2000.

[11] F. Liao, J. Wang, and G. Yang. Reliable robust flight tracking control: an lmi approach. *IEEE Trans. Control Systems Technology*, 10(1):76–89, 2000.

[12] M. Mahmoud, J. Jiang, and Y. Zhang. *Active Fault Tolerant Control Systems: Stochastic Analysis and Synthesis*. Springer-Verlag, Berlin, 2003.

[13] M. Mrugalski and M. Witczak. State-space gmdh neural networks for actuator robust fault diagnosis. *Advances in Electrical and Computer Engineering*, 12(3):65–72, 2012.

[14] M. Mrugalski, M. Witczak, and J. Korbicz. Confidence estimation of the multi-layer perceptron and its application in fault detection systems. *Engineering Applications of Artificial Intelligence*, 21(8):895–906, 2008.

[15] H. Noura, D. Theilliol, J. Ponsart, and A. Chamseddine. *Fault-tolerant Control Systems: Design and Practical Applications*. Springer-Verlag, Berlin, 2003.

[16] Z. Qu, C. M. Ihlefeld, J. Yufang, and A. Saengdeejing. Robust fault-tolerant self-recovering control of nonlinear uncertain systems. *Automatica*, 39(10):1763–1771, 2003.

[17] M. Witczak. Toward the training of feed-forward neural networks with the D-optimum input sequence. *IEEE Transactions on Neural Networks*, 17(2):357–373, 2006.

[18] M. Witczak. *Modelling and Estimation Strategies for Fault Diagnosis of Non-linear Systems*. Springer-Verlag, Berlin, 2007.

[19] M. Witczak, J. Korbicz, and M. Luzar. A lmi-based strategy for $\mathcal{H}_\infty$ fault estimation of non-linear systems: Application to the multi-tank system. In *Proceedings of the 13th European Control Conference, Strasbourg, France*, pages 270–275, 2014.

[20] M. Witczak, A. Obuchowicz, and J. Korbicz. Genetic programming based approaches to identification and fault diagnosis of non-linear dynamic systems. *International Journal of Control*, 75(13):1012–1031, 2002.

[21] M. Witczak and P. Pretki. Design of an extended unknown input observer with stochastic robustness techniques and evolutionary algorithms. *International Journal of Control*, 80(5):749–762, 2007.

[22] M. Witczak, V. Puig, and S. de Oca. A fault-tolerant control strategy for non-linear discrete-time systems: application to the twin-rotor system. *International Journal of Control*, 86(10):1788–1799, 2013.

[23] M. Witczak, R. Rybski, and J. Kaczmarek. Impedance measurement with the d-optimum experimental conditions. *IEEE Transactions on Instrumentation and Measurement*, 58(8):2535–2543, 2009.

[24] Y. Zhang and J. Jiang. Bibliographical review on reconfigurable fault-tolerant control systems. In *IFAC Symposium Fault Detection Supervision and Safety of Technical Processes, SAFEPROCESS*, pages 265–276, Washington, D.C., USA, 2003.

European Workshop on Advanced Control and Diagnosis IOP Publishing
Journal of Physics: Conference Series **570** (2014) 082003 doi:10.1088/1742-6596/570/8/082003

Feedback Linearization approach for Standard and Fault Tolerant control: Application to a Quadrotor UAV Testbed

J. Ghandour, S. Aberkane, J-C. Ponsart

University of Lorraine, CRAN, UMR 7039, Campus Sciences, B.P. 70239, 54506 Vandoeuvre-lès-Nancy Cedex, France

CNRS, CRAN, UMR 7039, France

Jean-Christophe.Ponsart@univ-lorraine.fr

Abstract. In this paper the control problem of a quadrotor vehicle experiencing a rotor failure is investigated. We develop a Feedback linearization approach to design a controller whose task is to make the vehicle performs trajectory following. Then we use the same approach to design a controller whose task is to make the vehicle enter a stable spin around its vertical axis, while retaining zero angular velocities around the other axis when a rotor failure is present. These conditions can be exploited to design a second control loop, which is used to perform trajectory following. The proposed double control loop architecture allows the vehicle to perform both trajectory and roll/pitch control. At last, to test the robustness of the feedback linearization technique, we applied wind to the quadrotor in mid flight.

1. Introduction

While the military and civil interest in Unmanned Arial Vehicles (UAVs) grows exponentially, the research centers race to publishing new and more efficient control techniques ([5], [16]). Among the various proposed control methods in the literature, one can cite for example LQ ([10]), H ([11], [12], [14]), backstepping ([7], [9], [15]) or sliding mode ([4], [13]) control techniques. In this paper, we propose to consider a nonlinear feedback linearization control ([2], [6]) in a fault tolerant control framework ([3]).

A UAV presents several basic advantages over a Manned Aerial Vehicle (MAV), including lower cost, prolonged continuous flight time, reduced radar signatures, smaller size, increased maneuverability, and the most important benefit is being less risky to crews.

Particular UAVs that have received much attention in the literature during the last ten years are quadrotors. This is mainly due to there suitability for a wide variety of UAV research applications and test-fly methods in control, fault diagnosis, fault tolerant control (FTC) as well as multi-agent based technologies in cooperative control, distributed control, formation flight, mobile wireless networks and communications.

During certain operations, a UAV cannot remains controlled from ground, for example in long continuous flights, in surveillance operations where multiple UAVs are in flight simultaneously, or behind the enemy line where signal transmission can put the stealth aspect of the UAV in jeopardy. This created the need for autonomous UAVs, which are programmed in a way to accomplish a given task without the need of human intervention at any point during the procedure. But with autonomous UAVs came another problem, these UAVs could perfectly handle themselves when fully operational, but a faulty sensor, or a damaged body, gets the system unstable, crashing the vehicle right away. Due to the high cost of vehicles used by the military and civil organisms around the world, a FTC system was needed to be installed on these vehicles, assuring the damaged vehicle returns to its base without further damages.

This paper is divided into several sections. At first, the quadrotor description and modeling is presented. Then, the control law based on feedback linearization is detailed. The last section presents simulation results done on Matlab/Simulink®.

2. Quadrotor description

The quadrotor is a simple to manufacture aircraft with vertical landing/take-off capability, hovering ability, and is greatly maneuverable. The figure 1 shows the four actuators in a cross configuration, producing three rotations, Roll-Pitch-Yaw, and a vertical translation, Lift.

The first problem that faces us when creating a hovering machine is the problem of torque reaction based on Newton's third law of motion. This problem is solved in a quadrotor by turning two rotors clockwise, while turning the others counter-clockwise, thus canceling the global torque applied on the quadrotor.

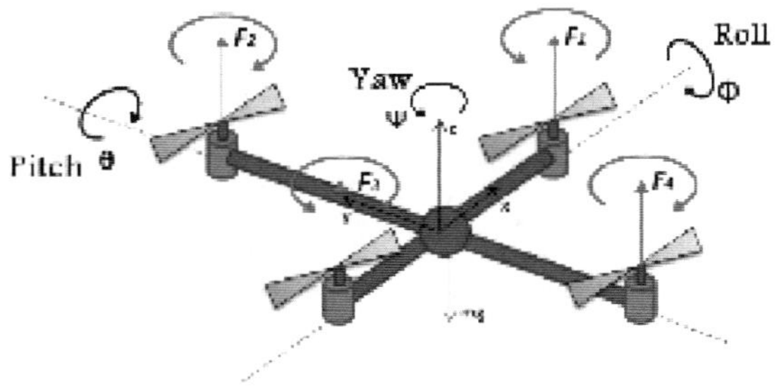

Figure 1. The Quadrotor

A relation exists between the vector $U=[U_z, U_\varphi, U_\theta, U_\psi]^T$ and $F=[F_1, F_2, F_3, F_4]^T$:

$$\begin{bmatrix} U_z \\ U_\varphi \\ U_\theta \\ U_\psi \end{bmatrix} = \begin{bmatrix} 1 & 1 & 1 & 1 \\ 0 & d & 0 & -d \\ -d & 0 & d & 0 \\ c & -c & c & -c \end{bmatrix} \begin{bmatrix} F_1 \\ F_2 \\ F_3 \\ F_4 \end{bmatrix} \tag{1}$$

where d is the distance from the epicenter of the quadrotor to the actuator axes, $c>0$ is the drag factor, F_i is the lift generated by the ith actuator, U_z is the total lift force generated by all four actuators and $U_\varphi, U_\theta, U_\psi$ are the couples generated respectively in the Roll, Pitch and Yaw directions.

The mathematical model describing the position and attitude of a quadrotor is based on that of a rotating rigid body with six degrees of freedom. Close to hovering conditions $(\varphi \approx 0 \text{ et } \theta \approx 0)$ one obtains ([14], [15]):

$$\begin{cases} m\,\ddot{x} = \left(C_\varphi S_\theta C_\psi + S_\varphi S_\psi\right) U_z - K_x\,\dot{x} \\ m\,\ddot{y} = \left(C_\varphi S_\theta S_\psi - S_\varphi C_\psi\right) U_z - K_y\,\dot{y} \\ m\,\ddot{z} = \left(C_\varphi C_\theta\right) U_z - m\,g - K_z\,\dot{z} \\ J_p\,\ddot{\varphi} = U_\varphi - \dot{\theta}\,\dot{\psi}\left(J_r - J_q\right) - \dot{\theta}\,J_T\,\Omega_T - K_p\,\dot{\varphi} \\ J_q\,\ddot{\theta} = U_\theta - \dot{\varphi}\,\dot{\psi}\left(J_p - J_r\right) - \dot{\varphi}\,J_T\,\Omega_T - K_q\,\dot{\theta} \\ J_r\,\ddot{\psi} = U_\psi - \dot{\varphi}\,\dot{\theta}\left(J_q - J_p\right) - K_r\,\dot{\psi} \end{cases} \tag{2}$$

where m is the mass of the quadrotor, J_p, J_q, J_r are the moments of inertia of the quadrotor, J_T is the moment of inertia for the engines and $\Omega_T = \Omega_1 - \Omega_2 + \Omega_3 - \Omega_4$ is the overall speed of the propellers wherein Ω_i is the speed of the ith propeller. $S_\bullet$ and $C_\bullet$ are the respective abbreviations of sin(.) and cos(.).

3. Controlling the vehicle

With 6 degrees of freedom, and 4 actuators, the quadrotor is an under-actuated system, making it impossible to control all of the states simultaneously. The controlled outputs are generally $\{x, y, z, \psi\}$, thus with stable pitch and roll angles, a quadrotor is able to reach a desired position in space and a desired yaw angle. As stated in [1] several UAV flight control algorithms already exist, including the Gain—Scheduled PID, adaptive control, Model Reference Adaptive Control (MRAC), Sliding Mode Control, Backstepping Control (BSC), Control allocation, Model Predictive Control (MPC), and Feedback Linearization (FL).

3.1. Feedback Linearization

3.1.1. Procedure's description

Feedback linearization is a common approach used in controlling nonlinear systems ([2], [6]). The approach involves coming up with a transformation of the nonlinear system into an equivalent linear system through a change of variables and a suitable control input. Feedback linearization may be applied to nonlinear systems as:

$$\begin{cases} \dot{x}(t) = f(x) + g(x)\,u(t) \\ y(t) = h(x) \end{cases} \tag{3}$$

where $x \in \Re^n$ is the state vector, $u \in \Re^p$ is the vector of inputs, and $y \in \Re^m$ is the vector of outputs. The goal is to develop a control input $u(t) = a(x) + b(x)\,v(t)$ that induces a linear input-output map between the new input v(t) and the output $y(t)$. An outer-loop control strategy for the resulting linear system can then be applied.

3.1.2. Procedure's application

In this section, we use the model (2) after neglecting the gyroscopic effects and the drag forces, while using the steps of the work in [2], to develop a feedback Linearization controller of a quadrotor ensuring that the position $\{x(t), y(t), z(t), \psi(t)\}$ tracks the desired trajectory $\{x_d(t), y_d(t), z_d(t), \psi_d(t)\}$ asymptotically.

Let $\xi = [x, y, z, \dot{x}, \dot{y}, \dot{z}, \varphi, \theta, \psi, \dot{\varphi}, \dot{\theta}, \dot{\psi}]^{\mathrm{T}}$ be the state vector, $U = [T, T_\varphi, T_\theta, T_\psi]^{\mathrm{T}}$ the control input, and $\eta(\xi) = [x, y, z, \psi]^{\mathrm{T}}$ the output vector. Being under-actuated, we cannot use a static but a dynamic output feedback control. By successive differentiation, a nonsingular input appears when we reach the forth derivative of x, y, and z, and the second derivative of ψ:

$$\begin{aligned} x^{(4)}(t) = \frac{U_z}{m}\big[& -\ddot{\psi} S_\psi S_\theta C_\varphi - \dot{\psi}^2 C_\psi S_\theta C_\varphi - \dot{\psi}\dot{\theta} S_\psi C_\theta C_\varphi + \dot{\psi}\dot{\varphi} S_\psi S_\theta S_\varphi + \ddot{\theta} C_\psi C_\theta C_\varphi - \dot{\theta}\dot{\psi} S_\psi C_\theta C_\varphi - \dot{\theta}^2 C_\psi S_\theta C_\varphi - \dot{\theta}\dot{\varphi} C_\psi C_\theta S_\varphi - \ddot{\varphi} C_\psi S_\theta S_\varphi \\ & + \dot{\varphi}\dot{\psi} S_\psi S_\theta S_\varphi + \dot{\varphi}\dot{\theta} C_\psi C_\theta S_\varphi - \dot{\varphi}^2 C_\psi S_\theta C_\varphi + \ddot{\psi} C_\psi S_\varphi - \dot{\psi}^2 S_\psi S_\varphi + \dot{\psi}\dot{\varphi} C_\psi C_\varphi + \ddot{\varphi} S_\psi C_\varphi + \dot{\psi}\dot{\varphi} C_\psi C_\varphi - \dot{\varphi}^2 S_\psi S_\varphi \big] \\ & + \frac{2\dot{U}_z}{m}\big[-\dot{\psi} S_\psi S_\theta C_\varphi + \dot{\theta} C_\psi C_\theta C_\varphi - \dot{\varphi} C_\psi S_\theta S_\varphi + \dot{\psi} C_\psi S_\varphi + \dot{\varphi} S_\psi C_\varphi\big] + \frac{\ddot{U}_z}{m}\big[C_\psi S_\theta C_\varphi + S_\psi S_\varphi\big] \end{aligned} \tag{4}$$

$$\begin{aligned} y^{(4)}(t) = \frac{U_z}{m}\big[& \ddot{\psi} C_\psi S_\theta C_\varphi - \dot{\psi}^2 S_\psi S_\theta C_\varphi + \dot{\psi}\dot{\theta} C_\psi C_\theta C_\varphi - \dot{\psi}\dot{\varphi} C_\psi S_\theta S_\varphi + \ddot{\theta} S_\psi C_\theta C_\varphi + \dot{\theta}\dot{\psi} C_\psi C_\theta C_\varphi - \dot{\theta}^2 S_\psi S_\theta C_\varphi - \dot{\theta}\dot{\varphi} S_\psi C_\theta S_\varphi - \ddot{\varphi} S_\psi S_\theta S_\varphi \\ & - \dot{\varphi}\dot{\psi} C_\psi S_\theta S_\varphi - \dot{\varphi}\dot{\theta} S_\psi C_\theta S_\varphi - \dot{\varphi}^2 S_\psi S_\theta C_\varphi - \ddot{\varphi} C_\psi C_\varphi + \dot{\varphi}^2 C_\psi S_\varphi + \dot{\psi}\dot{\varphi} S_\psi C_\varphi + \ddot{\psi} S_\psi S_\varphi + \dot{\psi}\dot{\varphi} S_\psi C_\varphi + \dot{\psi}^2 C_\psi S_\varphi \big] \\ & + \frac{2\dot{U}_z}{m}\big[\dot{\psi} C_\psi S_\theta C_\varphi + \dot{\theta} S_\psi C_\theta C_\varphi - \dot{\varphi} S_\psi S_\theta S_\varphi + \dot{\psi} S_\psi S_\varphi - \dot{\varphi} C_\psi C_\varphi\big] + \frac{\ddot{U}_z}{m}\big[S_\psi S_\theta C_\varphi - C_\psi S_\varphi\big] \end{aligned} \tag{5}$$

$$z^{(4)}(t)=\frac{U_z}{m}\left[-\ddot{\theta}S_\theta C_\varphi-\dot{\theta}^2C_\theta C_\varphi+\dot{\theta}\dot{\varphi}S_\theta S_\varphi-\ddot{\varphi}C_\theta S_\varphi+\dot{\varphi}\dot{\theta}S_\theta S_\varphi-\dot{\varphi}^2C_\theta C_\varphi\right]+\frac{2\dot{U}_z}{m}\left[-\dot{\theta}S_\theta C_\varphi-\dot{\varphi}C_\theta S_\varphi\right]+\frac{\ddot{U}_z}{m}\left[C_\theta C_\varphi\right] \tag{6}$$

with $\ddot{\varphi}=\frac{U_\varphi}{J_p}$, $\ddot{\theta}=\frac{U_\theta}{J_q}$ and $\ddot{\psi}=\frac{U_\psi}{J_r}$. Let $\varsigma=\left[U_z,\dot{U}_z\right]^{\mathrm{T}}$ and $P=\left[x,y,z\right]^{\mathrm{T}}$ thus we can write:

$$\begin{pmatrix} P^{(4)}(t) \\ \psi^{(2)}(t) \end{pmatrix}=l(\xi,\varsigma)+J(\xi,\varsigma)\,\tilde{u}(t) \tag{7}$$

With $\tilde{u}=\left[\ddot{U}_z,U_\varphi,U_\theta,U_\psi\right]^{\mathrm{T}}$, $l=\left[l_1,l_2,l_3,l_4\right]^{\mathrm{T}}$ and $J=\begin{bmatrix} J_{11} & J_{12} & J_{13} & J_{14} \\ J_{21} & J_{22} & J_{23} & J_{24} \\ J_{31} & J_{32} & J_{33} & J_{34} \\ J_{41} & J_{42} & J_{43} & J_{44} \end{bmatrix}$ where

$$\begin{aligned} l_1(t)=\frac{U_z}{m}\Big[&-\dot{\psi}^2C_\psi S_\theta C_\varphi-\dot{\psi}\dot{\theta}S_\psi C_\theta C_\varphi-\dot{\psi}\dot{\varphi}S_\psi S_\theta S_\varphi-\dot{\theta}\dot{\psi}S_\psi C_\theta C_\varphi-\dot{\theta}^2C_\psi S_\theta C_\varphi-\dot{\theta}\dot{\varphi}C_\psi C_\theta S_\varphi+\dot{\varphi}\dot{\psi}S_\psi S_\theta S_\varphi-\dot{\varphi}\dot{\theta}C_\psi C_\theta S_\varphi-\dot{\varphi}^2C_\psi S_\theta C_\varphi \\ &-\dot{\psi}^2S_\psi S_\varphi+2\dot{\psi}\dot{\varphi}C_\psi C_\varphi-\dot{\varphi}^2S_\psi S_\varphi\Big]+\frac{2\dot{U}_z}{m}\left[-\dot{\psi}S_\psi S_\theta C_\varphi+\dot{\theta}C_\psi C_\theta C_\varphi-\dot{\varphi}C_\psi S_\theta S_\varphi+\dot{\psi}C_\psi S_\varphi+\dot{\varphi}S_\psi C_\varphi\right] \end{aligned} \tag{8}$$

$$\begin{aligned} l_2(t)=\frac{U_z}{m}\Big[&-\dot{\psi}^2S_\psi S_\theta C_\varphi+\dot{\psi}\dot{\theta}C_\psi C_\theta C_\varphi-\dot{\psi}\dot{\varphi}C_\psi S_\theta S_\varphi+\dot{\theta}\dot{\psi}C_\psi C_\theta C_\varphi-\dot{\theta}^2S_\psi S_\theta C_\varphi-\dot{\theta}\dot{\varphi}S_\psi C_\theta S_\varphi-\dot{\varphi}\dot{\psi}C_\psi S_\theta S_\varphi-\dot{\varphi}\dot{\theta}S_\psi C_\theta S_\varphi-\dot{\varphi}^2S_\psi S_\theta C_\varphi \\ &+\dot{\psi}^2C_\psi S_\varphi+2\dot{\psi}\dot{\varphi}S_\psi C_\varphi+\dot{\varphi}^2C_\psi S_\varphi\Big]+\frac{2\dot{U}_z}{m}\left[\dot{\psi}C_\psi S_\theta C_\varphi+\dot{\theta}S_\psi C_\theta C_\varphi-\dot{\varphi}S_\psi S_\theta S_\varphi+\dot{\psi}S_\psi S_\varphi-\dot{\varphi}C_\psi C_\varphi\right] \end{aligned} \tag{9}$$

$$l_3(t)=\frac{U_z}{m}\left[-\dot{\theta}^2C_\theta C_\varphi+\dot{\theta}\dot{\varphi}S_\theta S_\varphi+\dot{\varphi}\dot{\theta}S_\theta S_\varphi-\dot{\varphi}^2C_\theta C_\varphi\right]+\frac{2\dot{U}_z}{m}\left[-\dot{\theta}S_\theta C_\varphi-\dot{\varphi}C_\theta S_\varphi\right] \tag{10}$$

$$l_4(t)=0 \tag{11}$$

$$\begin{aligned} &J_{11}=\frac{1}{m}\left[C_\psi S_\theta C_\varphi+S_\psi S_\varphi\right],\quad J_{12}=\frac{U_z}{mJ_p}\left[-C_\psi S_\theta S_\varphi+S_\psi C_\varphi\right],\quad J_{13}=\frac{U_z}{mJ_q}\left[C_\psi C_\theta C_\varphi\right],\quad J_{14}=\frac{U_z}{mJ_r}\left[-S_\psi S_\theta C_\varphi+C_\psi S_\varphi\right], \\ &J_{21}=\frac{1}{m}\left[S_\psi S_\theta C_\varphi-C_\psi S_\varphi\right],\quad J_{22}=\frac{U_z}{mJ_p}\left[-S_\psi S_\theta S_\varphi-C_\psi C_\varphi\right],\quad J_{23}=\frac{U_z}{mJ_q}\left[S_\psi C_\theta C_\varphi\right],\quad J_{24}=\frac{U_z}{mJ_r}\left[C_\psi S_\theta C_\varphi+S_\psi S_\varphi\right], \\ &J_{31}=\frac{1}{m}\left[C_\theta C_\varphi\right],\quad J_{32}=\frac{U_z}{mJ_p}\left[-C_\theta S_\varphi\right],\quad J_{33}=\frac{U_z}{mJ_q}\left[-S_\theta C_\varphi\right],\quad J_{34}=0,\quad J_{41}=J_{42}=J_{43}=0,\quad J_{44}=\frac{1}{J_r} \end{aligned} \tag{12}$$

By taking $\tilde{u}(t)=J^{-1}\left(-l+v(t)\right)$, we linearize the system and obtain:

$$\begin{aligned} x^{(4)}(t)&=\eta_1^{(4)}(t)=v_1(t) \qquad & z^{(4)}(t)&=\eta_3^{(4)}(t)=v_3(t) \\ y^{(4)}(t)&=\eta_2^{(4)}(t)=v_2(t) \qquad & \psi^{(2)}(t)&=\eta_4^{(2)}(t)=v_4(t) \end{aligned} \tag{13}$$

Remark 1: It should be noted that the determinant of the matrix J is $\frac{C_\varphi}{J_pJ_qm^2}$ therefore; J is nonsingular when $\left(-\frac{\pi}{2}<\varphi<\frac{\pi}{2}\right)$, which is satisfied generally.

For trajectory tracking ([1]), with $e_i=\left(\eta_{id}-\eta_i\right)$, we can use:

$$\begin{aligned} v_1(t)&=\eta_{1d}^{(4)}(t)+\sum_{j=1}^{4}C_{j-1}^{1}e_1^{j-1} \qquad & v_3(t)&=\eta_{3d}^{(4)}(t)+\sum_{j=1}^{4}C_{j-1}^{3}e_3^{j-1} \\ v_2(t)&=\eta_{2d}^{(4)}(t)+\sum_{j=1}^{4}C_{j-1}^{2}e_2^{j-1} \qquad & v_4(t)&=\eta_{4d}^{(2)}(t)+\sum_{j=1}^{4}C_{j-1}^{4}e_4^{j-1} \end{aligned} \tag{14}$$

3.2. Feedback Linearization Approach to Fault Tolerance

An effective fault recovery is achieved after two major successive procedures. A Fault Detection and Isolation (FDI), is a fast algorithm that indicates to the system the sensor, actuator, or body part that has been damaged. Following the FDI decision, the system switches to the adequate FTC algorithm capable of stabilizing and maneuvering the damaged quadrotor. This paper does not cover the FDI procedure. Simulations to the post-FDI stage were made by manually indicating the damaged actuator, and started with initial conditions chosen far from the hover equilibrium point, representing conditions after the accomplishment of the FDI. In other words, we assume in this work that the FDI decision is available with some delay. We don't consider here the case of false alarms, missed detection or incorrect fault isolation.

The fault tolerant controller is developed following a double control loop architecture in which an inner and faster controller has the task to regulate the attitude angles and the altitude of the vehicle, while an outer and slower controller has the aim of modifying the desired values of the attitude angles in order to perform trajectory tracking. The control law of the whole system is computed by using the same steps of the work in [3].

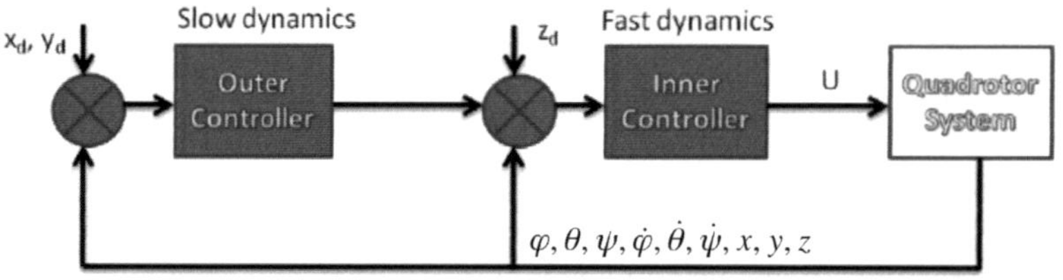

Figure 2. The double control loop architecture

Let $\chi = \left[\varphi, \theta, \psi, \dot{\varphi}, \dot{\theta}, \dot{\psi}, x, y, z, \dot{x}, \dot{y}, \dot{z}\right]^{T}$ the state vector. The dynamic equations (2), after neglecting the gyroscopic effects due to the propeller's rotation, can be rewritten in a state-space form:

$$\begin{cases} \dot{\chi}_1 = \chi_4 + \chi_5 S_{\chi_1} \dfrac{S_{\chi_2}}{C_{\chi_2}} + \chi_6 C_{\chi_1} \dfrac{S_{\chi_2}}{C_{\chi_2}} \\ \dot{\chi}_2 = \chi_5 C_{\chi_1} - \chi_6 S_{\chi_1} \\ \dot{\chi}_3 = \dfrac{1}{C_{\chi_2}}\left[\chi_5 S_{\chi_1} + \chi_6 C_{\chi_1}\right] \\ \dot{\chi}_4 = \dfrac{1}{J_p}\left[-K_p \chi_4 - \chi_5 \chi_6 \left(J_r - J_q\right) + U_\varphi\right] \\ \dot{\chi}_5 = \dfrac{1}{J_q}\left[-K_q \chi_5 - \chi_4 \chi_6 \left(J_p - J_r\right) + U_\theta\right] \\ \dot{\chi}_6 = \dfrac{1}{J_r}\left[-K_r \chi_6 - \chi_4 \chi_5 \left(J_q - J_p\right) + U_\psi\right] \end{cases} \qquad \begin{cases} \dot{\chi}_7 = \chi_{10} \\ \dot{\chi}_8 = \chi_{11} \\ \dot{\chi}_9 = \chi_{12} \\ \dot{\chi}_{10} = \dfrac{1}{m}\left[\left(C_{\chi_1} S_{\chi_2} C_{\chi_3} + S_{\chi_1} S_{\chi_3}\right) U_z - K_x \chi_{10}\right] \\ \dot{\chi}_{11} = \dfrac{1}{m}\left[\left(C_{\chi_1} S_{\chi_2} S_{\chi_3} - S_{\chi_1} C_{\chi_3}\right) U_z - K_y \chi_{11}\right] \\ \dot{\chi}_{12} = \dfrac{1}{m}\left[C_{\chi_1} C_{\chi_2} U_z - K_z \chi_{12} - mg\right] \end{cases} \tag{15}$$

In case of failure on M1: the control input vector is chosen as $u = \left[U_z, U_\varphi, U_\psi\right]^T$, where $U_\theta = \dfrac{d}{2}\left(U_z + \dfrac{U_\psi}{c}\right)$.

In case of failure on M2: the control input vector is chosen as $u = \left[U_z, U_\theta, U_\psi\right]^T$, where $U_\varphi = \dfrac{-d}{2}\left(U_z - \dfrac{U_\psi}{c}\right)$.

In case of failure on M3: the control input vector is chosen as $u = \left[U_z, U_\varphi, U_\psi\right]^T$, where $U_\theta = \dfrac{-d}{2}\left(U_z + \dfrac{U_\psi}{c}\right)$.

In case of failure on M4: the control input vector is chosen as $u = \left[U_z, U_\theta, U_\psi\right]^T$, where $U_\varphi = \dfrac{d}{2}\left(U_z - \dfrac{U_\psi}{c}\right)$.

3.2.1. Inner Control Loop

From the state-space form it can be seen that the dynamics of the state variables $\chi_1, \chi_2, \chi_4, \chi_5, \chi_6, \chi_9\ \chi_{12}$, which we will call $\bar{\chi}$ for notation simplicity, can be written as $\dot{\bar{\chi}} = f(\bar{\chi}) + g(\bar{\chi})u$.

The dynamics of the states χ_1, χ_2 and χ_9 can be similarly written as:

$$\begin{bmatrix} \dot{\chi}_1 \\ \dot{\chi}_2 \\ \dot{\chi}_9 \end{bmatrix} = \begin{bmatrix} \chi_4 + \chi_5 S_{\chi_1} \dfrac{S_{\chi_2}}{C_{\chi_2}} + \chi_6 C_{\chi_1} \dfrac{S_{\chi_2}}{C_{\chi_2}} \\ \chi_5 C_{\chi_1} - \chi_6 S_{\chi_1} \\ \chi_{12} \end{bmatrix} = \hat{f}(\overline{\chi}) \qquad (16)$$

which, in this case, is independent of the input vector u. This property becomes useful when calculating the second derivative of $[\chi_1, \chi_2, \chi_9]^T$:

$$\begin{bmatrix} \ddot{\chi}_1 \\ \ddot{\chi}_2 \\ \ddot{\chi}_9 \end{bmatrix} = \frac{d\hat{f}(\overline{\chi})}{dt} = \frac{\partial \hat{f}(\overline{\chi})}{\partial \overline{\chi}} \dot{\overline{\chi}} = \frac{\partial \hat{f}(\overline{\chi})}{\partial \overline{\chi}} f(\overline{\chi}) + \frac{\partial \hat{f}(\overline{\chi})}{\partial \overline{\chi}} g(\overline{\chi}) u \qquad (17)$$

Denoting the Jacobian matrix with $J(\overline{\chi}) = \dfrac{\partial \hat{f}(\overline{\chi})}{\partial \overline{\chi}}$, the previous dynamics can be written as:

$$\begin{bmatrix} \ddot{\chi}_1 \\ \ddot{\chi}_2 \\ \ddot{\chi}_9 \end{bmatrix} = J(\overline{\chi}) f(\overline{\chi}) + J(\overline{\chi}) g(\overline{\chi}) u \qquad (18)$$

It can be proved that matrix $J(\overline{\chi}) g(\overline{\chi})$ is invertible if and only if $\chi_2 \neq \arctan\left(\dfrac{\pm J_r l}{2 J_p d} C_{\chi_1}\right)$ and $\chi_1 \neq \arctan\left(\dfrac{\pm J_r l}{2 J_p d}\right)$. In most practical scenarios this condition is satisfied. In our case, with the parameters adopted for the simulation, the invertibility condition is guaranteed as long as the pitch and roll angles are limited to 80° which is a value never reached by the quadrotor during non-aerobatic flight. Let $\chi_{id}, \dot{\chi}_{id}, \ddot{\chi}_{id}$ be the desired values for $\chi_i, \dot{\chi}_i, \ddot{\chi}_i$ and define the ith error as $e_i = \chi_i - \chi_{id}$. If the control inputs are chosen as:

$$u^* = -\left(J(\overline{\chi}) g(\overline{\chi})\right)^{-1} J(\overline{\chi}) f(\overline{\chi}) + \left(J(\overline{\chi}) g(\overline{\chi})\right)^{-1} \left(\begin{bmatrix} \ddot{\chi}_{1d} \\ \ddot{\chi}_{2d} \\ \ddot{\chi}_{9d} \end{bmatrix} - \begin{bmatrix} Ci_1^1 e_1 \\ Ci_2^1 e_2 \\ Ci_9^1 e_9 \end{bmatrix} - \begin{bmatrix} Ci_1^2 \dot{e}_1 \\ Ci_2^2 \dot{e}_2 \\ Ci_9^2 \dot{e}_9 \end{bmatrix} \right) \qquad (19)$$

where Ci_i^j are positive constants, then the error dynamics can be written as

$$\begin{bmatrix} \ddot{e}_1 \\ \ddot{e}_2 \\ \ddot{e}_9 \end{bmatrix} + \begin{bmatrix} Ci_1^1 e_1 \\ Ci_2^1 e_2 \\ Ci_9^1 e_9 \end{bmatrix} + \begin{bmatrix} Ci_1^2 \dot{e}_1 \\ Ci_2^2 \dot{e}_2 \\ Ci_9^2 \dot{e}_9 \end{bmatrix} = \begin{bmatrix} 0 \\ 0 \\ 0 \end{bmatrix} \qquad (20)$$

which yields stable second-order dynamics providing exponential decay of e_i for i=1, 2, 9 (respectively the roll, pitch and altitude errors).

3.2.2. Outer Control Loop

An outer control loop's task is to generate the desired values for the roll and pitch angles, in a way to reach a desired position in the horizontal plane. The constants Ci and Co represent the natural frequency of the subsystem's dynamics, subsystems described by the state variables χ_1, χ_2, χ_9 and $\chi_7, \chi_8, \chi_{10}$ χ_{11}. To avoid compromising the stability of the system, we choose $Co << Ci$ which makes the outer control loop act much slower than the inner control loop. Under the presumption that χ_1 and χ_2 are small angles, χ_{1d} and χ_{2d} are chosen close to zero as:

$$\begin{bmatrix} \chi_{1d} \\ \chi_{2d} \end{bmatrix} = \frac{m}{U_z}\begin{bmatrix} S_{\chi_3} & -C_{\chi_3} \\ C_{\chi_3} & S_{\chi_3} \end{bmatrix}\begin{bmatrix} \dot{\chi}_{10d} + \frac{K_x}{m}\chi_{10} - Co_1^1 e_7 - Co_1^2 e_{10} \\ \dot{\chi}_{10d} + \frac{K_y}{m}\chi_{11} - Co_2^1 e_8 - Co_2^2 e_{11} \end{bmatrix} \tag{21}$$

where Co_i^j are positive constants, then error dynamics for the horizontal displacements in closed loop is :

$$\begin{bmatrix} \ddot{e}_x + Co_1^1 e_x + Co_1^2 \dot{e}_x \\ \ddot{e}_y + Co_2^1 e_y + Co_2^2 \dot{e}_y \end{bmatrix} = \begin{bmatrix} 0 \\ 0 \end{bmatrix} \tag{22}$$

which is asymptotically stable.

From a physical point of view, it means that when one of the rotor fails, the inner control law stabilizes roll, pitch and altitude, while the outer control law exploits the near hover condition to slowly change the pitch and roll angles in order to move the vehicle to a desired position in space.

4. Results and Discussions

Several simulations have been run using the Matlab and Simulink® softwares in order to validate the theoretical results. The model used to represent the quadrotor is preferred to be as realistic as possible, since eventually this model will be replaced by the quadrotor itself. On the other hand, the model used to calculate the control techniques tends to be as simplified as possible, with an acceptable error, in order to generate a simple and fast control algorithm.

In order to test the efficiency of the Feedback Linearization Control technique at handling and maneuvering the quadrotor, we needed a smooth trajectory with no sharp turns, and minimal acceleration value. A rectifiable polynomial trajectory that responds to the requirements was developed:

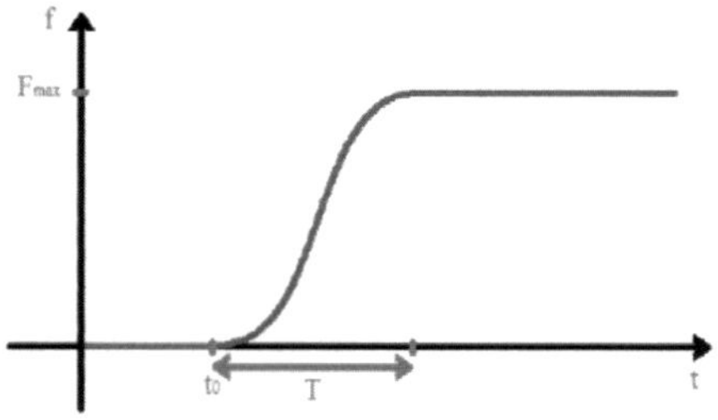

Figure 3. Polynomial equation

The polynomial equation shown in figure 3 starts ascending at t_0 with a speed and acceleration equal to 0, and after a T period of time, it gets to the desired $F_{\max}$ height with a speed and acceleration equal to 0 again:

$$f(t) = \frac{6F_{\max}}{T^5}(t-t_0)^5 - \frac{15F_{\max}}{T^4}(t-t_0)^4 + \frac{10F_{\max}}{T^3}(t-t_0)^3 \tag{23}$$

with $f(t_0) = \dot{f}(t_0) = \ddot{f}(t_0) = \dot{f}(t_0+T) = \ddot{f}(t_0+T) = 0$ and $f(T) = F_{\max}$.

4.1. Feedback Linearization Control

In order to verify the effectiveness and the efficiency of the proposed Feedback Linearization control technique, an application to quadrotor helicopter is conducted by simulation. The simulations were done with a Fixed-step size equal to 0.001s, and with the "ode5 (Dormand-Prince)" solver. The nominal parameters for a quadrotor helicopter are: m = 2 kg, $J_x = J_y = J_z/2$ = 1.2416 Nm.s²/rad, d = 0.2 m, c = 0.01 m, J_T = 2.8e^{-5}Nm.s²/rad, $K_x = K_y = K_z = 10^{-2}$ N.s/m and $K_\varphi = K_\theta = K_\psi = 10^{-3}$ Nm.s/rad . The proposed Feedback Linearization control law requires the knowledge of certain derivatives. In order to avoid analytical derivation difficulties, we estimate them by using the $\dot{v} = \frac{\Delta v}{\Delta t}$ differentiation approximation. Where Δv is the change in the value to be derived and Δt is the change in time since the previous simulation time step. The quadrotor is initially in hover flight and the initial conditions are $U = [T, T_{\varphi}, T_{\theta}, T_{\psi}]^T = [mg, 0, 0, 0]^T$.

Three modules consist the simulation system. The first being a nonlinear system module where we can find all the differential equations (2). The second is the controller module that includes the algorithm based on the control technique being tested. The third is an initialization module that sets the essential parameters to run the simulation.

By looking into the equations (14), we deduce that we have 3 fourth order integrators, and one second order integrator. In order to stabilize the outputs, a pole placement method will be employed. For example, the first equation of (14) where v_1 is the input and O_1 is the output, can be represented as:

$$\begin{cases} \begin{bmatrix} \dot{x} \\ x^{(2)} \\ x^{(3)} \\ x^{(4)} \end{bmatrix} = \begin{bmatrix} 0 & 1 & 0 & 0 \\ 0 & 0 & 1 & 0 \\ 0 & 0 & 0 & 1 \\ 0 & 0 & 0 & 0 \end{bmatrix} \begin{bmatrix} x \\ \dot{x} \\ x^{(2)} \\ x^{(3)} \end{bmatrix} + \begin{bmatrix} 0 \\ 0 \\ 0 \\ 1 \end{bmatrix} v_1 = A_1 \begin{bmatrix} x \\ \dot{x} \\ x^{(2)} \\ x^{(3)} \end{bmatrix} + B_1 v_1 \\ O_1 = \begin{bmatrix} 1 & 0 & 0 & 0 \end{bmatrix} \begin{bmatrix} x \\ \dot{x} \\ x^{(2)} \\ x^{(3)} \end{bmatrix} \end{cases} \tag{24}$$

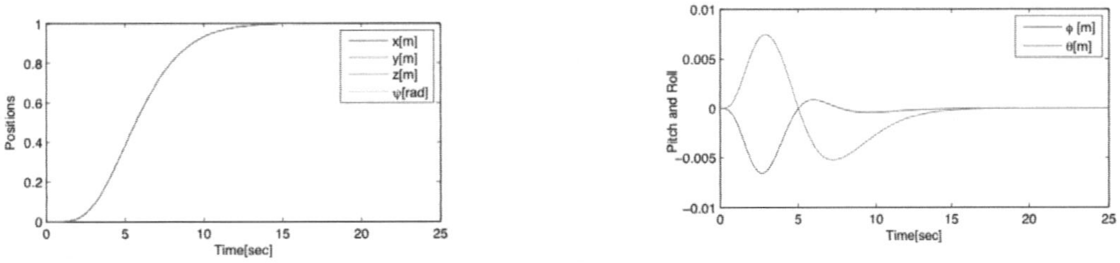

Figure 4. Positions

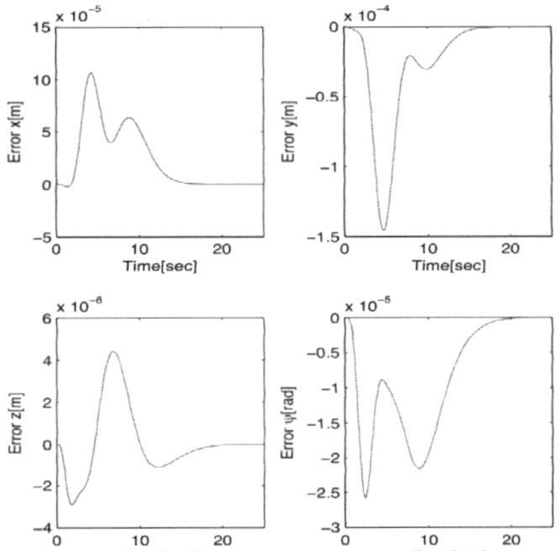

Figure 5. Errors

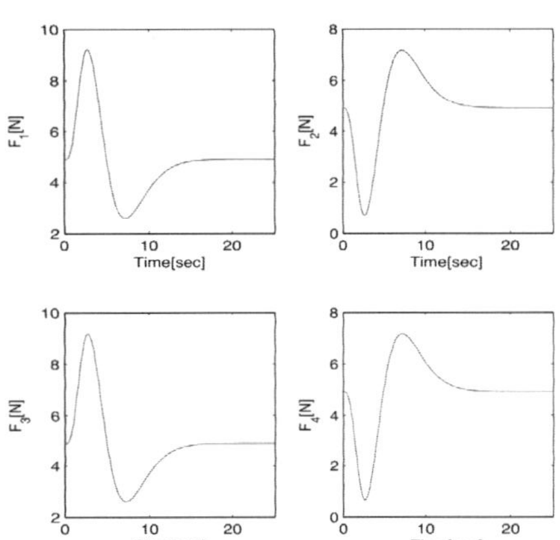

Figure 6. Forces

4.2. Fault Tolerant Feedback Linearization Control results

To simulate a free fall recovery after few milliseconds of the M_2 failure, with the same nominal parameters for a quadrotor, the same Fixed-step size, and the same solver as in section 4.1, the initial conditions are set in a way that $x = 2m,\ y = 5m,\ z = 7m,\ \varphi = 0.0001 rad,\ \theta = \text{-}0.0001 rad$ $\psi = 0 rad,\ \dot{x} = -0.1 m.s^{-1},\ \dot{y} = 0.1 m.s^{-1},\ \dot{z} = -2 m.s^{-1},\ \dot{\varphi} = \dot{\theta} = \dot{\psi} = 0 rad.s^{-1}$. At first, the desired values for the state variables are set to $x_d = 3m,\ y_d = 4m,\ z_d = 10m$, once the quadrotor has reached the desired lateral and longitudinal position, The altitude desired is then set to 10 (1-*LP*) where *LP* (landing procedure) is the step response of the following transfer function:

$$I(s) = \frac{1}{(s+1)^6} \tag{25}$$

where s is the Laplace variable.

The time needed to reach the desired set points is caused by the fact that the outer controller tends to force the attitude angles to differ from zero as long as the desired position is not reached, while the inner control tends to force those angles to zero. The simulation has been therefore run for 200 s, which is a sufficient time to reach the landing point with stable Pitch and Roll angles.

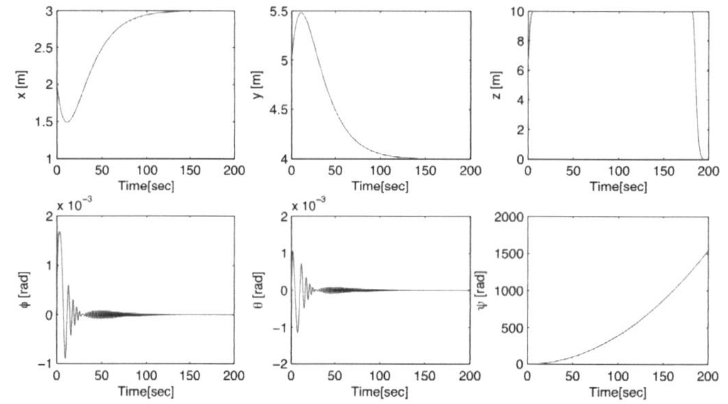

Figure 7. Positions

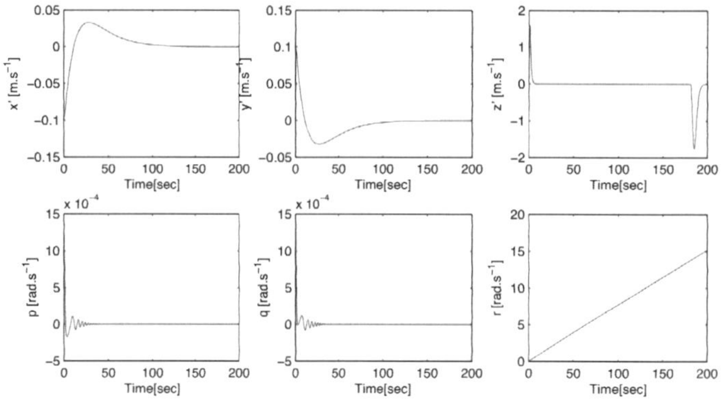

Figure 8. The Quadrotor's speed

As it can be seen in figure 7 the attitude angles φ and θ are stabilized, the oscillations are due to the presence of the outer controller. The ψ angle does not reach a steady state value, but increases according to an approximate linear law, this behavior is actually not a surprise since the controllability of the ψ variable has been sacrificed in the design phase of the fault tolerant controller. The linear positions reach their desired values, even if the regulation of z is much faster than that of x and y due to the fact that altitude regulation is operated by the inner and faster controller. The angular and linear velocities are all stabilized as it can be seen in figure 8, with the exception of $\dot{\psi}$, for the reasons already stated, and $\dot{z}$ which becomes negative during the landing procedure.

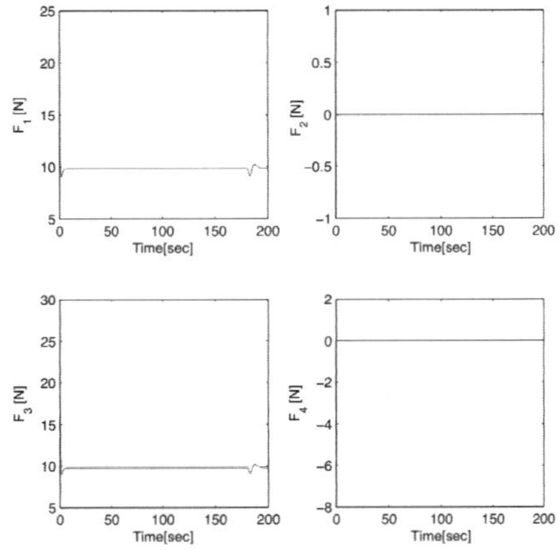

Figure 9. Forces

4.3. Wind disturbance

To test the robustness of the Feedback Linearization technique, the same simulation in section 4.1 was repeated but this time a wind force was introduced in the form of a transfer function's response (25) to the following square pulse:

$$P = \begin{cases} 1 & \text{for } 5 < t < 25 \\ 0 & \text{everywhere else} \end{cases} \tag{26}$$

We can see that during the 20 to 25 seconds time interval, while wind is still blowing through the quadrotor (figure 10), x, y, z, φ, θ, ψ are stabilized (figure 11), the errors are set to zero (figure 12) and the forces generated by the actuators are acceptable (figure 13). When the first 40 seconds of the simulation pass, the wind forces are totally eliminated, and we can see that all errors are set to zero again, x, y, z and ψ are stabilized, while φ and θ tend towards zero. This indicates a good robustness of the FL.

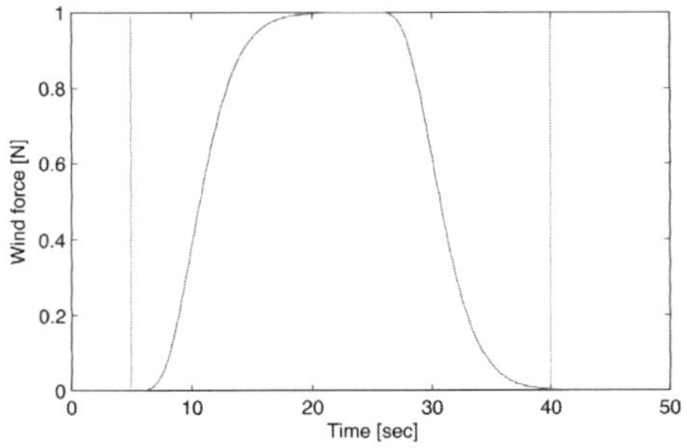

Figure 10. Wind Forces

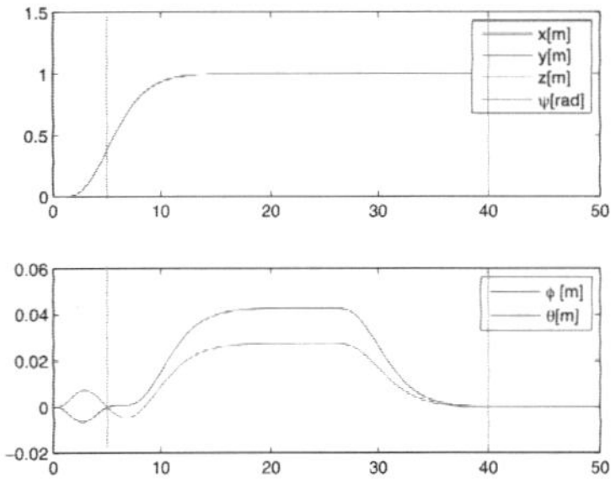

Figure 11. Positions

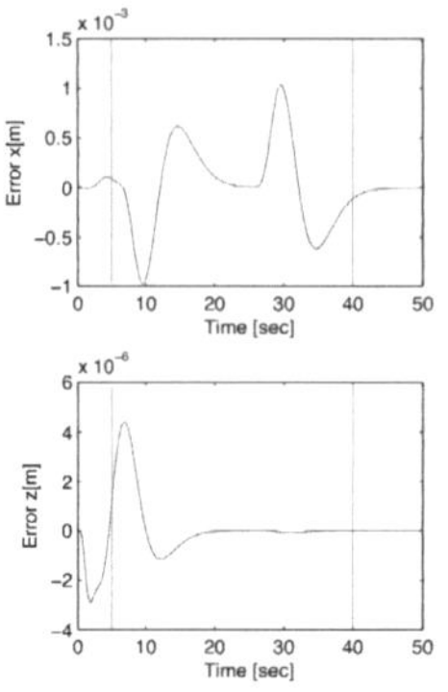

Figure 12. Errors

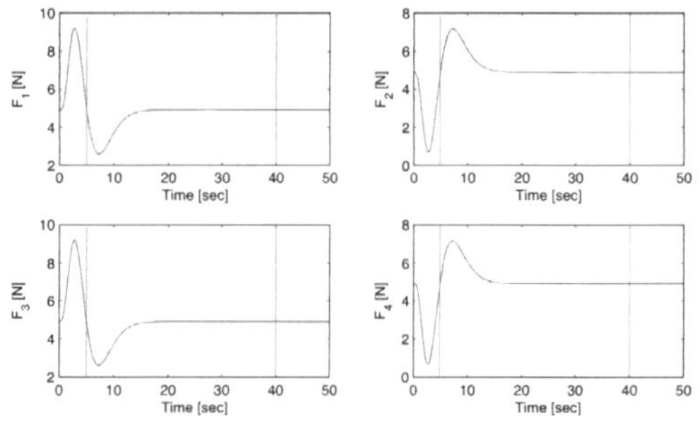

Figure 13. Forces

5. Conclusion and Perspective

It has been proven in this paper that a UAV can be automated using a Feedback Linearization approach, in a way that, without human intervention, it can reach any desired point in space, even in a faulty environment. The presence of tolerable errors in figure 5 could be explained by the differentiation approximation used to avoid analytical derivation difficulties, and by the fact that we used a simplified model to build the control technique. Last but not least, the Feedback Linearization technique proved to have excellent robustness in section 4.3.

References

[1] J. D'Azzo and C.H. Houpis, "Linear control system analysis and design, conventional and modern", *McGraw-Hill Series in Electrical and Computer Engineering*, 1995.

[2] A.J. Fossard and D. Normand-Cyrot (Eds.). Nonlinear Systems, *Vol. 3*: Control, *Springer*, 1996.

[3] A. Freddi, A. Lanzon, S. Longhi, "A feedback linearization approach to fault tolerance in quadrotor vehicles", in: Preprints of the 18th IFAC World Congress, Milano, Italy, 28 August–2 September 2011.

[4] L. Fridman A. Benallegue, A. Mokhtari. High-order sliding-mode observer for a quadrotor uav. International journal of robust and nonlinear control, 18 :427–440, 2007.

[5] G. M. Hoffmann, H. Huang, S. L. Wasl, and C. J. Tomlin, "Quadrotor helicopter flight dynamics and control: Theory and experiment", in Proc. of the AIAA Navigation, and Control Conference, 2007.

[6] A. Isidori, "Nonlinear Control Systems", *3rd Edition, Springer*, 1995.

[7] H. Khebbache. Tolérance aux défauts via la méthode backstepping des sytèmes non linéaires. Master's thesis, Université Ferhat abbas de setif, Algérie, 2012

[8] Quanser official site : Control Challenges - Unmanned Systems - Qball-X4 http://www.quanser.com/Products/qball

[9] H. Ramirez Rodriguez et al., "Integral Sliding Mode Backstepping Control of Quadrotors for Robust Position Tracking", ICUAS'13 International Conference on Unmanned Aircraft Systems, Atlanta, USA, May 28-31 2013.

[10] F.R. Rubio G.V. Raffo, M.G. Ortega. An integral predictive/nonlinear hinf control structure for a quadrotor helicopter. Automatica, 46 :29–39, 2009.

[11] I. Sadeghzadeh et al., "Linear Parameter Varying Control Synthesis: State Feedback versus H∞ Technique with Application to Quadrotor UAV", ICUAS'14 International Conference on Unmanned Aircraft Systems, Orlando, USA, May 27-30 2014.

[12] R. Siegwart S. Bouabdallah, A. Noth. PID vs LQ control techniques applied to an indoor micro quadrotor. IEEE Inteligent robotics system, 2004.

[13] R. Siegwart S. Bouabdallah. Backstepping and sliding-mode techniques applied to an indoor micro quadrotor. In International conference on robotics and automation, 2005.

[14] J.F. Whidborne A.K. Cooke I.D. Cowling, O.A. Yakimenko. A prototype of an autonomous controller for a quadrotor uav. European Journal of Control, 2007.

[15] X.B. Zhang, Y.M. Zhang, C.-Y. Su, Y. Feng, "Fault-tolerant control for quadrotor UAV via backstepping approach", in: The 48th AIAA Aerospace Sciences Meeting, Orlando, Florida, USA, 47 January 2010

[16] Y.M. Zhang et al., "Development of advanced FDD and FTC techniques with application to an unmanned quadrotor helicopter testbed", Journal of the Franklin Institute. (2013), http://dx.d oi.org/10.1016/j.jfranklin.2013.01.009

Sensor Fault-Tolerant Control of a Drivetrain Test Rig via an Observer-Based Approach within a Wind Turbine Simulation Model

Sören Georg, Stefan Heyde and Horst Schulte

HTW Berlin, School of Engineering I, Control Engineering, Wilhelminenhofstr. 75A, 12459 Berlin, Germany

E-mail: horst.schulte@htw-berlin.de

Abstract. This paper presents the implementation of an observer-based fault reconstruction and fault-tolerant control scheme on a rapid control prototyping system. The observer runs in parallel to a dynamic wind turbine simulation model and a speed controller, where the latter is used to control the shaft speed of a mechanical drivetrain according to the calculated rotor speed obtained from the wind turbine simulation. An incipient offset fault is added on the measured value of one of the two speed sensors and is reconstructed by means of a Takagi-Sugeno sliding-mode observer. The reconstructed fault value is then subtracted from the faulty sensor value to compensate for the fault. The whole experimental set-up corresponds to a sensor-in-the-loop system.

1. Introduction

Due to the growing size and complexity of wind turbines, there is an increasing demand for developing intelligent fault diagnosis and fault-tolerant control strategies to avoid turbine shut-downs that are caused by minor (non-critical) faults. Such strategies are particularly important for offshore wind turbines, which are more difficult and in bad weather conditions even impossible to access, such that a regular maintenance and replacement of faulty system parts is impeded. The fault diagnosis schemes that are nowadays routinely employed in wind turbines are typically on the level of the supervisory control and include sensor comparison, model comparison and thresholding [1]. These strategies enable a safe turbine operation, but no intelligent fault-tolerant control. Typically, the turbines are shut down even in case of faults that are in principle non-critical, like a fault of a redundant sensor. The gains in safety therefore often entail a reduced availability and thus a loss of produced energy.

Several advanced fault diagnosis schemes have been applied in wind turbine simulations in recent years. A good overview can be found in [2], where the results of a wind turbine FDI (fault detection and isolation) competition are summarised. In the presented solutions, all simulations were conducted with a reduced-order benchmark model introduced in [3, 4]. Most of these solutions rely on the evaluation of residuals, a standard technique in FDI methods [5]. The downside of evaluating residuals for FDI is that they yield no estimates of the faults including their magnitude and "direction". This information, however, is important for a full fault analysis and vital for fault-tolerant control.

The competition was later extended to fault-tolerant control (FTC). Of the presented solutions, only few include fault estimation, namely [6], [7] and [8].

The FDI benchmark model from [3], which features only drivetrain degrees of freedom but no blade and tower dynamics, is not a realistic wind turbine simulation model. In a later stage of the FDI/FTC competition [9], the aero-elastic code FAST by NREL [10] was thus used as a simulation model along with the turbine parameters of the NREL 5 MW reference turbine [11]. From among the presented solutions, only [12] includes fault estimation, though only for pitch dynamics parameter faults.

There is therefore still a need for further research on reliable and effective FDI and FTC algorithms that include estimation/reconstruction of faults and are validated in standard wind turbine simulation packages like FAST. The authors of this paper have already published the application of a Takagi-Sugeno sliding mode observer to wind turbine actuator and sensor fault reconstruction and fault-tolerant control [13, 14, 15]. This approach is capable of simultaneously reconstructing several faults (even with significantly differing orders of magnitude) and yields promising results.

What has not yet been conducted, and to the authors' knowledge has not been published either by other authors, is a validation of FDI/FTC algorithms on a real test rig, let alone on a real wind turbine. In this paper, a first step in this direction is presented in form of a sensor-in-the-loop set-up, where the Takagi-Sugeno sliding mode observer for fault reconstruction is implemented on a dSpace rapid control prototyping system alongside a wind turbine simulation model. The calculated (and scaled) rotor speed from the wind turbine model is then used as the setpoint for a closed-loop controller to control the shaft speed of a mechanical drivetrain. The measured value of the speed sensor on the load side of the drivetrain is then imposed with an additive fault and fed back into the wind turbine model and the observer. The reconstructed fault can then be subtracted from the "faulty" sensor value to achieve a near, though not full, fault-tolerance.

This paper is organised as follows. In Section 2, the combined simulation and experimental set-up is described. In Section 3, the fault reconstruction concept by means of the Takagi-Sugeno sliding mode observer is introduced. In Section 5, results are presented and Section 6 gives a conclusion and outlook.

2. Experimental and Real-Time Simulation Set-Up

A schematics of the combined experimental and simulation set-up highlighting the sensor-in-the-loop structure is depicted in Figure 1.

A reduced-order dynamic wind turbine (WT) model with four degrees of freedom (torsional drivetrain with rotor and generator rotational angles, tower-top and blade-tip deflection) as presented in [16] and [17] was implemented in Simulink and compiled on a dSpace rapid control prototyping system to be run in real-time. In parallel to the wind turbine model, a rotor blade pitch controller [14] was implemented. In Figure 1, the block "WT Model & Cntrl & Observer" subsumes both the wind turbine model and the pitch controller as well as the observer for fault reconstruction (see Section 3). As wind speed input, a time-series of turbulent wind (at a mean wind speed of 18 m/s), generated with the software TurbSim [18] by NREL, was used. In order to regulate the shaft speed of the drivetrain test rig, a separate controller (state-space with integral action) was implemented (see block "Motor Cntrl") and also compiled on the dSpace system. The rotor speed output of the wind turbine model scaled with a factor of 97 (this corresponds to the gearbox ratio of the NREL 5 MW turbine) was set as demanded speed for the shaft speed controller. The output of the speed controller is a demanded torque (T_M in Figure 1) for the motor of the servo drive, which is itself internally controlled via a Beckhoff industrial PC and frequency converter (not explicitly depicted in Figure 1). The mechanical drivetrain (see Figure 2) consists of a shaft with a flat steel spring in the middle to obtain a torsional-flexible shaft, and an inertia disc at the end. This drivetrain test rig was initially developed

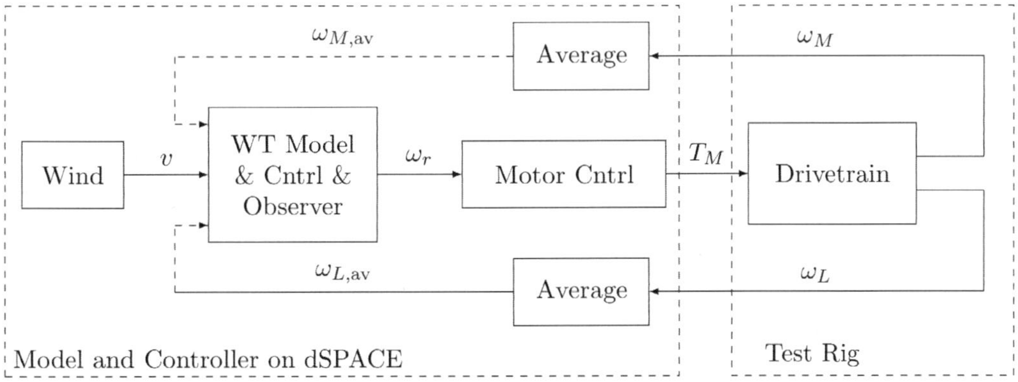

Figure 1: Schematics of the sensor-in-the-loop set-up with wind turbine model, controller and observer running on a dSpace rapid control prototyping system (left side) and the mechanical drivetrain test rig (right side). ω_r: calculated (and scaled) rotor speed from the wind turbine model. T_M: demanded motor torque. ω_M, ω_L: measured shaft speed sensor values at motor and load side, respectively. The right block 'Test Rig' consists of the mechanical drivetrain and a Beckhoff servo-motor controlled with a Beckhoff industrial PC and frequency converter. The signal line for the averaged motor speed $\omega_{M,av}$ is drawn dashed since this signal is not utilised further in the wind turbine model.

Figure 2: Mechanical drivetrain test rig - Control Engineering Laboratory of HTW Berlin

for designing a controller to reduce torsional vibrations, not with the intention to use it as a simulator for a wind turbine drivetrain (see also the outlook in Section 6). Two incremental encoders by Pepperl+Fuchs are used to measure the shaft speed at both the motor (ω_M) and the inertia/load side (ω_L) of the flat steel spring. To calculate both angular speeds, the standard method of counting the number of passing increments per cycle time (set to 0.001 s) was applied in this work. Even though the resolution of the encoders is quite high (1024 increments), at the angular speeds tested in the experiments around 120 rad/s, the measured/calculated speed values may still differ by +/- 6 rad/s. These differences occur if one increment more or less is counted per cycle, which easily happens due to the drivetrain vibrations. Ideally, a separate counter module, running at an independent high frequency (several MHz), should be used to measure the time between two increments. In this way, it would be possible to obtain a smoother speed measurement. However, such a configuration was not feasible in the presented system set-up. Therefore, the two speed measurements were filtered using a standard mean value calculation (over the last 20 timesteps) before they were fed back into the wind turbine simulation model (although the averaged motor speed $\omega_{M,av}$ was not utilised further in the wind turbine model).

3. Sliding Mode Observer Based Fault Reconstruction

In this section, the fault reconstruction method applied in this work is presented.
Edwards and Spurgeon have shown that sliding mode observers can be effectively employed for direct fault reconstruction [19, 20, 21]. The way this works is by evaluating the so-called equivalent output injection signal, which corresponds to the average behaviour of the nonlinear switching term used to establish and maintain the sliding motion.

In this work, a Takagi-Sugeno sliding mode observer (TS SMO), which is a nonlinear multiple-model extension of the Edwards-Spurgeon observer, is used. The methods for fault reconstruction can thus be employed in a similar way. For an introduction to Takagi-Sugeno (TS) models [22] see for example [23]. The TS SMO was first introduced in [24, 25]. In [13], an extension was presented, which involves a weighted switching action that allows a simultaneous reconstruction of faults of significantly differing orders of magnitude, for example pitch angle and generator torque actuator faults in wind turbines. The reason for using a nonlinear TS sliding mode observer is that the wind turbine model itself is nonlinear. By modelling the system nonlinearities within the TS structure, the switching term is solely needed for reconstructing faults, not for compensating deviations between the underlying nonlinear and a linearised model.

3.1. Takagi-Sugeno Sliding Mode Observer

In the most general form, the TS SMO design is based on a nonlinear TS system subject to external disturbances $\boldsymbol{\xi}$, actuator faults $\mathbf{f}_a$ and sensor fault $\mathbf{f}_s$, which is essentially the TS modified form of the basic state-space model normally assumed for FDI problems [5]:

$$\dot{\mathbf{x}} = \sum_{i=1}^{N_r} h_i(\mathbf{z})\left(\mathbf{A}_i\mathbf{x} + \mathbf{B}_i\mathbf{u} + \mathbf{D}_i\boldsymbol{\xi} + \mathbf{F}_i\mathbf{f}_a\right), \quad \tilde{\mathbf{y}} = \mathbf{y} + \mathbf{f}_s = \mathbf{C}\mathbf{x} + \mathbf{f}_s, \tag{1}$$

where N_r denotes the number of distinct nonlinearities. The TS membership functions h_i fulfill the condition $\sum_{i=1}^{N_r} h_i = 1$ and the premise variables $\mathbf{z}$ may depend on system states, inputs, and external variables [23].

Transformed Form of the TS SMO The TS SMO is designed in a transformed system, where the measurable system states $\mathbf{y}$ and the non-measurable system states $\mathbf{x}_1$ are separated. This separation is achieved via a transformation $\mathbf{T}_c = [\mathbf{N}_c \; \mathbf{C}]^T$, where $\mathbf{N}_c$ denotes the null-space of $\mathbf{C}$. With a series of transformations $\mathbf{T}_i = \bar{\mathbf{T}}_{L,i} \tilde{\mathbf{T}}_{DE,i} \mathbf{T}_c$ (see [20, 24, 25] for a description of the transformation matrices and the transformed system matrices), the TS system is brought into a structure where the uncertainties and faults only act on the measurable system states [24]:

$$\dot{\mathbf{x}}_1 = \sum_{i=1}^{N_r} h_i(\mathbf{z}) \left(\boldsymbol{\mathcal{A}}_{11,i} \mathbf{x}_1 + \boldsymbol{\mathcal{A}}_{12,i} \mathbf{y} + \boldsymbol{\mathcal{B}}_{1,i} \mathbf{u} \right), \tag{2}$$

$$\dot{\mathbf{y}} = \sum_{i=1}^{N_r} h_i(\mathbf{z}) \left(\boldsymbol{\mathcal{A}}_{21,i} \mathbf{x}_1 + \boldsymbol{\mathcal{A}}_{22,i} \mathbf{y} + \boldsymbol{\mathcal{B}}_{2,i} \mathbf{u} + \boldsymbol{\mathcal{D}}_{2,i} \boldsymbol{\xi} + \boldsymbol{\mathcal{F}}_{2,i} \mathbf{f}_a \right) + \dot{\mathbf{f}}_s, \tag{3}$$

The structure of the transformed system matrices is as follows: $\boldsymbol{\mathcal{A}}_i = \mathbf{T}_i \mathbf{A}_i \mathbf{T}_i^{-1} = \begin{bmatrix} \boldsymbol{\mathcal{A}}_{11,i} & \boldsymbol{\mathcal{A}}_{12,i} \\ \boldsymbol{\mathcal{A}}_{21,i} & \boldsymbol{\mathcal{A}}_{22,i} \end{bmatrix}$, $\boldsymbol{\mathcal{B}}_i = \mathbf{T}_i \mathbf{B}_i = \left[\boldsymbol{\mathcal{B}}_{1,i}^T \; \boldsymbol{\mathcal{B}}_{2,i}^T\right]^T$, $\boldsymbol{\mathcal{D}}_i = \mathbf{T}_i \mathbf{D}_i = \left[\mathbf{0}^T \; \boldsymbol{\mathcal{D}}_{2,i}^T\right]^T$, $\boldsymbol{\mathcal{F}}_i = \mathbf{T}_i \mathbf{F}_i = \left[\mathbf{0}^T \; \boldsymbol{\mathcal{F}}_{2,i}^T\right]^T$. For a stable observer to exist, the following three existence conditions have to be fulfilled [20, 24, 25]:

Condition 1. The disturbances and actuator faults are unknown but bounded: $\|\boldsymbol{\xi}^T(t) \; \mathbf{f}_a^T(t)\|^T \le \Xi$. Furthermore, individual bounds exist: $\|\boldsymbol{\xi}(t)\|^T \le \Xi_\xi$ and $\mathbf{f}_a(t) \le \Xi_{f_a}$. The sensor faults and their derivatives are assumed to be bounded, too: $\|\mathbf{f}_s\| \le \Psi$, $\|\dot{\mathbf{f}}_s\| \le \Psi_d$. Moreover, the system states and inputs are assumed to be bounded.

Condition 2. Let q be defined as the number of columns of $[\mathbf{D}_i \; \mathbf{F}_i]$. Then, the condition $q = \operatorname{rank}(\mathbf{C}[\mathbf{D}_i \; \mathbf{F}_i]) = \operatorname{rank}[\mathbf{D}_i \; \mathbf{F}_i]$ must be fulfilled. Furthermore, it must hold that $p > q$, where p is the number of measurable system states.

Condition 3. All invariant zeros of $(\mathbf{A}_i, [\mathbf{D}_i \; \mathbf{F}_i], \mathbf{C})$ must lie in $\mathbb{C}_-$.
The TS sliding mode observer in transformed form is given by

$$\dot{\hat{\mathbf{x}}}_1 = \sum_{i=1}^{N_r} h_i(\mathbf{z}) \left(\boldsymbol{\mathcal{A}}_{11,i} \hat{\mathbf{x}}_1 + \boldsymbol{\mathcal{A}}_{12,i} \hat{\mathbf{y}} + \boldsymbol{\mathcal{B}}_{1,i} \mathbf{u} - \boldsymbol{\mathcal{A}}_{12,i} \tilde{\mathbf{e}}_y \right), \tag{4}$$

$$\dot{\hat{\mathbf{y}}} = \sum_{i=1}^{N_r} h_i(\mathbf{z}) \left(\boldsymbol{\mathcal{A}}_{21,i} \hat{\mathbf{x}}_1 + \boldsymbol{\mathcal{A}}_{22,i} \hat{\mathbf{y}} + \boldsymbol{\mathcal{B}}_{2,i} \mathbf{u} - (\boldsymbol{\mathcal{A}}_{22,i} - \boldsymbol{\mathcal{A}}_{22}^s) \tilde{\mathbf{e}}_y + \boldsymbol{\nu} \right), \tag{5}$$

where $\tilde{\mathbf{e}}_y := \hat{\mathbf{y}} - (\mathbf{y} + \mathbf{f}_s)$ denotes the output error including sensor faults and $\boldsymbol{\mathcal{A}}_{22}^s$ is a stable design matrix. In this work, the discontinuous switching term necessary to establish and maintain a sliding motion is given by

$$\boldsymbol{\nu} = -\rho \frac{\mathbf{P}_2 \mathbf{W} \tilde{\mathbf{e}}_y}{\|\mathbf{P}_2 \mathbf{W} \tilde{\mathbf{e}}_y\|}, \quad (\tilde{\mathbf{e}}_y \neq \mathbf{0}). \tag{6}$$

Here, $\mathbf{P}_2$ is the symmetric, positive definite solution of the Lyapunov equation $\mathbf{P}_2 \boldsymbol{\mathcal{A}}_{22}^s + \boldsymbol{\mathcal{A}}_{22}^{s\,T} \mathbf{P}_2 = -\mathbf{Q}_2$, where $\mathbf{Q}_2$ is a symmetric positive definite design matrix. Expression (6) differs from the discontinuous switching terms in [20] and [24, 25] in that the output

Journal of Physics: Conference Series **570** (2014) 082004 doi:10.1088/1742-6596/570/8/082004

error vector $\tilde{\mathbf{e}}_y$ is modified with a diagonal weighting matrix $\mathbf{W}$ and that a diagonal gain matrix $\boldsymbol{\rho} = \mathrm{diag}\,(\rho_1 \cdots \rho_p)$ is used instead of a scalar gain factor. The elements of the weighting matrix are the reciprocal values of the estimated maximum values of the output vector: $\mathbf{W} = \mathrm{diag}\,(W_1 \cdots W_p) = \mathrm{diag}\,(1/\,|y_{\mathrm{max,1}}| \cdots 1/\,|y_{\mathrm{max,p}}|)$. The modified switching function (6) was introduced in [13] to enable a simultaneous reconstruction of both pitch angle and generator torque actuator faults, which have significantly differing orders of magnitude.

Equivalent Output Injection Signal Similarly to the equivalent control action in sliding mode control structures, for sliding mode observers the so-called equivalent output injection signal describes the average behaviour of the discontinuous component $\boldsymbol{\nu}$ [21].

The equivalent output injection signal is a measure for the effort to maintain the sliding motion [21] and can be approximated to arbitrary precision by introducing a small enough positive scalar δ into the discontinuous component (6):

$$\boldsymbol{\nu}_{\mathrm{eq}} = -\boldsymbol{\rho}\,\frac{\mathbf{P}_2\,\mathbf{W}\,\tilde{\mathbf{e}}_y}{\|\mathbf{P}_2\,\mathbf{W}\,\tilde{\mathbf{e}}_y\| + \delta}\,, \tag{7}$$

3.2. Fault Reconstruction

If faults occur in the system, a higher effort needs to be exerted to maintain the sliding motion, which is then reflected in the equivalent output injection signal $\boldsymbol{\nu}_{\mathrm{eq}}$. In [20, 21], direct fault reconstruction methods for actuator as well as sensor faults are described based on evaluating this signal. For a TS sliding mode observer, these methods can be adapted and used in a similar fashion [24, 25]. If no sensor faults are present and neglecting disturbances, actuator faults can be reconstructed using the relation

$$\hat{\mathbf{f}}_a = \left(\sum_{i=1}^{N_r} h_i\,(\mathbf{z})\,\boldsymbol{\mathcal{F}}_{2,i}\right)^{+} \boldsymbol{\nu}_{\mathrm{eq}}\,, \tag{8}$$

where $(\cdot)^{+}$ denotes the left pseudo-inverse.

For sensor fault reconstruction, the method introduced by Tan and Edwards [26], where sensor faults can be treated as actuator faults within an augmented system, is adapted for TS systems and the fault reconstruction then works analogously to (8). The basic idea is to filter the system outputs and then to augment the system with the filtered outputs.
Consider a system with sensor faults $\mathbf{f}_s$ in r selected output channels:

$$\begin{aligned} \dot{\mathbf{x}} &= \sum_{i=1}^{N_r} h_i\,(\mathbf{z})\,(\mathbf{A}_i\,\mathbf{x} + \mathbf{B}_i\,\mathbf{u})\,, \\ \tilde{\mathbf{y}} &= \mathbf{C}\,\mathbf{x} + \mathbf{N}\,\mathbf{f}_s\,, \end{aligned} \tag{9}$$

where the matrix $\mathbf{N} \in \mathbb{R}^{p\times r}$ $(n \geq p \geq r)$ (with n the number of system states and p the number of outputs) is used to sort the components of $\mathbf{f}_s$ into the channels of the respective output vector. Introducing a delay matrix $\mathbf{A}_f = \mathrm{diag}\left(-\frac{1}{\tau_1}\ \cdots\ -\frac{1}{\tau_p}\right)$, with the individual delay time constants τ_i $(i \in \{1,\,\ldots,\,p\})$ for each output, the dynamics of the filtered output vector $\mathbf{y}_f$ is given by:

$$\begin{aligned} \dot{\mathbf{y}}_f &= -\mathbf{A}_f\,\mathbf{y}_f + \mathbf{A}_f\,\tilde{\mathbf{y}} \\ &= -\mathbf{A}_f\,\mathbf{y}_f + \mathbf{A}_f\,\mathbf{C}\,\mathbf{x} + \mathbf{A}_f\,\mathbf{N}\,\mathbf{f}_s\,. \end{aligned} \tag{10}$$

Both systems (9) and (10) can be combined into an augmented system with system vector $\mathbf{x}_a$, where the sensor faults now appear as actuator faults:

$$\begin{pmatrix}\dot{\mathbf{x}}\\ \dot{\mathbf{y}}_f\end{pmatrix} = \sum_{i=1}^{N_r} h_i(\mathbf{z}) \left(\underbrace{\begin{bmatrix}\mathbf{A}_i & \mathbf{0}\\ \mathbf{A}_f\,\mathbf{C} & -\mathbf{A}_f\end{bmatrix}}_{=:\mathbf{A}_{a,i}} \underbrace{\begin{pmatrix}\mathbf{x}\\ \mathbf{y}_f\end{pmatrix}}_{=:\mathbf{x}_a} + \underbrace{\begin{bmatrix}\mathbf{B}_i\\ \mathbf{0}\end{bmatrix}}_{=:\mathbf{B}_{a,i}} \mathbf{u} \right) + \underbrace{\begin{bmatrix}\mathbf{0}\\ \mathbf{A}_f\,\mathbf{N}\end{bmatrix}}_{=:\mathbf{F}_a} \mathbf{f}_s \tag{11}$$

$$\mathbf{y}_f = \underbrace{\begin{bmatrix}\mathbf{0} & \mathbf{I}_{p\times p}\end{bmatrix}}_{=:\mathbf{C}_a} \begin{pmatrix}\mathbf{x}\\ \mathbf{y}_f\end{pmatrix} . \tag{12}$$

In analogy to (4) and (5), the TS sliding mode observer in transformed form for the augmented system (11)-(12) is given by

$$\dot{\hat{\mathbf{x}}}_{a,1} = \sum_{i=1}^{N_r} h_i(\mathbf{z}) \left(\boldsymbol{\mathcal{A}}_{a,11,i}\,\hat{\mathbf{x}}_{a,1} + \boldsymbol{\mathcal{A}}_{a,12,i}\,\hat{\mathbf{y}}_f + \boldsymbol{\mathcal{B}}_{a,1,i}\,\mathbf{u} - \boldsymbol{\mathcal{A}}_{a,12,i}\,\tilde{\mathbf{e}}_y\right) , \tag{13}$$

$$\dot{\hat{\mathbf{y}}}_f = \sum_{i=1}^{N_r} h_i(\mathbf{z}) \left(\boldsymbol{\mathcal{A}}_{a,21,i}\,\hat{\mathbf{x}}_{a,1} + \boldsymbol{\mathcal{A}}_{a,22,i}\,\hat{\mathbf{z}}_f + \boldsymbol{\mathcal{B}}_{a,2,i}\,\mathbf{u} - (\boldsymbol{\mathcal{A}}_{a,22,i} - \boldsymbol{\mathcal{A}}_{22}^{s})\,\tilde{\mathbf{e}}_y + \boldsymbol{\nu}\right) , \tag{14}$$

where $\tilde{\mathbf{e}}_y = \hat{\mathbf{y}}_f - \mathbf{y}_f$ denotes the error vector, which is now calculated from the filtered output components of the TS SMO and the sensor faults can then be reconstructed like actuator faults using the pseudo-inverse of the TS sum of the transformed form of the augmented matrices $\boldsymbol{\mathcal{F}}_{a,2,i}$:

$$\hat{\mathbf{f}}_s = \left(\sum_{i=1}^{N_r} h_i(\mathbf{z})\,\boldsymbol{\mathcal{F}}_{a,2,i}\right)^{+} \boldsymbol{\nu}_{\text{eq}} . \tag{15}$$

3.3. Design of the TS Sliding Mode Observer

The TS SMO design in this work was based on a further reduction of the 4-DOF wind turbine model containing only the drivetrain degrees of freedom [13, 14]. This proved a suitable choice for achieving good fault reconstruction results.

4. Fault-Tolerant Control

By subtracting the reconstructed sensor faults $\hat{\mathbf{f}}_s$ from the faulty outputs, a fault-tolerance can (in principle) be achieved, depending on the quality of the fault reconstruction:

$$\mathbf{y}_{\text{corr}} = \mathbf{y}_{F,s} - \hat{\mathbf{f}}_s = \mathbf{C}\,\mathbf{x} + \left(\mathbf{f}_s - \hat{\mathbf{f}}_s\right) . \tag{16}$$

For a stability analysis of this structure in the case of wind turbine simulations, please refer to [15]. As will be seen in the next section, even in the case with no speed sensor fault, the TS SMO reconstructs small fault magnitudes. To avoid feeding back these erroneous reconstructions, a threshold was applied, such that only reconstructed magnitudes greater than this threshold were considered for the fault compensation calculation (16).

5. Results

Five measurements over a 120 s period were conducted:

1. Fault-free case
2. Load side (ω_L) sensor fault with a final fault value of ...
   2.1 ... 0.2 rad/s - no fault compensation
   2.2 ... 0.2 rad/s - with active fault compensation
   2.3 ... 0.8 rad/s - no fault compensation
   2.4 ... 0.8 rad/s - with active compensation

The faults were initially fixed to 0 and then gradually increased with a linear ramp to the final fault values of 0.2 respectively 0.8 rad/s. These values are in relation to the magnitude of the calculated rotor speed within the wind turbine simulation model. In relation to the measured angular speed of the drivetrain, the fault values would be 0.2 rad/s $\cdot$ 97 = 19.4 rad/s respectively 0.8 rad/s $\cdot$ 97 = 77.6 rad/s.

The results are shown in Figures 3 and 4, where the three cases (fault-free (blue lines), fault without compensation (green lines), fault with active compensation (red lines)) are depicted together to see the effect of the faults and the fault compensation. Their are small time offsets between the three cases, because the measurements were started manually.

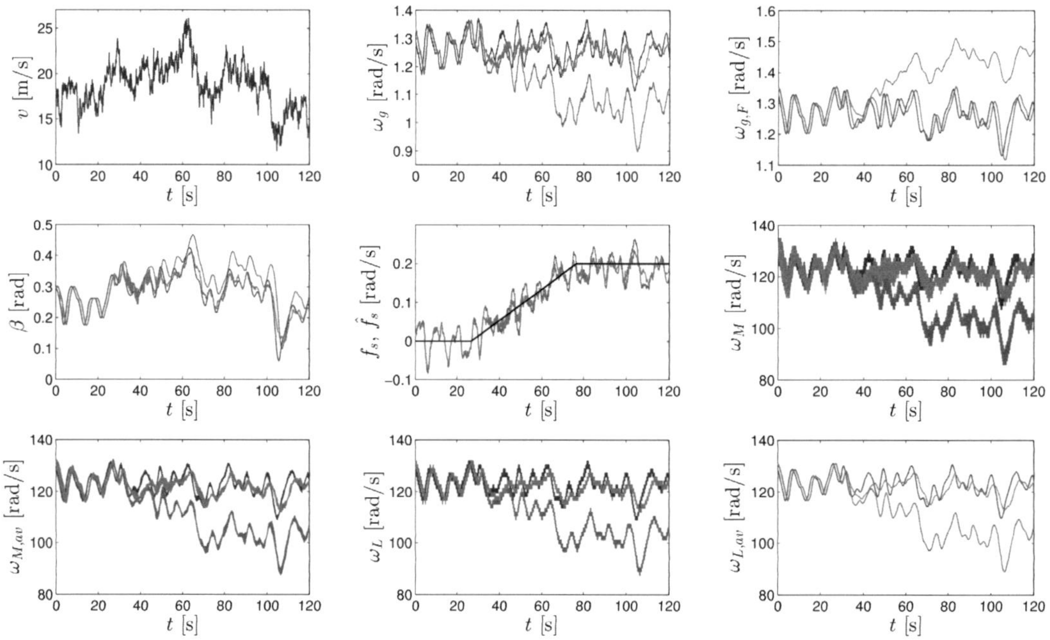

Figure 3: Simulation and measurement results for an incipient additive fault on the load side speed sensor (final fault value: 0.2 rad/s). Blue lines: fault-free case; green lines: faulty case without fault compensation; red lines: faulty case with active fault compensation. In the central subfigure, the black line shows the actual fault value for the two faulty measurements.

The results of the measurements with the smaller fault value of 0.2 rad/s are shown in Figure 3. The central middle subfigure shows the true fault value (black line) and the reconstructed fault values in the case without (green) and with (red) fault compensation. One can see that even with a true fault value of 0, the TS SMO reconstructs (erroneous) fault values with abolute values up to 0.07 rad/s. For this reason, a threshold value of 0.05 rad/s was defined, such that in the case with active fault compensation, only reconstructed fault values with abolute values greater than this threshold were considered to be fed back via the fault-tolerant control algorithm (16). Even though the largest values of the erroneous reconstructions are greater than this threshold, the value of 0.05 rad/s was chosen to achieve a compromise between a reasonably quick fault compensation and the avoidance of erroneous compensations.

The measured sensor values and their average values are shown in the last 4 subfigures (denoted ω_M, $\omega_{M,av}$, ω_L and $\omega_{L,av}$). The fault was added on the averaged measurement $\omega_{L,av}$ of the load side speed of the drivetrain and then used in the wind turbine simulation model ($\omega_{g,F}$: 3rd subfigure). One can see that in the faulty case without fault compensation, this value remains the same (neglecting the small time-offset) as the in the fault-free case, because the pitch controller within the WT model adjusts the pitch angle such as to control the generator speed around its setpoint. Due to this, the physical shaft speed of the drivetrain decreases (see green line in 2nd subfigure).

By contrast, in the case with active fault compensation, the fault-free behaviour is nearly recovered, though not perfectly, because the fault reconstruction is not perfect. The rising fault ramp and the mean fault value are quite well reconstructed by the TS sliding mode observer. However, there is quite a large variation around the mean fault value of 0.2 rad/s. Compared to the results in simulation-only studies presented in [15], the reconstruction quality is thus decreased. Still, it is sufficient to achieve a fairly well working fault-tolerant control.

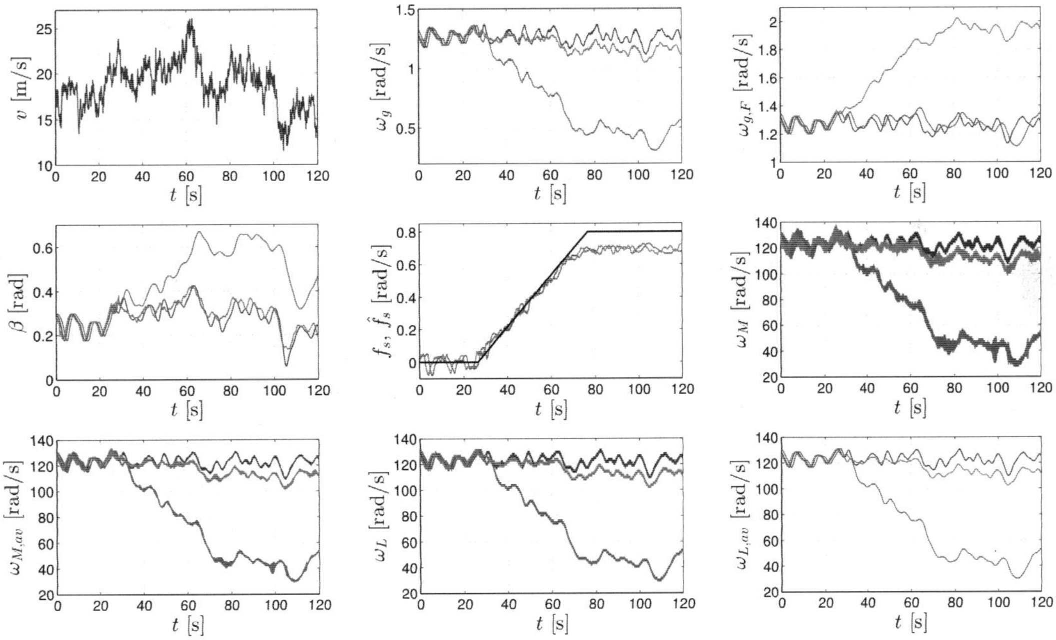

Figure 4: Simulation and measurement results for an incipient additive fault on the load side speed sensor (final fault value: 0.8 rad/s). Blue lines: fault-free case; green lines: faulty case without fault compensation; red lines: faulty case with active fault compensation. In the central subfigure, the black line shows the actual fault value for the two faulty measurements.

The results of the measurements with the larger fault value of 0.8 rad/s are shown in Figure 4. This fault value corresponds to more than 60 % of the demanded rotor speed 1.267 rad/s within the wind turbine simulation model and thus constitutes a very significant fault magnitude. From the central subfigure it can be seen that in this case, there is a discrepancy of about 0.1 rad/s between the actual final fault value of 0.8 rad/s and the mean of the reconstructed value. Due to this discrepancy, the physical shaft speed in the faulty case with active fault compensation is also slightly decreased compared to the fault-free case. When comparing this result to the case without fault compensation, however, one can see that a significant improvement can be achieved with the fault-tolerant control scheme.

6. Conclusion and Outlook

The investigations presented in this paper are a first step to a practical validation of the TS sliding mode observer for fault reconstruction, with focus on the wind turbine application. It could be shown that the fault-tolerant control strategy achieves a significant improvement of the controlled behaviour of the sensor-in-the-loop set-up in the presence of faults.

For future work, a modification of the drivetrain test rig is planned, for the following reasons. The system set-up with the shaft speed controller in between the wind turbine simulation model and the drivetrain is still somewhat cumbersome. Moreover, it leads to an additional time-lag within the system caused by the transient behaviour (rise time) of the shaft speed controller. Instead of the inertia disc at the one end of the drivetrain, it would be better to install a generator (with the same power as the driving motor). In such a set-up, the motor torque could be adjusted directly from the calculated (and scaled) aerodynamic rotor torque within the wind turbine model. The calculated generator torque from the torque controller within the wind turbine model would then be set as a demanded counteracting torque for the generator. In this way, not only sensor but also generator torque actuator faults could be modelled and reconstructed. Furthermore, the time-lag introduced by the shaft speed controller would not be present, such that the fault reconstruction quality should also be improved. The modified drivetrain with motor and generator would constitute a more realistic scenario for a drivetrain of a real wind turbine.

Acknowledgments

This work was conducted within a research project funded by the German Federal Ministry of Education and Research under grant no. 17N1411.

References

[1] Johnson K E and Fleming P A 2011 *Mechatronics* **21** 728–736
[2] Odgaard P F and Stoustrup J 2012 Results of a Wind Turbine FDI Competition *IFAC Symposium on Fault Detection, Supervision and Safety of Technical Processes* (Mexico City, Mexico) pp 102–107
[3] Odgaard P F, Stoustrup J and Kinnaert M 2009 Fault Tolerant Control of Wind Turbines - a benchmark model *IFAC Symposium on Fault Detection, Supervision and Safety of Technical Processes* (Barcelona, Spain) pp 155–160
[4] Odgaard P F, Stoustrup J and Kinnaert M 2013 *IEEE Transactions on Control Systems Technology* **21** 1168–1182
[5] Chen J and Patton R J 1999 *Robust Model-Based Fault Diagnosis for Dynamic Systems* (Kluwer Academic Publishers)
[6] Yang X and Maciejowski J M 2012 Fault-tolerant model predictive control of a wind turbine benchmark *IFAC Symposium on Fault Detection, Supervision and Safety of Technical Processes* (Mexico City, Mexico) pp 337–342
[7] Rotondo D, Nejjari F, Puig V and Blesa J 2012 Fault Tolerant Control of the Wind Turbine Benchmark using Virtual Sensors/Actuators *IFAC Symposium on Fault Detection, Supervision and Safety of Technical Processes* (Mexico City, Mexico) pp 114–119
[8] Sami M and Patton R J 2012 An FTC Approach to Wind Turbine Power Maximisation via T-S Fuzzy Modelling and Control *IFAC Symposium on Fault Detection, Supervision and Safety of Technical Processes* (Mexico City, Mexico) pp 349–354
[9] Odgaard P F and Johnson K E 2013 Wind turbine fault detection and fault tolerant control - an enhanced benchmark challenge *American Control Conference* (Washington, D.C., USA) pp 4447–4452

[10] Jonkman J M and Buhl Jr M L 2005 FAST User's Guide Tech. rep. NREL/EL-500-38230, National Renewable Energy Laboratory, Golden, Colorado
[11] Jonkman J, Butterfield S, Musial W and Scott G 2009 Definition of a 5-MW Reference Wind Turbine for Offshore System Development Tech. rep. NREL/TP-500-38060, National Renewable Energy Laboratory, Golden, Colorado
[12] Sheibat-Othman N, Othman S, Benlahrache M and Odgaard P F 2013 Fault detection and isolation in wind turbines using support vector machines and observers *American Control Conference* (Washington, D.C., USA) pp 4459–4464
[13] Georg S and Schulte H 2014 Takagi-Sugeno Sliding Mode Observer with a Weighted Switching Action and Application to Fault Diagnosis for Wind Turbines *Intelligent Systems in Technical and Medical Diagnostics* (*Advances in Intelligent Systems and Computing* vol 230) ed Korbicz J and Kowal M (Springer-Verlag Berlin Heidelberg) pp 41–52
[14] Georg S and Schulte H 2013 Actuator Fault Diagnosis and Fault-Tolerant Control of Wind Turbines using a Takagi-Sugeno Sliding Mode Observer *International Conference on Control and Fault-Tolerant Systems* (Nice, France) pp 516–522
[15] Georg S and Schulte H 2014 *Journal of Physics: Conference Series* **524** **(012053)**
[16] Bianchi F D, De Battista H and Mantz R J 2007 *Wind Turbine Control Systems - Principles, Modelling and Gain Scheduling Design* Advances in Industrial Control (Springer-Verlag, London Limited)
[17] Georg S, Schulte H and Aschemann H 2012 Control-Oriented Modelling of Wind Turbines Using a Takagi-Sugeno Model Structure *IEEE International Conference on Fuzzy Systems* (Brisbane, Australia) pp 1737–1744
[18] Kelley N NWTC Computer-Aided Engineering Tools (TurbSim by Neil Kelley, Bonnie Jonkman). http://wind.nrel.gov/designcodes/preprocessors/turbsim
[19] Edwards C and Spurgeon S K 1994 *International Journal of Control* **59**
[20] Edwards C and Spurgeon S K 1998 *Sliding Mode Control: Theory and Applications* (Taylor & Francis, Boca Raton)
[21] Edwards C, Spurgeon S K and Patton R J 2000 *Automatica* **36** 541–553
[22] Takagi T and Sugeno M 1985 *IEEE Transactions on Systems, Man, and Cybernetics* **15** 116–132
[23] Tanaka K and Wang H O 2001 *Fuzzy Control Systems Design and Analysis: A Linear Matrix Inequality Approach* (John Wiley & Sons, Inc)
[24] Gerland P, Groß D, Schulte H and Kroll A 2010 Robust Adaptive Fault Detection Using Global State Information and Application to Mobile Working Machines *International Conference on Control and Fault-Tolerant Systems* (Nice, France) pp 813–818
[25] Gerland P, Groß D, Schulte H and Kroll A 2010 Design of Sliding Mode Observers for TS Fuzzy Systems with Application to Disturbance and Actuator Fault Estimation *IEEE Conference on Decision and Control* (Atlanta, USA) pp 4373–4378
[26] Tan C P and Edwards C 2003 *International Journal of Robust and Nonlinear Control* **13** 443–463

AUTHOR INDEX